METHODS IN
PLANT ECOLOGY

METHODS IN
PLANT ECOLOGY

EDITED BY

S.B.CHAPMAN

A HALSTED PRESS BOOK

JOHN WILEY & SONS

NEW YORK

© 1976 Blackwell Scientific Publications
Osney Mead, Oxford
85 Marylebone High Street, London W1M 3DE
9 Forrest Road, Edinburgh
P.O. Box 9, North Balwyn, Victoria, Australia

First published 1976

Library of Congress Cataloging in Publication Data

Main entry under title:

Methods in plant ecology.

    Includes index.
    1. Botany—Ecology. 2. Botany—Methodology.
I. Chapman, S. B.
QK901.M47   581.5'028   75–37570
ISBN 0 470 01371 0

Printed in Great Britain

CONTENTS

List of Contributors   vii

Acknowledgments   ix

1   Introduction   1
    S.B. CHAPMAN

2   History of Vegetation   5
    K.E. BARBER

3   Description and Analysis of Vegetation   85
    F.B. GOLDSMITH and CAROLYN M. HARRISON

4   Production Ecology and Nutrient Budgets   157
    S.B. CHAPMAN

5   Physiological Ecology and Plant Nutrition   229
    P. BANNISTER

6   Site and Soils   297
    D.F. BALL

7   Climatology and Environmental Measurement   369
    R. PAINTER

8   Chemical Analysis   411
    S.E. ALLEN, H.M. GRIMSHAW, J.A. PARKINSON,
    C. QUARMBY and J.D. ROBERTS

9   Data Collection Systems   467
    C.R. RAFAREL and G.P. BRUNSDON

Index   507

LIST OF CONTRIBUTORS

S.E. ALLEN Institute of Terrestrial Ecology, Merlewood Research Station, Grange-over-Sands, Cumbria, England

D.F. BALL Institute of Terrestrial Ecology, Bangor Research Station, Bangor, Wales

P. BANNISTER Department of Biology, University of Stirling, Stirling, Scotland

K.E. BARBER Department of Geography, University of Southampton, Southampton, England

G.P. BRUNSDON Institute of Hydrology, Howbery Park, Wallingford, Berkshire, England

S.B. CHAPMAN Institute of Terrestrial Ecology, Furzebrook Research Station, Wareham, Dorset, England

F.B. GOLDSMITH Department of Botany and Microbiology, University College, London, England

H.M. GRIMSHAW Institute of Terrestrial Ecology, Merlewood Research Station, Grange-over-Sands, Cumbria, England

CAROLYN M. HARRISON Department of Geography, University College, London, England

R. PAINTER Institute of Hydrology, Howbery Park, Wallingford, Berkshire, England; now at: National Water Council, Goring, Reading, Berkshire, England

J.A. PARKINSON Institute of Terrestrial Ecology, Merlewood Research Station, Grange-over-Sands, Cumbria, England

C. QUARMBY Institute of Terrestrial Ecology, Merlewood Research Station, Grange-over-Sands, Cumbria, England

C.R. RAFAREL Institute of Terrestrial Ecology, Furzebrook Research Station, Wareham, Dorset, England

J.D. ROBERTS Institute of Terrestrial Ecology, Merlewood Research Station, Grange-over-Sands, Cumbria, England

ACKNOWLEDGEMENTS

Acknowledgements and thanks are due to a number of persons and organizations for collaboration or permission to use material in the production of this book. The following publishers are thanked for their permission to reproduce illustrative material:

Edward Arnold (Figs. 3.7, 3.14); Cambridge University Press (Fig. 2.4); W.H. Freeman & Co (Fig. 2.7); Longmans (Fig. 2.5); McGraw-Hill Book Co (U.K.) (Fig. 5.19); Ejnar Munksgaard (Figs. 2.1, 2.2); Oxford University Press (Fig. 3.2); Pergamon Press (Fig. 4.16); Sidgwick & Jackson (Fig. 3.9); Thames & Hudson (Fig. 2.9).

The officers of the following societies, or organizations are thanked for permission to reproduce material from their journals or publications: American Association for the Advancement of Science (Fig. 4.6); Antiquity (Fig. 2.8); Ecological Monographs (Fig. 3.20); Field Studies Council (Figs. 2.3, 2.11); Institute of British Geographers (Fig. 2.15); International Biological Programme (Fig. 4.8, Table 3.4); *Journal of Ecology* (Figs. 3.11, 3.12, 3.16, 3.18, 3.19, 3.23, 3.24, 5.4, 5.5, 5.24, 5.25, 5.27, 5.28, 6.8, Tables 4.1, 4.2); *Journal of Hydrology* (Figs. 4.19, 4.20); Oikos (Figs. 4.15, 4.18c); Societas Biologia Fennica Vanamo (Fig. 2.12); United States Department of Agriculture (Fig. 6.7).

Thanks are due to Dr S.R.J. Woodell for details of the litter trap shown in Fig. 4.8, to Dr D.P. Birch for Fig. 2.14, and to Dr M.A. Geyh of the Niedersachsisches Landesamt Für Bodenforschung, Hannover for radiocarbon dates quoted in Chapter 2. Acknowledgement must be made to Dr R.B. Painter & Mr P. Stevens for their contributions to the section dealing with soil moisture that is included in Chapter 6. Dr M. Hornung is thanked for helpful discussion in connection with the preparation of Chapter 6. The Director of the Institute of Hydrology is thanked for his permission to use photographs and material reproduced in Chapter 9. Finally the editor would like to thank the many people who have helped and encouraged him during the production of this book, and for the typing assistance provided by Mrs M. Osborne.

CHAPTER 1

INTRODUCTION

S.B. CHAPMAN

It is perhaps relevant that the introduction to a book entitled *Methods in Plant Ecology* should begin by looking at some of the developments and changes that have taken place since the word OECOLOGY was used in 1886. Since Haekel's introduction of the term for 'the science of treatment of the reciprocal relations of organisms and the external world' a great many changes both in the approaches of workers in the field and in the variety of techniques and methods at their disposal have taken place. Whilst ECOLOGY is now used and misused in a number of different ways it should still be possible to recognize the widest possible meaning of the word but for the purposes of a book such as this to retain the original biological meaning for the study of organisms and their relationship with other individuals and their environment.

Only a few years ago a leading ecologist was heard to comment that 'ecology cannot really be taught but can only be learnt', and some people may still sympathize with this point of view. It is also perhaps significant that a book published in 1934 entitled *Field Studies in Ecology* ends with a description of 'an ecology box' that in addition to a few extra items 'provides all that is required for good, accurate work'. These two separate items serve to illustrate the way that ecology has developed in recent years to a stage where available methods and techniques have increased in number, diversity, complexity, and in many cases also in cost to a degree that many people find alarming.

In 1939 Sir Arthur Tansley delivered a somewhat prophetic presidential address to the British Ecological Society in which he commented upon techniques in ecology:

'It is obvious that the progress of science depends very closely on technique, on the intelligent application of existing techniques and on the devising of new ones. The invention of a new technique frequently leads to very rapid advance in knowledge and to the opening up of completely new fields. Every well defined branch of biology develops its own techniques, so that beginners in the subject can use them from the outset; and if the worker is good enough he can apply them intelligently to new objects and perhaps go on to devise important modifications or completely new methods. In our subject

1

there are a few important techniques which are distinctively ecological. I have already pointed out that the ecologist has to borrow and adapt the techniques belonging to different specialized branches of science.'

Much of this remains very true today although there are now perhaps rather more techniques and methods that ecology might claim for its own. Since Tansley's statement the techniques that the ecologist can now borrow from other disciplines have also increased in number and diversity and it would seem to a casual observer that the ecologist need only select a suitable method to obtain the information needed to solve his problems. Many ecologists may no doubt wish that this were the case; but others would sadly miss the challenge of attempting to find ways of solving apparently insoluble problems.

There are a number of traps and pitfalls that await the noncritical user of methods and techniques especially if the assumptions upon which they are based are either not understood or, even worse, are simply ignored. So often one hears this criticism made regarding the use of statistical methods but many other techniques used by the ecologist are equally prone to misuse, and it should be emphasized that it is essential to check the suitability or accuracy of a method for a particular purpose. This is especially important when using a method for the first time, or under conditions that differ from those for which it was originally intended, and it is important that the basic principles involved in a method should be understood for the satisfactory application of the method.

Yet another danger that awaits the unwary is the addictive nature of apparatus and methodology. This affliction would appear to strike in one of two ways, firstly, there is the worker so addicted to new methods and to apparatus that he cannot see or find the time to approach real problems; secondly, there is the person who is so attached to an old and faithful technique that he is quite unwilling to try out or to develop anything new however useful it may be to his work. In each case the net result is often very much the same, a technique in search of a problem rather than a problem in search of a suitable technique to provide relevant information.

If prior thought is given to defining one's particular problem and into exactly what needs to be measured, a suitable technique or experimental approach can be selected. It is at this stage that consideration must be given to any modifications that may be needed to make a method suitable for one's own particular purpose. Problems will arise where no suitable technique exists and the initial research problem may well be to develop a method capable of providing a particular type of information or making some type of estimation.

In many cases there will be alternative methods to choose from, and because of the variability associated with ecological problems, it may be more appropriate to employ relatively simple pieces of apparatus at a number of sites than a more sophisticated and expensive piece of machinery at one or only few individual sites. Very often the lower, but often adequate, accuracy of a particular method will be more than compensated for by the ability to obtain readings from a greater number of sites. Alternatively, a simpler but replicable method may be used to provide preliminary data to assess the degree of variation in some factor and to allow a more sophisticated piece of apparatus to be sited and employed to the greatest advantage at some later stage.

Many ecologists have invested in some form of recorder that produces a continuous recording of an environmental factor only to find out later that all they required was an average or integrated result. It is always surprising to find how quickly such continuous recordings can accumulate and how much time is required to extract the relevant information. Very often some form of integrating recorder can provide all the information that is required thus saving a great deal of time. It must be admitted that with the development of more sophisticated forms of data logging and use of computers much of these data processing problems can or will be eliminated, but this approach can be expensive and will be beyond the reach of many ecologists. It is therefore still wise council to suggest that plenty of thought be given to the type of information that is required, the degree of accuracy that is needed, the variability of the parameter that is to be measured and therefore choose the most suitable method or technique for the purpose.

It would be very difficult to find a group of ecologists who could agree upon the limits of ecology; even if they were willing to attempt to define the limits, which is most unlikely, they would almost certainly not agree. To one ecologist the work of another might appear 'too physiological' whilst his own work might be considered to be almost non-biological by someone else. Despite these feelings about someone else's work it is hoped that neither would go so far as to dismiss the other's interests as not being ecology. However great the difficulty that such a group of ecologists might have in defining the limits of their subject it is of little consolation to someone attempting to organize a book that is intended to deal with ecological methods, where limits and artificial divisions do have to be made. The approach adopted must therefore represent something of a compromise and it is hoped that not too many people will be offended when they find that a favourite method or field of ecology has been omitted, or only briefly mentioned.

The approaches of the authors of the different sections in this book are governed by the nature of their particular subjects. In some sections the methods have been described in greater detail than in others. In some cases the provision of detailed methodology would have taken up too much space, or the technique may already be fully described in some readily available source. In some sections it has been felt that more use can be served by outlining the principles of methods available and to provide starting points and key references that will enable the potential users to develop procedures suitable for their own particular research problems.

One of the results of the International Biological Programme (I.B.P.) has been the publication of a number of excellent handbooks that deal with methodology in many branches of ecology and allied fields. However, it has been strongly suggested that there is still the need for a single volume that deals with *Methods in Plant Ecology*. This present volume is an attempt to produce such a book and is intended primarily to be of use to students at University and Technical College level, but it is hoped that it will be of some value to students at other levels.

REFERENCES

BRACHER R. (1934) *Field Studies in Ecology*. Arrowsmith, Bristol.
TANSLEY A.G. (1939) British ecology during the past quarter century: the plant community and the ecosystem. *J. Ecol.* **27**, 513–530.

CHAPTER 2

HISTORY OF VEGETATION

K.E. BARBER

1   **Introduction**   5
2   **Field techniques**   7
2.1   Sediments   7
   2.1.1   Site location and
      evaluation   8
   2.1.2   Sampling from sections   10
   2.1.3   Sampling from boreholes   12
   2.1.4   Sample storage   22
2.2   Other field techniques   23
   2.2.1   Woodland structure   23
   2.2.2   Indicator species   24
   2.2.3   Physical features   26
3   **Dating techniques**   27
3.1   Relative dating   27
   3.1.1   Pollen dating   27
   3.1.2   Dating by stratigraphy   29
3.2   Chronometric or absolute dating   30
   3.2.1   Radiocarbon dating   30
   3.2.2   Other radiometric
      methods   33
   3.2.3   Dating by Palaeomagnetism 34
   3.2.4   Varve dating   35
   3.2.5   Dendrochronology   35
   3.2.6   Lichenometric dating   38
   3.2.7   Dating hedges   38

4   **Laboratory techniques**   39
4.1   Miscellaneous techniques   39
4.2   Plant macrofossils   41
   4.2.1   Wood remains   42
   4.2.2   Seeds and fruits   43
   4.2.3   Bryophytes   44
   4.2.4   Other plant macrofossils   46
4.3   Pollen and spores   47
   4.3.1   Introduction   47
   4.3.2   Laboratory equipment   48
   4.3.3   Extraction of pollen and
      spores   49
   4.3.4   Absolute pollen counts   52
   4.3.5   Identification and
      counting   56
   4.3.6   Presentation of results   60
   4.3.7   Interpretation   61

5   **Documentary records**   62
5.1   Scope   62
5.2   Written evidence of past
   vegetation   63
   5.2.1   Agricultural and land use
      records   63
   5.2.2   Direct floral records   67
5.3   Maps and photographic records   68

6   **References**   69

1 INTRODUCTION

The ecology of existing plant communities can only be fully understood with reference to their history. The use of historical studies in clarifying the status and distribution of various species and communities is now commonplace, and with the greater sophistication and range of such studies the value of the knowledge so gained is widely recognized.

Studies such as that of Oldfield (1970a), on Blelham Bog National Nature

Reserve, Walker (1970) on the climax of hydroseres in the British Isles, and Peterken (1969) on the vegetation in Staverton Park are good examples of what has been achieved recently using the historical approach. More generally, the use of pollen analysis and associated techniques have transformed our knowledge of the vegetational history and indeed climatic history of much of the world throughout the Quaternary.

However, the accuracy of our reconstructions is dependent upon a great many interrelated factors, from the statistical and other assumptions inherent in the techniques themselves to the fragmentary nature of many of the deposits investigated. These problems have only recently attracted the close attention of historical ecologists but, especially in pollen analysis, some results of this questioning have been published (Janssen, 1970; Tauber, 1967; Rybnickova & Rybníček, 1971; Birks & West, 1973).

These investigations have not overturned the general picture of ecological history: rather they are adding fascinating detail and contributing to our aim of characterizing and explaining that history more precisely.

Two important factors must be borne in mind by anyone contemplating research in historical ecology. Firstly, a prerequisite for any such work is a sound understanding of present-day ecological situations and processes. Most of the techniques can be performed by a well-trained technician, but the interpretation of results requires experience with modern analogues of the past situation and a knowledge of the autecology of the species involved, especially when they are used as 'indicator' species. Secondly, because of the diversity of techniques that can be brought to bear on a problem in this field, the investigator must be aware of advances in subjects as diverse as palaeomagnetism and aerial photography.

This chapter divides the techniques used in historical ecology into four main classes: field techniques, dating techniques, laboratory analysis techniques, and documentary sources of information. These divisions are mainly for convenience of thought and are not meant to be definitive subdivisions of the subject. They also reflect, in their length and detail, something of the relative importance of each method and its applicability in a variety of situations. Hence pollen and spore analysis is dealt with at some length as it has been carried out all over the world and can give us unbroken sequences of vegetational change over thousands of years. In a number of ways pollen grains are ideal fossils. They are produced in millions as microscopic particles by almost all vegetation, most are widely dispersed and accumulate in peat and lake muds from which their recognizable outer shells can be extracted. By counting a large number from various levels of the sediment and expressing each taxon as a percentage of the whole sample one may construct a pollen diagram (Fig. 2.13) showing the percentages of different pollen types changing through the depth of the deposit. These changes can be dated both relatively, by the pollen assemblages themselves, and by chronometric dating (Section 3), and the vegetational history of the area may then be reconstructed by interpreting the pollen diagram using one's knowledge of present-day plant ecology, climatic change, land use history, etc. Many such studies, particularly recent ones, have included pollen diagrams which extend up to the modern surface so that the changes at the top of the diagram can be related to the present distribution of vegetation. This greatly increases the value and interest of such work to ecologists dealing with present communities and the general interrelationships of ecology and Quaternary palaeobotany have been commented on by West (1964) and Proctor (1973) amongst others.

Compared with the section on pollen analysis that on documentary records is much shorter and this is an example of the differences in applicability of the various techniques which may be used in historical ecology—pollen analysis may be applied world-wide without changing one's general approach but with documentary records one comes across difficulties such as the translation of old monastic records from Europe. They also relate to only a relatively small area so there are few general 'rules' in this work. A number of techniques have had to be omitted. These include the study of the inorganic deposits of glaciation, land and sea level changes and oxygen isotope analysis. Though it is often essential for the historical ecologist to make use of such studies and at times to work in such fields himself, space precludes any consideration of these techniques, which may be followed up by reference to texts such as West (1968).

2 FIELD TECHNIQUES

2.1 SEDIMENTS

This section covers the selection and sampling of sites capable of yielding fossiliferous material for the analyses described in Section 3.2. One point above all must be borne in mind during all field and laboratory work on fossil material, and that is the possibility of contamination by younger or older material, or both. Some contamination is easy to recognize and allow for. The presence of microspores derived from older sediments which have been eroded and washed into a lake (Birks, 1970), or the inclusion of 'exotic' (i.e. non-native) pollen in mid-post-glacial samples are examples of this. However, some contamination must go unnoticed and is undetectable. Usually this does not affect our interpretations greatly as a few oak pollen grains which are not contemporaneous with a pollen spectra containing a 'true' 60% oak contribution to the tree pollen total is neither here nor there, but the detection of a few beech pollen grains in an early post-glacial sample is more problematical. All one can do is to work with the care and cleanliness of a surgeon and to cultivate a suspicious mind!

2.1.1 Site location and evaluation

In choosing sites for, say, pollen analysis, much will depend on whether the historical ecologist is working on a particular problem, such as the autecology of an aquatic plant, or working on a particular area of phytogeographical importance. The distribution and frequency of deposits varies with the geological history of the area. In most areas where one is trying to relate present vegetation to its immediate predecessors the investigator will be looking for extant or recently drained mires and lakes.

In most cases the location of such sites is a straightforward matter. Topographic, geological and, if they exist, soil maps are an obvious first source, although the cartographic symbols do not always differentiate sufficiently between different types of organic deposit. Sites such as kettleholes are too small to be picked out by a closed contour on a one-inch or equivalent map but a careful survey of 1:25,000 or 1:10,560 sheets will often yield results. Any available air photographs should be studied looking especially for signs of disturbance by man and signs of flowage of the peat in the case of raised bogs.

Surveys by official bodies and natural history societies can be of great use as a guide to likely sites and historical work on such sites is all the more valuable when the present ecology is known to be of interest (*vide* Oldfield, 1970a). The local unpublished knowledge of appropriate government scientific officers, amateur naturalists, farmers and others has led to many a successful investigation; a maxim for all ecologists could well be: 'Seek out the oldest inhabitant', and the public bar of a local inn should not be forgotten as a source of ecological information!

Finally, there are the 'chance' sources of sites which are usually connected with some excavation work—motorway construction cuttings and boreholes, trenches for the laying of sewers and gas pipelines, sand and gravel pits and building excavations. A number of famous sites fall into this category: the interstadial sites of the English Midlands investigated by Coope, Shotton and others; the Bronze Age silts and muds exposed in a gas pipeline trench in Kent (Godwin, 1962); the last interglacial deposits from Trafalgar Square (Franks, 1960); and perhaps most fortuitous of all the interglacial deposits of Histon Road, Cambridge and the Devensian deposits found during excavations for a lecture block in Sidgwick Avenue, Cambridge (see Plate 41 in Sparks & West, 1972). If, as in the latter case large, recognizable remains are discovered then such sites are almost certain to be brought to the notice of the scientific community, but in a number of instances potentially valuable material is lost. It is incumbent upon all historical ecologists to keep an eye on any holes in the ground in their area and to make their interest known to the contractors who, in the author's experience, are cooperative if the scientific value of organic layers, buried soils and the like is explained to them.

Given a choice of sites in any one area, there are a number of criteria which may be used to select the most suitable site.

Faegri & Iversen (1964, p. 54) give a number of rules to follow:

1.   Prefer a lake deposit to peat.
2.   Avoid too small or too large lakes (ideal magnitude ca. 5,000 m²).
3.   Choose lakes that are not surrounded by extensive bogs.
4.   Choose a sample from a protected place where there has been no wind or current action.
5.   Avoid places where brooks enter a basin—risk of contamination, oxidation, and currents.

While they recognize that much useful work can and has been done on peat samples, recent advances in pollen analysis have led to a lessening of the emphasis on their first rule. Research into the transport and redeposition of sediments in lakes (Nichols, 1967; Davis, 1968), and the improved interpretation of pollen spectra from peats (Rybnickova & Rybníček, 1971) have led to an increased awareness of the relative merits of such deposits and of the importance of sediment stratigraphy. There are indeed serious shortcomings in certain types of lake deposits as sources of pollen sequences (Nichols, 1967), and Davis (1968) has pointed out the fact that pollen in lacustrine deposits may well have been disturbed and redeposited several times before becoming permanently incorporated in the sediment.

In a recent theoretical study on 'scale and complexity' problems in palaeoecology, Oldfield (1970b) notes that 'fewer assumptions or uncertainties *about dispersal in particular* occur if ombrogenous bog sites are used rather than lake sites'. This point is well brought out by Fig. 16 in West (1971) which shows six pathways leading to eventual incorporation of pollen in a lake sediment but only two where a raised bog is concerned. However, the rate of accumulation of peat, which may be formed by different species within the same profile, is rather more variable than that of a lake sediment and as it has not yet proved easy to perform 'absolute' pollen counts on peat deposits, both types of deposit have their own particular advantages and disadvantages. The most difficult sediments to work on are probably those of valley mires, where the worst features of both lakes and ombrogenous bogs can be found together. The peat may be woody, coarsely fibrous or very sloppy and all three types are found in the same profile in many of the valley mires of southern England. Added to this such mires have inflows and outflows to complicate and disturb the stratigraphy by erosion and deposition of silt, sand and gravel (and pollen!); their bottom profiles are often irregular, usually being old river courses; their summer water levels may fall far below the surface causing oxidation of the annual pollen fall; and finally their changing surface vegetation contributes substantial amounts of pollen. This last factor is a very difficult one to deal with, though some headway has been made by using detailed macrofossil analysis to indicate those taxa which should be excluded from an auxiliary pollen diagram in an attempt to interpret the regional vegetational history (Rybnickova & Rybníček, 1971). Many herbs are important 'agricultural indicators' in pollen analysis but the pollen of fen species may be indistinguishable from those of dry land species in the same family.

Despite these problems the great majority of pollen diagrams show broadly similar vegetational changes within the same area and we can be confident that our methods and interpretations are not too far from the truth; the studies mentioned above show refinements of detail rather than complete reinterpretations. Disturbance of deposits by natural or artificial means remains the major limiting factor in historical ecology based on analyses of sediments; the investigator must remain aware of this possibility throughout the field-work and subsequent analysis and interpretation of his results.

2.1.2 Sampling from sections

The two great advantages of sampling from an exposed section are that the lateral as well as vertical variations in the sediment can be easily seen and that contamination can be much reduced.

The first step in examining an exposed section in raised bog peat, or cliff face, is not to 'clean it up' with knife or spade as is often implied in the literature. The weathered surface may contain much valuable information which once erased will not reappear for some weeks. Hand cut peat sections may show up algal pool muds as sharp lines due to differential shrinkage, *Eriophorum* fibres and *Calluna* twigs stand out from the general surface and have a 'bleached' appearance, and the junctions between peats of different types and humifications are more sharply defined. The section should be photographed in colour incorporating a scale for subsequent measurement. The colour rendering of certain film emulsions make them more suitable than others for differentiating between the various shades of brown in peat. Colour transparencies can be very useful in checking drawn sections by projecting them onto graph paper at an appropriate scale.

A pencil drawing of the section is usually made by pegging a metre square frame with 10 cm wire divisions onto the section and transferring the stratigraphy onto graph paper. A system of sediment classification such as that devised by Troels-Smith (1955) may not be suitable when investigating a single material such as *Sphagnum* peat, in that all the symbols used will be very similar to each other, although for heterogeneous deposits the Troels-Smith system is preferable as it describes the sediment in terms of its colour, elasticity, stratification and composition without presuming on its origin. For rather more homogeneous sediments one may work out a system suitable for the deposit and for the purpose of the work. For *Sphagnum* peats the divisions used by Walker & Walker (1961) based mainly on humification and identifiable plant remains are quite adequate and easily modified to suit particular circumstances. Casparie (1972) successfully used a similar system in his monumental work on Bourtanger Moor. In mixed deposits such as clay, succeeded by sedge peat, woody peat, and *Sphagnum* peat it may be advantageous to use a recognized system of symbols combined with the Munsell system of recording colour. The recording of colour is often omitted because of the changes which occur when organic materials dry out and oxidize. Examples of these systems are given in Figs. 2.1 to 2.4.

Defining boundaries between sediment types can often be difficult. In borehole work it is a simple matter to record a few centimetres of transitional sediment but in an exposed section such transitions can make the resulting diagram very confusing. Somewhat arbitrary decisions must therefore be made and a simple code of line thicknesses, or solid and broken lines, used to define major and minor, or less sharp, boundaries. Special attention should be given to all sharply defined boundaries in sediments; they may indicate a period of erosion, non-deposition or human interference with the deposit.

The recognition of some major plant remains is quickly learnt by practice but many field identifications require checking by the use of microscopic characters. The identification of plant macrofossils is dealt with in Section 4.2.

Once the section has been drawn and photographed, sampling can be considered. The number and position of section profiles to be sampled will vary with the particular investigation. To investigate the vegetational history of an area a monolith of sediment for pollen analysis should be taken which extends right through that part of the deposit which can best be judged to have had a continuous sequence of accumulation—that is, avoiding sharp stratigraphic breaks, sloping contacts and dry oxidized layers.

When the detailed ecological history of a growing mire is the problem under investigation far more samples may be taken, and correlated in time by radiocarbon dates and comparative pollen diagrams, so that simultaneous stages of development and vegetation may be established. Walker & Walker (1961) used this method for a preliminary survey of a number of Irish bogs; more recently Casparie (1972) and the author (unpublished) have looked in detail at bog-growth in this way and have been able to reconstruct simultaneous phases of development.

Samples are generally and most satisfactorily taken in monolith boxes, $50 \times 10 \times 10$ cm, constructed of aluminium sheet 1·5 mm thick. When the monolith position has been selected, the labelled box is lightly pressed into place, then removed to leave an outline on the section. This can then be cut around with a sharp knife—to sever fibrous remains—before the box, lined with aluminium foil, is pressed home.

The monolith is removed with a spade, covered tightly with aluminium foil and sealed in a polythene bag. Such monoliths, especially of acid raised bog peat may be stored for years at room temperature but are safer from deterioration if kept in a freezer. The stratigraphy should be checked in the laboratory soon after collection as a safeguard against colour change and shrinkage though. In detailed work the face checked must be that on the bottom of the tin and the inclusion of an aluminium foil liner in the box makes this much easier. Hammond (1974) has described a method for the preservation of peat monoliths for permanent display.

In unconsolidated sediments such as sands, lacquer sections may be taken by the method described by West (1968, p. 366) but this is only useful for demonstration specimens. Samples for pollen analysis from soil (Dimbleby, 1961) are usually taken as small separate samples from a cleaned section, working from the base of the section to avoid contamination.

In considering open sections one should be aware of the advantage of being able to get a three-dimensional look at the stratigraphy. Cutting a peat face back in slices of, say, 20 or 40 cm is not a difficult job and by doing so, leaving each new section to weather for a couple of weeks; recording, sampling and cutting back again and finally recording the side walls of the pit the author has been able to reconstruct the development of a hummock and hollow complex in a peat block 2·40 m × 1·40 m. Stewart & Durno (1969) have demonstrated the usefulness of a similar sampling technique.

2.1.3 Sampling from boreholes

A number of well tried and tested samplers for peat and mud are now available commercially or can be made up in institutional workshops following the detailed plans published in scientific journals.

There are three basic types of instrument in common use for organic sediments; the Hiller pattern, the 'Russian' type, and samplers providing a full core of the material (Figs. 2.5 & 2.6). Various other screw and mechanical-driven augers may have to be used for inorganic sediments but such sediments are not likely to be used by investigators of relatively recent ecological history; the reader is referred to West (1968) for further details, from which the following table of comparisons of the different samplers is taken. (Table 2.1)

The principles on which these samplers operate are quite simple; their successful use in the field is less straightforward. The author's preference is to use a Russian type sampler for all terrestrial organic deposits and a Mackereth pneumatic sampler for underwater sediments, on the grounds of ease of use and relative freedom from sediment contamination or distortion.

The Russian-type sampler was first described in English by Jowsey (1966). The sampler consists of a semi-circular chamber which rotates round a finned anchor plate thus enclosing a virtually undisturbed 'half-core' of sediment. The nose, chamber and head are joined solidly together and the anchor plate pivoted between head and nose sections as shown (Fig. 2.5). Though it must turn freely on these pivots it should be as close a fit as is consistent with this requirement. Since 1966 various modifications have been made without altering the basic design and the author's present sampler has a strengthened chamber of

Limnic sediment.

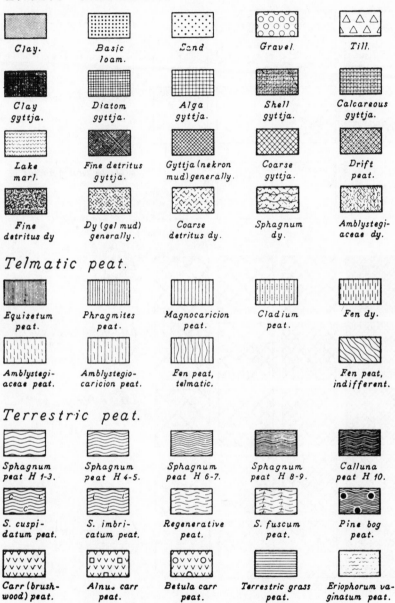

Clay.　　Basic loam.　　Sand　　Gravel.　　Till.

Clay gyttja.　　Diatom gyttja.　　Alga gyttja.　　Shell gyttja.　　Calcareous gyttja.

Lake marl.　　Fine detritus gyttja.　　Gyttja (nekron mud) generally.　　Coarse gyttja.　　Drift peat.

Fine detritus dy　　Dy (gel mud) generally.　　Coarse detritus dy.　　Sphagnum dy.　　Amblystegiaceae dy.

Telmatic peat.

Equisetum peat.　　Phragmites peat.　　Magnocaricion peat.　　Cladium peat.　　Fen dy.

Amblystegiaceae peat.　　Amblystegiocaricion peat.　　Fen peat, telmatic.　　Fen peat, indifferent.

Terrestric peat.

Sphagnum peat H 1-3.　　Sphagnum peat H 4-5.　　Sphagnum peat H 6-7.　　Sphagnum peat H 8-9.　　Calluna peat H 10.

S. cuspidatum peat.　　S. imbricatum peat.　　Regenerative peat.　　S. fuscum peat.　　Pine bog peat.

Carr (brushwood) peat.　　Alnus carr peat.　　Betula carr peat.　　Terrestric grass peat.　　Eriophorum vaginatum peat.

Figure 2.1 Sediment symbols in common use: originally proposed by Faegri & Gams. (From Faegri & Iversen, 1966.) The 'H' figures refer to peat humification on von Post's scale; Davies (1944) gives this in full. H1-3 is little humified with recognizable plant remains; H8-9 is almost structureless black peat.

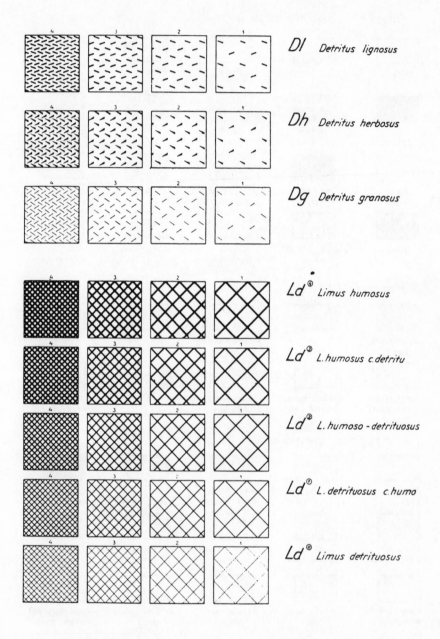

Figure 2.2 Sediment symbols from the Troels-Smith (1955) system. (From Faegri & Iversen, 1964.)

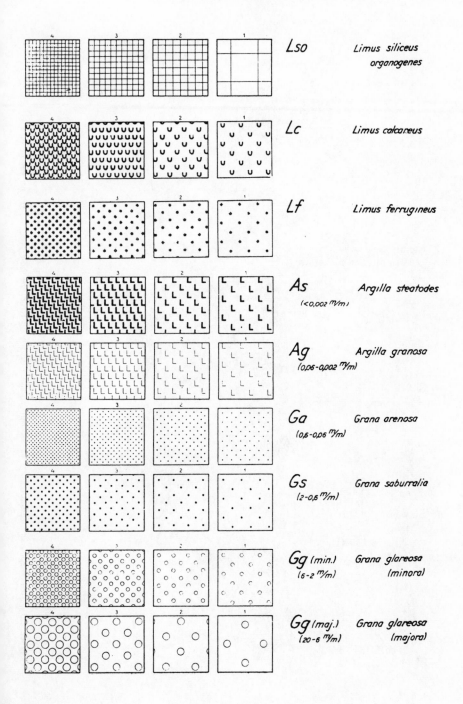

Lso   Limus siliceus organogenes

Lc   Limus calcareus

Lf   Limus ferrugineus

As   Argilla steatodes
$(<0,002 \, ^m/m)$

Ag   Argilla granosa
$(0,06-0,002 \, ^m/m)$

Ga   Grana arenosa
$(0,6-0,06 \, ^m/m)$

Gs   Grana saburralia
$(2-0,6 \, ^m/m)$

Gg (min.)   Grana glareosa
$(6-2 \, ^m/m)$ (minora)

Gg (maj.)   Grana glareosa
$(20-6 \, ^m/m)$ (majora)

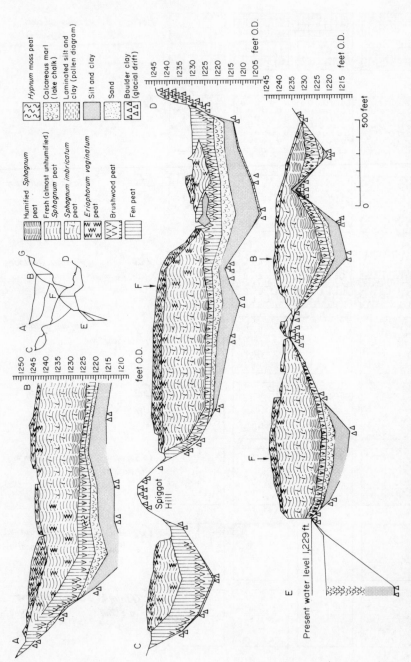

Figure 2.3 Stratigraphy of Malham Tarn Moss. Full sections drawn up from borehole records. (From Pigott & Pigott, 1959.)

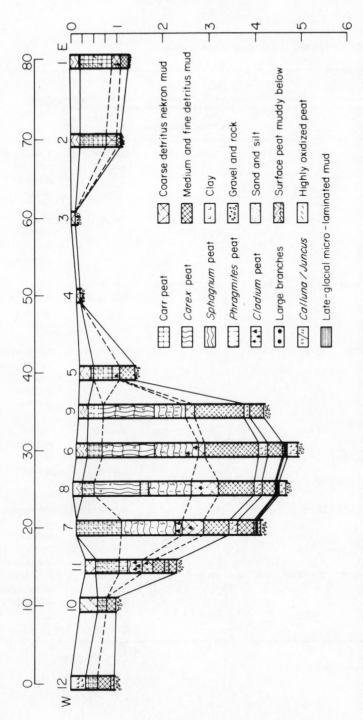

Figure 2.4 Stratigraphy of Boring Transect C, Blelham Bog. Individual boreholes shown with possible correlations. (From Oldfield, 1970a.)

Table 2.1 Comparison of some hand-operated samplers.

Type	Sample		Defor-mation of sediment	Com-pression of sediment	Suitability for sediments	
	Length	Width			Good for	Bad for
Hiller	50 cm	3 cm	much	none	peat, compact mud	loose organic sediment, inorganic sediment
Russian	50 cm	5 cm	none	none	peat, mud	loose organic sediment, inorganic sediment
Dachnowski	30 cm	5 cm	little	little	non-fibrous peat, mud	fibrous peat, loose organic sediment
Livingstone	50 cm	4 cm	little	some	non-fibrous peat, mud	fibrous peat, loose organic sediment, inorganic sediment
Punch	50 cm	6 cm	little	some	compact organic sediment, clay silt, fine sand	fibrous peat, loose organic sediment
Screw auger	25 cm	4 cm	much	much	compact organic sediment, clay and silt	soft organic and inorganic sediment
Other augers	10–30 cm	5–10 cm	much	little	compact organic sediment, clay, silt, sand, gravel	soft organic and inorganic sediment

From West (1968), p. 102.

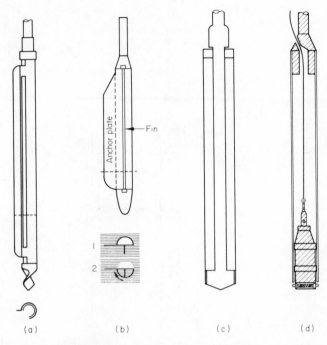

(a)   (b)   (c)   (d)

Figure 2.5 Peat samplers in common use. (a) Hiller, (b) Russian, (c) Dachnowski, (d) Livingstone.
For details of operation see text. (From West, 1968.)

stainless steel 5 mm thick and a shortened nose of 100 mm to allow basal deposits to be sampled more easily (Fig. 2.6). Tolonen (1966) has made similar modifications. In use the sampler is simply pushed down to the desired depth and turned through 180° by means of a T-handle.

The anchor plate is held stationary by the sediment because of its area and the chamber, with a sharpened cutting side, encloses the sediment by rotating as shown in the diagram (Fig. 2.5). While the sampler is usually shown taking two quarter cylinders of sediment (one for field examination and one for laboratory analysis) by inserting it with the fin on the outside, it is preferable to take a full half-core for laboratory use. The advantages of this are that any disturbance of the sediment by the fin is avoided (Jowsey, 1966) and it enables one to use commercially available plastic drainpipe tubes, cut longitudinally into 500 × 50 mm sections to contain the core samples. It is still possible to investigate the sample in some detail in the field; with careful handling and wrapping the whole tube in foil it is possible to preserve even soft, watery surface peat and the larger sample allows microfossil sub-samples to be taken from the centre of the half-core. For field demonstrations of the stratigraphy throughout a whole deposit this instrument is, of course,

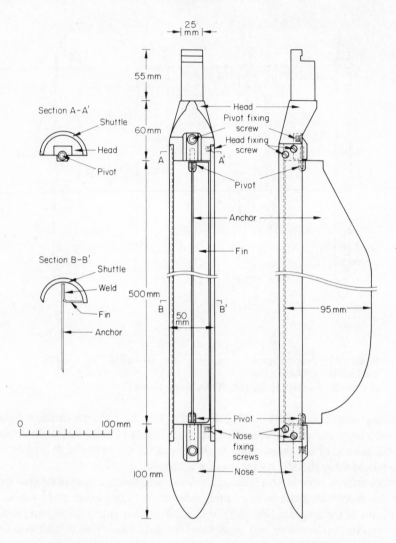

25
mm

Section A–A'
Shuttle
Head
Pivot

55 mm

60 mm

Head
Pivot fixing
screw
Head fixing
screw

A   A'

Pivot

Anchor

Fin

Section B–B'
Shuttle
Weld
Fin
Anchor

500 mm

B   B'

50
mm

95 mm

0   100 mm

Pivot
Nose
fixing
screws
Nose

100 mm

Figure 2.6 Russian peat sampler as modified by the author.

unsurpassed as the samples in their plastic containers may be laid end to end on the ground before a class.

The one disadvantage of the Russian sampler is that it must not be turned while being pushed into the sediment, and it does mean that getting through woody or mineral layers can sometimes be difficult. Experience in the valley mires of southern England has shown that this can almost always be overcome by gently raising the sampler a few centimetres and employing two or three persons in a concerted push. The sampler must then be turned carefully in case wood or pebbles jam between anchor plate and chamber, with the danger of twisting the sampler.

The Hiller sampler, the 'classical peat borer' as West puts it, is still in common use despite its disadvantages when compared with the Russian type. It is usually made with chambers of 350 or 500 mm, and consists of two tubes of stainless steel, the inner one of 30 mm or more inside diameter with a 20 mm slit down one side, the outer one with a curved and sharpened flange set at an angle down the side of a similar slit (Fig. 2.5). Most have a screw nose as shown but it is an advantage to have provision for replacing this with a simple cone head for soft deposits (Thomas, 1964) as this will lead to less disturbance of the sediment. The Hiller is pushed down to the desired depth maintaining some clockwise pressure on the T-handle to keep the flange in the closed position and is then twisted three to four times counter-clockwise. The flange cuts into the sediment opening the slits and material is forced into the chamber. Distortion is therefore inevitable and because of the two close-fitting tubes rotating together, it is easily jammed by sandy deposits.

Despite their disadvantages Hiller samplers are still routinely used because of their ability to penetrate some stiff muds, clays and woody sediments where they may have something of an edge over the Russian sampler because of their screw action. The basic design has been modified by Thomas (1964) to provide removable liners allowing whole core samples to be retained, obviating the need for sub-sampling in the field.

Endfilling corers of the Livingstone and Dachnowski type have been extensively reviewed in existing publications (Wright *et al.*, 1965; Deevey, 1965). These accounts contain a mine of useful information won by experience and go into the details of using these corers in the field. The principle of using the Livingstone corer is shown in Fig. 2.7, and Table 2.1 shows the sizes in common use and their suitability for various types of deposit.

The Mackereth-type sampler for subaqueous deposits also makes use of a piston but the driving force is supplied by compressed air. This elegant and ingenious device, conceived and developed by the late F.J.H. Mackereth, is fully described in its large form, for cores up to 8 m long, in Mackereth (1958) and, as a 'mini-corer' taking 1 m cores which can include the sediment—water interface, in Mackereth (1969). Both papers include engineering details which would enable the corers to be made by institutional workshops.

A simple large-diameter corer, without a piston, for work on fairly dry bog peats, has been described in detail by Smith *et al.* (1968). The sampler is operated by hand, assisted by a simple rig and chain-hoists, and with the use of plastic liners very good recovery rates of cores 1·5 m long and of 12 cm diameter have been achieved. Compression, distortion and contamination are also very slight. The corer was designed primarily to obtain samples for radiocarbon dating and deserves serious consideration for this purpose; it is certainly simpler than digging a pit to obtain monoliths or repeated sampling with a smaller diameter sampler.

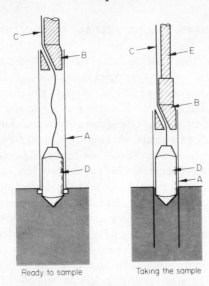

Ready to sample   Taking the sample

Figure 2.7 Principle of operation of the Livingstone corer. A: coring tube. B: adaptor. C: cable. D: piston. E: extension rod. (From Deevey (1965) in Handbook of Paleontological Techniques, edited by Bernhard Kummel and David Raup. W.H. Freeman and Company. Copyright © 1965.)

Finally, mention should be made of the sampler used by Hubbard & Stebbings (1968) in their investigation of the stratigraphy of Keysworth Marsh. With a simple polished steel tube 120 cm × 5·3 cm they obtained cores of mud a metre long. The end of the tube was slightly 'pinched' to retain the cores and a sharp cutting edge cleanly severed the plant material in the mud instead of pushing it to one side or causing it to form a plug in the tube.

2.1.4 Sample storage

The basic idea in storing samples of fossil material is to maintain the field conditions of fossilization. The main factor in the field is the waterlogging of the peat or lake sediment and maintaining this condition in the laboratory is fairly easy for short periods of time, that is, a few months. Aluminium monolith tins or plastic core tubes, covered closely with aluminium foil and then with polyethylene will do temporarily. Polyethylene is, of course, permeable, but it does reduce evaporation; its main function is to protect the foil and exclude dust and pollen from the atmosphere. Aluminium foil eventually corrodes in contact with damp acid peat and should be renewed every few months. No wrapping material is entirely satisfactory; foil has the advantage of being cheap and easily available, and it is easy to mark with felt-tipped pen.

For longer periods of storage, freezing of complete monoliths or cores is the best method and does not destroy stratification. Smaller samples, in glass tubes with cork or polypropylene stoppers, may also be frozen (if there is room for expansion of the material) or the tubes sealed with Parafilm or dipped in wax.

2.2 OTHER FIELD TECHNIQUES

This section covers a somewhat mixed bag of techniques, distinguished from sedimentary methods because they do not usually provide the long sequential patterns that one can obtain from sediments. They are based upon evidence from living vegetation and its non-fossilized litter and are applicable mainly to woodland communities. Shorter-lived communities such as sand dunes and heathlands, which may involve spatial zonation and marked species changes with time are not considered here—an excellent account of the short term ecological changes in heathland is given by Gimingham (1972) who includes some consideration of the structure and ageing of *Calluna* stands.

New techniques and uses for field observations are being constantly developed and in Britain the Historical Ecology Discussion Group organized by ecologists at the Monks Wood Experimental Station, Abbots Ripton, Cambridgeshire meets to discuss current research. The objective of this informal group, open to anyone interested, is to bridge the gap between ecologists, historians, archaeologists, archivists and historical geographers. Since its inception in 1969 it has held more than a dozen meetings on subjects such as old woodland and grassland floras, the ecological use of archives, and the historical ecology of specified areas.

2.2.1 Woodland structure

Everyone is aware that woodlands exhibit different structures which are a reflection of the wood's history in terms of its management, or lack of it, the efficiency of regeneration, the mixture of species, changes in the soil profile, etc. The determination of tree age is basic to an understanding of this structure and is dealt with in Section 3.2.5. Other details of the woodland that are visible in the field and which give information on its history include the broad age classes of the trees (seedling, sapling, mature, over-mature, etc.) their form, the presence or absence of regeneration, the depth of litter and the ground flora.

Tubbs (1968) in discussing the history of the New Forest uses three broad age-classes, A, B, and C, which can be more or less easily appreciated by the observer in the field and are a great aid in discussion. Peterken (1969) provides no less than 14 diagrams of the species and generations of trees and shrubs in Staverton Park and uses these to illustrate the woodland history and patterns of regeneration such as the contraction in the range of holly due to regeneration failure. Tree form is also of importance in indicating disturbance by man or his animals, for example the old pollarded oaks and beeches of the

New Forest, most of which date prior to 1698 when pollarding was forbidden, demonstrate a previous form of management carried out to increase the grazing for deer in a Royal Hunting Forest. Fairly distinct growth forms which occur naturally in woody plants growing in an exposed environment have been described in Peterken and Hubbard's study (1972) of the holly wood on the shingle of Dungeness. Using these growth forms they erect a succession of types to explain the history of the wood. This sort of reasoning from growth form and structure is often, though less overtly, used in our interpretation of woodland history and many examples can be gleaned from the literature of arguments which involve relative age, 'protection' of one species by another (e.g. oak 'protecting' holly seedlings at its base), canopy closure and its later opening, invasion by introduced species (e.g. sycamore in the British Isles) and grazing by domesticated and wild animals. Excellent examples of arguments along these lines are contained in the studies by Pigott (1969) and Merton (1970) of woodlands on limestone, and that of Kassas (1951) of fen woodland.

It must always be remembered that such studies can be misleading in the absence of absolute age determinations (Section 3.2.5) or management records of some sort. Age determinations of trees by ring counts on stumps or by increment corings often show surprising ages for trees growing in a poor habitat, ages which would not be accurately assessed by simply observing girth and height, and the converse can be true for specimens in an optimum habitat. The evidence of such straightforward management as planting in rows at equal distances and, perhaps, thinning, can be detected in many woods in long-settled areas such as the British Isles though after a century or so of neglect when disease, felling by wind, etc., have taken their toll early management will be much less evident. Where records exist the task of the historical ecologist is much easier though descriptions, as against records, of notable woods such as Wistman's Wood on Dartmoor may conflict (Anderson, 1954) and can make the task more difficult.

2.2.2 Indicator species

A blaze of bluebells or foxgloves in an area of recently felled woodland is one of the most striking indicators of ecological change but lasts only a short time in the natural state. A more lasting indicator of change is the gorse *Ulex europaeus*. Tubbs & Jones (1964) reported that gorse distribution in the New Forest was largely controlled by soil disturbance due to man in the form of old enclosure banks and the now abandoned margins of cultivation made at the time of the Napoleonic Wars. Old grasslands may also be recognized by their general species richness and in a survey of many sites, Gulliver (pers. comm.) found 'a group of several species that had not been able to re-invade following reseeding operations between 5–15 years previously, and these species may be considered as indicator species of undisturbed grassland sites. Most of the proposed indicator species were very rare, and hence of limited use for interpolation, but a few are fairly common e.g. *Conopodium majus, Filipendula ulmaria, Galium verum* and *Rhinanthus minor* agg.'. T. Wells (pers. comm.) has used one of the rarer species of this group, *Orchis morio,* to provisionally date three meadows in Huntingdonshire (now in Cambridgeshire), giving them an undisturbed life span of at least 250 years.

Indicator species in woodlands have been treated by a number of authors, notably Pigott (1969), Tittensor & Steele (1971), Pollard (1973) and Rose (in Peterken, 1969). Pigott (1969) considered the field-layer of *Tilia*-rich woodlands on the Derbyshire limestone to be related to the age-structure of the woodland as well as to the nature of the soil. *Mercurialis perennis* for example, is said to be indicative of older even-aged woods, whereas *Convallaria majallis* is more characteristic than any other species of uneven-aged woods and is especially associated with stands of *Tilia.* It is dangerous to rely on only one or two species, as these authors recognize, for other factors influence the distribution of plants (e.g. Martin (1968) on *Mercurialis*). The use of groups of species rather than individual species is therefore preferable in this type of study. This is well illustrated by Tittensor & Steele (1971) in their studies of the ground flora of the Loch Lomond oakwoods, based on Tittensor's earlier studies (1970a,b) of the history of the woods. They found a complexity of interacting factors including soil type, grazing and trampling which could account for most of the variation in the herb layer but that undocumented effects of man on the trees themselves could also have affected the situation. Pollard (1973) was able to show that some old hedges in the vicinity of Monks Wood Experimental Station are woodland relics managed as boundary hedges rather than planted hedges, herb species such as *Mercurialis perennis, Endymion non-scriptus* and *Anemone nemorosa* being particularly characteristic of the old 'woodland' hedges (see also Section 3.2.7).

Rose's work on epiphytic lichens in British woodlands promises to be a very useful tool in elucidating woodland history, especially as regards the incidence of disturbance, or conversely, the continuity of woodland cover on any one site. It must be remembered that other factors are important in the distribution of lichens, these include atmospheric pollution and woodland management practices such as coppicing and drainage. Rose, having visited some 4,000 or so sites, has concluded that primaeval forest in Britain had something like 120 species of lichen per km^2 and that the closest approximation to this total today is found in the New Forest (Rose & James, 1974). Twenty to twenty-five species per kilometre is the average for coppiced woods in drier parts of Britain; Staverton Park (Peterken, 1969) contained 64 species, including a number characteristic of ancient woodland, and Rose was able to conclude that the woodland there '. . . is almost certainly a relic of the ancient forest cover of this part of England and that there has been here a fair degree of permanence of forest condition and tree cover, and lack of any major clear felling and replanting'. Clearly studies of the lichen flora can offer ecologists a most useful measure for assessing the amount of disturbance in a community especially when allied to documentary and other evidence of woodland continuity.

2.2.3 Physical features

The recognition of old field-systems and boundaries, of disturbed soil profiles, and of ancient trackways and habitations should be included in every ecologist's field observations when in inhabited or once-inhabited areas. Even in non-inhabited areas such as tundra and alpine areas physical features such as rock-falls and fossil periglacial features and other events which occurred in the past may be of importance in the present distribution of plants and communities. So diverse are these phenomena that no complete coverage can be attempted here but a few examples may be noted.

Dimbley (1962) used pollen analysis of soils from beneath prehistoric and historic banks and monuments of various sorts in his important study of the origin of lowland heaths in Britain and used the results to ascribe an anthropogenic origin to these heaths. Visual examination of these buried soil profiles is often striking enough—a brown forest soil below a burial mound with well-developed podsols all around—but pollen analysis of the buried soil and the horizons above it give even more ecological information. Man-made physical features can therefore give valuable evidence not only in indicating land enclosed or used for some other purpose but also in preserving part of the old land surface.

Hewlett (1973) used field and documentary evidence to great effect in reconstructing a historical landscape around Otford in Kent. Using 'hedge-dating' as a basic field technique (Section 3.2.7), he also made use of a physical feature, the 'evolved bank', similar to a lynchet. These 'evolved banks' occur when a hedge boundary is established across a slope; ploughing and soil-creep on the upslope side causes soil to accumulate against the hedge, while on the downslope side the soil slips away. Hewlett investigated over 150 of these 'evolved banks' and found a rough relationship between height and age—the highest being nine feet—similar to that between the number of species in a hedge and its age. Obviously there are a number of complicating factors such as slope angle, frequency of ploughing, etc., but the technique certainly seems to work in the Otford area and may be applicable elsewhere.

It is common knowledge that many vegetation patterns owe their origin to physical features such as soil depth, microtopography and the like, but one physical process of some interest to the historical ecologist, because it is no longer active in most areas but has an effect on the vegetation, is that of periglaciation. In East Anglia especially, the permafrost conditions of the last glaciation, ending only some 10,000 years ago, have left their mark in 'chalkland patterns', 'Breckland polygons' composed mainly of sand over chalk or chalky till, and ice-mound features (Sparks & West, 1972). Here some knowledge of the physical processes involved is very valuable in explaining the present vegetation pattern—for example the alternate stripes of *Calluna*-dominated and more basic grass-dominated vegetation on slopes in the Breckland (Watt *et al.*, 1966).

Finally one might mention the usefulness of some knowledge of the history of man's physical artifacts. Changes in the style of buildings and houses (Smith & Yates, 1968) can give useful field clues of settlement establishment and change which must, however, be followed up by documentary work if records exist, and the same is true of the study of field systems and the like. The linking of field and documentary work in studies of the history of the landscape is very clearly shown in the admirable works of Professor W.G. Hoskins (1955, 1967), which have been recently followed by Lambert (1971) on the Netherlands, and Williams (1974) on South Australia.

3 DATING TECHNIQUES

In any historical study it is vital to be able to place events in their correct chronological sequence and to be able to fix at least some of these events with reference to the scale of calendar years. In the wider field of Quaternary palaeoecology there has been much controversy recently concerning the length of the glacial/interglacial cycle (Cooke, 1973) and the whole relationship between lithostratigraphy, biostratigraphy, and absolute chronology is in a state of flux (Royal Society, 1969), but in terms of the historical ecology of the past few hundred or even few thousand years this need not concern us.

Dating techniques may be either 'relative' or 'absolute'. The former method involves placing an event or series of events in an existing temporal framework and is the basis of stratigraphic geology, archaeological excavations, etc. 'Absolute' dating involves the more or less direct measurement of time and so is often referred to as chronometric dating—a more precise term. Time may be measured by the decay or build-up of radioactive isotopes or by counting the growth increments of a tree amongst other methods, and while there are a number of factors which make our datings more or less precise these factors do not discredit the methods as a whole but rather enable us to make successive approximations to the truth.

3.1 RELATIVE DATING

3.1.1 Pollen dating

In its simplest form this means referring a stage in a particular pollen diagram to a widespread event from a number of pollen diagrams. This technique was used before radiocarbon dating became available and is still widely used today but dates derived by this method should always be confirmed by radiocarbon dates. This is especially important since it has now been shown just how time-transgressive pollen zone boundaries can be. Smith & Pilcher (1973) demonstrate that the rise in *Alnus* pollen percentages in the British Isles at the start of the Atlantic period of the post-glacial transgresses no less than 2,000 years, whereas this important vegetational boundary is conventionally taken to be around 5500 B.C. Fortunately the *Ulmus* decline, signifying the beginning of the Sub-Boreal period and the start of Neolithic agriculture seems to be synchronous within the limits of resolution of radiocarbon dating (3000 B.C.), though here again individual sites may differ and this very difference is of great archaeological and palaeoecological importance. The moral clearly is that all pollen diagrams should have a number of radiocarbon dates. But these are expensive and there are insufficient analytical facilities to satisfy the demand. The next best thing is for a series of securely dated regional standard pollen diagrams to be constructed with well defined pollen assemblage zones. This has been advocated by West (1970), and examples published by Godwin *et al.* (1957) Hibbert *et al.* (1971), and Pennington *et al.* (1972). Other pollen sequences within the same general area can then be compared with this standard diagram and 'dated' by correlation of vegetation boundaries. Comparisons can be made by eye or by numerical techniques (Gordon & Birks, 1974). It goes without saying that a familiarity with the general vegetational history and ecology of an area is essential before this sort of comparison can be attempted. Pennington (1974) provides a good summary of British vegetational history and Wright (1971) serves as an introduction to the North American sequences. Many areas of the world still lack this basic framework, although the situation is changing quickly.

There are situations where radiocarbon dating is not available or not very reliable in providing 'fixed' points—where contamination of the sample by older or younger carbon cannot be excluded or where the sample is so recent that the standard deviation of the radiocarbon date makes the date a poorer estimate of age than can be had by other means. In the latter case (often of particular interest to an ecologist studying the present-day communities) the correlation between known historical events and pollen diagram changes is even better than the radiometric date. Oldfield (1963, 1969) has exploited this procedure in dating and explaining the recent ecological history of the south-east Lake District—events such as the dissolution of the monasteries around A.D. 1540, the great extension and intensification of agriculture at around A.D. 1800 in response to the Napoleonic War and the reafforestation of country estates with pine, elm and beech can be clearly seen in many of these diagrams which come right up to the modern sediment surface. The dating of the upper part of the author's diagram from Bolton Fell Moss is also based on this principle (Fig. 2.13), and Mitchell (1965), Moore (1968) and Turner (1970) have used similar arguments with success. In the United States the European settlement and clearance dates can be used similarly (e.g. McAndrews, 1966).

Where it is suspected that radiocarbon samples will be contaminated due to root penetration or carbonate-rich ground water, it may still be useful to have a radiocarbon date. In the case of rootlet penetration the radiocarbon date will be too recent but will at least give a 'minimum date', the 'real' age will have to be interpolated from unaffected sites in the same area (Oldfield, 1963, p. 30). Beyond the range of radiocarbon dating, that is, older than 60–70,000 years, relative dating by pollen assemblage zones is pretty well the only method of dating organic deposits and the characterization of the various interglacial stages in Britain (West, 1970, and other papers) is the basis here. These zones may then be related to chronometrically dated marine or volcanic deposits though this is a difficult and largely unresolved task (Shackleton & Turner, 1967).

3.1.2 Dating by stratigraphy

This somewhat unsatisfactory title is used to cover the relative, and in favourable circumstances the 'absolute' age one can attribute to an event by its occurrence within a particular stratum, or in relation to a particular 'marker' such as a dated volcanic ash fall or inwash layer in a lake sediment. 'Pollen dating' is, of course, dating by means of biostratigraphy but was felt to be sufficiently distinctive and important to deserve its own section.

Certain features of sediment stratigraphy have long been used as indicators of age. The change from minerogenic sediment, indicating exposed mineral soils and probably solifluction, to organic sediment in lake basins is a fair indicator of the change from the last Late-Devensian cold period, the Younger Dryas, to the post-glacial. This is not to say, however, that the late-glacial/post-glacial boundary should always be drawn at this junction as biological indicators of change may not coincide with the lithological change and, in general, biostratigraphical indicators are to be preferred (Watts, 1963). Bradbury & Waddington (1973) made great use of stratigraphic evidence in dating their core from Shagawa Lake, Minnesota. The core base was radiocarbon-dated at A.D. 15 ± 125 years and the top 33 cm, too recent for radiocarbon, was dated by such indicators as the beginning of haematite inwash due to iron ore mining (A.D. 1889), the sharp rise in phosphorus in the profile due to phosphates in detergents (A.D. 1948) and its decline due to a remodelled sewage plant (A.D. 1963). This direct link between stratigraphic evidence and recent human activity could be useful elsewhere.

Archaeological finds in peat and other sediments have long been used for dating. Relative age based on the typology of artifacts was of course the whole basis of archaeological dating before the availability of chronometric techniques. Many techniques are now available besides radiocarbon, including fission-track dating, obsidian dating, fluorine dating of bones and thermoluminescence of pottery. Accounts of these and other techniques are conveniently gathered together in Brothwell & Higgs (1969). One area where archaeological and ecological evidence intertwine is the Somerset Levels of England where, prior to radiocarbon dating, Godwin made great use of archaeological dating (Godwin, 1948, 1955). Features such as the Neolithic and Bronze Age trackways, the establishment, occupation and abandonment of Iron Age Lake Villages and Romano–British hoards, as well as a number of smaller finds, all buried in peat, gave a chronological framework which was not significantly altered by subsequent radiocarbon dating and a consistent relationship between archaeological events and stratigraphy from 3500 B.C. to A.D. 400 has been built up. (Godwin, 1960; Dewar & Godwin, 1963; Coles *et al.,* 1970.)

One of the major errors that can be involved in dating by stratigraphy is that an object or artifact may be introduced into a stratum of a different age. For example, biological remains may be washed out of an older stratum and redeposited in one much younger, or the artifact may be dug into or pushed down into the sediment. Usually the error in such cases is so gross as to be detectable but there must be cases where for example a spear-head had penetrated a few centimetres into the sediment and no exact correlation can be made between its depth and the depth of the contemporary surface. Dewar & Godwin (1963) discuss a number of finds in relation to stratigraphy and it is apparent that in general the age relationships can be satisfactorily determined when all such factors are taken into account, and especially when radiocarbon dating is available.

3.2 CHRONOMETRIC OR ABSOLUTE DATING

3.2.1 Radiocarbon dating

There are a number of clear accounts of the theory and technique of radiocarbon dating such as Libby (1965), West (1968) and Willis (1969). What is of immediate concern to the historical ecologist is how to get ^{14}C assays done if he needs them, how to collect the samples and how to interpret the resulting dates.

Radiocarbon is produced in the upper atmosphere by the bombardment of nitrogen atoms with cosmic ray particles to produce the radio isotope of carbon (^{14}C) and this forms $^{14}CO_2$ which mixes freely in extremely small proportions with $^{12}CO_2$ and is taken up by plants and animals and dissolves in water. The radiocarbon content of live plants and animals is a function of the concentration of radiocarbon in the atmosphere, and when the organism dies exchange with the atmosphere ceases and the ^{14}C in the organism decays away at a more or less accurately known rate. Half of the ^{14}C is lost in $5,730 \pm 30$ years—its 'half-life'—though the earlier 'Libby standard' of 5,568 years is still used in reporting dates to the journal *Radiocarbon*. The measurement of this residual ^{14}C activity, its comparison to modern standards and the calculations of the date of the sample are the job of the radiocarbon laboratory, who will take into account the various technical sources of error (West, 1968; Willis, 1969) and report the result in 'radiocarbon years B.P.' ('before present'—which is taken as A.D. 1950 before large amounts of radiocarbon were introduced into the atmosphere by nuclear explosions).

The date reported will also have an 'error' of plus or minus so many years—for example '$1,170 \pm 50$ radiocarbon years B.P.'. This 'error' should really be referred to as the 'statistical uncertainty' (Vogel, in Neustupny, 1970a) because it is calculated on the counting statistics of the random decay of ^{14}C and usually refers to one standard deviation. In the example given this means that there is a 68% probability that the true sample age lies between 1,120 and 1,220 years B.P. If the estimate of the age of a sample made prior to radiocarbon assay (and the laboratory will ask for such an estimate) lies outside this standard deviation then the investigator who submitted the sample must revise his previous ideas and/or consider afresh the possibilities of contamination.

The reason for the use of the term 'radiocarbon years' is that it became increasingly apparent throughout the 1960s that one of the fundamental assumptions of the technique—that ^{14}C in the atmosphere had remained more or less constant over millenia—was not valid. Small deviations back to about A.D. 750 were previously known from counts made of the ^{14}C activity of annual rings of *Sequoia gigantea* (Willis *et al.*, 1960), but larger deviations became apparent when *Pinus aristata* (Bristlecone Pine) was used to correlate dendrochronological dates with radiocarbon dates further back in time. Between about 0 A.D./B.C. and the present time these deviations are very small but become progressively larger until at 5000 B.C. the ^{14}C date is about 800 years 'too young' according to the Bristlecone Pine calendar (Fig. 2.8). This discovery has led to a great outpouring of papers and much controversy, especially in archaeology where events dated by reference to some sort of written history stay the same, while events dated by radiocarbon may now appear to have moved back in time. Olsson (1970) edited the proceedings of a symposium on this major change in radiocarbon dating; Neustupny (1970b) has provided a useful summary from the archaeological point of view and Burleigh *et al.* (1973) discuss some of the major points, with a useful bibliography. As things stand at the moment ^{14}C assays will continue to be reported without correction and in terms of the 5,568 years half-life until there is an agreed tree-ring calibration—the laboratory supplying the dates will give advice on this.

Returning to the practical side of radiocarbon dating, the question of how

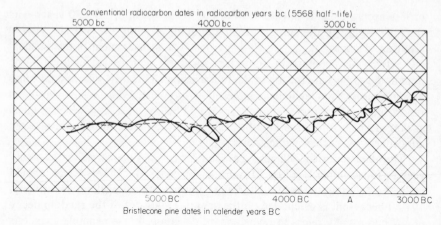

Conventional radiocarbon dates in radiocarbon years bc (5568 half-life)

5000 bc   4000 bc   3000 bc

5000 BC   4000 BC   A   3000 BC

Bristlecone pine dates in calender years BC

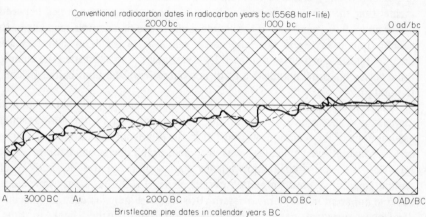

Conventional radiocarbon dates in radiocarbon years bc (5568 half-life)

2000 bc   1000 bc   0 ad/bc

A   3000 BC   A₁   2000 BC   1000 BC   OAD/BC

Bristlecone pine dates in calendar years BC

Figure 2.8 Tree ring calibration of radiocarbon dates. The smoothed curve (broken line) is based on the averaging of dates from three laboratories; the solid line shows the calibration curve of one of these laboratories. (From Burleigh *et al.*, 1973.)

to obtain radiocarbon dates on one's samples resolves itself into either interesting a laboratory in the research being done so that the dates are obtained free, or having enough money to pay for them! The former is always preferable and is the most usual way that individual workers are able to obtain dates. The laboratory making the assay will give instructions on sampling and upon the amount of sample necessary, which varies according to the amount of carbon in the material. Usually 5 g of carbon are needed; with peat, organic mud or wood this may be about 30 g dry weight but with a partly inorganic lake deposit as much as 300 g may be required.

Any contamination which may have escaped notice is most easily seen if a series of dates is performed from the same site. Modern contamination, making a sample too young, is the most common and in a sample of 'infinite' age even 0·1% contamination with modern ^{14}C will give a date of 57,000 years B.P. and 5% contamination will reduce a true age of 50,000 years B.P. by half, to 24,300 years B.P. (Shotton, 1967). A number of other examples like this can be seen in the laboratory reports in *Radiocarbon*. Coope *et al.* (1971) have also detected such errors, in one case due to percolation of sewage effluent. Measurement of such old material, very near the limit of radiocarbon dating, can only be reliably done in the very best laboratories which specialize in such dates; generally the limit is 40,000 years. With much younger material, such as medieval samples of 600 years B.P. true age, 1% contamination will give an apparent age of 540 years, which will probably be within the statistical uncertainty figures which are given with each age determination — for example 600 ± 60 years—and may therefore be undetectable. Five per cent contamination with modern ^{14}C though, will be obvious in reducing this date to an apparent age of 160 years B.P., while 10% contamination will give such a sample a modern apparent age (Sparks & West, 1972, p. 258).

Contamination with older carbon will give a date some years older than the true age; 50% replacement with infinitely old carbon will give an apparent age older by one half-life—5,730 years. This 50% contamination would, of course, be noticeable; a more likely amount of contamination such as 5% or 10% may not be noticeable unless the date was part of a series or at variance with other evidence as to the true age of the sample. A well-worked out example of 'hard-water error' is given by Shotton (1972). He describes how wood and small twigs gave dates in agreement with the palynological evidence and other dates elsewhere, whereas a black mud from the same time-horizon, probably algal and therefore capable of synthesizing tissue from carbonate in hard water, gave dates in a parallel series 1,700 years older than the wood dates.

Radiocarbon dating has been of inestimable value in confirming some of our ideas and changing others in historical ecology; indeed some would go further and replace conventional stratigraphic chronology with one based on radiometric dates (Vita-Finzi, 1973) though this is probably premature in the present state of knowledge (compare Godwin, 1960 with Smith & Pilcher, 1973).

3.2.2 Other radiometric methods

^{14}C is not the only radioactive isotope useful for dating purposes though the historical ecologist may be forgiven for thinking so from this and various other accounts in the literature. Uranium-series dating, based on the longer-lived isotopes in the decay series from uranium to lead, is of little direct use to the historical ecologist though of great interest to the Quaternary stratigrapher. These isotopes are used to date ocean sediment cores, corals and molluscs in the general range 0–400,000 years B.P., far beyond the range of radiocarbon, and together with potassium-argon dating of volcanic rocks have given us an 'absolute' time-scale for the whole of the Quaternary Period. This is not of course perfect as yet; there are a number of assumptions and uncertainties in the methods but these are being actively researched and hold out great hope of an agreed chronology in the future. Further details may be had from Broecker (1965), Shotton (1967), West (1968), Vita-Finzi (1973), Brothwell & Higgs (1969) and Szabo & Collins (1975).

3.2.3 Dating by palaeomagnetism

This is a form of 'indirect absolute' dating in that the palaeomagnetic changes recorded in the stratigraphy can be used as 'absolute' dates once the changes have been firmly dated in a standard section. That the magnetic field has reversed several times in geological history is now common knowledge and has been a major factor in proving and dating continental drift; that the magnetic declination is changing can be seen from the notes on any topographic map and compass readings have to be compensated for it. Both of these changes are of use in dating.

The last major reversal occurred some 700,000 years ago and this and earlier changes have been used in constructing broad chronologies of the whole of the Quaternary Period (Cox *et al.,* 1965, Cooke, 1973). On a much finer time-scale than this though are the declination changes recorded in lake sediments by Mackereth (1971). He was able to show that the sediment had acquired sufficient magnetisation in the direction of the ambient field at the time of deposition and that this remanent magnetisation was stable. Using his own design of pneumatic corer (Section 2.1.3) to obtain undisturbed cores of the upper metre of sediment in Lake Windermere, Mackereth showed that the 'swings' recorded since A.D. 1580 in observatories, and particularly the 'maximum Westerly excursion' at about A.D. 1820, were recorded in the sediments. A six-metre core from Lake Windermere was also analysed in this way and radiocarbon dated. The 'swings' from this long core, back to the end of the Allerød interstadial, appeared to have a frequency of 2,700 years, and on the basis of these data 'magnetic ages' were calculated and showed an excellent agreement with radiocarbon ages. In a further paper Mackereth collaborated with physicists (Creer *et al.,* 1972) in developing the method with a view to more widespread applications, and with similar work on archaeological remains (Aitken, 1970) we may look forward to a great improvement in the precision of our chronologies. The method has already been used with some success in the preliminary studies of Lough Neagh sediments (O'Sullivan *et al.,* 1973), and it may prove invaluable for dating recent deposits contaminated by inwash of ancient carbon and is presumably a means of dating sediments from hard water lakes where radiocarbon dates would be in error. Unfortunately there does not appear to be any remanent magnetism in peat (Mackereth, personal communication).

3.2.4 Varve dating

Two types of varve or annual lamination occur in suitable lakes and are one of the best absolute dating methods—they are also useful in calibrating tree ring and radiocarbon dates (Stuiver, 1970). The first type of varve to be found, and the best known, is that laid down in glacial lakes where the spring and summer meltwater deposits a coarse light-coloured layer of sand and clay which is succeeded by the settling of a thinner layer of finer material as the meltwater decreases or stops in autumn and winter. Work on these varves (also known as rhythmites, pioneered by de Geer (1912, 1934, 1940) in Sweden and Sauramo (1932) in Finland, enabled a chronology of the glacial retreat stages to be drawn up long before the advent of radiocarbon dating, by which method the chronology has been largely confirmed.

The second type of lamination which is thought to be annual in origin is of organic origin in meromictic lakes where the thermal stratification is not disturbed at depth allowing undisturbed sedimentation. The spring and summer sediments are light in colour being rich in diatoms; the autumn and winter layers are dark with organic material. Turner (1970) has investigated such sediments from the Hoxnian interglacial deposits at Marks Tey, Essex and has been able to estimate the total length of the interglacial as 30–50,000 years (much less than other estimates) and the length of the different zones of vegetational succession. At Lake of the Clouds in Minnesota, Craig (1972) also worked on laminated sediments and was able to use some 9,400 annual laminae to date the pollen assemblages. Watts (1973) has made use of both the above studies in his excellent review of the rates of change and stability of vegetation over time; clearly such lakes are of great importance in plant ecological studies and it is unfortunate that so few of them exist. West (1968) shows examples of both types of laminae (Plates 3 and 4) and discusses further the problems of correlation between different sites of varved clays, and of the formation of 'false' varves due to storms.

3.2.5 Dendrochronology

That trees form growth rings and that these can be used to determine age and past growth conditions has been known and used by ecologists for many years, but as Bannister (1969) points out the 'coming of age' of dendrochronology occurred on the 22nd June, 1929 when Douglas succeeded in connecting up a centuries-long 'floating chronology' from archaeological material in the American South-West with a chronology built up from contemporary trees. The principle is illustrated in Fig. 2.9, and depends upon

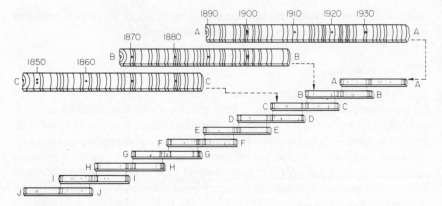

Figure 2.9 Chronology building. A: radial sample from a living tree cut after the 1939 growing season; B-J: specimens taken from old houses and successively older ruins. The ring patterns match and overlap back into prehistoric times. (From Bannister, 1969, after Stallings.)

the matching of abnormal rings, preferably narrow ones, to extend the chronology beyond the range of existing trees. The laboratory of Tree-Ring Research at the University of Arizona in Tuscon is the pre-eminent centre for this work, and for related studies in dendroecology and dendroclimatology. A major review of these studies is that by Fritts (1971), and Bannister (1969) summarizes the world-wide effort to produce long chronologies and gives many references. Smith and his co-workers are also working on a master dendrochronological curve for ^{14}C calibration (Burleigh *et al.,* 1973). The usefulness of long tree-ring chronologies to check and calibrate radiocarbon dating has already been mentioned (Section 3.2.1).

These long chronologies are not relevant to the ecologist investigating the history of a particular area of woodland. A number of recent studies have relied on age determinations gained by counting rings either in a core obtained by a special hollow-tube corer, an 'increment corer' (Pigott, 1969, Merton, 1970) or on a stump (Peterken & Tubbs, 1965, Peterken & Hubbard, 1972). Relying on girth or height to assess the age of individual trees is not at all reliable because of variation. Yarwood (1974) found diameters varying between less than 8 cm to over 28 cm in a plot of 85×85 m within a 40-year-old plantation of *Pinus sylvestris*. Growth rings may also be absent, or double rings formed, within one year (Kozlowski, 1971) and assymetrical growth also makes for difficulties in the use of increment cores. Complete transverse sections of the trunk are therefore to be preferred but are only obtainable by destroying the object of study!

Merton (1970) considers the errors that may arise from ages based on cores and summarizes these as: '(1) the omission of partial rings; (2) the failure to distinguish rings; (3) heart rot; (4) failure to penetrate to the centre of the tree; (5) the difficulty in estimating the time taken to reach 140 cm'—which was breast height, at which the cores were taken. He points out that partial rings are not uncommon in slow-growing species with diffuse-porous wood, but such rings, or intra-annual rings, are unlikely in most species growing in temperate or cold regions (Fritts, 1971). In semi-arid areas, such as the American South-West, or with trees at their climatic limits such complications are more likely and Fritts gives three complex physiological models attempting to show how reduced ring-growth may arise. Cross-dating of different trees (Stokes & Smiley, 1968) can help in the recognition of partial or intra-annual rings.

Heart-rot will be evident in the field but a minimum age can be assigned to such a tree by coring further up the stem and estimating the time taken to reach this height. As Merton (1970) and others point out, the time taken to reach the height at which cores are taken can be a fairly important source of error; the variation in this time can be considerable between shaded and unshaded conditions.

The practical details of taking a core are usually omitted from papers such as those mentioned above but there can be difficulties for a novice. The Swedish Pressler corer (see also Chapter 4, Section 3.3.2) is the usual tool and is made in seven sizes, the most convenient being the No. 3, 20 cm long. It consists of three parts; handle, core-tube and spring-steel extractor, both the latter being stored in the handle. The core-tube is locked into the handle to form a T-shape and screwed into the tree, aiming for the centre. When it is judged that the centre of the tree has been passed the corer is rapidly screwed or jerked in the reverse direction for a couple of turns to break the core off (a snapping sound is often heard), and unscrewed from the tree. The hole should then be plugged with a wooden peg coated with fungicide. The core is extracted with the trough-like length of spring-steel provided and examined to see if a complete radius has been obtained. The point where the curves of the rings reverse is usually easily seen, most trees forming fairly wide rings in their youth, and if any have been missed due to the tree having an eccentric centre the number missing can be estimated by 'projecting the arcs of the innermost rings into complete circles and dividing the distance to the constructed centre by the mean thickness of the oldest rings in the core' (Merton, 1970). If this gives a result greater than seven missing rings Merton recommends re-coring. The result is usually much less than seven rings and checks by the author, taking three or four cores from the same tree, showed the technique to be accurate to within one or two rings.

The Pressler corers were intended originally for softwoods and certainly work very well on these but difficulties may be experienced with some hardwoods such as yew (*Taxus baccata*). One further point to watch with all old trees is that if rotten heartwood is penetrated the corer may be difficult to extract as the screw will have nothing to grip on.

For counting rings a hand-lens is not really ideal; it is better to wrap and label the cores and examine them in the laboratory with a low-power binocular microscope or a 'travelling microscope' with a measuring scale. The cores may be ring-counted in their untreated form but where the rings are to be accurately measured and cross-correlated the cores must be mounted in lengths of slotted wood and sectioned or sanded.

Further details on sampling and analysis can be found in Bannister (1969) and Stokes & Smiley (1968), who mention more specialized techniques to deal with particular problems such as very hard wood or archaeological material, and show how ring series may be correlated and expressed graphically. An earlier paper by Kassas (1951) is also a good example of the way detailed tree-ring data can be used in historical ecology.

3.2.6 Lichenometric dating

Recent papers by Benedict (1967, 1968), Denton & Karlen (1973) and Karlen (1973) have established that reliable dates can be assigned to late-Holocene moraines by measurement of the size of lichen thalli growing on the morainic material. The principle is that once morainic material has been deposited, or a rock surface uncovered, it will be colonized by *Rhizocarpon* which grows radially outward so that the largest diameter lichen found is the oldest and represents a minimum age for the surface. The establishment of a lichen growth curve in an area is a necessary first step (Benedict, 1967) and this is achieved by measuring the largest-diameter lichen on various dated surfaces—gravestones, old mine-workings, moraines dated from old maps and air photographs and railway embankments—and this growth-curve is extended by radiocarbon dating. The basis of this is fully discussed and justified in Denton & Karlen (1973); their lichen-growth curve extends back through the Little Ice-Age to A.D. 1550 with some certainty, and is extrapolated back to some 8,000 years B.P. It certainly seems to be a valid and extremely useful dating method which could possibly be employed on surfaces other than moraines and aid in the reconstruction of the historical ecology of an area.

3.2.7 Dating hedges

This technique, mentioned in Section 2, has been much discussed of late by historical ecologists in Britain and the general theory clarified by its originator (Hooper, 1970). Working from the observation that the number of shrub species in a hedgerow could not be explained solely by edaphic, climatic or agricultural factors, he formulated the 'Hooper hedgerow history hypothesis' that the rate of colonization of a hedge is about one new species per century. There are local variations due to personal or local traditions of planting mixed hedges for example, but in the Huntingdon/Northamptonshire border area Hooper was able to predict the age of a hedge from the number of species in a 30 yard length according to the formula: 'age in years $= 99 \times$ no. species $- 16$'. A fair literature is now accumulating on this intriguing topic, some of which is detailed in Hooper (1970). Pollard (1973) has provided an alternative explanation for some hedgerows rich in species, suggesting that they are the managed relics of former woodland. He substantiates this with a detailed study of the boundary hedge of Monks Wood and a survey of other hedges in Huntingdon and Peterborough. The importance of documentary evidence in this work can not be overemphasized and this is clearly a fruitful field for collaboration between ecologists and local historians.

4 LABORATORY TECHNIQUES

This section is concerned with the isolation and identification of plant macrofossils and microfossils, particularly pollen grains and spores, from the sediments obtained by the methods dealt with in Section 2.1. There are also many other methods in more or less general use which throw light on the historical ecology of certain areas or plant communities, and these are summarized below.

4.1 MISCELLANEOUS TECHNIQUES

A number of chemical techniques have been used on sediments; most of these are modifications of the methods used on non-fossil material (see Chapter 8). Mattson & Koulter-Anderson (1954), Chapman (1964) and Ingram (1967) provide a good introduction to the methods, results and literature of chemical analyses on peat; the work of Mackereth (1965, 1966) on the chemistry of lake sediments may be followed up by reference to Pennington & Lishman (1972) and Pennington *et al.* (1972) where these techniques are exploited to great effect in the ecological history of lake basins in northern Britain. Some analyses, such as those for iodine, seem to show a correlation with past climatic conditions; others, such as carbon, show something of the status of the soils in the lake basin, while the degradation products of plant pigments (Fogg, 1965; Swain, 1965) give some information on the productivity of the lakes.

Analyses of diatoms and of various animal remains are somewhat specialist undertakings. The taxonomic knowledge necessary is not easily acquired and because of the great number of species which may be involved an extensive reference collection of modern examples is usually essential. This also applies to plant macrofossil and pollen analyses but for plant ecologists such material is more easily come by from herbaria or field collecting. When one is dealing with, for example, beetle remains in recent sediments from Britain the individuals that are found may range from present day arctic to mediterranean species and together with the fact that the fossils are usually fragments of the whole insect means that such work is often centred on one or two specialist laboratories. Workers from those laboratories may be called in to advise on such fossils and may well undertake analyses on sites of importance. As a guide to the problems and potentialities of diatom analyses reference should be made to Round (1957, 1961, 1964), Haworth (1969), and Crabtree (1969), who also cite the most important works on identifying diatoms. The most recent advance in diatom analyses is that of counting the values on an 'absolute' basis—numbers per unit wet volume—as developed by Battarbee (1973). Using this technique fossil and modern production figures can be compared. Beetle analyses owe much to the pioneering work of Coope (1961, 1965) and have led to revisions of our ideas on climates derived from plant evidence (Coope & Sands, 1966; Coope *et al.,* 1971; Osborne, 1972).

Non-marine molluscs are similar to beetles in giving information on climate and local environmental conditions, and in being a somewhat specialist area of palaeoecological research demanding a reference collection of modern specimens. B.W. Sparks has drawn attention to the usefulness of these fossils (1961, 1964) and published a number of papers where mollusc evidence has been used together with evidence from plant fossils (Sparks & West, 1959; Sparks & Lambert, 1961; West & Sparks, 1960). The best account of the practical problems involved is in Sparks & West (1972), and keys for land snails are given by Evans (1972) and by Macan (1969) for fresh and brackish-water snails as well as good introductions to the species likely to be found. Isolation techniques for diatoms, beetles and molluscs are given in Appendix 1 of West (1968).

Various other animals may be found in lake and bog sediments and in soil and litter. Frey (1964, 1965) provides some insight into just how many groups and species may be encountered; a good grounding in invertebrate zoology seems essential for any serious work in this field. There is, though, a group of the Protozoa which are important and easy to recognize in sediments—testate rhizopods. The tests or shells of these creatures are usually identifiable to species level and those which occur in peat have been successfully used in characterizing the hydrological/vegetational state of bog surfaces in the past. Frey (1964) summarizes much of the earlier work and gives references to a number of papers; since then one may note the thorough studies by Tolonen (1966a,b, 1971), combining rhizopod, stratigraphic and pollen analyses on Finnish bogs, and the work of Heal (1961, 1962, 1963 and 1964) on the modern ecology of rhizopods. Paulson (1952–53), Tolonen (1966a and Fig. 2.12) and de Graaf (1956–58) contain drawings of the main species which help identification. Most of the smaller tests may be counted at the same time as a pollen count (provided the pollen preparation procedure used does not destroy the tests by, for example, demanding hydrofluoric acid digestion of silica), but ideally rhizopods should be prepared and counted separately. Tolonen (1966a) gives a simple method for this.

Opal phytoliths, microscopic particles of silica laid down in plant cells, especially those of grasses, also have some use in historical ecology. Occasionally recognized in pollen preparations which have not been treated with hydrofluoric acid, they are abundant in the right circumstances (buried grassland soils, for instance) and their potential has recently been reviewed by Rovner (1971). Though not sufficiently differentiated as yet to allow species determination they appear to be useful in indicating a change from woodland to grassland, which would be a great help in areas where pollen-bearing sediments are absent. One can also foresee their use in tracing the presence of a particular species or genus. Rovner (1971) includes some drawings of typical specimens, as well as many references and a preparation technique and Dimbleby (1967) has two plates illustrating phytoliths from thirteen grass species.

4.2 PLANT MACROFOSSILS

When one considers the variety of parts of widely differing plants that can be incorporated into a sediment the prospect of identifying and constructing some sort of meaningful analysis of macrofossils is a daunting one. This partially explains the 'decline' of macrofossil studies once pollen analysis became established, but there has been a resurgence of interest recently (Watts & Winter, 1966; H.H. Birks, 1973). Most pollen-analytical studies have some regard to macrofossils in the sediments, usually listing the more noticeable seeds such as *Menyanthes* or *Potamogeton*, and of course, the general components of the profile analysed for pollen are shown at the side of the diagram, but in some cases the presence of macrofossils can clinch an argument. The finding of macrofossils at Ballybetagh (Jessen & Farrington, 1938) and in the Lake District (Pennington, 1947) are early examples of this sort of evidence, showing that tree-birches, rather than the dwarf birch *Betula nana,* reached these areas during the Late-Devensian interstadial (the Allerød, Zone II, of Godwin, 1956).

Whereas pollen grains are all of the same order of magnitude—15–200 microns, most falling in the 25–40 micron class—macrofossil remains can vary from whole tree trunks, to leaves and whole plants of mosses down to seeds of less than a millimetre in size. Collecting, preparation and identification procedures differ accordingly though some general guidelines can be laid down. Collecting follows the same pattern as for microfossils; there is less likelihood of contamination but the same precautions should be taken and again it is preferable to sample from sections rather than boreholes. This applies especially if peat stratigraphic studies are being undertaken—the stratigraphy of ombrogenous peat reflects the mosaic of plant communities which developed on the old bog surfaces and this stratigraphy can be so complex as to make borehole data confusing and misleading (Section 2.1.2).

A major review of macrofossil work in general by C.A. Dickson (1970), and the recent book *Bryophytes of the Pleistocene* by J.H. Dickson (1973) are essential reading for all contemplating such research. Dimbleby (1967) covers the field from an archaeological standpoint and Godwin (1956) is still a most useful text. The following short accounts of techniques draws largely upon the above.

4.2.1 Wood remains

A large stump of pine exposed in a peat-cutting, with its bark still adhering and with pine needles and cones in the surrounding peat is no problem at all to identify but the 'instant' identification of small fragments of wood brought up in a peat sampler is a temptation to be resisted. Despite the strictures of Godwin (1956) and Dickson (1970) many wood remains are still 'recognized' like this; the cutting of transverse, radial and tangential sections for microscopic examination is the only way to be sure as even bark characteristics can be misleading (e.g. the similar bark of *Betula, Corylus* and *Alnus*) except to the experienced investigator.

Sectioning by hand is a rapid and easy task with the soft, wet wood of peat deposits (as long as it is not too soft when aqueous wax embedding is the best method) and gum chloral can be used to mount the sections for examination. Hypochlorite (bleach) may also be used to clear sections for observation. Charcoal and old dry wood are more difficult and are even more of a specialist undertaking. Dimbleby (1967, Chapter 8) is quite the best account of these techniques and includes 33 excellent photomicrographs of charcoal. Needless to say a reference collection of sections and access to publications such as Jane's *The Structure of Wood* (1956) and the illustrated key to the microscopic structure of hardwoods by Brazier & Franklin (1961) are essential; Clifford's key for fossil wood, a fold-out page in Godwin (1956) is also useful in that it concentrates on those features most likely to have been preserved. A detailed study of the preparation of decayed wood for microscopical examination, not mentioned by the above authors, is given by Wilcox (1964). Although primarily intended for wood decayed by pathogens, this account does cover practical items such as the use of the sledge or sliding microtome and various staining techniques which may be of use to the historical ecologist studying wood remains. A note of caution though, for all those contemplating this task; unlike seeds, fruits and most moss remains, fragments of wood show great variability. They may be pieces of root, twig, branch or main stem, and may have been growing very slowly or in a distorted manner so that comparison with 'ideal' reference slides and photomicrographs is not easy. As Dimbleby (1967) says: '... the job is not one for the inexperienced'.

4.2.2 Seeds and fruits

These are again a specialist study but one much more safely undertaken than the study of wood as the seeds in a vial of modern material may be exactly matched with fossils some thousands of years old, allowing for some morphological variability due to state of ripening, etc.

If at all possible the mud, peat or silt sample should only be washed with water on a sieve or nest of sieves of different meshes. The material can be stirred carefully with the fingers or a variable sprinkler head of the sort used on hosepipes may be left to play over the sample. Treatment with hydrochloric or nitric acid, or sodium hydroxide, at appropriate dilutions, may be necessary to fragment the material (Dickson, 1970) but plain water should be tried first for simplicity and the avoidance of softening and darkening plant remains as happens with sodium hydroxide. Dilute nitric acid is better than sodium hydroxide in that it leaves the plant material cleaner and firmer and the gas bubbles liberated tend to make some of the seeds and fruits float to the top of the container (Godwin, 1956). Mesh size of the sieves used will vary according to the material; obviously the smallest seeds must not be washed down the sink! 125–150 microns will catch everything identifiable but with coarse material (that is, most peats) such a small mesh will quickly clog up and a wider mesh sieve of, say, 250 or even 500 microns, must be used above the fine one and the material examined in two fractions.

Using a stereomicroscope with easily varied magnification from × 8 to × 80 or so (a 'stereozoom' instrument allowing continuous viewing is best) the material is then scanned in a shallow trough and identifiable remains picked out with a sable-hair brush from which most of the hairs have been removed with a razor-blade. Different types are segregated into the compartments of palette dishes and identified either when wet, or drying on damp filter paper, or dry. (West, 1968, Appendix 1; Dickson, 1970).

There is no substitute for experience and a reference collection in identifying fruits and seeds. A number of atlases and short monographs exist and are referenced in Dickson (1970) who goes through a number of the families found as fossils, pointing out the main diagnostic characters. The best of these atlases, by Katz *et al.* (1965) on the Quaternary seeds and fruits of the USSR has about 1,000 species illustrated mainly by line drawings but is unfortunately out of print, as well as being available only in Russian. Similarly Beijerinck's (1947) atlas, with Dutch text and key, and Bertsch's (1941) keys, in German, are not easy guides for the beginner. Workers in North America are well served with the 824 plates of 600 species in Martin & Barkley's *Seed Identification Manual* (1961), though this deals only with 'living' seeds and the fossil characteristics are different in a number of cases. Berggren's atlas (1969) of the Cyperaceae is a model of what such a work should be, with keys to genera and species, detailed descriptions, plates, an index of the type specimens used and even a colour chart. It is intended to issue further volumes to cover about 2,000 species and when complete there is no doubt that this will be one of the standard works on the subject. Use of the scanning electron microscope will also increase our knowledge of critical genera as the work on fossil Ericaceae seeds by Huckerby *et al.* (1972) has demonstrated.

4.2.3 Bryophytes

Moss remains are here treated separately because they are such important constituents of the sediments an historical ecologist may deal with and indeed in ombrogenous peat formations they form the bulk of the peat. Their importance has recently been given full credit in J.H. Dickson's *Bryophytes of the Pleistocene* (1973). Only the most important and recognizable Sphagnaceae will be dealt with here; Dickson gives extensive details of other species and genera.

The special attention given over the years to *Sphagnum* species is reflected in the number of papers dealing with mire growth and stratigraphy and its relationship to climatic change. Since Osvald's classic study of the Komosse bogs (1923), his early work in Britain (1949) and that of Godwin & Conway (1939) on Tregaron bog in Wales, papers by Kulczynski (1949), Godwin (1954), Ratcliffe & Walker (1958), Walker & Walker (1961), Overbeck (1963), Hansen (1966), and Casparie (1969) have speculated on the historical ecology of this genus. One particular species, *Sphagnum imbricatum,* has, as Dickson (1973) puts it: '. . . occasioned more comment from British ecologists than any other species considered in this book'. It was formerly abundant on mires, forming most of the peat, but is now restricted to the north and west and a similar distribution pertains in northwest Europe as a whole. Explanations fall broadly into the categories of anthropogenic extinction due to burning, grazing and drainage (Pigott & Pigott, 1963) or climatic/trophic (Godwin, 1956; Morrison, 1959; Green, 1968).

The techniques for investigating the historical ecology of the Sphagnaceae could hardly be simpler. Walker & Walker (1961) cut 1 cm thick slices from their monoliths, macerated them in 20 cc of 2% sodium hydroxide and sieved through a 150 micron sieve. The filtrate gave samples for pollen analysis, used to correlate levels in the peat stratigraphy, and the material remaining on the sieves was assessed on a 5-point scale and presented as depth/frequency histograms for the various *Sphagnum* species. The author, working on the same problem of peat growth and the succession of *Sphagnum* species in

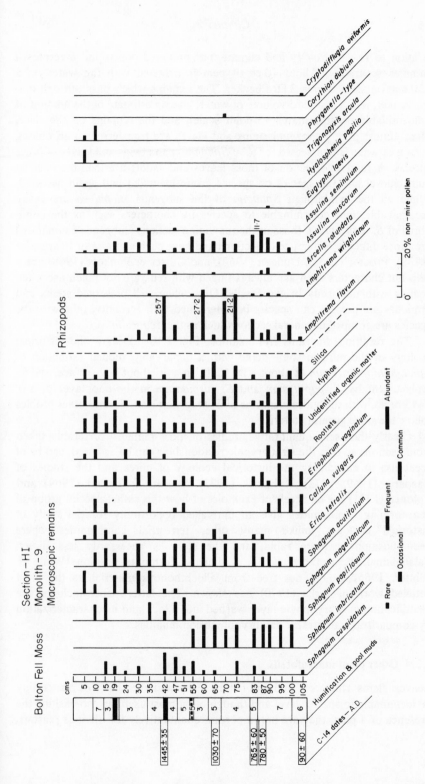

Figure 2.10 Macrofossil diagram from raised bog peat.

relation to bog hydrology and climate, has prepared pollen and macrofossil samples separately. About 10 cc of peat is irrigated with tap water on a 250 micron sieve over a 3 litre beaker. The remains which pass through into the beaker, into a standard volume of water, give an estimate of the amount of unidentifiable organic matter (5-point scale), and the remains on the sieve, often almost pure *Sphagnum* stems and leaves are transferred to an oblong trough of water and assessed for total *Sphagnum* and twigs and leaves of other species. A random selection of moss leaves and shoots are then mounted in gum chloral and the species or species group identified and again assessed. While all the cymbifolian *Sphagna* of the subgenus *Inophloea* are easily recognizable, and determinable to species by characters such as the comb fibrils of *S. imbricatum,* those of the remaining sections (subgenus *Litophloea*) are more difficult. Reference collections and keys such as those of Fearnsides (1938), Proctor (1955), Duncan (1962), and Cöster & Pankow (1968) are a help but characters such as: 'always tinged with red', are not much use when dealing with uniformly brown fossils! Only when well preserved stems and branches are present can species be determined with certainty, otherwise the species are grouped in general sections such as the *Acutifolia.*

The resulting diagrams (for example Fig. 2.10) all show the historical ecology of the mire in some detail with a remarkably sudden extinction of *Sphagnum imbricatum* at about A.D. 1270, on a dried out bog surface, and its replacement by *S. papillosum* and *S. cuspidatum* in the pool-layer of A.D. 1445 or so, with the subsequent rise of *S. magellanicum.* Several other profiles follow the same general pattern.

Clearly then semi-quantitative studies of moss remains, preferably taken from sections so that the stratigraphic relationships can be assessed, can be of great use in elucidating the historical ecology of mires and the studies of Casparie (1969, 1972), Rybniček (1973), Stewart & Durno (1969) and Tolonen (1971) are all excellent examples of how this rather special group of peat-forming communities give an unrivalled opportunity for the study of historical ecology, unlike many other terrestrial communities where decomposition is more rapid, and the story is necessarily less direct. Palaeolimnological studies are in a similarly fortunate position (Watts & Winter, 1966) though not free from allochthonous material as the most detailed work of Birks (1970) has shown. The latter paper includes the identification of a bryophyte layer washed into a loch and its characterization by comparison with present day bryophyte communities.

4.2.4 Other plant macrofossils

Leaves, stems, rhizomes, bud and catkin scales and even hairs of plants may be identifiable to species level and though often most useful in establishing the presence of a plant they are not important quantitatively for obvious reasons.

Dickson (1970) gives numerous examples of identifiable leaves, of which the best known examples are probably those of the Arctic willows and the birches, and those of *Phragmites* in reedswamp peats. *Eriophorum vaginatum* may be easily determined in the field by the blackened, crescentric spindles within the leaf bases, and the papery rhizomes of *Scheuchzeria palustris* are commonly found in the flooding horizons of the peat in Somerset (Plate XII, Godwin, 1956). The occurrence of megaspores and sporangia of pteridophytes and the oosporangia of the Characeae are also dealt with by Dickson (1970) and no doubt there are various other parts of any number of species yet to be found and recognized (for example, the recognition of plant cuticular patterns by Stewart & Follet, 1966).

The specialized field of domesticated plants and wild species used by man has its own problems, such as the extinction of the early types of cereals and the identification of grain from impressions in mud walls or pottery. This subject is reviewed in two papers in Brothwell & Higgs (1969) by Helbaek, the pioneer of these studies in Europe, and Yarnell, with reference to the Americas, and has received monograph treatment in Renfrew's *Palaeoethnobotany* (1973).

4.3 POLLEN AND SPORES

4.3.1 Introduction

Nowhere in the whole field of historical ecology has the 'knowledge explosion' been so marked as in pollen and spore analysis. Godwin's 1934 paper entitled 'Pollen analysis—problems and potentialities of the method', following on the pioneer quantitative work of Von Post and Erdtman, lit the fuse and the technique seems to have resulted in an almost exponential curve of publications since the 1950's. No attempt will be made here to summarize the great amount of work on the basic methodology of the subject—pollen production, dispersion, preservation, representivity, etc.—for which the following key studies should be consulted.

Experimental work on differential pollen production and dispersion has been covered by a number of authors, notably Andersen (1967), Davis (1963), Faegri (1966), O'Sullivan (1973a), Ritchie & Lichti-Federovich (1967) and Tauber (1965, 1967a,b, 1970). More recently two sessions of the Quaternary Plant Ecology Symposium at Cambridge 1972, were devoted to pollen dispersal, sedimentation and representation (Birks & West, 1973). These covered such processes in temperate, tropical and arctic and alpine areas and include citations to almost all the modern literature in this field; because of their importance they are referenced in full in the list at the end of this chapter (papers by: Janssen, Peck, Crowder & Cuddy, Pennington, Andersen, Berglund, Flenley, Birks. All 1973).

A brief review of the more theoretical aspects of 'pollen-analytically based palaeoecology' by Oldfield (1970), draws attention to areas of research which would repay further investigation. These include the relationship between pollen source strength and dispersal distance and the statement that 'work both contemporary and fossil based on areas where distinctive biotopes coincide with characteristic and pollen analytically well represented plant communities could be particularly valuable'. Examples are given of this kind of work and one might add the more recent studies of O'Sullivan (1973b) and Tinsley & Smith (1974).

The linking of contemporary and past pollen deposition is becoming increasingly common in order to explain some of the features of the fossil pollen diagram. Salmi (1962) and Turner (1970) are good examples of this, applied to raised bogs; Birks' study (1970) of inwashed pollen spectra is a particularly well worked out example for lake sediments; O'Sullivan (1973b) interpreted his pollen diagrams from mor humus layers with the help of surface samples, and Webb (1973) compares modern and pre-settlement pollen from Michigan, USA. The work of Rybnickova & Rybniček (1971) on 'the determination and elimination of local elements in pollen spectra from different sediments' has already been mentioned (Section 2.1.1).

4.3.2 Laboratory equipment

Pollen analysis is not an especially cheap technique to fund; even a fairly simple laboratory set-up must have a fume cupboard, centrifuge, hotplate, low-power and high-power binocular microscope, as well as the usual glassware, reagent dispensers, etc. Perhaps though the first purchase should be Faegri's very readable and informative *Textbook of Pollen Analysis* (third edition, 1975; second edition, Faegri & Iversen, 1964). This is the standard text, covering all aspects of pollen preparation and with a reliable key to the pollen types though as with *all* keys it should not be relied on alone. Brown (1960) is also a useful 'cookbook' to have to hand and includes recipes for dealing with coal, shales, oil, asphalt and other somewhat exotic sediments. Part III of Kummel & Raup's *Handbook of Paleontological Techniques* (1965) contains 14 papers on palynological techniques, including such specialities as the cutting of ultra-thin sections of pollen. Erdtman's *Handbook of Palynology* (1969) is a diverting compendium to dip into while the tubes are in the water bath, but by no means a straightforward textbook.

The first essential of all the physical equipment in the pollen-analytical laboratory is that it should be clean. The fume cupboard must be capable of drawing up the fumes of heavy organic liquids (e.g. toluene, acetone) and capable of withstanding attack by such noxious chemicals as hydrofluoric acid. The test tubes used for the actual preparation have a lot to put up with—HCl, H_2SO_4 and HF as well as organic solvents—polypropylene tubes are safer and last longer than glass though they should be replaced regularly. The sieves used should be completely creviceless (e.g. 'Endecotts'), and inspected for damage before each preparation.

The centrifuge need not be especially sophisticated but must reach speeds of over 2,500 r.p.m. and a facility for interchanging swing-out heads to accommodate 15 ml or 50 ml tubes gives added flexibility.

The microscope, the heart of the system, should be as good as can be afforded. The basic essentials are optics capable of about × 300 for routine counting and × 1000 for critical identifications—that is, an oil immersion of very high numerical aperture (n.a. usually 1·30) with as good a colour correction as one can afford, a planapochromat being desirable. A mechanical stage is necessary to move the slides in a controlled manner and to be able to relocate traverses and particular grains. Phase contrast equipment and photomicrographic facilities may be added later if they cannot be afforded at the outset, though phase contrast is essential for advanced work.

4.3.3 Extraction of pollen and spores

Few things make a palynologist happier than to see through his microscope an almost pure suspension of fossil pollen, all the obscuring matrix of plant material and silica having been cleared by various means—disaggregation and dispersal, chemical solution, density separation, etc. This ideal can be approached in certain finely divided sediments, of one chemically-soluble type, such as dy (gel-mud) soluble in potassium hydroxide, or where the plant matrix is relatively fresh and may be sieved and dissolved away, such as little-humified *Sphagnum* peat. One must not, though, aim to produce such preparations from coarse, heterogeneous sediment by applying every technique in the book. Differential destruction and collapse of pollen and spores is well known (Hafsten, 1960; Praglowski, 1970).

The following preparation schedule, based on a variety of sources and commonly used in British laboratories, is an effective and safe technique. Its main use is for Flandrian peats and muds; for soils Dimbleby (1957, 1961) should be consulted, for loess, Frenzel (1964), and for other sediments Gray (1965) and the texts mentioned earlier.

The basis of the technique is to remove carbonates when present with hydrochloric acid, humic colloids by sodium or potassium hydroxide, cellulose by acetylation with acetic anhydride and sulphuric acid (this is not 'acetolysis' though often referred to as that—see Gray, 1965, p. 544), and silica by hydrofluoric acid. Oxidation of lignin is not usually necessary (*Sphagnum* contains no lignin) and if used must be done with the greatest care (Faegri & Iversen, 1964). The polleniferous residue is then stained and mounted in a medium such as silicone fluid or glycerol.

POLLEN PREPARATION SCHEDULE

SOLUTION OF CARBONATES AND HUMIC COMPOUNDS

1. Using a clean spatula place a small quantity of sediment ($c. \frac{1}{2}$ cm^3) in a 15 ml test tube, $\frac{2}{3}$ filled with dilute HCl.
2. When reaction, if any, is complete centrifuge at 3,000 rpm for 3–4 minutes, ensuring first that tubes are filled to the same level. This applies thoughout the schedule.
3. Carefully decant, i.e. pour away liquid from tube, retaining residue. Do it in one smooth movement.
4. Add a few ccs of 10% KOH to test tube, mix on vortex mixer,* top up KOH to within 3 cm of top of tube and place in boiling water bath in fume cupboard for 10 minutes.
5. Using a *little* distilled water, wash residue through a fine (180 micron) sieve sitting over a 400 ml beaker. Clean test tubes with distilled water and refill with contents of beaker. N.B. Be especially careful in keeping sieves, beakers and test tubes in correct number order.
6. Centrifuge and decant.
7. Add distilled water, centrifuge and decant. Repeat procedure until brown stain removed from supernatant liquid.

HYDROFLUORIC ACID TREATMENT

N.B. HF burns are painful and slow to heal. Rubber gloves must be worn and face protected. It is unnecessary for highly organic samples.

(a) Add a small amount of distilled water to residue and mix thoroughly.
(b) Add $c.$ 2″ of HF to test tube and place in beaker of boiling water or water bath for $c.$ 30 minutes. (Time required will vary greatly with quantity of silicates present; for very siliceous samples repeat steps a & b.)
(c) Centrifuge and carefully decant either into a collector vessel or into an acid proof sink in the fume cupboard, copiously flushing with water during and after.
(d) Add dilute HCl, mix and put in boiling water bath for 3–5 minutes—do not boil HCl. (This removes colloidal silica, etc.)
(e) Centrifuge and decant.
(f) Add distilled water, mix, centrifuge and decant. Repeat.

ACETYLATION

8. Add glacial acetic acid, stir, and centrifuge. Decant into fume cupboard sink with water running during and after.
9. Repeat.

---

* Vortex mixers are far superior to mixing with glass rods and eliminate risk of cross-contamination.

10. Make up acetylation mixture, freshly, just before it is required. Using a measuring cylinder mix acetic anhydride and conc. sulphuric acid in proportions of 9:1 by volume. Measure out acetic anhydride, using automatic pipette first, then add conc. H_2SO_4 carefully, stirring to prevent heat build-up. Stir again just before adding mixture to test tubes.
11. Add a few cc to sample, mix and fill tube about $\frac{1}{2}$ or $\frac{2}{3}$ full.
12. Put in boiling water bath for 1–3 minutes. (Stirring is unnecessary unless acetylation is carried on for longer than a few minutes—never leave glass rods in tubes as steam condenses on the rods and runs down into the mixture reacting violently.)
13. Centrifuge and decant *into large beaker of water in fume cupboard.*
14. Add glacial acetic acid, mix, centrifuge and decant.
15. Add distilled water, mix, centrifuge and decant.

MOUNTING—(A)—IN GLYCEROL

16. Add distilled water, 1–2 drops of safranin (red stain) and mix; centrifuge and decant.
17. Carefully drain excess water from the tube.
18. Add 3(–6) drops glycerol and *mix thoroughly with residue.* Amount to use will be determined by amount of residue and concentration of pollen in residue. Test by preparing trial slide—better to have too little glycerol than too much.
19. Using small spatula or rod transfer a small drop of material on a clean glass slide and cover with a cover slip. (18 × 18 mm:No. 0.)
    When concentration is right prepare 3–4 slides for each sample.
    Label with sample number, depth, site, etc.
20. Seal cover slip with nail varnish—taking care not to get polliniferous material onto the varnish brush.

MOUNTING—(B) IN SILICONE FLUID

This is a longer, more refined technique and has the important advantage of not altering pollen grain size overmuch. Also, as silicone fluid is practically non-volatile the slides never dry out and do not need sealing.

16. Add distilled water, mix, centrifuge and decant.
17. Add 1 cc distilled water, 5 cc 100% ethanol and 1–2 drops of safranin. Mix, top up with ethanol, centrifuge and decant.
18. Add 100% ethanol, mix, centrifuge and decant.
19. Add *c.* 1 cc of toluene and pour from test-tubes into *labelled* glass vials. Repeat to wash all pollen from test tube and balance vials carefully. Seal with polythene tops.
20. Carefully lower vials into centrifuge using forceps, and spin at about 750 rpm for 10 minutes.
21. Carefully decant into beaker in fume cupboard.

22. Add silicone fluid (viscosity MS 200/2,000 cs.), 2–6 drops, and *mix* well–pollen
     grains may otherwise clump together.
23. Allow excess toluene to evaporate—24 hours in fume cupboard.
24. Make up slides. Fix cover slip with four blobs of varnish at corners—no need to
     seal.

Vials and slides of silicone fluid preparations may be kept for years without
deterioration, indeed so great is this advantage over glycerol, glycerine jelly
and other mounting media, together with the significant advantage of not
altering pollen grain size, that silicone fluids (which may be obtained in various
viscosities from Hopkin & Williams, Chadwell Heath, Essex) should be used in
preference to all others (Andersen, 1965). Praglowski's (1970) observations on
grain collapse do not seem to be borne out by the appearance of fossil grain
preparations with silicone fluids, probably due to the generally small size of
most grains encountered fossil. Grains also eventually swell and lose their
characteristics after a time in glycerine jelly (Faegri, 1975) so that quick and
easy as this method is (West, 1968, Appendix 1) it is not recommended or
included here. Regarding stains, it is important to note that overstaining will
seriously hinder identifications; in severe cases the grains become almost
completely opaque. Like Brown (1960) the author has experimented with a
number of stains and concluded that although safranin O and basic fuchsin
are adequate, Bismarck brown has the advantage of not appearing to
overstain. Some workers do not use stain at all; the slight 'burning' effect of
acetylation and the good contrast of silicone fluid being deemed sufficient.

4.3.4 Absolute pollen analysis

The foregoing section dealt with a method for the preparation of pollen for
relative pollen counts; that is grains of one taxon are expressed as percentage
of the total grains counted and the discussion in the following sections is based
largely on this method. In an unstable vegetational situation where there occur
massive changes in absolute pollen production (e.g. the immigration of trees in
a late-glacial tundra) or in species coverage (e.g. alder immigration into a mire

Figure 2.11 Some characteristic pollen grains and spores present in the Malham
deposits. a, *Pinus* (pine); b, *Betula pubescens* or *B. verrucosa* (birch tree); c,
Betula nana-type (dwarf birch); d, *Corylus* (hazel); e, *Taxus* (yew); f, *Juniperus*
(juniper); g, *Ulmus* (elm); h, Gramineae (grasses); i, *Quercus* (oak); j, *Fraxinus*
(ash); k, *Alnus* (alder); l, *Salix* (willow); m, Ericaceae (heather); n, Cyperaceae
(cotton grass and sedges); o, *Plantago lanceolata* (ribwort plantain); p, *Tilia
cordata* (lime); q, *Artemisia* (wormwood); r, Compositae, *Taraxacum*-type; s,
Potamogeton (pondweeds); t, Umbelliferae, *Heracleum*-type (hogweed); u,
Helianthemum (rockrose); v, *Sphagnum* pore; w, *Selaginella selaginoides*
microspore. (From Pigott & Pigott, 1959.) NB: For remarks on identification of
Betula nana pollen see Birks (1968).

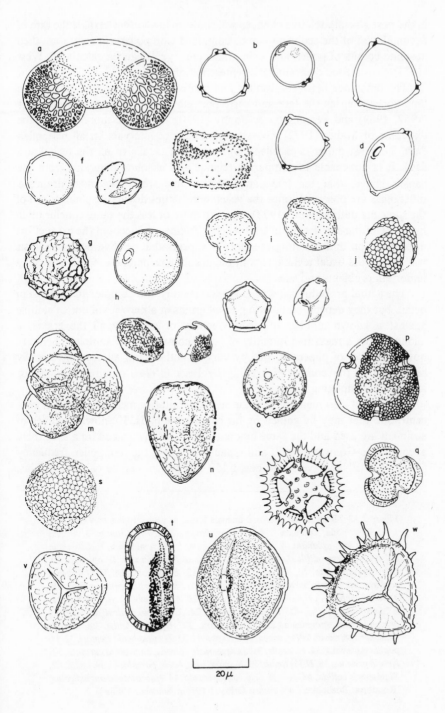

20μ

in the post-glacial), relative changes will show up in all other taxa. If the rate of accumulation of the sediment can be combined with pollen concentration then absolute counts of pollen influx (in grains cm^{-2} yr^{-1}) can be calculated. West (1971) gives a good diagrammatic representation of this.

The difference between absolute and relative diagrams can be dramatic in time zones such as the late and early post-glacial as the diagrams of Davis (1967, 1969) and Pennington & Bonny (1970) show, and more recently the diagrams of Maher (1972) who shows confidence intervals on all his pollen data. This last paper is excellent, with very full details of all the techniques used. It is noticeable in comparing relative and absolute diagrams from the papers above, that the late-glacial is the time when the most dramatic differences are observed; once the vegetation 'settles down' the changes are of the same magnitude. Sims (1973) comes to more or less the same conclusion in his study of the elm decline in East Anglia. Taking into account the extra effort in preparation and counting therefore it appears that relative pollen analysis will remain the usual technique with the absolute technique used on especially important problems and sites.

The actual procedures involved are too lengthy and complex to describe in detail, but they depend on counting fossil grains in a known volume of sample against a known number of 'exotic' pollens or counting all the grains in aliquots from a measured quantity of liquid containing the sample. There are now a number of papers on this by various authors but all are conveniently referenced in the comparative study by Peck (1974), required reading for anyone contemplating this work.

One might also mention here the semi-quantitative measure of pollen concentration that may be gained by far simpler methods. If known volumes of sediment are used and the same amount of final residue placed on a slide then, according to Conway (1947) and Chapman (1964) the tree pollen frequency (TPF = 1,000/No. traverses to count 150 tree pollen) can be calculated. This,

---

Figure 2.12 Recent and subfossil animals found in peat samples of Varrassuo. 1–43 Rhizopoda, Testacea 1. *Amphitrema flavum*, 2. *A. wrightianum*, 3. *Arcella rotundata* var. *aplanata*. 4. *Assulina muscorum*, 5. *A. seminulum*, 6, *Corythion dubium*, 7. *Cryptodifflugia oviformis*, 8. *Diffludia bacillifera*, 9. *D. oblonga*, 10. *Euglypha alveolata*, 11. *E. laevis*, 12. *E.* sp. 13. *E. strigosa*, 14. *Hyalosphenia subflava*, 15. *Nebela parvula*, 16. *N. militaris*, 17. *Phryganella hemisphaerica*, 18. *Trinema enchelys*, 19. *A—C. Arcella artocrea*, 20. *A. cantinus*, 21. *A. cantinus*, 22. *A discoides*, 23. A—D. *Bullinula indica*, 24. *Centropyxis aculeata*, 25. *C. aerophila* var. *spagnicola*, 26. *C. laevigata*, 27. *Difflugia leidyi*, 28. A—B. *Heleopera petricola*, 29. *H. rosea*, 30. *H. sphagni*, 31. *Hyalosphenia elegans*, 32. *H. papilio* (anomal.), 34. *H. ovalis*, 35. *Lesquereusia spiralis*, 36. *Nebela carinata*, 37. *Nebela griseola*, 38. *N.* (?) *tincta*, 39. *N. marginata*, 40. *N. parvula*, 41. *N. tincta*, 42. *Plagiopyxis callida*, 43. A—B. *Trigonpyxis arcula*, 44. *Habrotrocha angusticollis*, (Rotatoria: Bdelloidea), 45. Acari: Oribatei. (From Tolonen, 1966a.)

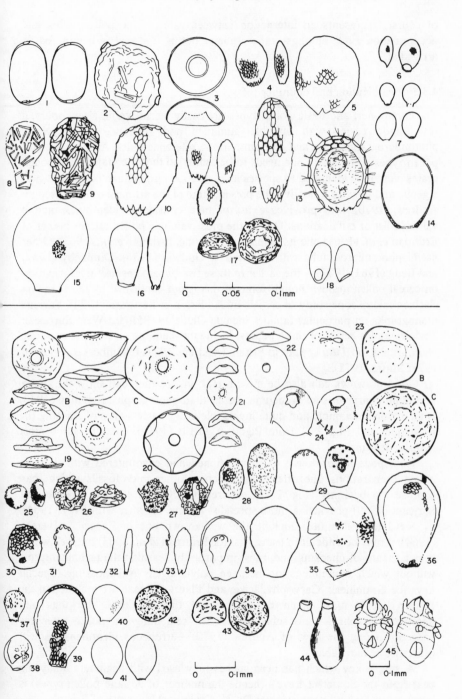

of course, represents an interaction between rate of tree pollen influx and accumulation rate, amongst other things, but is a useful indicator, easily arrived at.

4.3.5 Identification and counting

The commoner, and basic pollen types are easily learnt within a few hours, a task made easier with suitable guidance and reference to drawings, photomicrographs, keys and, most important, reference slides. As more difficult pollen types come along one needs to be reminded that the majority of pollen grains cannot be identified to species level. A good number can be run down to a genus and most to a family. The present state of the art has been assessed by Andrew (1970) who provides a list of taxa considered identifiable and a valuable list of publications. One of the most valuable pollen atlases is that of Erdtman *et al.* (1961) and its companion volume, Erdtman *et al.* (1963). Other useful photomicrographs will be found in Godwin (1956), Dimbleby (1967) and Beug (1961, 1971)—the earlier of these two papers has useful information on cereal pollen grains, though some of this is not confirmed by the scanning electron microscope study of cereals by Andersen & Bertelsen (1972). Various monographs on particular families include Oldfield (1959) on West European Ericales and a series of three papers, including scanning electron microscope photographs, on *Tilia* species in Britain (Andrew, 1971; Chambers & Godwin, 1971; Mittre, 1971).

Good drawings can also be most useful; indeed they can show the shape and main characteristics sometimes better than photographs. Erdtman (1943) is an excellent example and quite invaluable in the early stages of learning the technique, and Fig. 2.11 from Pigott & Pigott (1959), contains some most realistic drawings.

Rhizopods have already been mentioned and encountered with these in pollen preparations one often comes across other microfossils which may confuse the beginner (Fig. 2.12). Andrew (1970) notes some of these but Sarjeant (1969) provides a more specialist review, together with Frey (1964).

Keys to pollen taxa include the well-known one in Faegri & Iversen (1964); it may be difficult to use at first (or Erdtman *et al.* 1963) but well worth it as a means of checking and, most importantly, learning the technical terms without which communication between palynologists is impossible. Special keys for Gramineae, Caryophyllaceae and Plantaginaceae are included in the former but it must again be emphasized that all keys, drawings and photomicrographs are no substitute for the real thing—a pollen reference collection along the lines of Andrew (1970)—erroneous identifications will otherwise be inevitable.

A simple key which has been successfully used with beginners analysing peat from the Somerset Levels (hence the number of aquatic pollen types) is given below by kind permission of Dr. K. Crabtree, University of Bristol.

POLLEN KEY FOR USE WITH BRITISH POST GLACIAL MATERIAL

1.	Single grain with air bladders	*Pinus*
	Single grain	4
	Tetrad	2
2.	Smooth or irregular surface pattern	3
	Regular elongate reticulate pattern single pore	*Typha latifolia*
3.	Tetrad relatively small and smooth surfaced	*Empetrum*
	Medium sized, irregular rough pattern	*Calluna*
	Large size, rounded tetrad	*Erica*
4.	No visible pore, bean shaped to oval	FILICALES
	A three pronged (triradiate) scar	5
	Single pore visible	6
	More than 1 pore or furrow visible	7
5.	Smooth walls	*Pteridium*
	Fine cloud pattern on walls	*Sphagnum*
	Very coarse large pattern on walls	*Osmunda*
		Botrychium, etc.
6.	Smooth wall, rounded oval in shape	GRAMINEAE
	Fine rough pattern, pore poorly defined, triangular to oval in shape	CYPERACEAE
	Clearly defined reticulate pattern:	
	(a) 2 walls, regular-shaped reticulum	*Sparganium*
	(b) 2 walls, elongate reticulum	*Typha angustifolium*
	(c) 1 ill defined wall, regular-shaped reticulum	*Potamogeton*
7.	More than 3 pores visible	8
	More than 3 furrows visible	9
	3 pores visible	11
	3 furrows visible	12
	Surface patterns obscuring any furrows and pores	10
8.	4–6 pores, cloud pattern	*Plantago* spp.
	7–12 pores cloud pattern	*Plantago lanceolata*
	4–7 pores, smooth wall, arci leading from pores	*Alnus*
	4 pores, smooth wall, large grain	*Carpinus*
	5 pores, elongate reticulate pattern on wall	*Ulmus*
	12–24 pores, smallish grain, thick wall	CHENOPODIACEAE
	8–20 pores, larger grain, pores membraneous	CARYOPHYLLACEAE
	4–5 large barrel pores smooth wall	*Myriophyllum*
9.	6 furrows very fine reticulate pattern	*Galium*
	6 furrows fine reticulate pattern larger grain	LABIATAE
10.	Oval grain with pointed spines	*Nuphar*
	Round grain with small blunt processes	NYMPHACEAE
	Grain thick walled with spines and star or crest shaped thickenings	*Taraxacum*
	Grain largish and rounded with coarse spines	(Thistle group) COMPOSITAE

11. Smooth wall (a) simple pore   *Corylus*
                 (b) Snake-head pore   *Betula*
                 (c) Double wall, small grain   *Filipendula*
     Thickened wall (a) Cylindrical shape   UMBELLIFERAE
                    (b) Thickened recessed pores   *Tilia*
                    (c) Very large grain, thickened   *Epilobium*
                         pores at corners
     Slight rough pattern, thinnish wall;
                    (a) pore with marked triangular furrow   *Fagus*
                    (b) small grain 3–4 pores with short   *Rumex acetosa*
                         furrow
                    (c) as above but longer furrow   *Rumex acetosella*
                    (d) Larger grain 3–4 pores and furrow   *Rumex* spp.
12. Rough surface pattern (spotty) (a) 1 wall   *Quercus*
                                    (b) 2 walls   *Ranunculus*
     Pattern of striae (a) overlapping   *Menyanthes*
                       (b) separate   *Acer*
     Reticulate pattern (a) Smallish oval grain   *Salix*
                        (b) Larger rounded oval   *Fraxinus*
                        (c) Large coarse reticulum   *Hedera*
                        (d) Rounded grain, reticulum   CRUCIFERAE
                             composed of fine dots
                        (e) Large rounded processes   *Ilex*
                             projecting from the
                             surface
     Thick tripartite walls (a) with spines   *Bellis* (COMPOSITAE)
                            (b) without spines but   *Artemisia*
                                 internal vertical (radial)
                                 thickenings
     Very large size, with thick pallisade and small   *Succisa*
     spines

In the actual counting it is important to be methodical and establish a routine so that grains are not counted twice or special grains lost by not being noted, with their coordinates on the slide for further study. Starting, therefore, at the same corner of the cover-slip one traverses along or down a particular line, identifying and counting all the grains encountered and noting them on a specially prepared scoring sheet.

The slide is moved at least $1\frac{1}{2}$ field diameters to the next traverse and counting continued until the particular pollen sum being used is reached. Note must be taken of Brookes & Thomas' (1967) demonstration of the non-random distribution of pollen on the slide which can seriously distort spectra counted from only the edge of the slide. To try and overcome this slides should not be too rich in pollen so that more or less the whole slide is counted. Sometimes, of course, 2–3 or more slides need to be counted though one should beware of achieving a result at all costs. Low pollen content may be a

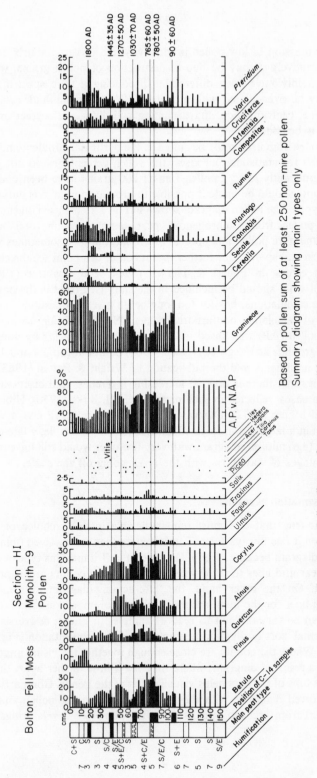

Figure 2.13 Pollen diagram from same monolith as macrofossil diagram (Fig. 2.10).

Based on pollen sum of at least 250 non-mire pollen
Summary diagram showing main types only

genuine expression of low pollen productivity and/or rapid matrix accumulation. Alternatively it may be due to destruction of pollen grains, which will almost certainly have been differential, producing a false spectrum, or the pollen may be over-diluted with mountant. (With silicone fluid, resuspension with toluene, centrifugation and decanting will allow the correct amount of mountant to be added.)

The pollen sum itself must be suited to the particular problem and must be high enough for statistical purposes; the pollen percentages do not generally change significantly after a pollen sum of 200 grains have been counted but higher counts should be made if possible to produce continuous curves of less frequent types. The pollen diagram shown in Fig. 2.13 is based on sums of 250 non-mire pollen from ombrogenous peat samples giving total sums (i.e. including the mire pollen and spores) of around 500 but sometimes well over 1,000 pollen and spores. For further discussion of this point see the chapter on 'Statistical errors' in Faegri & Iversen (1964) and Crabtree (1968). The question of what should be included in the sum itself (that is the basis of the percentage calculations) has for too long beeen a vexed question. The custom in Britain of calculating non-arboreal pollen (NAP) and *Corylus* against 150 arboreal pollen (AP) was based on the exclusion of *Corylus* by von Post on fallacious grounds as Faegri & Iversen (1964) point out (pp. 84 and 134–35). With their arguments and the publication of Wright & Patten (1963) entitled 'The pollen sum', there is now an increasing awareness and discussion of the calculation bases, reflected in such papers as H.H. Birks (1970), Hibbert *et al.* (1971) and Maher (1972). Diagrams based solely on % AP give a very false impression in a largely non-forested landscape—for example, a late-glacial or sub-recent (agricultural) diagram with very few trees would still have individual AP percentages of, say, 80% with the NAP going off the scale.

4.3.6 Presentation of results

While it is true that one must remain flexible in one's choice of diagram presentation it has always seemed unfortunate that the resolved ('blacked-in') saw-edge diagram became so popular, at least in Britain. This type of diagram, without clear grid-lines to show where the individual counts are, is potentially misleading. On the other hand, the 'interaction' diagram of symbols and connecting lines for the main arboreal pollens, commonly used in continental Europe, can be very difficult to read. (Examples in Faegri & Iversen, 1964.) For a 'normal' sort of diagram a simple bar graph presentation (Fig. 2.13) is preferred. Where the counts are close enough together the visual impression is almost of a resolved diagram; where they are a little farther apart then a line joining the tops of the bars helps visualization of the trends. Comparison of the earlier resolved diagrams with the later bar diagrams of pollen analyses by F.A.Hibbert from archaeological sites in the Somerset Levels brings out this point clearly (Coles & Hibbert, 1968; Coles *et al.*, 1970). Presentation is largely a matter of personal taste; the perfect diagram does not exist and one can always see improvements that could be made to one's own diagrams after the second and third re-drawing.

4.3.7 Interpretation

To attempt interpretation of pollen diagrams without a good working knowledge of the ecology of plant communities and of certain critical species is to invite erroneous conclusions. This is less likely to occur in diagrams from the 'well-worked' temperate regions than to diagrams from areas like the tropics. In such cases notes on the state of ecological knowledge should be included in the text, as in Morrison (1968) in Uganda and Kershaw (1971) in Australia. Interpretation naturally takes account of what has gone before, building from the basic outline of workers such as Godwin (1940) in Britain and Deevey in the Northeastern United States (Davis, 1965). The theoretical and experimental researches mentioned in Section 4.3.1 must also be applied to the problem.

The Assemblage Zone concept (Pollen Assemblage Zone = PAZ) has recently come into use and a series of local and regional zonations have been formally described; these largely replace the old 'Godwin zones' though they will undoubtedly continue for ease of reference. Cushing (1967) published the first PAZ's for North America and the advantages of this method were soon realized and introduced to Britain, where they had already been discussed. Turner & West (1968) and West (1970) proposed chronozones as broad divisions that could be easily subdivided. When securely dated, PAZ's can be correlated with these chronozones. H.H. Birks (1970) established Local PAZ's in her pollen diagram from Abernethy Forest, northeast Scotland and Hibbert *et al.* (1971) reported the 'first in a series [of ^{14}C dated PAZ's] from Flandrian sites throughout the British Isles which it is hoped will form a secure basis for the definition and dating of the major pollen assemblage zones in the country'. A recently discovered site in the New Forest that includes the whole of the Flandrian could similarly become a standard pollen diagram for that ecologically important area (Barber, 1975b). For diagrams covering the very recent past PAZ's are of limited usefulness; the pollen diagram from Bolton Fell Moss (Fig. 2.13) and other diagrams from the same time period, may be interpreted as records of agricultural as well as ecological history and often show too many variations due to localized human activity to be used as correlative tools for a large area. Pollen analyses from archaeological sites need to be very carefully interpreted with the aid of the archaeologist and any other specialists involved. Barber (1975a) provides an example of such an analysis from a medieval urban situation.

Numerical aids to interpretation should not be overlooked. Mosimann's

paper (1965) 'a collection of statistical tools . . . ready for use' by the pollen
analyst is probably the best known and is made considerably more useful to
the less numerate of us by its worked examples! Westenberg (1947a,b, 1967) is
also an authority on these pollen statistical problems. Computers have been
used in the interpretation, comparison and plotting of diagrams (Gordon &
Birks, 1972, 1974; Damblon & Schumacker, 1972). A punch card data bank
of Quaternary plant fossil records (Deacon, 1972) has helped yield a numerical
analysis of the past and present flora of the British Isles (Birks & Deacon,
1973) and shows promise for further development.

5 DOCUMENTARY RECORDS

5.1 SCOPE

This section must inevitably be a scrapbook of sources, so wide is the
variability in type and time-span of material. In time it could be said that
'ecological' records go back to information on the harvests of the early Near
Eastern civilizations. The type of material may be visual—paintings,
photographs, maps and diagrams—or written, in the form of monastic
records, farm and estate records, old floras and now that ecological studies
have been published over some decades they themselves become historical
documents. One thing stands out in considering the sources quoted below;
they are all to some degree local, sometimes intensely so. The message must be
that there are few general rules to follow and each study must be approached
individually. Simply knowing where to look for information is perhaps one of
the most important generalizations that can be made, and one can draw a
distinct contrast between the Old and New Worlds. In Britain the Public
Record Office, London, is one of the richest sources of information together
with the County Archives and old-established libraries such as the Bodleian in
Oxford. In particular areas there may be substantial amounts of estate papers
still in the private hands of large landowners and often uncatalogued. In the
USA, New Zealand, etc., the records start with the first settlement by
Europeans but it was the Land Surveys of the 19th century which began the
really systematic gathering of information, much of it of direct ecological
relevance. Virtually all of the data are available for scrutiny in the National
Archives, Washington D.C. and N.Z. Lands and Survey Department.
Paradoxically one can encounter great difficulty in building up a picture of the
landscape in the past from a number of scattered sources (many of which do
not refer to vegetation directly) in the longest-settled areas, whereas in the
USA detailed maps of pre-settlement vegetation can be built up relatively
easily.

What follows is an introductory guide to the literature—general books on landscape evolution, historical ecology papers which have used documentary evidence of one sort or another, ecological studies which can now be regarded as 'historical' themselves, and information on old maps and photographs.

5.2 WRITTEN EVIDENCE OF PAST VEGETATION

5.2.1 Agricultural and land-use records

The outstanding examples in this category are the Land Surveys of the 19th century already mentioned. Gordon (1966) is a good example of what can be done with the information, producing a map of the natural vegetation of the State of Ohio in no less than 10 categories. Birch (1971) and Rankin & Davis (1971) have used the surveys for parts of Illinois and Alabama respectively. Birch was even able to construct maps of tree distribution and size for the survey period 1807–1818, before any permanent settlement disturbed the pattern (Fig. 2.14). Potzger *et al.* (1956) did similar work on the survey records of Indiana.

As already mentioned the situation is somewhat different in Britain—much more difficult but intriguing and full of interest. Certain prolific writers are well-known and their books are valuable introductions to any case study. Professor H.C. Darby's work on the *Domesday Geography of England* (1952–1967 in five volumes with other authors contributing) and his other numerous writings on this theme are justly famous (Darby, 1973). The main information upon the vegetation is on the amount of woodland remaining at the time of the Domesday Inquest in 1086 and its subsequent clearance (Darby, 1950, 1951), both papers referring to more detailed work and generally on a county scale.

Professor W.G. Hoskins, pre-eminent in the study of English local history, gives a number of sources in his *Making of the English Landscape* (1955) and has built on this in two further informative volumes (1959, 1967). He is now editor of a county series under the same title. Professor Maurice Beresford's chapter on English parks in his *History on the Ground* (1971) is of interest in pointing out some old parklands (*vide* Peterken, 1969), and recent monographical treatment of the 'reclamation' of two of the wilder parts of England, Exmoor (Orwin & Sellick, 1970) and the Somerset Levels (Williams, 1970) would be essential reading for any historical ecologist working in those areas.

The theme of woodland clearance and enclosure is, in detailed studies, a rich vein for the historical ecologist. The following papers, chosen from the last few years' issues of one geographical journal, illustrate the point.

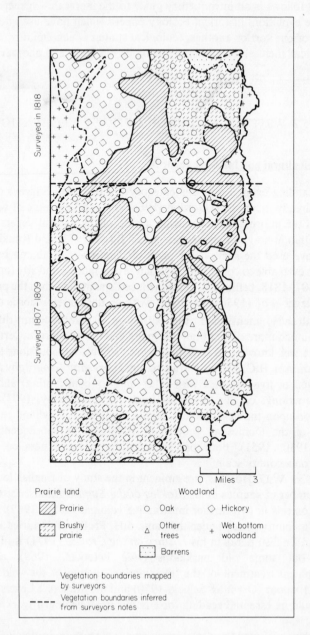

Figure 2.14 The vegetation of Edwards County, Illinois, in the early 19th century. (From Birch, 1971.)

(a) Wightman (1968) on the pattern of vegetation in the Vale of Pickering area *c.* A.D. 1300; Fig. 2.15 is an example of the sort of map that can be produced by this careful work. Various species are also mentioned.

(b) Brandon (1969) on medieval clearances in the East Sussex Weald; includes chronology and extent of woodlands *c.* A.D. 1500.

(c) Williams (1970) on enclosure and reclamation in 18th and 19th centuries, examining particularly the sources and their reliability.

(d) Nicholls (1972) on the evolution of Needwood Forest; refers especially to forest laws (cf. Tubbs, 1968).

(e) Williams (1972) on enclosure in Somerset; an example with particularly full documentation.

Other papers to note on this theme are those of Donkin (1960) on the activities of the Cistercians in the Royal Forests, and Yates (1964) on 'waste-edge' settlements in Sussex, Needwood Forest, Dartmoor and Shropshire. On woodland management *per se* the following papers are good guides to the sources available; Roden (1968) on medieval management of the Chiltern woodlands and Tubbs (1964) on the pre-1700 coppicing system in the New Forest. Rackham (1974a, b), on the detailed history of an ancient wood and on the history of oak in Britain, are also excellent examples.

The close linking of documentary and pollen evidence has already been mentioned as a dating procedure (Section 3.1), but it is far more than just this. One sees a praiseworthy trend in two recent studies where a pollen analyst and a historian have joined forces: Tallis & McGuire (1972) on the pollen and documentary (court rolls, accounts, charter, wills, etc.) evidence for woodland clearance in part of Lancashire and Roberts *et al.* (1973) on the land use of Weardale in which pollen and documentary evidence is welded together in a very satisfactory manner. In all pollen/documentary work one must bear in mind the 'resolution' of the pollen diagram, that is the closer the spectra the more detailed the events that may be demonstrated, but there is also a threshold factor operating in that by no means every historical event will show up as a change in the pollen frequencies. The pollen diagram (Fig. 2.13) from Bolton Fell Moss in northern Cumbria seems to have quite good resolution in showing something like the following succession.

(a) Pre-Roman largely forested landscape up to *c.* A.D. 90.

(b) Roman clearance—the site is only 5 km north of Hadrian's Wall.

(c) Post-Roman decline and forest regeneration over some 600 or so years.

(d) Norse/early Medieval clearance—a short period.

(e) Medieval decline—the Border wars and raids.

(f) Late Medieval sudden clearance and much more extensive agriculture.

(g) Peak agricultural activity at time of Napoleonic Wars.

(h) Re-afforestation and modern estate management.

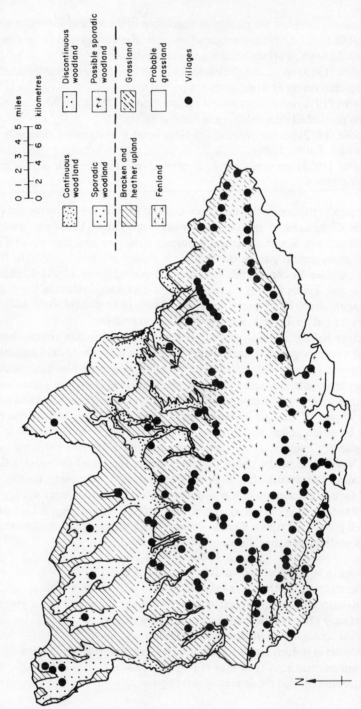

Figure 2.15 Probable vegetation pattern, Vale of Pickering, c. A.D. 1300. The large black circles indicate medieval settlements and give some indication of the major areas in which the vegetation pattern was interrupted by village arable.

The full details of this work are yet to be reported but it can be seen from the diagram that much ecological information can be extracted, and matched with independent documentary results, especially in the upper 20–30 cm (e.g. Nicholson, 1777; Tate, 1943). Oldfield (1963, 1969) has published similar correlations from southern Cumbria. One may also link work on pollen/documentary and peat stratigraphic data with climatic change (Barber, in prep.) and Brandon (1971) gives correlations between weather conditions and harvests in Sussex—all leading on to the equally intriguing and important field of recent climatic change from documentary sources (Lamb, 1966; Manley, 1958, 1964; Ladurie, 1972).

As with all historical studies, information increases as one nears the present and for those interested in the ecological information to be derived from agricultural surveys of the 18th and 19th centuries the following source-guides give an insight into what is available. Grigg (1967) is the best survey that I know and includes copious notes (refers to the period 1770–1850). Cox & Dittmer (1965) survey the tithe files of the mid-19th century (see Prince, 1959 for tithe maps), and Franklin (1953) and Fussell (1964a,b) deal with grassland cultivation, pre-1700 to the 20th century. While thinking of agriculture we should recall the great effects grazing animals have on vegetation and how the countryside has been modified for game. Sheail's excellent *Rabbits and their History* (1971) shows just how great an influence 80 million mouths (and teeth!) can have on the vegetation of huge areas, while *The Book of Duck Decoys* by Sir Ralphe Payne-Gallwaye (1886) includes old decoy locations and diagrams, as well as dates of abandonment (and therefore the beginning of the hydrosere) interspersed with stories of outrageously anthropomorphic ducks!

5.2.2 Direct floral records

Any vegetation survey or record of a particular plant species is potentially an historical document. The records of Diver mentioned in the next section are paralleled by the published and unpublished records of Goode (1948). Both of these workers provided detailed information on the flora of Dorset and their manuscript records are fortunately housed in a government research station—there must be many more unpublished records which are not yet quite so safe and available for comparison with the present situation.

Published records which can be used to follow changes in the flora abound in the reports of local natural history societies and in old floras. A guide to local floras in Britain is provided by Wanstall (1963). Use may also be made of the *Atlas of the British Flora* (Perring & Walters, 1962) and of the records of the Botanical Society of the British Isles (and their counterparts elsewhere) to follow the history of particular species—a good introduction to such studies is provided by Perring (1970a,b); the latter paper surveys the decline of rare species in Britain over the last seventy years. The role of fire in the ecological history of the natural conifer forests of Western and Northern America was the subject of a symposium in 1972. Papers included one by Heinselman (1973) who used written evidence and growth ring data, and by Swain (1973) who linked evidence from lake sediments to the written history of recent times but also extended the record back into the prehistoric period. Many of the other papers contain data which will become useful historical information for workers in the future.

Records of introduced species of trees into the British Isles may be found in Mitchell (1974) and older works such as Mongredien (1870). Salisbury's *Weeds and Aliens* (1964) covers the history and spread of many common British herbs—Chapters 2 and 3 on weed floras of the past and recent dispersals of species are of especial interest to the historical ecologist.

5.3 MAPS AND PHOTOGRAPHIC RECORDS

Landscape and plant community changes are dramatically highlighted by maps and air or ground photographs. The accuracy of old maps and nautical charts has been assessed by Carr (1962), Harley (1972) and Harley & Phillips (1964). The development of aerial photography, with reference to archaeology and therefore crop-marks and other signs of soil disturbance is the subject of a book by Deuel (1969); while Howard (1971) *Aerial Photo-Ecology,* is a standard work on modern techniques, together with publications such as the *Transactions of the Symposium on photo interpretation* (1962).

Three good examples of change photographed from ground level are to be found in Godwin *et al.* (1974), Tansley (1968) and the memoir to Captain Cyril Diver (Merrett, 1971). Godwin's photographs, taken in 1928 and 1944, show invasion of *Rhamnus frangula* into Cladio-Molinietum at Wicken Fen, while Plates 123 and 124 in Tansley, taken 13 years apart, bear witness to the rapidity of the spread of *Spartina townsendii.* Diver's work in the 1930s on South Haven Peninsula, Dorset, was extraordinarily comprehensive and enabled the very same views to be rephotographed in 1971; the changes, especially in tree growth, are thus particularly clearly seen. Old and new maps of the South Haven Peninsula show similar changes in physiography. Another good example of the vegetation as it was in the recent past is provided by Küchler's (1964) map and photographs relating to the USA.

The ecological usefulness of standard topographic maps is fairly limited. Woodland presence and boundary changes are the most obvious information gained, though Green & Pearson (1968) found previous editions of the Ordnance Survey maps of Wybunbury Moss an aid in assessing the extent of man's interference with the plant communities. Also, the Second Land Utilisation Survey, progressing at present, may be compared with the First

LUS of the 1930s, though again one is limited by the categories of vegetation defined to rather large scale changes. In New Zealand, Johnston (1961) has drawn attention to the vegetation detail on manuscript maps of the Canterbury area dating from the mid- to late-19th century; he discusses the accuracy and amount of material on each and has compiled a series of 1:63360 vegetation maps.

Maps and air photographs prepared for a specific purpose are therefore more rewarding than many state agency records. Godwin *et al.*'s (1974) maps of Wicken Fen and Watt's permanent quadrat exclosures of grassland on Lakeheath Warren (1962) are models of their kind and the Cambridge University Collection of air photographs of Blelham Bog aided Oldfield's (1970) interpretation of some of the man-made features there. One could go on multiplying the number of references but the point has already been made: documentary records, written, cartographic and photographic can be crucial to the success of any study in historical ecology.

6 REFERENCES

AITKEN M.J. (1970) Dating by archaeomagnetic and thermoluminescent methods. *Phil. Trans. R. Soc. A* **269**, 77–78.

ANDERSEN S.TH. (1960) Silicone oil as a mounting medium for pollen grains. *Danmarks geol. Unders. ser.* **4**, 4, 1. 24 pp.

ANDERSEN S.TH. (1965) Mounting media and mounting techniques, in B. Kummel & D. Raup (eds.), *Handbook of Paleontological Techniques,* pp. 587–598. San Francisco & London, Freeman.

ANDERSEN S.TH. (1967) Tree-pollen rain in a mixed deciduous forest in south Jutland (Denmark). *Rev. Palaeobotan. Palynol.* **3**, 267–275.

ANDERSEN S.TH. (1973) The differential pollen productivity of trees and its significance for the interpretation of a pollen diagram from a forested region. In H.J.B. Birks & R.G. West (eds.), *Quaternary Plant Ecology,* pp. 109–116. Oxford, Blackwell Scientific Publications.

ANDERSEN S.TH. & BERTELSEN F. (1972) Scanning electron microscope studies of pollen of cereals and other grasses. *Grana* **12**, 79–86.

ANDERSON M.L. (1954) The ecological status of Wistman's Wood, Devonshire. *Trans. bot. Soc. Edin.* **36**, 195–206.

ANDREW R. (1970) The Cambridge pollen reference collection. In D. Walker & R.G. West (eds.), *Studies in the Vegetational History of the British Isles,* pp. 225–231. Cambridge University Press.

ANDREW R. (1971) Exine Pattern in the pollen of the British species of *Tilia. New Phytol.* **70**, 683–686.

BANNISTER B. (1969) Dendrochronology. In D. Brothwell & E. Higgs (eds.), *Science in Archaeology,* 2nd edition, pp. 191–205. London, Thames & Hudson.

BARBER K.E. (1975a) Pollen Analysis: pit 101 High Street B. In C.P.S. Platt & R. Coleman-Smith (eds.), *Excavations in Medieval Southampton.* Leicester University Press.

BARBER K.E. (1975b) Vegetational history of the New Forest: a preliminary note. *Proc. Hants. Field Club.* **30** (for 1973).

BARTLEY D.D. (1966) Pollen analysis of some lake deposits near Bamburgh in Northumberland. *New Phytol.* **65,** 141–156.

BATTARBEE R.W. (1973) Preliminary studies of Lough Neagh sediments II. Diatom analysis from the uppermost sediment. In H.J.B. Birks & R.G. West (eds.), *Quaternary Plant Ecology,* pp. 279–288. Oxford, Blackwell Scientific Publications.

BELL F.G. (1970) Late Pleistocene floras from Earith, Huntingdonshire. *Phil. Trans. R. Soc. B* **258,** 347–378.

BEIJERINCK W. (1947) *Zadenatlas der Nederlandsche Flora.* Wageningen, Veenman.

BENEDICT J.B. (1967) Recent glacial history of an alpine area in the Colorado Front Range, USA I. Establishing a lichen-growth curve. *J. Glaciol.* **6,** 817–832.

BENEDICT J.B. (1968) Recent glacial history of an alpine area in the Colorado Front Range, USA II. Dating the glacial deposits. *J. Glaciol.* **7,** 77–87.

BERESFORD M. (1971) *History on the Ground.* London, Methuen.

BERGGREN G. (1969) *Atlas of Seeds and small fruits of Northwest-European plant species, with morphological descriptions.* Part 2: Cyperaceae. Stockholm, Swedish Natural Science Research Council.

BERGLUND B.E. (1973) Pollen dispersal and deposition in an area of Southeastern Sweden—some preliminary results. In H.J.B. Birks & R.G. West (eds.), *Quaternary Plant Ecology,* pp. 117–130. Oxford, Blackwell Scientific Publications.

BERTSCH K. (1941) *Fruchte and Samen. Handbucher der praktischen vorgeschicts forschung,* I. Stuttgart, Enke.

BEUG H.-J. (1961) *Leitfaden der Pollenbestimmung fur Mitteleuropa und angrenzende Gebiete.* LI. Stuttgart, Fisher.

BEUG H.-J (1971) *Leitfaden der Pollenbestimmung fur Mitteleuropa und angrenzende Gebiete.* LII/III. Stuttgart, Fisher.

BIRCH B.P. (1971) The environment and settlement of the prairie-woodland transition belt—a case study of Edwards County, Illinois. *Southampton Research Series in Geogr.* **6,** 3–30.

BIRKS H.H. (1970) Studies in the vegetational history of Scotland. I: A pollen diagram from Abernethy Forest, Inverness-shire. *J. Ecol.* **58,** 827–846.

BIRKS H.H. (1973) Modern macrofossil assemblages in lake sediments in Minnesota. In H.J.B. Birks & R.G. West (eds.), *Quaternary Plant Ecology,* pp. 173–189. Oxford, Blackwell Scientific Publications.

BIRKS H.J.B. (1968) The identification of Betual nana pollen. *New Phytol.* **67,** 309–314.

BIRKS H.J.B. (1970) Inwashed pollen spectra at Loch Fada, Isle of Skye. *New Phytol.* **69,** 807–821.

BIRKS H.J.B. (1973) Modern pollen rain studies in some Arctic and Alpine environments. In H.J.B. Birks & R.G. West (eds.), *Quaternary Plant Ecology,* pp. 143–168. Oxford, Blackwell Scientific Publications.

BIRKS H.J.B. & DEACON J. (1973) A numerical analysis of the past and present flora of the British Isles. *New Phytol.* **72,** 877–902.

BIRKS H.J.B. & WEST R.G. (eds.) (1973) *Quaternary Plant Ecology.* Oxford, Blackwell Scientific Publications.

BONNY A.P. (1972) A method for determining absolute pollen frequencies in lake sediments. *New Phytol.* **71,** 393–405.

BRADBURY J.P. & WADDINGTON J.C.B. (1973) The impact of European settlement on Shagawa Lake, Northeastern Minnesota, USA. In H.J.B. Birks & R.G. West (eds.), *Quaternary Plant Ecology,* pp. 289–307. Oxford, Blackwell Scientific Publications.

BRANDON P.F. (1969) Medieval clearances in the East Sussex Weald. *Trans. Inst. Br. Geogr.* **48,** 135–153.

BRANDON P.F. (1971) Late-medieval weather in Sussex and its agricultural significance. *Trans. Inst. Br. Geogr.* **54,** 1–18.

BRAZIER J.D. & FRANKLIN G.L. (1961) Identification of hardwoods—a microscope key. *Forest Prod. Res. Bull.* **46,** 1–96.

BRITISH NATIONAL COMMITTEE FOR GEOLOGY (1968) International Geological Correlation Programme: United Kingdom Contribution. London, The Royal Society.

BROECKER W.S. (1965) Isotope geochemistry and the Pleistocene climatic record. In H.E. Wright & D.G. Frey (eds.), *The Quaternary of the United States,* pp. 737–753. Princeton University Press.

BROOKES D. & THOMAS K.W. (1967) The distribution of pollen grains on microscope slides. Part I: The non-randomness of the distribution. *Pollen et Spores* **9,** 621–629.

BROTHWELL D. & HIGGS E. (1969) *Science in Archaeology,* 2nd Edition. London, Thames & Hudson.

BROWN C.A. (1960) *Palynological techniques.* Baton Rouge, Louis. 188 pp. Private pub.

BURLEIGH R., SWITSUR V.R. & RENFREW C. (1973) The radiocarbon calendar calibrated too soon? *Antiquity* **47,** 309–317.

CARR A.P. (1962) Cartographic record and historical accuracy. *Geogr. J.* **47,** 135–146.

CASPARIE W.A. (1969) Bult-und Schlenkenbildung in Hochmoortorf: (zur Frage des Moorwachstums-Mechanismus). *Vegetatio, Acta Geobotanica* **19,** 146–180.

CASPARIE W.A. (1972) *Bog development in Southeastern Drenthe (The Netherlands).* The Hague, Junk.

CHAMBERS T.C. & GODWIN H. (1971) Scanning electron microscopy of *Tilia* pollen. *New Phytol.* **70,** 687–692.

CHAPMAN S.B. (1964a) The ecology of Coom Rigg Moss, Northumberland. I. Stratigraphy and present vegetation. *J. Ecol.* **52,** 299–313.

CHAPMAN S.B. (1964b) The ecology of Coom Rigg Moss, Northumberland. II. The chemistry of peat profiles and the development of the bog system. *J. Ecol.* **52,** 315–321.

COLES J.M. & HIBBERT F.A. (1968) Prehistoric roads and tracks in Somerset, England. I. Neolithic. *Proc. prehist. Soc.* **34,** 238–258.

COLES J.M., HIBBERT F.A. & CLEMENTS C.F. (1970) Prehistoric roads and tracks in Somerset, England. II. Neolithic. *Proc. prehist. Soc.* **36,** 125–151.

CONOLLY A.P. & DAHL E. (1970) Maximum summer temperature in relation to the modern and Quaternary distributions of certain Arctic-Montane species in the British Isles. In D. Walker & R.G. West (eds.), *Studies in the Vegetational History of the British Isles,* pp. 159–223. Cambridge University Press.

CONWAY V.M. (1947) Ringinglow Bog, near Sheffield. Part I. Historical. *J. Ecol.* **34,** 149–181.

COOKE H.B.S. (1973) Pleistocene chronology: long or short? *Quaternary Research* **3,** 206–220.

COOPE G.R. (1961) On the study of glacial and interglacial insect faunas. *Proc. Linn. Soc. Lond.* **172,** 62–65.

COOPE G.R. (1965) Fossil insect faunas from Late Quaternary deposits in Britian. *Advmt Sci., Lond.* **21,** 564–575.

COOPE G.R. & SANDS C.H.S. (1966) Insect faunas of the last glaciation from the Tame Valley, Warwickshire. *Proc. R. Soc. B,* **165,** 389–412.

COOPE G.R., MORGAN A. & OSBORNE P.J. (1971) Fossil Coleoptera as indicators of climatic fluctuations during the last glaciation in Britain. *Palaeogeography, Palaeoclimatol., Palaeoecol.* **10,** 87–101.

COSTER I. & PANKOW H. (1968) Illustriester Schussel zur Bestimmung einiger Mitteleuropaischer *Sphagnum*-Arten. *Wiss. Z. Univ. Rostock* **415,** 285–323.

COX A., DOELL R.R. & DALRYMPLE G.B. (1965) Quaternary paleomagnetic stratigraphy. In H.E.Wright & D.G. Frey (eds.), *The Quaternary of the United States,* pp. 817–830. Princeton University Press.

COX E.A. & DITTMER B.R. (1965) The tithe files of the mid-nineteenth century. *Agric. Hist. Rev.* **13,** 1–16.

CRABTREE K. (1968) Pollen analysis. *Sci. Prog., Oxf.* **56**, 83–101.

CRABTREE K. (1969) Post Glacial diatom zonation of limnic deposits in North Wales. *Mitt. Internat. Verein. Limnol.* **17**, 165–171.

CRAIG A.J. (1972) Pollen influx to laminated sediments: a pollen diagram from northeastern Minnesota. *Ecology* **53**, 46–57.

CREER K.M., THOMPSON R., MOLYNEUX L. & MACKERETH F.J.H. (1972) Geomagnetic secular variation recorded in the stable magnetic remanence of recent sediments. *Earth Planet. Sci. Letters* **14**, 115–127.

CROWDER A.A. & CUDDY D.G. (1973) Pollen in a small river basin: Wilton Creek, Ontario. In H.J.B. Birks & R.G. West (eds.), *Quaternary Plant Ecology*, pp. 60–78. Oxford, Blackwell Scientific Publications.

CUSHING E.J. (1967a) Evidence for differential pollen preservation in late Quaternary sediments in Minnesota. *Rev. Palaeobotan. Palynol.* **4**, 87–101.

CUSHING E.J. (1967b) Late-Wisconsin pollen stratigraphy and the glacial sequence in Minnesota. In E.J. Cushing & H.E. Wright (eds.), *Quaternary Paleoecology*, pp. 59–88. New Haven & London, Yale University Press.

CUSHING E.J. & WRIGHT H.E. (eds.) (1967) *Quaternary Paleoecology*. New Haven & London, Yale University Press.

DALE M.B. & WALKER D. (1970) Information analysis of pollen diagrams. I. *Pollen et Spores* **12**, 21–37.

DAMBLON F. & SCHUMACKER R. (1972) New prospects for study of palynological data: the use of computers. *Pollen et Spores* **13**, 609–614.

DARBY H.C. (1950) Domesday woodland. *Econ. Hist. Rev.* Series 2, **3**, 21–43.

DARBY H.C. (1951) The clearing of the English woodlands. *Geography* **36**, 71–83.

DARBY H.C. *et al.* (1952–1967) *The Domesday Geography of England.* 5 vols. Cambridge University Press.

DARBY H.C. (ed.) (1973) *A New Historical Geography of England.* Cambridge University Press.

DAVIES E.G. (1944) Figyn Blaeu Brefi: a Welsh upland bog. *J. Ecol.* **32**, 147–166.

DAVIS M.B. (1963) On the theory of pollen analysis. *Am. Jnl Sci.* **261**, 897–912.

DAVIS M.B. (1965) Phytogeography and palynology of northeastern United States. In H.E. Wright, Jr. & D.G. Frey (eds.), *The Quaternary of the United States*, pp. 377–401. Princeton University Press.

DAVIS M.B. (1967) Pollen accumulation rates at Rogers Lake, Connecticut, during Late- and Postglacial-time. *Rev. Palaeobotan. Palynol.* **2**, 219–230.

DAVIS M.B. (1968) Pollen grains in lake sediments. Redeposition caused by seasonal water circulation. *Science* **162**, 796–799.

DAVIS M.B., BREWSTER L.A. & SUTHERLAND J. (1969) Variation in pollen spectra in lakes (1), *Pollen et Spores* **11**, 557–571.

DAVIS R.B., BRUBAKER L.A. & BEISWENGER J.M. (1971) Pollen grains in lake sediments: pollen percentages in surface sediments from Southern Michigan. *Quaternary Research*, 450–467.

DEACON J. (1972) A data bank of Quaternary plant fossil records. *New Phytol.* **71**, 1227–1232.

DEEVEY E.S. (1965) Sampling lake sediments by use of the Livingston sampler. In B. Kummel & D. Raup (eds.) *Handbook of Paleontological Techniques*, pp. 520–529. San Francisco, Freeman.

DE GEER G. (1912) A geochronology of the last 12,000 years. *C.r. XI int. geol. Congr. (Stockholm)* **1**, 241–258.

DE GEER G. (1934) Geology and geochronology. *Geogr. Annlr.* **16**, 1–52.

DE GEER G. (1940) Geochronologia Suecica Principles. *K. Svenska VetenskAcad. Handl. Ser.* **3**, 18, 6.

DENTON G.H. & KARLEN W. (1973) Holocene climatic variations—their pattern and possible cause. *Quaternary Research* 3, 155–205.

DEUEL L. (1973) *Flights into Yesterday*. London, Penguin.

DEWAR H.S.L. & GODWIN H. (1963) Archaeological discoveries in the raised bogs of the Somerset levels. *Proc. prehist. Soc.* 29, 17–49.

DICKSON C.A. (1970) The study of plant macrofossils in British Quaternary deposits. In D. Walker & R.G. West (eds.), *Studies in the Vegetational History of the British Isles*, 233–254. Cambridge University Press.

DICKSON J.H. (1967) The British moss flora of the Weichselian Glacial. *Rev. Palaeobotan. Palynol.* 2, 245–253.

DICKSON J.H. (1973) *Bryophytes of the Pleistocene*. Cambridge University Press.

DIMBLEBY G.W. (1961) Soil pollen analysis. *Jl Soil Sci. 12*, 1, 1–11.

DIMBLEBY G.W. (1962) The Development of British Heathlands and their soils. *Oxford Forestry Memoirs* No. 23.

DIMBLEBY G.W. (1965) Post-Glacial changes in soil profiles. *Proc. R. Soc. B* 161, 355–362.

DIMBLEBY G.W. (1967) *Plants & Archaeology*. London, Baker.

DONKIN R.A. (1960) The Cistercian settlement and the English Royal Forests. *Citeaux* XI, 1–33.

DUNCAN U.K. (1962) Illustrated Key to Sphagnum mosses. *Trans. & Proc. bot. Soc. Edinb.* 39, 290–301

ERDTMAN G. (1943) *An Introduction to Pollen Analysis*. Waltham, Mass., Chronica Botanica.

ERDTMAN G. (1969) *Handbook of Palynology*. Copenhagen, Munksgaard.

ERDTMAN G., BERGLUND B. & PRAGLOWSKI J. (1961) An introduction to a Scandinavian Pollen Flora. *Grana Palynologica* 2, 3–92.

ERDTMAN G., PRAGLOWSKI J. & NILSSON S. (1963) *An Introduction to a Scandinavian Pollen Flora*, Vol. II. Stockholm, Almquist & Wiksell.

EVANS G.H. (1972) The diatom flora of the Hoxnian deposits at Marks Tey, Essex. *New Phytol.* 71, 379–386.

EVANS J.G. (1973) *Land Snails in Archaeology*. London, Seminar Press.

FAEGRI K. (1966) Some problems of representivity in pollen analysis. *Palaeobotanist* 15, 135.

FAEGRI K. (1975) *Textbook of Pollen Analysis*, third edition. Copenhagen, Munksgaard.

FAEGRI K. & IVERSEN J. (1964) *Textbook of Pollen Analysis*, second edition. Copenhagen, Munksgaard.

FEARNSIDES M. (1938) Graphic keys for the identification of *Sphagna*. *New Phytol.* 37, 409–424.

FLENLEY J.R. (1973) The use of modern pollen rain samples in the study of the vegetational history of tropical regions. In H.J.B. Birks & R.G. West (eds.), *Quaternary Plant Ecology*, pp. 131–142. Oxford, Blackwell Scientific Publications.

FOGG G.E. (1965) (In discussion of Mackereth 1965). *Proc. R. Soc. B* 161, 353–354.

FRANKLIN T.B. (1953) *British Grasslands from the Earliest Times to the Present Day*. London, Faber & Faber.

FRANKS J.W. (1960) Interglacial deposits at Trafalgar Square, London. *New Phytol.* 59, 145–152.

FRANKS J.W. & PENNINGTON W. (1961) The Late- and Post-glacial Deposits of the Esthwaite Basin, North Lancashire. *New Phytol.* 60, 27–42.

FRENZEL B. (1964) Pollenanalyse von Losen. *Eiszeit. Gegenw.* 15, 5–39.

FREY D.G. (1964) Remains of animals in Quaternary lake and bog sediments and their interpretation. *Arch. Hydrobiol., suppl. Ergebn. Limnol.*, No. 2, 1–114.

FREY D.G. (1965) Other invertebrates—an essay in biogeography. In H.E. Wright, Jr. & D.G. Frey (eds.), *The Quaternary of the United States*, pp. 613–633. Princeton University Press.

FRITTS H.C. (1971) Dendroclimatology and Dendroecology. *Quaternary Research* **1**, 419–449.

FUSSELL G.E. (1964a) The Grasses and grassland cultivation of Britain. I. Before 1700. *J. Br. Grassld Soc.* **19**, 49–54.

FUSSELL G.E. (1964b) The Grasses and grassland cultivation of Britain. II. 1700–1900. *J. Br. Grassld Soc.* **19**, 212–217.

GIMINGHAM C.H. (1972) *Ecology of Heathlands*. London, Chapman & Hall.

GODWIN H. (1934) Pollen-analysis—problems and potentialities of the method. *New Phytol.* **33**, 278, 325.

GODWIN H. (1940) Pollen analysis and forest history of England and Wales. *New Phytol.* **39**, 370–400.

GODWIN H. (1948) Studies of the Post-Glacial history of British vegetation: X Correlation between climate, forest composition, prehistoric agriculture and peat stratigraphy in sub-Boreal and sub-Atlantic peats of the Somerset Levels. *Phil. Trans. R. Soc. B* **233**, 275–286.

GODWIN H. (1954) Recurrence surfaces. *Danm. Geol. Unders. II R* **80**, 22–30.

GODWIN H. (1955) Studies of the Post-Glacial history of British vegetation. XIII. The Meare Pool region of the Somerset Levels. *Phil. Trans. R. Soc. B* **239**, 161–190.

GODWIN H. (1956) *History of the British Flora*. Cambridge University Press.

GODWIN H. (1958) Pollen-analysis in mineral-soil: an interpretation of a podzol pollen-analysis by Dr. G.W. Dimbleby. *Flora* **146**, 321.

GODWIN H. (1960) Radiocarbon dating and Quaternary history in Britain. *Proc. R. Soc. B* **153**, 287–320.

GODWIN H. (1962) Vegetational History of the Kentish chalk downs as seen at Wingham and Frogholt. *Veroff. Geobot. Inst. Rubel, Zurich* **37**, 83–99. (Festschrift Franz Firbas).

GODWIN H., CLOWES D.R. & HUNTLEY B. (1974) Studies in the ecology of Wicken Fen. V. Development of fen carr. *J. Ecol.* **62**, 197–214.

GODWIN H. & CONWAY V.M. (1939) The ecology of a raised bog near Tregaron, Cardiganshire. *J. Ecol.* **27**, 315–359.

GODWIN H., WALKER D. & WILLIS E.H. (1957) Radiocarbon dating and post-glacial vegetational history: Scaleby Moss. *Proc. R. Soc. B* **147**, 352–366.

GOOD R. (1948) *A Geographical Handbook of the Dorset Flora*. Dorchester, Dorset Nat. Hist. Soc.

GORDON R.B. (1966) *Map: Natural vegetation of Ohio at the time of the earliest land surveys*. Ohio Biological Survey, Ohio State University.

GORDON A.D. & BIRKS H.J.B. (1972) Numerical methods in Quaternary palaeoecology I. Zonation of pollen diagrams. *New Phytol.* **71**, 961–979.

GORDON A.D. & BIRKS H.J.B. (1974) Numerical methods in Quaternary palaeoecology. II. Comparison of pollen diagrams. *New Phytol.* **73**, 221–249.

GRAAF FR. DE (1956) Studies on Rotatoria and Rhizopoda from the Netherlands. I. Rotatoria and Rhizopoda from the 'Grote Huisven'. *Biol. Jaarb.* **23**, 145–217.

GRAY J. (1965a) Palynological Techniques. In B. Kummel & D. Raup (eds.), *Handbook of Paleontological Techniques*, pp. 471–481. San Francisco, Freeman.

GRAY J. (1965b) Extraction Techniques. In B. Kummel & D. Raup (eds.), *Handbook of Paleontological Techniques*, pp. 530–587. San Francisco, Freeman.

GREEN B.H. (1968) Factors influencing the spatial and temporal distribution of *Sphagnum imbricatum* Hornsch ex Russ. in the British Isles. *J. Ecol.* **56**, 47–58.

GREEN B.H. & PEARSON M.C. (1968) The ecology of Wybunbury Moss, Cheshire. I. The present vegetation and some physical, chemical and historical factors controlling its nature and distribution. *J. Ecol.* **56**, 245–268.

GRIGG D.B. (1967) The changing agricultural geography of England: a commentary on the sources available for the reconstruction of the agricultural geography of England, 1770–1850. *Trans. Inst. Br. Geogr.* **41**, 73–96.

HAFSTEN U. (1959) Bleaching + HF + acetolysis: a hazardous preparation process. *Pollen et Spores* **1**, 77–79.

HAMMOND R.F. (1974) The preservation of peat monoliths for permanent display. *J. Soil Sci.* **25**, 63–66.

HANSEN B. (1966) The raised bog Draved Kongsmose. *Bot. Tidsskr.* **62**, 2–3, 146–185.

HARLEY J.B. (1972) Maps for the local historian. A guide to the British sources. London, National Council of Social Service (for the Standing Conference for Local History).

HARLEY J.B. & PHILLIPS C.W. (1964) The historian's guide to Ordnance Survey maps. London, National Council of Social Service (for The Standing Conference for Local History).

HARNISCH O. (1949) Alterer und jungerer Sphagnumtorf. Eine Rhizopodenanalytische Studie an nordwest Europaischen Hochmooren. *Biol. Zb.* **68**, 398–412.

HAVINGA A.J. (1967) Palynology and pollen preservation. *Rev. Palaeobotan. Palynol.* **2**, 81–98.

HAWORTH E.W. (1969) The diatoms of a sediment core from Blea Tarn, Langdale. *J. Ecol.* **57**, 429–440.

HEAL O.W. (1961) The distribution of testate amoebae (Rhizopoda: Testacea) in some fens and bogs in northern England. *J. Linn. Soc. (Zool.)* **44**, 369–382.

HEAL O.W. (1962) The abundance and microdistribution of testate amoebae (Rhizopoda: Testacea) in *Sphagnum. Oikos* **13**, 35–47.

HEAL O.W. (1964) Observations on the seasonal and spatial distribution of Testacea (Protozoa: Rhizopoda) in *Sphagnum. J. Anim. Ecol.* **33**, 395–412.

HEINSELMAN M.L. (1973) Fire in the virgin forests of the Boundary Waters Canoe Area, Minnesota. *Quaternary Research* **3**, 329–382.

HELBAEK H. (1969) Palaeo-ethnobotany. In D. Brothwell & E. Higgs (eds.), *Science in Archaeology*, pp. 206–214. 2nd edition. London, Thames & Hudson.

HEWLETT G. (1973) Reconstructing a historical landscape from field and documentary evidence: Otford in Kent. *Agric. Hist. Rev.* **21**, 94–110.

HIBBERT F.A., SWITSUR V.R. & WEST R.G. (1971) Radiocarbon dating of Flandrian pollen zones at Red Moss, Lancashire. *Proc. R. Soc. B* **177**, 161–176.

HOOPER M. (1970) Dating hedges. *Area* **4**: 63–65.

HOSKINS W.G. (1955) *The Making of the English Landscape*. London, Hodder & Stoughton. (Published with revisions, Penguin 1970).

HOSKINS W.G. (1959) *Local History in England*. London, Longmans. (Second edition, 1972.)

HOSKINS W.G. (1967) *Fieldwork in Local History*. London, Faber & Faber.

HOWARD J.A. (1971) *Aerial Photo-Ecology*. London, Faber & Faber.

HUBBARD J.C.E. & STEBBINGS R.E. (1968) *Spartina* marshes in southern England. VII. Stratigraphy of the Keysworth Marsh, Poole Harbour. *J. Ecol.* **56**, 707–722.

HUCKERBY E., MARCHANT R. & OLDFIELD F. (1972) Identification of fossil seeds of *Erica* and *Calluna* by scanning electron microscopy. *New Phytol.* **71**, 387–392.

HUSTEDT F. (1930) *Die Susswasser—flora Mitteleuropas 10. Bacillariophyta (Diatomeae)*. Ed. A. Pascher.

INGRAM H.A.P. (1967) Problems of hydrology and plant distribution in mires. *J. Ecol.* **55**, 711–725.

INTERNATIONAL SOCIETY FOR PHOTOGRAMMETRY (eds.) (1962) Transactions of the Symposium on Photo Interpretation. Int. Arch. Photogramm. 14. Delft. Waltman.

IVERSEN J. (1964) Retrogressive vegetational succession in the post-glacial. *J. Ecol.* **52**, (Suppl.) 59–70.

IVERSEN J. (1969) Retrogressive development of a forest ecosystem demonstrated by pollen diagrams from fossil mor. *Oikos* **12**, 35–49.

JANE F.W. (1956) *The Structure of Wood*. London, Black.

JANSSEN C.R. (1959) *Alnus* as a disturbing factor in pollen diagrams. *Acta Bot. Neerl.* **8**, 55–58.

JANSSEN C.R. (1966) Recent pollen spectra from the deciduous and coniferous-deciduous forests of Northeastern Minnesota: a study in pollen dispersal. *Ecology* **47**, 804–25.

JANSSEN C.R. (1970) Problems in the recognition of plant communities in pollen diagrams. *Vegetatio, Acta Geobotan.* XX 187–198.

JANSSEN C.R. (1973) Local and regional pollen deposition. In H.J.B. Birks & R.G. West (eds.), *Quaternary Plant Ecology,* pp. 31–42. Oxford, Blackwell Scientific Publications.

JESSEN K. & FARRINGTON A. (1938) The bogs at Ballybetagh, near Dublin with remarks on late-glacial conditions in Ireland. *Proc. R. Ir. Acad.* **44B,** 205–260.

JOHNSTON W.B. (1961) Locating the vegetation of early Canterbury: a map and the sources. *Trans. R. Soc. New Zealand. (Botany)* 1, 5–15.

JOWSEY P.C. (1966) An improved peat sampler. *New Phytol.* **65,** 245–248.

KARLEN W. (1973) Holocene glacier and climatic variations, Kebnekaise Mountains, Swedish Lapland. *Geogr. Ann.* **55A,** 29–63.

KASSAS M. (1951) Studies in the ecology of Chippenham Fen II. Recent history of the fen, from evidence of historical records, vegetational analysis and tree-ring analysis. *J. Ecol.* **39,** 19–33.

KATZ N.J., KATZ S.V. & KIPIANI M.G. (1965) *Atlas and Keys of Fruits and Seeds Occurring in the Quaternary Deposits of the USSR* (In Russian). Moscow, Nauka.

KERSHAW A.P. (1971) A pollen diagram from Quincan Crater, North-east Queensland, Australia. *New Phytol.* **70,** 669–681.

KOZLOWSKI T.T. (1971) Growth and development of trees, Vol. 2: Cambial growth, root growth and reproductive growth. New York, Academic Press.

KUCHLER A.W. (1964) Potential natural vegetation of the conterminous United States. (Manual to accompany map). *Am. Geogr. Soc. Spec. Pub.* **36,** 39 pp. 116 plates.

KULCZYNSKI S. (1949) Peat bogs of Polesie. *Mem. de l'Acad. Polon. des Sciences et des Lettres,* Serie B, 1–356.

KUMMEL B. & RAUP O. (eds.) (1965) *Handbook of Paleontological Techniques.* San Francisco, Freeman.

LADURIE E. LE ROY (1972) *Times of Feast, Times of Famine. A History of Climates Since the Year 1,000.* London, Allen & Unwin.

LAMB H.H. (1966) *The Changing Climate.* London. Methuen.

LAMBERT A.M. (1971) *The Making of the Dutch Landscape.* London, Academic Press.

LIBBY W.F. (1955) *Radiocarbon Dating.* Chicago University Press. 2nd ed.

MACAN T.T. (1969) *A Key to the British Fresh- and Brackish-water Gastropods.* Freshwater Biological Association Sci. Pub. No. 13. 3rd Ed.

MCANDREWS J.H. (1966) Postglacial history of prairie, savanna and forest in Northwestern Minnesota. *Torrey Bot. Club Mem.* **22,** 72 pp.

MCCRACKEN E. (1971) *The Irsih Woods since Tudor Times. Distribution and Exploitation.* Newton Abbot, David & Charles.

MACKERETH F.J.H. (1958) A portable core sampler for lake deposits. *Limnol. Oceanogr.* **3,** 181–191.

MACKERETH F.J.H. (1965) Chemical investigation of lake sediments and their interpretation. *Proc. R. Soc. B* **161,** 295–309.

MACKERETH F.J.H. (1966) Some chemical observations on post-glacial lake sediments. *Proc. R. Soc. B* **250,** 165–213.

MACKERETH F.J.H. (1969) A short-core sampler for subaqueous deposits. *Limnol. Oceanogr.* **14,** 145–151.

MACKERETH F.J.H. (1971) On the variation in direction of the horizontal component of remanent magnetisation in lake sediments. *Earth Planet Sci. Letters* **12,** 332–338.

MAHER L.J. (1972) Absolute pollen diagram of Redrock Lake, Boulder County, Colorado. *Quaternary Research* **2,** 531–553.

MANLEY G. (1959) Temperature trends in England 1698–1957. *Archiv fur Met. Geophys. und Biokl.* Serie B9, 413–433.

MANLEY G. (1964) The evolution of the climatic environment. In J.B. Sissons & J.W. Watson (eds.), *The British Isles: A Systematic Geography,* pp. 152–176. London, Nelson.

MARTIN A.C. & BARKLEY W.D. (1961) *Seed Identification Manual.* Berkeley, University of California Press.

MARTIN M.H. (1968) Conditions affecting the distribution of *Mercurialis perennis* L. in certain Cambridgeshire woodlands. *J. Ecol.* **56,** 777–793.

MATTSON E. & KOULTER-ANDERSSON E. (1954) Geochemistry of a raised bog. *Kongl. Lantbrukshogskolsrs Ann.* **21,** 321–366.

MERCER J.H. (1972) Chilean glacial chronology 20,000–11,000 carbon-14 years ago: some global comparisons. *Science* **176,** 1118–1120.

MERRETT P. (ed.) (1971) *Captain Cyril Diver (1892–1969): A memoir.* Furzebrook Research Station, Natural Environment Research Council.

MERTON L.F.H. (1970) The history and status of woodlands of the Derbyshire limestone. *J. Ecol.* **58,** 723–744.

MITCHELL A. (1974) *A Field Guide to the Trees of Britain and Northern Europe.* London, Collins.

MITCHELL G.F. (1965) Littleton Bog, Tipperary: an Irish agricultural record. *J. Roy. Soc. Antiquaries of Ireland* **95,** 121–132.

MITCHELL G.F., PENNY L.F., SHOTTON F.W. & WEST R.G. (eds.) (1973) A correlation of Quaternary deposits in the British Isles. *Geol. Soc. Lond.,* Special Report No. 4, 99 pp.

MITTRE V. (1971) Fossil pollen of *Tilia* from the East Anglian Fenland. *New Phytol.* **70,** 693–697.

MONGREDIEN A. (1870) *Trees and Shrubs for English Plantations.* London, Murray.

MOORE P.D. & CHATER E.H. (1969) The changing vegetation of west-central Wales in the light of human history. *J. Ecol.* **57,** 361–379.

MORRISON M.E.S. (1959) The ecology of a raised bog in Co. Tyrone, Northern Ireland. *Proc. R. Ir. Acad.* **60B,** 291–308.

MORRISON M.E.S. (1966) Low-latitude vegetation history with special reference to Africa. In *World Climate 8000–0 B.C.:* Royal Met. Soc. pp. 142–148.

MORRISON M.E.S. (1968) Vegetation and climate in the uplands of south-western Uganda during the later Pleistocene period: I Muchoya Swamp, Kigezi district. *J. Ecol.* **56,** 363–385.

MOSIMANN J.E. (1965) Statistical methods for the pollen analyst: multinomial and negative multinomial techniques. In B. Kummel & D. Raup (eds.), *Handbook of Paleontological Techniques,* pp. 636–673. San Francisco & London, Freeman.

NEUSTUPNY E. (1970a) A new epoch in Radiocarbon Dating. *Antiquity XLIV* 38–45.

NEUSTUPNY E. (1970b) The Accuracy of Radiocarbon Dating. In I.U. Olsson (ed.) *Radiocarbon Variations and Absolute Chronology.* New York/Stockholm, Wiley.

NICHOLLS P.H. (1972) On the evolution of a forest landscape. *Trans. Inst. Br. Geogr.* **56,** 57–76.

NICHOLS H. (1967) The suitability of certain categories of lake sediments for pollen analyses. *Pollen et Spores* **9,** 615–620.

OLDFIELD F. (1959) Pollen morphology of some of the Western European Ericales. *Pollen et Spores* **1,** 19–48.

OLDFIELD F. (1963) Pollen analysis and man's role in the Ecological History of the South-East Lake District. *Geogr. Ann.* **45,** 23–40.

OLDFIELD F. (1969) Pollen analysis and the history of land-use. *Advmt Sci., Lond.* **25,** 298–311.

OLDFIELD F. (1970a) The ecological history of Blelham Bog National Nature Reserve. In D. Walker & R.G. West (eds.), *Studies in the Vegetational History of the British Isles,* pp. 141–157. Cambridge University Press.

OLDFIELD F. (1970b) Some aspects of scale and complexity in pollen-analytically based palaeoecology. *Pollen et Spores* **12,** 163–171.

OLSSON I.U. (ed.) (1970) *Radiocarbon Variations and Absolute Chronology.* (Proc. 12th Nobel Symposium: Uppsala 1969). New York/Stockholm, Wiley.

ORWIN C.S. & SELLICK R.J. (1970) *The Reclamation of Exmoor Forest.* 2nd Edition. Newton Abbot, David & Charles.

OSBORNE P.J. (1972) Insect faunas of late Devensian and Flandrian age from Church Stretton, Shropshire. *Phil. Trans. R. Soc. B* **263.** 327–367.

O'SULLIVAN P.E. (1973a) Contemporary pollen studies in a native Scots Pine ecosystem. *Oikos* **24,** 143–150.

O'SULLIVAN P.E. (1973b) Pollen analysis of Mor humus layers from a native Scots pine ecosystem, interpreted with surface samples. *Oikos* **24,** 259–272.

O'SULLIVAN P.E., OLDFIELD F. & BATTARBEE R.W. (1973) Preliminary studies of Lough Neagh sediments I. Stratigraphy, chronology and pollen analysis. In H.J.B. Birks & R.G. West (eds.), *Quaternary Plant Ecology,* pp. 267–278. Oxford, Blackwell Scientific Publications.

OVERBECK F. (1963) Aufgaben botanisch-geologischer Moorforschung in Nordwestdeutschland. *Ber. Deutsch. bot. Gesell.* **76,** 2–12.

OSVALD H. (1923) Die Vegetation des Hochmoores Komosse. *Svensk. Vaxtsoc. Sallsk. Handl.* 1.

OSVALD H. (1949) Notes on the vegetation of British and Irish Mosses. *Acta Phytogeog. Suecica,* **26,** 7–62.

PAULSON B. (1952) Some rhizopod associations in a Swedish mire. *Oikos* **4,** 151–165.

PAYNE-GALLWAYE R. (1886) *The Book of Duck Decoys.* London, John van Voorst.

PECK R.M. (1972) Efficiency tests on the Tauber trap used as a pollen sampler in turbulent water flow. *New Phytol.* **71,** 187–198.

PECK R.M. (1973) Pollen budget studies in a small Yorkshire catchment. In H.J.B. Birks & R.G. West (eds.), *Quaternary Plant Ecology,* pp. 43–60. Oxford, Blackwell Scientific Publications.

PECK R.M. (1974) A comparison of four absolute pollen preparation techniques. *New Phytol.* **73,** 567–587.

PENNINGTON W. (1947) Lake sediments: pollen diagrams from the bottom deposits of the north basin of Windermere. *Phil. Trans. R. Soc.* **B233,** 137–175.

PENNINGTON W. (1969) The usefulness of pollen analysis in interpretation of stratigraphic horizons, both Late-glacial and Post-glacial. *Mitt. Internat. Verein. Limnol.* **17,** 154–164.

PENNINGTON W. (1970) Vegetation history in the north-west of England: a regional synthesis. In D. Walker & R.G. West (eds.), *Studies in the Vegetational History of the British Isles,* pp. 41–80. Cambridge University Press.

PENNINGTON W. (1973) Absolute pollen frequencies in the sediments of lakes of different morphometries. In H.J.B. Birks & R.G. West (eds.) *Quaternary Plant Ecology,* pp. 79–104. Oxford, Blackwell Scientific Publications.

PENNINGTON W. (1974) *The History of British Vegetation.* Second edition. London, English University Press.

PENNINGTON W. & BONNY A.P. (1970) An absolute pollen diagram from the British late-glacial. *Nature, Lond.* **226,** 871–873.

PENNINGTON W. & LISHMAN J.P. (1971) Iodine in lake sediments in Northern England and Scotland. *Biol. Rev.* **46,** 279–313.

PENNINGTON W., HAWORTH E.Y., BONNY A.P. & LISHMAN J.P. (1972) Lake sediments in Northern Scotland. *Phil. Trans. R. Soc.* **264B,** 193–294.

PERRING F.H. (ed.) (1968) *Critical Supplement to the Atlas of the British Flora.* London, Nelson.

PERRING F.H. (ed.) (1970a) The flora of a changing Britain. *Bot. Soc. Br. Isles,* Conference Report No. 11.

PERRING F.H. (1970b) The last seventy years. In *The flora of a changing Britain*. pp. 129–135. *Bot. Soc. Br. Isles,* Conference Report No. 11.

PERRING F.H. & WALTERS S.M. (ed.) (1962) *Atlas of the British Flora*. London, Nelson.

PERRY P.J. (1969) H.C. Darby and historical geography: a survey and review. *Geogr. Zeitschr.* **57,** 161–177.

PETERKEN G.F. (1969) Development of Vegetation in Staverton Park, Suffolk. *Field Studies* **3**(1), 1–39.

PETERKEN G.F. & HUBBARD J.C.E. (1972) The shingle vegetation of southern England: the holly wood on Holmstone Beach, Dungeness. *J. Ecol.* **60,** 547–572.

PETERKEN G.F. & TUBBS C.R. (1965) Woodland regeneration in the New Forest, Hampshire, since 1650. *J. appl. Ecol.* **2,** 159–170.

PHILLIPS L. (1972) An application of fluorescence microscopy to the problem of derived pollen in British Pleistocene deposits. *New Phytol.* **71,** 744–762.

PIGOTT C.D. (1969) The status of *Tilia cordata* and *T. platyphyllos* on the Derbyshire limestone. *J. Ecol.* **57,** 491–504.

PIGOTT C.D. & PIGOTT M.E. (1959) Stratigraphy and pollen analysis of Malham Tarn and Tarn Moss. *Field Studies* **1,** 1–18.

PIGOTT C.D. & PIGOTT M.E. (1963) Late-glacial and Post-glacial deposits at Malham, Yorkshire. *New Phytol.* **62,** 317–324.

PILCHER J.R. (1973) Pollen analysis and radiocarbon dating of a peat on Slieve Gallion, Co. Tyrone, N. Ireland. *New Phytol.* **72,** 681–689.

POLLARD E. (1973) Hedges VII: Woodland relic hedges in Huntingdon and Peterborough. *J. Ecol.* **61,** 343–352.

POTZGER J.E., POTZGER M.E. & MCCORMICK J. (1956) The forest primeval of Indiana as recorded in the original US land surveys and an evaluation of previous interpretations of Indiana vegetation. *Butler Univ. bot. studies* **13,** 95–111.

PRAGLOWSKI J. (1970) The effects of pre-treatment and the embedding media on the shape of pollen grains. *Rev. Palaeobotan. Palynol.* **10,** 203–208.

PRINCE H.C. (1959) The Tithe Surveys of the mid-19th C. *Agric. Hist. Rev.* **7,** 14–26.

PROCTOR M.C.F. (1955) Key to the British species of *Sphagnum*. *Trans. Brit. Bryol. Soc.* **2,** 552–560.

PROCTOR M.C.F. (1973) Summing up: an ecologists's viewpoint. In H.J.B. Birks & R.G. West (eds.), *Quaternary Plant Ecology*, pp. 313–314. Oxford, Blackwell Scientific Publications.

RACKHAM O. (1974a) *Hayley Wood: Its History and Ecology*. Cambridgeshire & Isle of Ely Naturalist's Trust Ltd.

RACKHAM O. (1974b) The oak tree in historic times. In M.G. Morris & F.H. Perring (eds.). *The British Oak: its History and Natural History*. Bot. Soc. Br. Isles: Conference Report No. 13.

RANKIN H.T. & DAVIS D.E. (1971) Woody vegetation in the Black Belt Prairie of Montgomery County, Alabama, in 1845–46. *Ecology* **52,** 716–719.

RANKINE W.F., RANKINE W.M. & DIMBLEBY G.W. (1960) Further excavations at a Mesolithic site at Oakhanger, Selborne, Hants. *Proc. prehist. Soc.* **26,** 246–262.

RATCLIFFE D.A. & WALKER D. (1958) The Silver Flowe, Galloway, Scotland. *J. Ecol.* **46,** 407–445.

RENFREW J.M. (1973) *Palaeoethnobotany*. London, Methuen.

RITCHIE J.C. & LICHTI-FEDEROVICH S. (1967) Pollen dispersal phenomena in Arctic- Subarctic Canada. *Rev. Palaeobotan. Palynol.* **3,** 255–266.

ROBERTS B.K., TURNER J. & WARD P.F. (1973) Recent forest history and land use in Weardale, Northern England. In H.J.B. Birks & R.G. West (eds.), *Quaternary Plant Ecology*, pp. 207–221. Oxford, Blackwell Scientific Publications.

RODEN D. (1968) Woodland and its management in the medieval Chilterns. *Forestry* **41,** 59–71.

ROSE F. & JAMES P.W. (1974) Regional studies on the British lichen flora. I: The corticolous and lignicolous species of the New Forest, Hampshire. *The Lichenologist* **6,** 1–72.

ROUND F.E. (1957) The Late-glacial and Post-glacial diatom succession in the Kentmere Valley deposit. *New Phytol.* **56,** 98–126.

ROUND F.E. (1961) The diatoms of a core from Esthwaite Water. *New Phytol.* **60,** 43–59.

ROUND F.E. (1964) Diatom sequences in lake deposits: some problems of interpretation. *Verh. int. Verein. theor. angew. Limnol.* **15,** 1012–20.

RÓVNER I. (1971) Potential of Opal Phytoliths for use in palaeoecological reconstruction. *Quaternary Research 1,* **3,** 343–359.

ROWLEY T. (1972) *The Shropshire Landscape.* London, Hodder & Stoughton.

ROYAL SOCIETY (1969) International Geological Correlation Programme: United Kingdom contribution. London.

RYBNIČEK K. (1974) A comparison of the present and past mire communities of Central Europe. In H.J.B. Birks & R.G. West (eds.), *Quaternary Plant Ecology,* pp. 237–261. Oxford, Blackwell Scientific Publications.

RYBNICKOVA E. & RYBNIČEK K. (1971) The determination and elimination of local elements in pollen spectra from different sediments. *Rev. Palaeobot. Palynol.* **11,** 165–176.

SALISBURY E. (1964) *Weeds and Aliens.* 2nd Edition. London, Collins.

SALMI M. (1962) Investigations on the distribution of pollens in an extensive raised bog. *Bull. Comm. geol. Finl.* **204,** 160–193.

SARJEANT W.A.S. (1969) Microfossils other than pollen and spores in palynological preparations. In Erdtman, G. *Handbook of Palynology,* pp. 165–208. Copenhagen, Munksgaard.

SAURAMO M. (1932) Studies on the Quaternary varve sediments in southern Finland. *Comm. geol. Finl. Bull,* 60.

SHACKLETON N.J. & TURNER C. (1967) Correlation between marine and terrestrial Pleistocene successions. *Nature, Lond.* **216,** 1079–1082.

SHEAIL J. (1971) *Rabbits and their History.* Newton Abbot, David & Charles.

SHOTTON F.W. (1967) The problems and contributions of methods of absolute dating within the Pleistocene period. *Q. J. geol. Soc., Lond.* **122,** 357–384.

SHOTTON F.W. (1972) An example of hard-water error in radiocarbon dating of vegetable matter. *Nature, Lond.* **240,** 460–461.

SIMS R.E. (1973) The anthropogenic factor in East Anglian vegetational history: an approach using A.P.F. techniques. In H.J.B. Birks & R.G. West (eds.), *Quaternary Plant Ecology,* pp. 223–236. Oxford, Blackwell Scientific Publications.

SMITH A.G., PILCHER J.R. & SINGH G. (1968) A large capacity hand-operated peat sampler. *New Phytol.* **67,** 119–124.

SMITH A.G. & PILCHER J.R. (1973) Radiocarbon dating and vegetational history of the British Isles. *New Phytol.* **72,** 903–914.

SMITH J.T. & YATES E.M. (1968) On the dating of English Houses from external evidence. *Field Studies 2,* **5,** 537–577.

SPARKS B.W. (1961) The ecological interpretation of Quaternary non-marine mollusca. *Proc. Limn. Soc. Lond.* **172,** 71–80.

SPARKS B.W. (1964) Non-marine Mollusca and Quaternary ecology. *J. Ecol.* **52** (suppl.) 87–98.

SPARKS B.W. & LAMBERT C.A. (1961) The Post-glacial deposits at Ahethorpe, Northamptonshire. *Proc. malac. Soc. Lond.* **34,** 302–315.

SPARKS B.W. & WEST R.G. (1959) The palaeoecology of the inter-glacial deposits at Histon Road, Cambridge. *Eiszeitalter Gegenw.* **10,** 123–143.

SPARKS B.W. & WEST R.G. (1972) *The Ice Age in Britain.* London, Methuen.

SPRING D. (1955) A great agricultural estate: Netherby under Sir James Graham, 1820–1845. *Agric. History XXIX,* **2,** 73–81.

STEERE W.C. (1965) The Boreal bryophyte flora as affected by Quaternary glaciation. In H.E. Wright & D.G. Frey (eds.), *The Quaternary of the United States,* pp. 485–495. Princeton University Press.

STEWART J.M. & DURNO S.E. (1969) Structural variations in peat. *New Phytol.* **68,** 167–182.

STEWART J.M. & FOLLET E.A.C. (1966) The electron microscopy of leaf surfaces preserved in peat. *Canad. Jl bot.* **44,** 421–427. 5 plates.

STOKES M.A. & SMILEY T.L. (1968) *An Introduction to Tree-ring Dating.* Chicago University Press.

STUIVER M. (1970) Tree ring, varve and carbon-14 chronologies. *Nature, Lond.* **228,** 454–455.

SUESS H.E. (1970) The three causes of the secular C14 fluctuations, their amplitudes and time constants. In I.U. Olsson (ed.) *Radiocarbon Variations and Absolute Chronology.* New York/Stockholm, Wiley.

SWAIN A.M. (1973) A history of fire and vegetation in northeastern Minnesota as recorded in lake sediments. *Quaternary Research* **3,** 383–396.

SWAIN F.M. (1965) Geochemistry of some Quaternary lake sediments of North America. In H.E. Wright Jr. & D.G. Frey (eds.), *The Quaternary of the United States,* pp. 765–781. Princeton University Press.

SZABO B.J. & COLLINS D. (1975) Ages of fossil bones from British interglacial sites. *Nature, Lond.* **254,** 680–682.

TALLIS J.H. & McGUIRE J. (1972) Central Rossendale: the evolution of an upland vegetation. I. The clearance of woodland. *J. Ecol.* **60,** 721–738.

TANSLEY A.G. (1968) *Britain's Green Mantle.* 2nd Edition, revised by M.C.F. Proctor. London, Allen & Unwin.

TATE W.E. (1943) A handlist of English Enclosure Acts and Awards. *Trans. Cumb. Westmor. Antiq. Arch. Soc.* NS **43,** 175–198.

TAUBER H. (1965) Differential pollen dispersal and the interpretation of pollen diagrams. With a contribution to the interpretation of the elm fall. *Danm. geol. Unders. (Ser. II)* **89,** 7–64.

TAUBER H. (1967) Differential pollen dispersion and filtration. In E.J. Cushing & H.E. Wright (eds.), *Quaternary Paleoecology,* pp. 131–141. Yale University Press.

TAUBER H. (1967) Investigations of the mode of pollen transfer in forested areas. *Rev. Palaeobotan. Palynol.* **3,** 277–286.

THOMAS K.W. (1964) A new design for a peat sampler. *New Phytol.* **63,** 422–425.

TINSLEY H.M. & SMITH R.T. (1974) Surface pollen studies across a woodland/heath transition and their application to the interpretation of pollen diagrams. *New Phytol.* **73,** 547–565.

TITTENSOR R.M. (1970a) History of the Loch Lomond oakwoods. I. Ecological history. *Scott. For.* **24,** 100–110.

TITTENSOR R.M. (1970b) History of the Loch Lomond oakwoods. II. Period of intensive management. *Scott. For.* **24,** 110–118.

TITTENSOR R.M. & STEELE R.C. (1971) Plant communities of the Loch Lomond oakwoods. *J. Ecol.* **59,** 561–582.

TOLONEN K. (1966a) Stratigraphic and rhizopod analyses on an old raised bog, Varrasuo in Hollola, South Finland. *Ann. Bot. Fenn.* **3,** 147–166.

TOLONEN K. (1966b) Soiden Kehityshistorian Tutkimusmenetelmista. *Eripainos Suo-lehdesta* n:o 6.

TOLONEN K. (1971) On the regeneration of North European bogs. I. Klaukkalan Isosuo in S. Finland. *Acta Agralia Fennica* **123,** 143–166.

TROELS-SMITH J. (1955) Characterization of unconsolidated sediments. *Danm. geol. Unders.* *4R* **3,** 10.

TROELS-SMITH J. (1960) Ivy, Mistletoe and Elm. Climatic Indicators—Fodder Plants. *Danm. Geol. Unders 4R* **4,** 4.

TUBBS C.R. (1964) Early encoppicements in the New Forest. *Forestry* **37,** 95–105.

TUBBS C.R. (1968) *The New Forest: An Ecological History.* David & Charles, Newton Abbot. 248 pp. illus.

TUBBS C.R. & JONES E.L. (1964) The distribution of gorse *(Ulex europaeus* L.) in the New Forest in relation to former land-use. *Proc. Hants. Field Club* **23,** 1–10.

TURNER C. (1970) The Middle Pleistocene deposits at Marks Tey, Essex. *Phil. Trans. R. Soc. B* **257**, 373–435.

TURNER C. & WEST R.G. (1968) The subdivision and zonation of inter-glacial periods. *Eiszeit. Gegenw.* **19**, 93–101.

TURNER J. (1964) Surface sample analyses from Ayrshire, Scotland. *Pollen et Spores* **6**, 583.

TURNER J. (1970) Post-Neolithic disturbance of British vegetation. In D. Walker & R.G. West (eds.), *Studies in the Vegetational History of the British Isles*, pp. 97–116. Cambridge University Press.

VAN DER HAMMEN T., WIJMSTRA T.A. & ZAGWIJN W.K. (1971) The floral record of the late Cenozoic of Europe. In K. Turekian (ed.), *Late Cenozoic Glacial Ages*, pp. 391–424. Yale University Press.

VITA-FINZI C. (1973) *Recent Earth History*. London, Macmillan.

WALKER D. (1966) The Late Quaternary history of the Cumberland Lowland. *Phil. Trans. R. Soc.* **B251**, 1–210.

WALKER D. (1970) Direction and rate in some British Post-glacial hydroseres. In D. Walker & R.G. West (eds.) *Studies in the Vegetational History of the British Isles*. Cambridge University Press.

WALKER D. (1972) Vegetation of the lake Ipea region, New Guinea Highlands. II. Kayamanda Swamp. *J. Ecol.* **60**, 479–504.

WALKER D. & WALKER P.M. (1961) Stratigraphic evidence of regeneration in some Irish bogs. *J. Ecol.* **49**, 169–185.

WANSTALL P.J. (ed.) (1963) *Local Floras*. Bot. Soc. Br. Isles, Conference Report No. 7.

WATT A.S. (1962) The effect of excluding rabbits from Grassland A (Xerobrometum) in Breckland, 1936–60. *J. Ecol.* **50**, 181–198.

WATT A.S., PERRIN R.M.S. & WEST R.G. (1966) Patterned ground in Breckland: structure and composition. *J. Ecol.* **54**, 239–258.

WATTS W.A. (1963) Late-glacial pollen zones in Ireland. *Ir. Geogr.* **4**, 367–376.

WATTS W.A. (1973) Rates of change and stability in vegetation in the perspective of long periods of time. In H.J.B. Birks & R.G. West (eds.), *Quaternary Plant Ecology*, pp. 195–206. Oxford, Blackwell Scientific Publications.

WATTS W.A. & WINTER T.C. (1966) Plant macrofossils from Kirchner Marsh, Minnesota—a paleoecological study. *Bull. geol. soc. Am.* **77**, 1339–1360.

WEBB T. (1973) A comparison of modern and presettlement pollen from southern Michigan (USA) *Rev. Palaeobotan. Palynol.* **16**, 137–157.

WEBB T. & BRYSON R.A. (1972) Late and Postglacial climatic change in the northern Midwest, USA: Quantitative estimates derived from fossil pollen spectra by multivariate statistical analysis. *Quaternary Research* **2**, 70–115.

WEST R.G. (1964) Inter-relations of ecology and Quaternary palaeobotany. *J. Ecol.* **52** (Suppl.), 47–57.

WEST R.G. (1968) *Pleistocene Geology and Biology*. London, Longmans.

WEST R.G. (1970) Pollen zones in the Pleistocene of Great Britain and their correlation. *New Phytol.* **69**, 1179–1183.

WEST R.G. (1971) *Studying the Past by Pollen Analysis*. Oxford Biology Reader No. 10. Oxford University Press.

WEST R.G. & SPARKS B.W. (1960) Coastal interglacial deposits of the English Channel. *Phil. Trans. R. Soc.* **B243**, 95–133.

WESTENBERG J. (1947a) Mathematics of pollen diagrams I. *Proc. K. ned. Akad. Wet.* **50**, 509–520.

WESTENBERG J. (1947b) Mathematics of pollen diagrams II. *Proc. K. ned. Akad. Wet.* **50**, 640–648.

WESTENBERG J. (1967) Testing significance of difference in a pair of relative frequencies in pollen analysis. *Rev. Palaeobotan. Palynol* **3**, 359–369.

WIGHTMAN W.R. (1968) The pattern of vegetation in the Vale of Pickering circa 1300 A.D. *Trans. Inst. Br. Geogr.* **45**, 125–142.

WILCOX W.W. (1964) Preparation of decayed wood for microscopical examination. US Forest Service Research Note FPL-056.

WILLIAMS M. (1970) *The Draining of the Somerset Levels.* Cambridge University Press.

WILLIAMS M. (1970) The enclosure and reclamation of waste in England and Wales in the eighteenth and nineteenth centuries. *Trans. Inst. Br. Geogr.* **51**, 55–70.

WILLIAMS M. (1972) The enclosure of waste land in Somerset, 1700–1900. *Trans. Inst. Br. Geogr.* **57**, 99–123.

WILLIAMS M. (1974) *The Making of the South Australian Landscape.* London, Academic Press.

WILLIS E.H. (1969) Radiocarbon Dating. In D. Brothwell & E. Higgs (eds.), *Science in Archaeology*, pp. 46–57. London, Thames & Hudson.

WILLIS E.H., TAUBER H. & MUNNICH K.O. (1960) Variations in the atmospheric radiocarbon concentration over the past 1300 years. *Radiocarbon* **2**, 1–4.

WRIGHT H.E. (1971) Late Quaternary vegetational history of North America. In K. Turekian (ed.) *Late Cenozoic Glacial Ages*, pp. 423–464. New Haven, Yale University Press.

WRIGHT H.E., CUSHING E.J. & LIVINGSTONE D.A. (1965) Coring devices for lake sediments. In B. Kummel & D. Raup (eds.), *Handbook of Paleontological Techniques*, pp. 494–520. San Francisco & London, Freeman.

WRIGHT H.E. & PATTEN H.L. (1963) The pollen sum. *Pollen et Spores* **5**, 445–450.

YARNELL R.A. (1969) Palaeo-ethnobotany in America. In D. Brothwell & E. Higgs (eds.), *Science in Archaeology*, pp. 215–229. 2nd Edition, London, Thames & Hudson.

YARWOOD S.M. (1974) Spatial aspects of growth and competition in a forest plantation. Unpub. B.Sc. Thesis, University of Southampton.

YATES E.M. (1965) Dark Age and Medieval settlement on the edge of wastes and forests. *Field Studies* **2**, 113–153.

CHAPTER 3

DESCRIPTION AND ANALYSIS OF VEGETATION

F. B. GOLDSMITH

and

C. M. HARRISON

1 **Introduction** 85
1.1 The dynamics of vegetation 88

2 **Description of vegetation** 90
2.1 Measures based on physiognomy 91
   2.1.1 Life-form 91
   2.1.2 Periodicity 94
   2.1.3 Stratification 95
2.2 Measures based on floristics 96
   2.2.1 Destructive measures 96
   2.2.2 Non-destructive measures 97
     Density 97
     Cover 97
     Frequency 100
     Basal area, girth and diameter 101
     Plotless measures 101
     Performance 103
   2.2.3 Sampling 103
     Representative 103
     Random 103
     Regular or systematic 104
     Restricted random 104
     Transects 104
     Stratified 105
     Number of quadrats 105
     Size of quadrats 106
     Shape of quadrats 107
     Permanent quadrats,
       photographic recording 107

     Minimal area 107
   2.2.4 Indices of diversity 109

3 **Analysis of vegetation** 109
3.1 Association and correlation 109
   3.1.1 Association 110
   3.1.2 Correlation 112
   3.1.3 Regression 113
3.2 Measures of non-randomness 113
   3.2.1 The detection of pattern 114
   3.2.2 Scale of pattern 116
3.3 Classification and mapping 120
   3.3.1 Small-scale (large area)
     classifications 120
     Divisive methods 121
     Agglomerative methods 123
   3.3.2 Large-scale (small area)
     classifications 129
     Divisive methods 130
     Agglomerative methods 135
   3.3.3 Vegetation mapping 137
   3.3.4 Use of aerial photography 139
3.4 Vegetational and environmental
   gradients 140
   3.4.1 Gradient analysis 141
   3.4.2 Ordination 141
   3.4.3 Trend surface analysis 148

4 **Choice of methods** 148
5 **References** 149

1 INTRODUCTION

Many ecologists view vegetation as a component of ecosystems which displays the effects of other environmental conditions and historic factors in an obvious and easily measurable manner. The careful analysis of vegetation is therefore used as a means of revealing useful information about other components of the ecosystem. In this way the research worker's study frequently proceeds from the description of vegetation in the field to the subsequent analysis of these records in the laboratory. The two phases of study reflect fundamentally different concepts of vegetation. The description is of 'real' vegetation while the analysis is concerned with forming 'abstractions' or 'generalized types' of vegetation which are simplifications of the complexities of the 'real' world. At the outset of the study the worker therefore has to decide which method of description and analysis best suits his purposes. The choice is a difficult one and the decision is affected by the purpose of study, the scale of enquiry, the botanical knowledge of the worker and the nature of vegetational variation itself (Webb, 1954).

Ecologists with a botanical training and studying a botanically-orientated problem generally prefer an approach based upon floristics, that is upon plant species composition. This necessitates a knowledge of the flora and may require considerable time and/or expertise, especially if working in an unfamiliar area. Their approach also depends upon whether the problem is autecological (single species) or synecological (communities), and orientated towards a production study or the identification of causality. Autecologists usually require a measure of abundance or performance of a species which can be easily supplied whereas synecologists, dealing with communities, become blighted by the problems associated with the nature of vegetational variation. An introductory text in plant ecology is provided by Willis (1973). Production ecologists require data about dry weight and calorific content which is very time consuming to collect and destructive. Also they are often concerned about the number of samples necessary and their most informative spatial arrangement.

Zoologists, on the other hand, being interested in vegetation as a matrix in which animals live and feed, are usually more interested in its structure, usually the degree of stratification, and habitat diversity (Elton & Miller, 1954; Elton, 1966).

Vegetation description is also an integral part of much resource survey work, especially in the preparation of inventories of timber and the assessment of range-land carrying capacity. Soil scientists, and to a lesser extent, geologists and climatologists, are interested in vegetation as an expression of the factors they study. Their research is often conducted on more general scales than that of the ecologists and requires classifications that produce mappable units (Birse & Robertson, 1967).

Foresters frequently use an assessment of species composition to indicate site potential and to help in selecting species for planting. In practice, however, this usually involves the establishment of indicator species which identify the main characteristics of the site (*Juncus, Sphagnum, Pteridium,* etc.) and objective methods for describing vegetation are usually restricted to research projects (van Groenewoud, 1965).

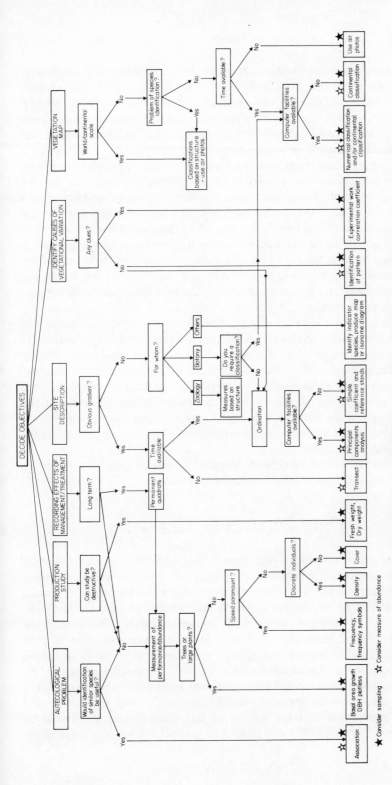

Figure 3.1 Some considerations to be used in choosing an appropriate method for describing or analysing vegetation.

The objective of this chapter is to review the range of methods available to the ecologist and to discuss the theoretical and practical problems involved with each and to indicate the kind of situation in which its use is appropriate. Full descriptions will not be given where reference to a readily available book or journal can provide the theoretical background and examples of the application of a method.

The primary objective in selecting a method should be its informativeness, but no ecologist has yet succeeded in establishing an objective test for comparing the information return from a range of techniques. Other criteria which should be used in selecting a method include relevance to the aims of the project, speed, accuracy, objectivity (reproducibility) and non-destructiveness (Moore *et al.,* 1970). In practice, the research worker usually has to make a compromise because the most objective method may not be the fastest, or the most accurate may be destructive. However, if these criteria are considered prior to sampling and analysing vegetation the final outcome should be proportionally more useful.

The first decision is to identify the objectives of the study. Although this sounds unnecessarily obvious it determines a series of decisions that must subsequently be made. Fig. 3.1 summarizes the kind of decision-making process the research worker should adopt. However, it must be emphasized that these are only guidelines. In practice, every problem is unique, and the selection of a method should be based on a detailed consideration of the characteristics of the problem. It is hoped, however, that Fig. 3.1 will provide an indication of the kinds of questions that should be asked and the decisions that are likely to be made.

Four textbooks are of general relevance to this chapter and should be consulted for reference (Greig-Smith, 1964; Kershaw, 1973; Kuchler, 1967; Shimwell, 1971).

1.1 THE DYNAMICS OF VEGETATION

A perplexing characteristic of vegetation is its dynamic nature; that is changes in the distribution and structure of components of vegetation which occur over time. For example, seasonal changes in vegetation are of universal occurrence but will affect the production ecologist and autecologist most directly. In addition, any descriptive measure of the vegetation will be affected by seasonality (Hope-Simpson, 1940) and for this reason the date of field observation should always be stated with the presentation of results. While the spatial implications of seasonal changes are likely to be easily recognized by the field ecologist, other spatial patterns in vegetation which are the product of temporal changes may not be so readily appreciated. For example several workers have noted that the pattern of individual species is affected by competition (Harper, 1967), and (Watt, 1947; 1964) related pattern in a heathland community to the competitive ability of a species and its age. Watt recognized four growth phases for *Pteridium* and *Calluna,* namely pioneer, building, mature and degenerate and he documented changes in the relative distribution of these two species and their associated growth phases. During the pioneer and degenerate phases of growth, the individual plant was least competitive, while it proved most aggressive during the building and mature phases. Such cyclical phases of development clearly affect the distribution of species, but also affect the structure and productivity of the community to which they belong (Barclay-Estrup & Gimingham, 1969). Since this type of vegetational pattern depends upon the life-history of the individual, the pattern will change most rapidly in communities dominated by short-lived species, e.g. 35 years for *Calluna* dominated heathland but 2–300 years in deciduous woodland. Nevertheless in an undisturbed condition the net effect of these cyclical processes is to create an observable vegetation mosaic, which over a long period remains the same. Many communities appear to exhibit these cyclical patterns of growth, for example heathland, grassland, and woodland (Watt, 1947), tundra (Billings & Mooney, 1959) and bogs (Oswald, 1923), and while most relate to the life-history of the individual dominant species, other examples seem to suggest that environmental factors such as fire, wind, frost-action and fluctuating water-tables can accentuate certain phases of the cycle (Ratcliffe & Walker, 1958, Burges, 1960; Anderson, 1967; Boatman & Armstrong, 1968).

In this respect therefore the pattern of vegetation may be a product of intrinsic change wrought by the plants themselves, may be caused by external environmental factors, or produced by a combination of both. The implications of these cyclical processes for the community ecologist are that in certain circumstances spatial changes are largely the result of temporal changes and not due to initial differences in the habitat conditions.

On a longer time-scale the pattern of the vegetation of an area may exhibit directional changes other than changes which maintain the *status quo*, as do the cyclical changes described above. These long-term changes are referred to as plant succession. During plant succession one community is replaced by another of different specific and structural composition, so that over a long period of time, several communities may succeed one another. The process is a continuous one although distinct stages may be described for the purpose of convenience. Eventually these seral communities are replaced by a stable climax community within which no further directional changes take place. In its original form the climax community was thought to be that community controlled by the climate of the region, that is the climatic climax (Clements, 1916) but Tansley (1935) later modified the concept to include all those communities which obtained relative permanence and stability in the landscape. It is within these stable communities that cyclical changes of regeneration will predominate. Successional communities on the other hand do not replace themselves but are invaded by new species and here spatial pattern in the vegetation is likely to result from competition rather than from cyclical processes.

The dynamics of vegetation are a complex subject and have been reviewed by Margalef (1968) and Odum (1969) who characterize succession in terms of increasing 'information' content. What is of especial interest to students of the description and analysis of vegetation however, is the relationship between the scale of spatial pattern and the scale of temporal change. Thus short-term daily and seasonal changes are of most relevance as an aid to understanding large-scale vegetational pattern, such as is encountered in autecological studies, cyclical changes such as those described by Watt (1947) are of most relevance to studies of within-community variation, while successional changes are probably most relevant, to small-scale studies of 'between' community variation (Harrison & Warren, 1970). Thus spatial scale and temporal scales are related but not in any absolute way, for cyclical process may also be relevant to autecological studies, and successional studies to large-scale community variation. It is for reasons such as these that any ecologist needs to consider the dynamic relationships of the vegetation which he is studying.

2 DESCRIPTION OF VEGETATION

Vegetation may be defined as an assemblage of plants growing together in a particular location and may be characterized either by its component species or by the combination of structural and functional characters that characterize the appearance, or physiognomy, of vegetation. This is an important distinction which is reflected by the range of methods available for describing vegetation. Structural or physiognomic methods do not demand species identification and are often considered more meaningful for small-scale (large area) studies and for habitat description for scientists of other disciplines. For example, zoologists favour descriptions of vegetation which they can interpret in terms of the niche, habitat and food resources of animals. Methods based on species composition or floristics are more useful for large-scale (small area) studies of a more detailed, botanical, nature. They are, however, used by the Continental European school of phytosociologists for vegetation classification and mapping on an extensive scale. It may be significant that these studies are very detailed and time-consuming. Thus they may be considered large-scale (small area) in character, although the synthesis of several decades work covers an extensive area. It appears that classificatory methods are initially devised for large-scale (small area) studies and are subsequently used for increasingly extensive studies. This has occurred with both the Continental methods and the quantitative ones such as association-analysis (q.v.).

2.1 MEASURES BASED ON PHYSIOGNOMY

Physiognomy is used to characterize an assemblage of plants but it is a measure which is surrounded in controversy and its usage in the literature is inexact. In general, physiognomy refers to the appearance of the vegetation; its height, colour and luxuriance; and to leaf size and shape. Although these characters may seem self-evident they tend to result from a combination of functional and structural characters (Fosberg, 1967). Functional characters are those which serve an adaptive role for survival in existing or past environments, as for example the evergreen or deciduous habit. Structural characters refer to the horizontal and vertical arrangement of components of the vegetation, as for example in the spacing between individuals and their vertical layering. Purely physiognomic characters which do not relate to either or both of these other characters are difficult to isolate. For example leaf size may be regarded as a functional adaptation to particular climatic conditions, or as a product of the age of the individual or as a result of shading when the plant is a member of an understorey. However, physiognomic characters have proved useful for describing vegetation and include life-form (*sensu* Du Rietz which are essentially descriptive), life-form (*sensu* Raunkiaer which are essentially functional), periodicity, and stratification.

2.1.1 Life-form

The definition and discussion provided by Du Rietz (1931) is perhaps the most comprehensive statement of life-form and most other authors who utilize life-form as a character for description base their classes upon those of Du Rietz. His categories reproduced in Table 3.1 are more or less self-explanatory as are the types of vegetation which can be recognized on this basis, for example, forest, woodland, scrub and grassland. The types are however very superficial and generalized so that only a gross description of vegetation from a large area can be achieved.

The life-form categories recognized by Raunkiaer (1934) provide one of the most widely used functional criteria for describing vegetation. He based his approach on the position with respect to the ground surface on the plant of the perennating bud, that is the bud from which the next seasons growth would be made. Five, broad, life-form classes were recognized namely phanerophytes, chamaephytes, hemicryptophytes, cryptophytes and therophytes, see Fig. 3.2. Phanerophytes bear their perennating buds freely in the air at varying heights at

Table 3.1   The life-form system of Du Rietz (1931)

A   Higher plants
  I. Ligniden (woody plants)
    (a) Magnoligniden (m)—Trees taller than 2 m
      1. Deciduimagnoligniden (md) deciduous
      2. Aciculimagnoligniden (ma) needleleaf evergreen
      3. Laurimagnoligniden (ml) other evergreens
    (b) Parvoligniden (p)—Shrubs 0·8 m to 2 m tall
      4. Deciduiparvoligniden (pd)
      5. Aciculiparvoligniden (pa)
      6. Lauriparvoligniden (pl)
    (c) Nanoligniden (n)—Under 0·8 m tall
    (d) Lianen (li)—Climbing plants
  II. Herbiden (herbs)
    (a) Terriherbiden—Terrestrial herbs
      9. Euherbiden (h) herbs
      10. Graminiden (g) grasses
    (b) Aquiherbiden—Water plants
      11. Nymphaeiden (ny) rooted with floating leaves *(Nymphaea)*
      12. Elodeiden (e) rooted without floating leaves *(Elodea)*
      13. Isoetiden (i) rooted, bottom rosettes *(Isoetes)*
      14. Lemniden (le) free floating, not rooted *(Lemna)*
B   Moose (Bryophytes)
      15. Eubryiden (b) all mosses and liverworts excluding *Sphagnum*
      16. Sphagniden (s)—*Sphagnum* spp.
C   Fletchten
      17. Lichens
D   Algen
      18. Algae
E   Pilze
      19. Fungi

least 25 cm above the ground. They are mostly woody plants, trees and shrubs and are sub-divided into classes according to height:

Megaphanerophytes	30 + m
Mesophanerophytes	8–30 m
Microphanerophytes	2–8 m
Nanophanerophytes	25 cm–2 m

Further subdivisions were also made on the presence or absence of protection of the bud and on whether or not the species is deciduous. *Chamaephytes* are also woody or semi-woody perennials bearing their buds close to the ground but less than 25 cm from the surface:

Suffrutescent or semi-shrubby forms
Passively decumbent forms
Actively creeping or stoloniferous forms
Cushion plants

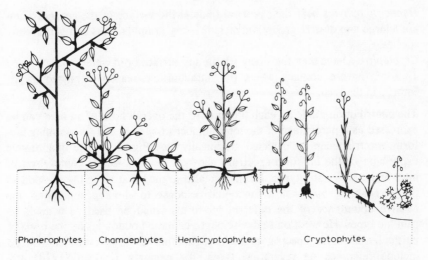

Phanerophytes | Chamaephytes | Hemicryptophytes | Cryptophytes

Parts of the plant which die in the unfavourable season are unshaded; persistent axes with surviving buds are black

Figure 3.2 Diagrammatic representation of Raunkiaer's life-forms (from Raunkiaer, 1934).

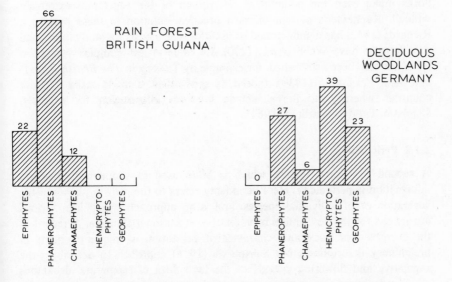

Figure 3.3 Life-form spectra for rain-forest in British Guiana and deciduous woodland in Germany.

Hemicryptophytes bear their renewal buds at the surface of the ground. They are a large and diverse group and include many graminaceous and herbaceous species.

Cryptophytes have their buds beneath the soil surface or in water.

Therophytes are annuals where the unfavourable season is passed as an embryo in the seed.

The contribution made by each life-form to the overall flora of an area can be expressed as a percentage of the total number of species and the resulting life-form spectrum can be depicted graphically (see Fig. 3.3). The quantitative expression of the life-form spectrum allows a more objective comparison of inter-regional floristic types. Raunkiaer also argued that since the position of the perennating bud was a functional response to climatic conditions, the relative abundance of the different life-forms could be used as a guide to climatic types. He went on to devise phyto-climatic boundaries on the basis of particular life-form spectra, and although other climatic classifications do include reference to vegetation types, for example Koppen (1928) and Thornthwaite (1948), the relationship between vegetation boundaries and climate must be extremely tenuous. The danger of a circular argument here is such that the value of these classifications must be questioned even on a global scale. This does not negate the use of Raunkiaer's life-form classification as a tool for the description of vegetation at other scales. It has been widely used by workers in many different environments. For example, in the tropics where floristic richness, structural complexity and the paucity of up-to-date published floras make even the preliminary description of the vegetation extremely difficult, Raunkiaer's system offers a practical solution to these problems. Richards (1952) has demonstrated its usefulness as a basis for study in tropical situations as have Webb *et al.* (1970). Its value in less complex vegetation types has also been established, for example by Tansley in *The British Islands and their Vegetation* (1939). Indeed its application is made easier in those countries where hand floras include life-form information for example, Clapham, Tutin & Warburg (1962).

2.1.2 Periodicity

A second functional feature which is often used to supplement life-form information is periodicity. Here periodicity refers to the growth phases of the vegetation or of individual species, and is an approach which has obvious attractions for the description of vegetation in seasonal climates. A record of the evergreen or deciduous character of vegetation is frequently made in preliminary descriptions, but Salisbury's (1916) approach to describing the vegetative and flowering periods of the herb flora of temperate, deciduous woodlands introduces a more dynamic element into these descriptions. He recognized three main growth phases, prevernal, vernal and aestival, and compiled phenological diagrams for the main species encountered. Other studies utilizing this approach describe the growth phases of Raunkiaer's life-forms (Preis, 1939). Periodicity records may be tedious to collect in the field, requiring a long field season for observation, but they can provide a useful basis for more detailed lines of enquiry.

2.1.3 Stratification

Periodicity is often related to the vertical and horizontal structure of the vegetation and in particular to the vertical stratification or layering of components of the vegetation. The recognition of more or less continuous layers of vegetation on the basis of height differences is a structural approach to vegetational description and is inherent in most life-form classifications. On a more local scale a structural approach can be used to simplify and describe the organization of complex vegetation types. Each layer is described in terms of height and in many instances floristic information is used to supplement this preliminary description. A technique pioneered in the tropics by Davies & Richards (1933) involves a complete visual representation of the stratification of the community in a profile diagram (see Fig. 3.4). These authors made

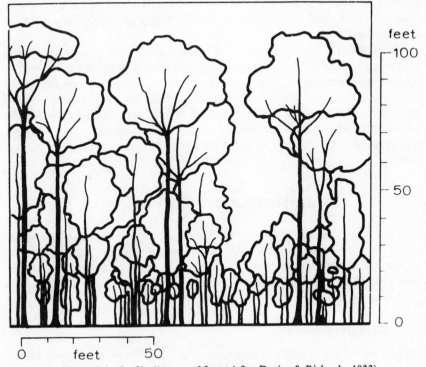

Figure 3.4 Profile diagram of forest (after Davies & Richards, 1933).

careful measurements of the main trees felled along a route through tropical rain-forest. In such an environment, characterized by species richness, luxuriance of growth and the overall complexity of the vegetation, a structural approach through stratification provides a useful means of supplementing life-form descriptions. The procedure, however, may be very destructive and expensive of time and effort. Tansley (1939) adapted this approach for use in the description of temperate deciduous woodlands but in this case estimated by visual inspection the height of the various layers *in situ*. Elton & Miller's (1954) classification of habitats which is designed primarily for field zoologists also adopts a structural approach through stratification charac-terized by differences in height.

This structural approach can also be applied to the root component of vegetation as well as to above ground parts. In this case, a trench dug through the soil exposes the roots which can then be cleaned by spraying with a jet of water. Weaver & Clements (1938) and Coupland & Johnson (1965) provide many examples of root stratification among prairie grasses and other workers have examined the inter-relationships among heath plants (Rutter, 1955) and between calcicole and calcifuge species (Grubb *et al.*, 1969). This approach through stratification is not an end in itself but it can provide a useful preliminary guide to plant/environment relationships which might be profitably examined later, and it provides a clear summary of structurally complex communities.

In summary all descriptions of vegetation based on physiognomic characters are essentially generalized but they provide a means of arriving at informative descriptions often based on easily measurable properties of the vegetation, relatively quickly. Such characteristics can be isolated even by relatively untrained observers (see Webb *et al.*, 1970) and it is for this reason, above all others, that physiognomic descriptions retain their popularity for reconnaissance surveys among many different users, and over a wide range of scales.

2.2 MEASURES BASED ON FLORISTICS

Detailed studies usually require an assessment of the species composition of an area. This may be accompanied by information about the amount or *abundance* of each species present at a site. It is useful to distinguish between abundance and *richness,* the latter being the number of species present on a particular area.

2.2.1 Destructive measures

Ideally, the ecologist requires data about the relative bulk of different species but the collection of such data requires destruction of the samples taken. This is undesirable if (1) further samples are required (if, for example, annual or long-term changes are being observed) or (2) if the area is of outstanding natural beauty or biological interest or (3) any of the species are rare.

The most commonly used measures are fresh weight and dry weight, which are self-explanatory. Fresh weight suffers the disadvantage that it varies with the moisture content of the plant; dry weight is determined after 24 hours drying at 100°C. These measures are often also referred to as *biomass* when the total vegetation, rather than individual species is considered, or as the *yield*, if expressed on a per unit area and/or per unit time basis. (See Chapter 4.)

2.2.2 Non-destructive measures

These have the advantages of repeatability and of causing minimal damage to the vegetation. There are several methods, some of which may be sub-divided, but none are Utopian in their efficacy (Brown, 1954; Greig-Smith, 1964; Kershaw, 1973). All involve the use of a sampling unit the size and number required will be dealt with later in the section.

Density

Density is defined as the number of individuals of a particular species per unit area. Counts are usually made in a number of quadrats, multiplied by the area under study and divided by the area sampled to give the density in the study area. This measure is independent of the size of the sampling unit and is said to be 'absolute'. It is much favoured by zoologists who rarely have difficulty in defining an individual, but application to many vegetation types is impractical. The determination of the density of trees, shrubs, tussocks of grasses or sedges, arable weeds and conspicuous individual herbs such as orchids is simple, but species which spread vegetatively, such as grasses and clover, are often almost impossible to deal with in this way. An individual grass plant is impossible to define in a permanent pasture and the counting of tillers is impractical unless the quadrats are extremely small.

Cover

Cover is defined as the proportion of the ground occupied by a perpendicular projection of the aerial parts of individuals of the species under consideration (Greig-Smith, 1964) and is usually expressed as a percentage. Because of the overlayering of different species, the total cover for an area may exceed 100%, and in the case of highly stratified forests may reach several hundred percent. It is also an absolute measure and may be recorded (1) by visual estimate (2) with single cover pins (often called point quadrats), or (3) with frames of pins or frames of crosswires.

The subjective estimation of cover simply involves a visual estimate. The percentage obtained may then be expressed as a figure indicating the range within which it falls. Such values are often called *Domin* values and are much used by phytosociologists on the continent of Europe (Bannister, 1966). The original use of the Domin scale involved information about cover and abundance but combined scales are unpopular amongst ecologists striving for objectivity and a simple cover scale is generally preferred.

Although several cover scales have been suggested (Shimwell, 1971) there are two in regular use Domin and Braun-Blanquet (Table 3.2).

Table 3.2　Cover scales in general use

Class	Domin	Braun-Blanquet	Hult-Sernander	Lagerberg-Raunkiaer
+	A single individual	Less than 1%	—	—
1	1–2 individuals	1–5	0–6·25	0–10
2	Less than 1%	6–25	6·5–12·5	11–30
3	1–4	26–50	13–25	31–50
4	4–10	51–75	26–50	51–100
5	11–25	76–100	51–100	—
6	26–33	—	—	—
7	34–50	—	—	—
8	51–75	—	—	—
9	76–90	—	—	—
10	91–100	—	—	—

These measures may be criticized on the grounds that they are 'pseudo-quantitative' and give little more information than frequency symbols (Tansley & Adamson, 1913), but they are easily and quickly used in the field.

Single cover pins may be used on low vegetation such as grassland to give a quick, easy and accurate measure of abundance. The pins should have zero cross-sectional area as exaggeration occurs with increasing diameter as is shown in Fig. 3.5. The exaggeration is greater for fine-leaved species than for broad-leaved ones. Bicycle spokes are often used as cover pins and are normally held vertically in the vegetation, although some workers have advocated the use of inclined pins (Warren Wilson, 1960). The pins may be held in a frame, and moved up and down through a series of holes in two parallel bars, one fixed vertically above the other (Fig. 3.6). Frames using cross-wires are similar in design (Fig. 3.6) except that the holes are larger or tubes in which wires are inserted are used and are viewed from above (Winkworth & Goodall, 1962).

The species touched by each pin between the canopy and ground level is normally recorded although some workers note the number of times each

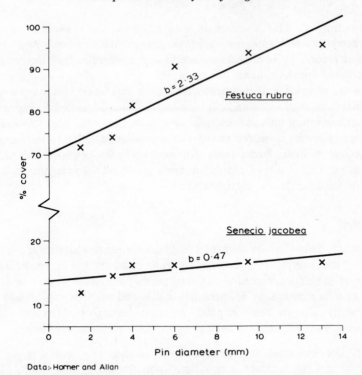

Data:- Horner and Allan

Figure 3.5 A comparison of the exaggeration of cover values with increasing pin diameter for two species.

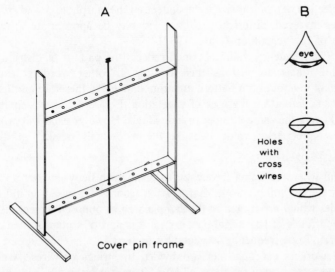

Figure 3.6 (A) Frame for holding pins for recording percentage cover, (B) to show the use of cross-wires.

species is touched. This is known as *cover repetition* and is likely to be more highly correlated with yield but suffers the disadvantage of being much more difficult to record. There appears to be no value in recording only the first, that is uppermost, species touched by each pin.

The use of cover is largely restricted to low vegetation such as grassland and the herb layer of woodlands. It becomes difficult to handle the pins and/or frames in vegetation such as tall heaths and scrub. Cover has been shown to be sufficiently sensitive to record changes in vegetation resulting from change in management or biotic fluctuations. Thomas (1960) for example, successfully used this measure to record changes in chalk grassland following the decline in rabbit populations due to myxomatosis.

Frequency

Frequency is defined as the chance of finding a species in a particular area in a particular trial sample. It is obtained by using quadrats and expressed as the number of quadrats occupied by a given species per number thrown or, more often, as a percentage. It is extremely quick and easy to record and has consequently always been popular amongst ecologists. However, these advantages must be weighed against two quite serious disadvantages;

1. The frequency is dependent upon quadrat size, and it is for this reason that the measure is considered non-absolute (Greig-Smith, 1964). Consequently care must be taken to select the optimum quadrat size (see below) and this must be stated whenever frequency values are reported. As the selection of quadrat size is dependent on the morphology and size of the species involved, different quadrat sizes may be appropriate to different strata of the vegetation (Cain, 1932).

2. The frequency value obtained reflects the pattern of distribution of the individuals as well as their density. In other words it expresses information about both pattern and abundance and therefore confuses two basic and important features of vegetation. Figure 3.7 shows three areas each having the same number of individuals. However each will produce a different frequency value because the individuals occur in different patterns.

It is useful when recording frequency to distinguish between root and shoot frequency (Greig-Smith, 1964), that is when records are based on individuals of a species which are rooted in the sample area as opposed to records based on the occurrence of any aerial part even if it is part of a plant rooted outside the quadrat. Root frequency is generally preferred.

Some workers use quadrats sub-divided by strings or wires (Archibald, 1949). These often take the form of 50×50 cm quadrats subdivided into 25 smaller quadrats. Frequency may then be expressed for each of these frames

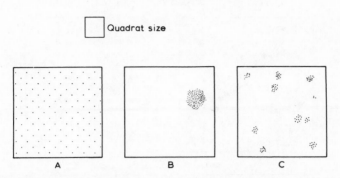

Figure 3.7 The dependence of percentage frequency on the pattern of distribution of individuals. With the same number of individuals present in each case and using the same size of quadrat, widely differing values will be obtained from the three communities (from Kershaw, 1973).

and may be useful if a single value is required for a small area. This is sometimes the case when collecting data for ordination or numerical classifications.

Basal area, girth and diameter

These are all measures used for species with large individuals such as tussocks, shrubs and trees. They have been extensively used by foresters and in association with data on density permit the prediction of timber yield. As their names imply they involve the measurement of diameter, girth or area at various heights of the plant. Ground level, as for basal area, is not popular because of the convoluted outline of many individuals due to buttresses, prop roots, etc. Consequently an arbitrary height (4 ft 3 in or 1·3 m and called breast height) is commonly used.

Girth of trees is often measured with special tapes (quarter girth tapes) which for commercially important species permit direct reference to tables for the estimation of timber yield.

Plotless measures

The practical difficulties of using quadrats and cover pins or marking out sample areas in forest and shrub have led to the use of plotless methods to estimate density (see Greig-Smith, 1964; Shimwell, 1971). This sampling design may also be used for collecting information about species composition, growth and environmental factors. Four different procedures have been suggested (Cottam, 1947; Cottam & Curtis, 1949, 1955, 1956) and are based on a number of random points (Fig. 3.8).

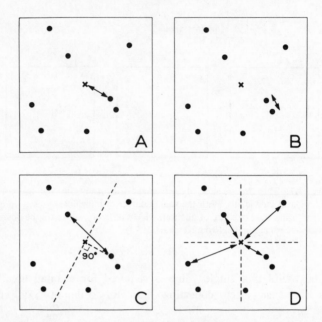

Figure 3.8 Plotless sampling methods: (A) closest individual method; (B) nearest neighbour method; (C) random pairs method; (D) point-centred quarter method.

1. Closest individual method.

Measurements are made from each random point to the nearest individual.

2. Nearest neighbour method.

Measurements are made from the individuals to their nearest neighbour.

3. Random pairs method.

Measurements are made from an individual to another on the opposite side of the sampling point.

4. Point-centred quarter method.

Distances are measured from the sampling point to the nearest individual in each quadrant.

From the mean value of each of these measurements the mean area and thus the density of the species may be calculated.

$$\text{Density} = \frac{\sqrt{\text{mean area}}}{2}$$

Pielou (1959, 1962) suggests that the correction factor in the denominator should in some circumstances be calculated empirically and has discussed the use of these measures in the study of pattern and of competition. Yarranton (1966) has used the intersections of herring netting as point quadrats to detect interspecific association and referred to this as a plotless method of sampling vegetation.

Performance

Various measures of vigour or performance have been used by ecologists and these are described by Chapman in Chapter 4.

2.2.3 Sampling

Both non-destructive and destructive measures described above require the use of some kind of sampling unit, usually quadrats. As we have seen, the choice of method may be a major problem, but even when this has been satisfactorily resolved there still remains the difficulty of deciding on the number, size and arrangement of samples. The choice of sampling method will again depend upon the nature of the problem, the morphology of the species, its pattern, and the time available. Here it is only possible to present the main alternatives and outline their theoretical and practical advantages and disadvantages. The research worker must always make the final decision within the context of his particular problem.

There are six main types of sampling arrangement from which to choose, they vary in their popularity but each is appropriate for some set of circumstances.

1. Representative (subjective or selective)

The quadrats are arranged subjectively to include representative areas or areas with some special feature, such as the species under study. In some circumstances practical considerations may render this the only arrangement possible, for example, where access is difficult or dangerous as on cliffs. Data collected from such a sampling pattern are not acceptable for statistical analyses involving the assessment of significance such as t-tests, F-tests, association, correlation or regression but are suitable for some multivariate techniques such as ordination.

2. Random

This is generally considered to be the 'ideal' method of sampling. Each sample by definition, must have an equal chance of being chosen and the samples should be positioned by using pairs of random numbers as distances along two axes positioned at right angles to each other. A variant of this type of approach, known as a random walk, may involve walking on a compass bearing for a random number of paces, sampling, changing direction and repeating the procedure. Random numbers may be taken from Statistical Tables such as those of Fisher & Yates (1963), Bingo sets, telephone directories, or playing cards. Throwing quadrats over one's left shoulder, or any other form of acrobatic, does not however achieve a random coverage of the sample area.

All statistical tests assume that data have been collected from a random arrangement of the sample units and, if the research worker is in any doubt, this approach should be adopted.

3. Regular or systematic

One of the problems with random sampling is that the cover of the area with sample points is not regular, some areas being under-sampled and others over-sampled. Sampling using a grid, however, achieves a regular arrangement and consequently the estimates derived are more accurate. But, their accuracy cannot be assessed and the data cannot be analysed statistically. If a statistical analysis is not required, then the method has the advantage that fewer sample points will be required compared with random sampling.

Care must be exercised if there is a phased pattern in the area under study. The classic example usually quoted is of broad ridge-and-furrow permanent pasture. If, in this situation, a regular sampling arrangement were used, and the phase of the sampling coincided with the phase of pattern in the vegetation, a very distorted result could be obtained. Similar problems could arise with dune ridges and slacks, and polygonal patterns in arctic vegetation.

4. Restricted random (partial random)

This is a compromise between systematic and random sampling and combines some of the advantages of each. The area under study is sub-divided and each sub-division then sampled at random. In this way each point in the area has a greater chance of being sampled and the data are suitable for statistical analysis. It is, however, more time-consuming than either 'parent' method because the area has to be marked out with a grid as well as necessitating the location of the random points.

5. Transects

Transects are a form of systematic sampling in which samples are arranged linearly and usually contiguously. They are very commonly used in studies but are really only appropriate to the investigation of gradients of change when they should be positioned at right angles to the zonation. For example, on salt-marshes, inter-tidal zones, successions on dunes, hydroseres, altitudinal gradients, from dry to wet heath, and across gradients of trampling intensity. Forests are often sampled by using nested plot techniques (Fig. 3.9).

Data from transects are invariably presented as frequency histograms for the species and environmental factors concerned, although 'profile diagrams'

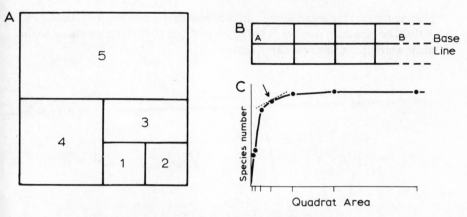

Figure 3.9 Quadrats may be conveniently arranged in the manner shown (A or B) to produce (C) a species-area relation curve which may be used to identify the minimal area of a plant community (from Shimwell, 1971).

after Davies & Richards (1933, see above) are often used to represent stratification in forests.

6. Stratified

This method of sampling has been more extensively used in disciplines other than ecology. It involves sub-dividing the field of study into relatively homogeneous parts and then sampling each sub-division according to its area, or some other parameter. For example, a mosaic of grassland and scrub could be divided into these two components and then each sampled, or a zonation with clearly demarcated zones is suitable for this type of sampling. Thus, it may be seen that the recognition of distinct communities is a form of stratified sampling. It may be that the imprecise nature of community boundaries has discouraged the use of this method of sampling in ecology but the advantages it offers are potentially considerable, particularly for relatively small scale surveys of large areas.

Number of quadrats

When considering the number of sampling units the research worker should adopt the general rule 'the more the better'. However, the objective of sampling, as opposed to recording everything, is to reduce the amount of labour and time involved. The actual number to be used is therefore a compromise between an ideal number, which is quite large, and a number which would take very little time to collect, which is quite few. Although the decision is relatively arbitrary a guide to the minimum number may be obtained from a simple test such as plotting the running mean or the variance against the number of quadrats (Fig. 3.10). The minimum number is the number of quadrats that correspond to the point where the oscillations damp down.

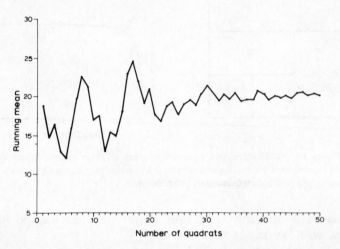

Figure 3.10 A simple test for estimating the minimum number of samples required. The running mean or variance is plotted against the number of quadrats.

Alternatively the number may be arrived at from a percentage basis. Assuming that the optimal area of the sampling unit has been decided (see below) the optimal number of quadrats can be chosen by deciding whether to sample 5%, 10% or 20% of the area under study. Some statistical tests require a minimum of thirty values and this may be considered a useful indication in the absence of other criteria. Another useful guide is that the variance within a sample area should be less than the variance between sample areas. If the data are relatively uniform fewer samples are required than if values vary widely. No absolute number can be suggested, however, because the range of heterogeneity encountered in the field is so wide and research projects vary in the degree of accuracy required.

Size of quadrats

The principal considerations in choosing the size of quadrat are the morphology of the species in the vegetation to be sampled and the homogeneity of the vegetation. Small quadrats are appropriate to the study of small plants, for example 10×10 cm or 25×25 cm quadrats may be suitable for chalk grassland, arable weeds and fixed dune grassland, and large quadrats for scrub and woodlands. The sampling unit for the collection of data for vegetation classification or ordination should be related to the homogeneity of the vegetation and be slightly larger than the 'minimal area' (*sensu* Poore, 1955, see below) of that vegetation.

Shape of quadrats

Quadrats by tradition are square. However, one of the causes of error, edge effect, is slightly reduced if the perimeter is reduced relative to the area, and this can be achieved by using round quadrats. Clapham (1932) however, advocates the use of rectangular quadrats, orientated parallel to the principal gradient of variation in the vegetation. However, the greater the length/breadth ratio the greater the edge effect. The difference that will be obtained with variously shaped quadrats is very small and not an important consideration.

Permanent quadrats, photographic recording

Long term changes in vegetation are best studied by means of permanent quadrats or permanent transects. Watt (1962) for example, used permanent quadrats to study the long term changes in grassland after the removal of rabbit grazing and Thomas (1960) used transects for a similar purpose. For large scale (small area) studies it is also possible to map individuals reasonably accurately using marked tapes from two reference pegs. Smaller scale studies require techniques of plane-tabling if accuracy is not of paramount importance or surveying using theodolite traverses. At the smallest scale (largest area) aerial photography or a series of old maps are recommended. Moore (1962) used the latter approach to study changes in the Dorset heathlands.

Direct photographic recording of vegetation with hand held 35 mm cameras may save time in the field but the determination of species abundance subsequently becomes more difficult. Modifications of this type of approach include the use of balloons to hold the camera several feet above the ground (Duffield & Forsythe, 1972). Chapman (pers. comm.) has developed this approach which might be called 'sub-aerial photography' by using a bipod and producing stereo-pairs of photographs. Automatic densiometric recording combined with computer data processing is being developed for aerial photographic recording and could in theory be used but has not yet been attempted.

Marking of experimental plots is usually effected by means of posts, pegs or coloured polythene tags. Wadsworth (1970) however, has recommended a method which is vandal-proof by using magnetic markers which are relocated with a magnetometer.

Minimal area

This is both a concept and a measure, or series of measures, which aim to reflect that tantalizing property of vegetation, its relative homogeneity at different scales. Minimal area is sometimes useful as a concept and a means of characterizing vegetation but very often it produces more confusion than clarification. Shimwell (1971, p. 15) defines minimal area as the smallest area which provides enough environmental space (environmental and habitat features) for a particular community type or stand to develop its true characteristics of species complement and structure. There are at least three working definitions of the subject, as outlined below.

1. Methods based on species composition

Species number may be plotted against sample area (Fig. 3.9), to produce a species-area relation curve. A small quadrat should be used initially and subsequent ones should double the sample area using a nested arrangement. The quadrats may be conveniently arranged in the manner shown in Fig. 3.9.

Species-area curves initially rise steeply, then level off and sometimes rise again as another phase of the vegetation is entered. The point at which the curve levels off is considered to be the minimal area of the community. It may be identified visually from the graph, as by most phytosociologists on the continent of Europe, or, at a predetermined gradient (Hopkins, 1955). This method is simple to use and effective, but it is generally preferred not to attach any special ecological significance to it, although it may be used as a preliminary guide for sample size in small scale vegetation surveys (Poore, 1955).

2. Methods based on species frequency

If frequency is estimated using an increasing quadrat size, the number of species with over 90% frequency at each quadrat size is referred to as the number of constants. The number of constants may then be plotted against quadrat size as described above and the minimal area is defined as that area in which the full number of constants occurs.

Archibald (1949) defines minimal area as that quadrat size which has a frequency value of 95% for at least one species. This is a somewhat arbitrary definition and not generally used.

3. Methods based on homogeneity

Goodall (1954) defines minimal area as being the smallest sample area for which the expected differences in composition between replicates are independent of their distance apart. In other words an area in which there is no pattern (see below, p. 113). This approach is of academic interest and is not usually used in practice.

2.2.4 Indices of diversity

Diversity is a concept that has been interpreted in different ways by different workers depending principally on their scale of study. (1) Williams (1964) describes several methods based on numbers of individuals in different populations, (2) MacArthur (1965), Mcbintosh (1967a) and Whittaker (1965, 1970) discuss methods based on richness, i.e. number of species per unit area, although this may be incorporated with a measure of the degree of dominance, and (3), the word is used to refer to the number of habitats, especially in the field of conservation management objectives.

Most of the methods referred to by Williams, based on the work of Simpson (1949) and others, concern the zoologist and are less appropriate to the characterization of vegetation. The species diversity of vegetation is highly correlated with latitude and altitude and animal species diversity with the degree of stratification of the vegetation. In spite of the confusion about the meaning of diversity, it is an important concept, principally because several workers have suggested that high diversity is related to stability (Leeuwen, 1966; Ratcliffe, 1971; Maarel, 1971; Whittaker, 1969). This discussion sometimes involves a consideration of Information Theory and Relation Theory to confirm that the two are related. However, it appears that the simplest measure of species diversity is richness, and that of habitat diversity is to enumerate habitats as defined by some recognized system (Elton & Miller, 1954; Elton, 1966; Society for the Promotion of Nature Reserves, 1969, etc.).

3 ANALYSIS OF VEGETATION

The techniques discussed above have been concerned with the description of vegetation on the ground. We now consider the techniques available to the ecologist to (1) compare areas, (2) relate vegetational variation to environmental factors, and (3) identify controlling factors. In any research or management problem concerning vegetation, reference should be made to both parts of this chapter; the first half to select the measure of abundance, and this second half to select the appropriate method of analysis, as well as to the choice of method diagram (Fig. 3.1).

3.1 ASSOCIATION AND CORRELATION

The distinction between association and correlation is made here in the statistical sense. Association concerns two attributes, usually species, which may be present or absent, that is qualitative data, whereas correlation concerns two variables which are quantitatively related.

3.1.1 Association

Association is used in ecology in either the abstract sense to refer to a characteristic assemblage of species comparable to a community, that appears as a unit of vegetation or, in the concrete sense, as a measure of the similarity of occurrence of two species. The former is discussed below (p. 120). In the concrete sense it is measured using a statistic, such as chi-squared (χ^2). Presence of the species should be recorded in randomly placed quadrats and the data arranged in the form of a contingency table such that (a) is the number of quadrats containing both species, (b, c) the number with only one and (d) the number with neither. The formula is as shown below:

2 × 2 contingency table

Species A

		+	−	
	+	a	b	$a+b$
Species B				
	−	c	d	$c+d$
		$a+c$	$b+d$	n

$$\chi^2 = \sum \frac{(\text{obs} - \text{exp})^2}{\text{exp}}$$

or

$$\chi^2 = \frac{(ad - bc)^2 n}{(a+b)(b+d)(c+d)(a+c)}$$

where the observed and expected values for each of the four cells of the contingency table are used, and n is the total number of quadrats.

The result can be examined for significance by reference to χ^2 tables. For 2 × 2 tables there is one degree of freedom. The result may indicate positive or negative association, depending on whether the observed number of joint occurrences is greater than expected or not. The result is dependent upon quadrat size because the data are of frequency type and should be interpreted with care. It is possible that a χ^2 value may be negative at one scale and positive at another for the same pair of species in the same area. If large numbers of positive χ^2 values are obtained it is probable that the quadrat size is too large.

When the number of samples is small such that any expected value is less than 5 it is advisable to use Yates' correction, which makes the observed values nearer to the expected by 0·5. Alternatively the formula below may be used which achieves the same result.

$$\chi^2 = \frac{n\{(|ad-bc|)^2 - \tfrac{1}{2}n\}^2}{(a+b)(c+d)(a+c)(b+d)}$$

Two species which are positively associated occur together more often than would be expected by chance. This does not necessarily indicate that the presence of one causes the occurrence of the other but more probably that the two are responding to a similar combination of environmental factors. This information is in itself useful but its real value is indicated when the γ^2 values for all the species in an area are arranged in a half-matrix (Fig. 3.11) as it permits the construction of constellation diagrams (Fig. 3.12). Agnew (1961) used this approach in his study of *Juncus effusus* in North Wales and found it

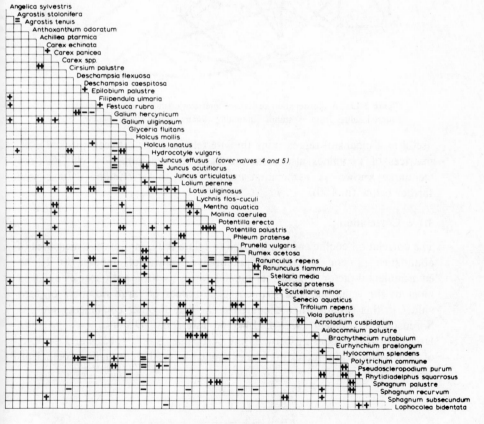

Figure 3.11 Complete chi-square matrix for 99 stands containing *Juncus effusus* showing positive and negative species relationships present (from Agnew, 1961).

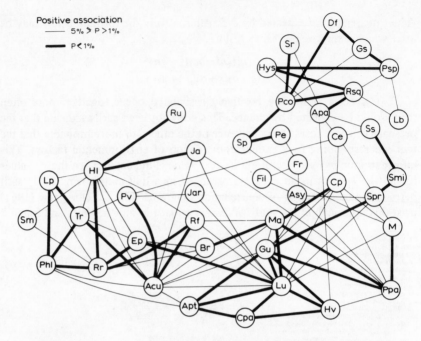

Figure 3.12 A constellation of species interrelationships based on positive chi-squared values from 99 stands containing *Juncus effusus* (from Agnew, 1961).

useful as a means of representing the interrelationships between species. Half-matrices of χ^2 values also form the basis for the numerical classificatory technique known as association-analysis and this technique is discussed further below (p. 130).

3.1.2 Correlation

The correlation coefficient can be used to relate the quantititative measures of abundance of one species to quantitative values of another or to an environmental factor. The former application is useful if both species occur in most of the quadrats. The frequent occurrences of large numbers of zero values in this situation has encouraged the use of other coefficients such as Sorenson's coefficient (see p. 144) or Interstand Distance (see p. 145). However, the relationship between the performance of a species and the level of an environmental factor may be examined using the correlation coefficient. The formula used is:

$$r = \frac{\sum\limits_{n}^{i=1}(x-\bar{x})(y-\bar{y})}{\sqrt{\sum\limits_{n}^{i=1}(x-\bar{x})^2 \sum\limits_{n}^{i=1}(y-\bar{y})^2}}$$

Where r is the correlation coefficient, x and y are the values for either species or environmental factors and $\bar{x}$ and $\bar{y}$ are their means, $n =$ number of quadrats. The significance of r may be obtained from tables using $n - 1$ degrees of freedom.

The correlation of a species with an environmental factor may be used to indicate possible causal factors but care should be taken when interpreting results as the environmental factor tested may itself be correlated with the causal factor. Thus, a highly significant correlation is no proof of cause-and-effect.

3.1.3 Regression

Techniques of regression are described in all statistical text books and the reader is recommended to refer to Bishop (1966) for a very simple introductory account.

Regression is usually used as a technique to fit a line to a series of co-ordinates on a graph and as a means of assessing the degree to which two or more variables are related. Ideally, one variable should be treated as the dependent or Y variable, and the other as the independent or X variable. The approach is comparable to the calculation of the correlation coefficient, except that with the regression coefficient only the variance of the dependent variable is taken into consideration. Normally, as in the study of the effect of nitrogen level on the abundance of a species, it is easy to identify the independent and dependent variables. Sometimes however there are no *a priori* reasons why one variable should be considered dependent and the other independent, for example the regression of one species against another. Whilst this may not matter, it should be realized that the regression coefficient of X on Y will differ from that of Y on X.

Regression analysis is appropriate in a situation where one continuous variable, for example the abundance of a species, is to be related to a series of levels of another variable, for example concentrations of a nitrogen fertilizer. This application most often arises in designed experiments and will not be discussed here (see Fisher, 1960). There are, however, circumstances in the analysis of vegetation in the field when regression is appropriate. One such example is the prediction of timber yield (Y variable) from a series of values for one or more (X) environmental variables. If several independent variables are involved methods of multiple regression are appropriate (Searle, 1966). An example of this type of approach has been used as the basis for the formulation of simple predictive models (see for example, Yarranton, 1969, 1971).

3.2 MEASURES OF NON-RANDOMNESS

Non-randomness in vegetation is often referred to as *pattern*. It may take the form of an aggregation of individuals known as *contagion*, or an even distribution referred to as *regularity*. The former is more common than either the latter or randomly distributed individuals. The departure from randomness interests the ecologist because it is a way of characterizing the vegetation or a particular species. Also because it must have a cause and it provides an opportunity for identifying the factors that control the distribution of a species.

The causes may be either *intrinsic*, i.e. a property of the plant, or *extrinsic*, due to environmental factors, or both. The former tends to occur at a smaller scale than extrinsically caused pattern and is less interesting to the ecologist. Intrinsically caused pattern may be the result of inefficient dispersal of seed referred to as reproductive pattern, or to the vegetative morphology of the species.

Pattern may be examined in terms of its occurrence, scale, intensity and degree of association between species. Gross expressions of pattern are of lower interest in the identification of the requirements of individual species as there are likely to be several species and several environmental factors correlated together. It is only in vegetation that is visually homogeneous that environmental correlations are likely to break down and a single controlling factor indentified.

3.2.1 The detection of pattern

Patterns may be detected by examining the departure of observed values from a Poisson distribution which assumes that individuals are randomly distributed. However, the Poisson distribution is only appropriate where there are few occurrences in relation to the total number possible, that is, where the mean number of individuals per quadrat is low.

If the data conform to this condition and the individuals are distributed at random (Fig. 3.13) then, by definition, the variance (V) equals the mean ($\bar{x}$) or, $V/\bar{x} = 1$.

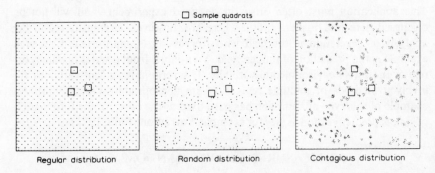

□ Sample quadrats

Regular distribution   Random distribution   Contagious distribution

Figure 3.13 Examples of (a) regular (b) random, and (c) contagious distribution of individuals.

If there is a tendency towards clumping or contagion, then, $V/\bar{x} > 1$ and if there is a tendency towards regularity, then, $V/\bar{x} < 1$. The variance: mean ratio may be calculated directly,

$$\text{variance} = \frac{\Sigma x^2 - \dfrac{(\Sigma x)^2}{n}}{n-1}$$

$$\text{mean} = \frac{\Sigma x}{n}$$

and tested for departure from the expected (unity) by a t-test:

$$t = \frac{\text{observed} - \text{expected}}{\text{standard error}}$$

where the standard error of variance/mean ratio $= \sqrt{2/(n-1)}$.

Pattern may also be detected by examining the departure of density data from a random (Poisson) distribution. This test is again applicable when the mean number of individuals per quadrat is very small. The observed number of quadrats containing 0, 1, 2, 3, 4, etc., individuals is first calculated and then the mean number of individuals per quadrat (m) permits the calculation of the comparable expected values for the Poisson series:

number of individuals per quadrat	0	1	2	3	4
expected number of quadrats in each category	e^{-m}	me^{-m}	$\dfrac{m^2}{2!}e^{-m}$	$\dfrac{m^3}{3!}e^{-m}$	$\dfrac{m^4}{4!}e^{-m}$

that is, $\dfrac{m^4}{4!}e^{-m}$ = expected number of quadrats with four individuals.

where 4! = factorial 4, i.e. $4 \times 3 \times 2 \times 1$

e = base to natural logarithms $= 2 \cdot 7183$

but e^{-m} = may be obtained directly from Greig-Smith (1964) p. 227.

For statistical reasons all the expected values should be greater than 5. In order to satisfy this condition it may be necessary to add together some of the expected values for the higher numbers of individuals per quadrat.

The departure of the observed values from the expected values may be tested by chi-squared, χ^2:

$$\text{where } \chi^2 = \frac{(\text{observed} - \text{expected})^2}{\text{expected}}$$

where the number of degrees of freedom = number of classes − 2. If the distribution of a species is shown to depart from randomness it may demonstrate a response to environmental heterogeneity or a biological characteristic of the species. The former may provide a means of identifying the environmental factor to which the species is responding and the latter may provide a measure of the inefficiency of dispersal of the plant or its capacity to spread vegetatively.

3.2.2 Scale of pattern

Pattern exists in vegetation at a variety of scales ranging from those reflecting different climatic zones of the earth to those reflecting local environmental factors that can only be detected by statistical methods. It is the latter which will be discussed in this section.

It is possible to examine the scale of pattern using either of the two techniques described above with different sized quadrats. However, the method generally referred to as 'pattern analysis' and developed by Greig-Smith and Kershaw (Greig-Smith, 1952, 1961; Kershaw, 1957, 1958, 1959) is more efficient. The data may be collected from a grid of nested quadrats or more usually from a transect of contiguous sampling units and any measure of abundance is appropriate (see however the discussion in Kershaw, 1957). The calculation required to analyse the pattern demonstrated by each species is a modified analysis of variance. The total sum of squares (variance) of the data is calculated and partitioned into its different component scales (block sizes).

The procedure is best explained by reference to Table 3.3 where the left hand column are the data as collected in the field. The transect may be any length that is a power of 2, for example, 32, 64, 128, or 256 units. The raw data are taken as block size 1 and increasing scales of pattern are examined by adding adjacent scores together in pairs to form larger block sizes. The blocking continues until there is only one value left. Thus for a transect of 32 basic units there will be 1, 2, 4, 8, 16 and 32 block sizes.

For each block size the values are squared and then summed giving Σx_1^2, Σx_2^2, Σx_4^2, Σx_8^2, Σx_{16}^2 and Σx_{32}^2. Each of these is divided by the appropriate block size, and the sum of squares for each block size is obtained from:

$$\frac{\Sigma x_1^2}{1} - \frac{\Sigma x_2^2}{2}, \quad \frac{\Sigma x_2^2}{2} - \frac{\Sigma x_4^2}{4}, \quad \frac{\Sigma x_4^2}{4} - \frac{\Sigma x_8^2}{8}, \quad \frac{\Sigma x_8^2}{8} - \frac{\Sigma x_1^2}{16}.$$

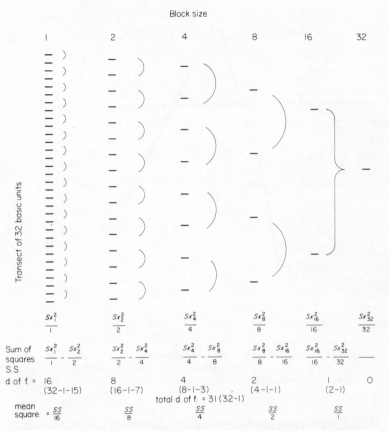

Block size

Table 3.3 The arrangement of the original data and its blocking in the calculation of pattern analysis.

The mean square is obtained by dividing the sum of squares by the corresponding number of degrees of freedom (number of observations minus one, minus the number of degrees of freedom already accounted for).

The mean square, a measure of variance, can then be plotted against the corresponding block size to obtain a *pattern analysis* graph (Fig. 3.14). This graph illustrates the intensity and scale of pattern for the species under study. There are problems involved in assessing the significance of the peaks (see Thompson, 1958) and it is generally preferred to construct several graphs from several sets of data and examine the peaks for consistency. When it is possible to assess significance, the confidence limits increase with increasing block size, so that a small peak at a small block size may be more significant than a higher peak at a larger block size.

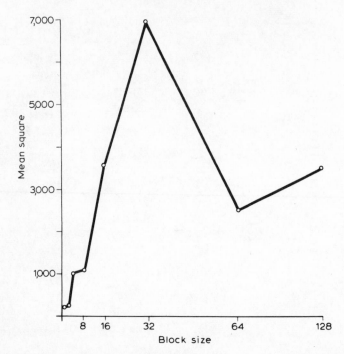

Figure 3.14 An example of a pattern analysis graph. Three scales of pattern in *Trifolium repens* shown by three peaks, at block size, 1, 4, and 32, on the graph of mean square/block size (from Kershaw, 1973).

Useful information can be obtained by comparing pattern analysis graphs for different species and for environmental factors. If a species has a consistent peak at the same block size as an environmental factor it may indicate a cause-and-effect relationship and this should be further studied experimentally.

The relationship between the patterns of pairs of species along the same transects may be further examined by means of the calculation of their covariance and correlation. The procedure to obtain the mean squares at different block sizes for each of the two species (V_A and V_B) is first followed. Then the data for the two species are bulked and the corresponding mean squares calculated (V_{A+B}).
Then,

$$2V_{A+B} = V_A + V_B + 2C_{AB}$$

Where C_{AB} is the covariance between the two species. It may be interpreted directly or converted into the correlation coefficient, r:
where

$$r = \frac{C_{AB}}{\sqrt{V_A V_B}}$$

The correlation coefficient should then be plotted against the block size for different species pairs. It may be positive at one block size and negative at another (Fig. 3.15) which may indicate that the two species are responding similarly at one scale but not at another.

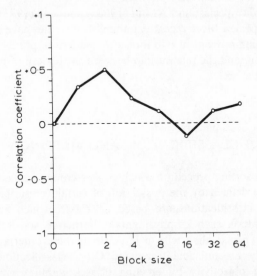

Figure 3.15 Graph to show correlation between two species at different block sizes. Note that the correlation coefficient is positive at block sizes 2 and negative at block size 16 (data from A.J. Morton).

Pattern analysis provides the ecologist with a technique for describing the spatial variability of species distribution as well as for investigating the environmental factors that control species distribution. The method is generally used in ecological studies of an academic nature being time-consuming in the collection of data due to the necessity of replicating transects, each containing several sample units. Environmental data are generally even more time-consuming to collect than vegetational data. Pattern analysis is applied to visually homogeneous vegetation as opposed to vegetational gradients, although the amount of 'textural' variation within a visually homogeneous area does not matter. When pattern analysis is used to identify causal factors the caution concerning hypothesis generation must be made, that is, the erection of an hypothesis is no proof of cause-and-effect. The method has considerable potential in the study of competition especially in the form of covariance analysis although examples of this application are few.

Examples of the application of pattern analysis are numerous and the reader is recommended to refer to Phillips (1954) for work revealing the morphology of the rhizome system of *Eriophorum angustifolium*. Anderson

(1961) relates the abundance of *Pteridium aquilinum, Calluna vulgaris* and *Vaccinium myrtillus* to soil oxygen diffusion rates and Agnew (1961) has conducted morphological studies on *Juncus effusus*; Greig-Smith (1961) for sand-dune species; Kershaw (1958) for work on *Agrostis tenuis,* (1959) for *Dactylis glomerata, Lolium perenne,* and *Trifolium repens,* (1962) for work on various Icelandic species; and Austin (1968a) and Hall (1971) for work on chalk grassland species. More recently, unpublished studies have examined the vertical stratification of tropical rain forest, the pattern of primary production and decomposition, and the relationship between the pattern of plant biomass and nutrients in the soil.

3.3 CLASSIFICATION AND MAPPING

Classification is a sorting procedure which creates groups or classes of similar objects which are defined by the possession of certain prescribed attributes. Some vegetation classifications are based on classes which are arbitrarily assigned using criteria such as physiognomy, structure and floristics. On a small scale (i.e. when dealing with a large area) these criteria do not lend themselves readily to statistical treatment. Other classifications achieve a greater degree of objectivity by erecting classes within which there is a minimum of variation so that the objects, together with their attributes, satisfy more rigorous, statistically based criteria. Such classifications are based upon floristic information and include the association-analysis of Williams & Lambert (1959, 1960) and a group of methods referred to as cluster analysis (Sokal & Sneath, 1963). They are primarily used for large-scale (small-area) studies of the variability of vegetation encountered *within* particular community types.

3.3.1 Small-scale (large area) classifications

Two basic approaches are available at this scale. The first proceeds by the sub-division of broad, inclusive classes to provide a large number of small, exclusive groups; for example the methods suggested by Fosberg (1967) and Ellenberg & Mueller-Dombois (1967) and earlier, by Drude (1913). The second is an agglomerative method which proceeds by recognizing a large number of small exclusive units which are subsequently joined together to form large, collective groups; for example, the method advocated by Braun-Blanquet (1932) and closely followed by Tuxen (1937) and to a lesser extent by Poore (1955). In either situation, however, the central objective of the classification is to provide a description of the vegetation of large regional areas within a single, unitary framework.

Divisive methods

Methods which first identify large major vegetation classes and then proceed by subdividing these into successively smaller classes generally employ either life-form criteria as the divisive character, for example the classifications of Rubel (1930), Du Rietz (1930, 1931) and Ellenberg (1956), or structural criteria, as for example in Fosberg (1961) and Drude (1913). In either case these main divisions correspond to the major vegetation formations of the world, for example, forest, woodland, grassland, etc. Subsequent divisions can be made on a variety of criteria such as physiognomy, structure, periodicity, function, habitat and floristics. Fosberg (1961, 1967) for example dismisses physiognomy as being inexact, floristic composition as requiring specialist knowledge, habitat as possibly involving a circularity of argument and recommends and adopts a system based on structure and function. His primary structural groups are based on spacing (open, closed and sparse) and a second division into formation classes (forest, scrub and grass) is based on vertical stratification (Table 3.4). The two formation groups within each formation class are based on function and are evergreen or deciduous and similarly the formations are based on function (orthophyll/sclerophyll; microphyll/mesophyll/megaphyll). This system is logically based and is providing· a useful framework for describing and processing I.B.P. check sheets. Descriptions of the method are available in Peterken (1967) and Kuchler (1967).

An alternative system to that of Fosberg has been prepared for UNESCO by a working party (Ellenberg & Mueller-Dombois, 1967) and is based on plant life-forms or physiognomy but includes ecological (habitat) criteria. There are seven major formation classes (Table 3.5). Each formation is subdivided into formation sub-classes, formation group, formation and sub-formation. This method has been less widely used than Fosberg's system probably because of the involvement of ecological criteria. This is unfortunate if the classification is to be used for the subsequent correlation of the vegetation classes with environmental factors because there is a risk of circular argument. However, practical trials have indicated that it is as easy to allocate vegetation in the field to one of these categories as it is any other of the numerous physiognomic systems (Goldsmith, 1974). Other workers such as Richards, Tansley & Watt (1940) suggest that 'vegetation should be primarily characterised by its own features not by habitat, indispensable as is the study of habitat for an understanding of its nature and distribution'. They argue that it is structure and composition which should form the basis of the description of plant communities. Indeed most early classifications were based upon physiognomic characters such as those reviewed above (see p. 91). Works such as those of Rubel (1930), Beard (1944), Cain & Castro (1959), Dansereau (1961) and Kuchler (1967) include examples of these physiognomic classifications. In addition a number of small-scale (large-area) classifications

Table 3.4   Fosberg extract table (from Fosberg, 1967)

■ Closed	▦ Absent closed
▨ Open	◺ Absent open
▒ Sparse	⬚ Absent sparse
☐ Absent	

1 Closed vegetation
A Forest
B Scrub
C Dwarf scrub
D Open forest with closed lower layers
E Closed scrub with scattered trees
F Dwarf scrub with scattered trees
G Open scrub with closed ground cover
H Open dwarf scrub with closed ground cover
I Tall savanna
J Low savanna
K Shrub savanna
L Tall grass
M Short grass
N Broad leaved herb vegetation
O Closed bryoid vegetation
P Submerged meadows
Q Floating meadows

2 Open vegetation
A Steppe forest
B Steppe scrub
C Dwarf steppe scrub
D Steppe savanna
E Shrub steppe savanna
F Dwarf shrub steppe savanna
G Steppe
H Bryoid steppe
I Open submerged meadows
J Open floating meadows

3 Sparse vegetation
A Desert forest
B Desert scrub
C Desert herb vegetation
D Sparse submerged meadows

Table 3.5   Ellenberg's seven formation classes

i Closed Forest
ii Woodlands
iii Fourrés (scrub)
iv Dwarf Scrub and related communities
v Terrestrial Herbaceous communities
vi Deserts and other sparsely vegetated areas
vii Aquatic plant formations

have been devised with particular problems in mind, for example, Warming's classification (1923) emphasizes plant/water relations, and is useful for the description of aquatic vegetation, and Raunkiaer's life-form classification (1934) emphasizes the relationship between plant growth and the severity of the climate. In comparison to these classifications and those early classifications based upon physiognomy, the more recent classifications of Fosberg, and Ellenberg & Mueller-Dombois, are best described as synthetic physiognomic methods which bring together into one system many of the useful characteristics of other more specialized classifications. For this reason they are more flexible and likely to satisfy the needs of small-scale (large-area) studies more readily.

Agglomerative methods

The second approach to small-scale vegetation description is an agglomerative method which is associated with the Zurich-Montpellier school of phytosociology under the guidance of Braun-Blanquet. This system proposed by Braun-Blanquet in 1928 (1932) has found widespread acceptance among Continental ecologists but has fewer followers among British and American workers. While its initial aim was to provide a classification framework for the vegetation of the world, realistically it has been most fruitfully used on regional and national scales. It is a field method based upon the detailed examination of the floristics of sample stands of vegetation, so although the physiognomic classifications can often be partially or wholly derived from aerial photographs, that of the Braun-Blanquet system cannot. The classification

Table 3.6 Taxonomic units of the Braun-Blanquet system of classification

Rank	Ending	Example
Class	-etea	Festuco-Brometea
Order	-etalia	Brometalia erecti
Alliance	-ion	Mesobromion
Association	-etum	Cirsio-Brometum
Sub-association	-etosum	Cirsio-Brometum caricetosum
Variant	—	(specific names used)

seeks to erect a hierarchy of units arranged from the simplest unit, the association, to the most complex, the vegetation circle. Successively higher units are the alliance, order and class. While all groups are based upon floristics, the higher units (orders and classes) tend to group also upon physiognomic criteria, and to some extent parallel the formations of the physiognomic classifications. Each association is a generalization derived from the examination of the floristic composition of numerous samples (relevées) and represents an 'ideal type' which is constantly being revised as new stands are described. The field procedure adopted for the selection of sample stands is well documented by Ellenberg (1956) and Moore (1962). Traditionally only those stands which are recognizably homogeneous in composition are selected for description. Their selection is often subjective, even though the 'minimal area' of an association is frequently advocated as providing the basis for selection (see above, p. 107). In the absence of any statistically based sampling design, the overall appearance of the vegetation is likely to dictate to the observer which stands are selected, and this in turn is influenced by the field dominants as much as it is by the experience of the observer. Sample selection is one of the most intractible problems facing the investigator wishing to work on a regional scale for, in the absence of any definite knowledge of the area and characteristics of the vegetation to be surveyed, even systematic or random sampling is difficult. At present, phytosociologists on the continent of Europe favour subjectively chosen sample plots of varying sizes to arrive at a representative inventory of the vegetation concerned, thus a plot $1-25$ m^2 is regarded as adequate for sampling herbaceous vegetation, while in small-scrub, a plot $25-100$ m^2 is preferred, and in forests, plots of $200-500$ m^2 are used for trees, with appropriate sizes substituted for the shrub and herb layers.

The floristic record for each plot includes a cover-abundance rating for each species on a five-point scale (see above) and each is assigned a sociability index, also derived from a five (or ten) point scale (see Table 3.7). The dual

Table 3.7   Sociability and cover-abundance classes

Sociability 1—growing once in a place, singly
          2—grouped or tufted
          3—in troops, small patches or cushions
          4—in small colonies, in extensive patches or forming carpets
          5—in great crowds or pure populations

---

value for each species together with information on various environmental factors relevant to the site, is entered into the field book. In the laboratory the floristic information is used to compile a large data table referred to as the raw table (see Table 3.8) in which lists from similar vegetation types are entered. By a process of sorting and continuous refinement, similar stands are grouped together on the basis of their floristic composition. This process of sorting relies heavily upon the recognition of potential differential species. These are species which may be mutually exclusive in their occurrence in the table or species which have joint occurrences over a range of sites. These species are used as the basis upon which several partially ordered tables are produced. A final orderly extract table (see Table 3.9) arranges the stands in groups which can be characterized by the presence or absence of certain more or less mutually exclusive species which are termed differential species (underlined in Table 3.9). At this stage the worker returns to the field to establish the validity of the groups which have been erected in the laboratory. The laboratory procedure is well documented in Becking (1957), Kuchler (1967), Shimwell (1971), Moore (1962), and Szafer (1966). The subsequent recognition of distinctive associations defined by character species can only be made after a considerable number of relevees has been described from a wide geographical area. Character species, which were originally thought to be almost exclusive to a particular association, represent a special case of the differential species used in the sorting of the data tables. Their selection for a particular association can only be made by understanding the concepts of constancy and fidelity. Here, constancy refers to the percentage occurrence of a species across a range of relevees and is established by reference to a five point presence scale and fidelity to the selective preference of a species for particular communities. Both can only be established once the full range of the species concerned and the communities in which they occur have been fully studied. Where high constancy is accompanied by a high degree of fidelity the species can be used as character species and these species can refer to the levels of the association, alliance and order. But although constancy and fidelity are related this is not always the case and for this reason a more flexible approach to definining character species has now been adopted. Becking (1957), Shimwell (1971) and Szafer (1966) illustrate how the concept of character species now embraces (i) absolute and relative character species, and (ii) regional and local character species. It nevertheless remains to be said that even these concepts require a fairly thorough knowledge of the vegetational variation on a regional scale. In western and central Europe and Scandinavia, this detailed knowledge has been compiled using the Braun-Blanquet method so that in several cases absolute character species can be assigned at both the level of the alliance and of the association. These character species are clearly documented in the continental publications, notably in those contributions to the journal *Vegetatio,* in Oberdorfer (1957) and Braun-Blanquet (1964). Elsewhere in the world, this information is not available and the establishment of character species will have to wait until vegetational descriptions have progressed more widely.

Criticisms of the Braun-Blanquet method by British workers include that of Pearsall (1924) with respect to minimal area, and that of Tansley & Adamson (1926). Poore's reassessment and modification of the method (1955) includes a positive attempt to reconcile the British tradition of reliance on the field dominants with the need for more accurate floristic description and comparisons. Moore (1962) also reiterated Poore's attempt to consider the Braun-Blanquet method as a reconnaissance technique worthy of serious attention. The modifications which Poore suggested have probably served to

Table 3.8 Raw table of saltmarsh community from Blakeney Point, Norfolk

	1	2	3	4	5	6	7	8	9	10	11	12	13	14	15	16	17	18	19	20	21	22
Aster tripolium	+	1	2	2	2	3	2	1	1	1	3	3	2	+	2	1	2	2	1	1	2	3
Halimione portulacoides				2	1			3	1	5			+			3	+	1	1	+	+	+
Limonium vulgare			+					+		1	+	+	+		+			+	1		1	+
Limonium humile			+	1		1	+				1	+	+					+		+		
Puccinellia maritima							2	2	2	2	2		4					+	3		1	+
Salicornia herbacea	2	2	4	4	3	4	2	1	2	+	2	4	1	2	3	1	3	2	1	1	1	2
Salicornia perennis								1	1				1			1			+			
Spartina anglica							1	+	1	+	+	1	+	+				+	2	1	+	3
Suaeda maritima	1	2	1	1			1	1	1	+	1	+	1	2	+	+	+	1	+	+	+	1
Triglochin maritima								+		1									1		+	
Bostrychia scorpioides			+	+	+	+	1	+	1	1	5	4	4		+	1	1	2	3		1	4
Enteromorpha spp.	3	+	+	+	+	+					1		+	2	+		+	+	+	+		
Enteromorpha intestinalis	+	+	+	+	+	+	+				+		+		+	+	+					
Fucus volubilis	1	2	2	1	1	+	+		+		+	1	+	2	1		+					
Pelvetia canaliculata						+							3						+			

Table 3.9 Orderly extract table of saltmarsh community from Blakeney Point, Norfolk

	46	13	24	19	45	21	36	9	7	11	44	23	33	49	50	18	22	34	35	48	12	30
Pucinellia maritima	4	4	3	3	3	3	2	2	2	2	2	1	1	1	1	1	±	±	±	±	±	
Limonium vulgare	1	+	2	1	1	1	+	1	+	+	+		+	2	1	+	+	+	1	+	+	1
Limonium humile	1	+			1	+	+		+	1	1			+	+	1	+		1	+	1	+
Triglochin maritima	4		1	1		+			+		1	2			+				+			
Spartina anglica	1	+	1	1	1	+	+	1	1	+	+	1	1	+	+	2	3	+	+	+	1	+
Bostrychia scorpioides	1	4	2	3	3	1	1	1	1	5	1	2	4	3	3	2	4	1	2	2	4	1
Halimione portulacoides		+	+	1	+	+	+	1			+	+				1	+	+				1
Salicornia perennis	+	1	+	+				1			2								1			
Pelvetia canaliculata											+						3		1		3	+
Fucus volubilis		+	+					+	+	+	+		+	+						+	+	1
Enteromorpha spp.	±	±		±						1				±		1	±				±	
Enteromorpha intestinalis		+						+	+						+				+			
Suaeda maritima	1	1	+	+	+	+	+	1	1	1	1	2	+	+	+	1	1	1	+	1	+	+
Salicornia herbacea	1	1	2	1	2	1	+	2	2	2	2	2	1	2	3	2	2	1	2	1	4	2
Aster tripolium	2	2	1	1	3	2	1	1	2	3	2	2	1	2	2	2	3	1	1	3	3	1

4	25	26	27	28	29	30	31	32	33	34	35	36	37	38	39	40	41	42	43	44	45	46	47	48	49	50	Frequency
1	1	1	1	2	1	1	1	1	1	1	1	2				1	2	1	3	2	2	3	2	1	3	2	2
		1	+	1	3						+		+					2	3	+	+		3				
	+			1			+	+	1	+						+		+	1	1			+	2	1		
		+	1	+			+	1	+	+	1					+	1	1	1	1	1	+	1	+	1		
			+			1	+	+	3							1	3	4	1	+	1	1					
+	2	2	2	2	2	+	2	1	1	2	+	1	1	3	2	1	2	2	2	2	1	1	1	2			3
																2			+	1	1						
		+			+	+		1	+	+	+	+	+			+	+	1	+	1	1		+	+	+		
	1	+	+	+	+	+	1	+	1	+	+	+	+			2	+	1	+	1	+	1	1	1	+	+	
						+			+												+			2	+		
				1	+		4	1	2	1	2					1	1	1	3	1	1	2	3	3			
5	+	+		+		1	+							+	1		2		1			+	+	+		1	
+				+	+	+	+	+								1	+						+				
+	2	1	2	+	+				+		+	1	2	1	1	1		+				+	+	+			
			+				+		1							+		+									

50	47	8	31	42	43	16	5	6	28	20	17	29	25	1	2	3	4	15	14	41	26	38	27	32	39	40
2	1	2	±																							
		+												+		+				+						
+		+	1	1				1							+		1				+					
	+	+	+	1					+		+								+	+		+				
1	+	+	1	1	1	+	+				1					+	+	+								
5	3	3	3	2	3	3	2	1	1	+	+	+														
1	1								1																	
		+		+																						
	1	1		+	+	2				+	+	1	2	2	1	1	2	1	2	1	1			2	1	
±	±	1		1		±	±			±	±	±	5	3	±	±	±	±	2	2	±	1	±	±		
	+	1	+	+	+	+	+			+	+	+	+	+	+	+							+			
1	1	+	1	+	+				+	+	+	+		1	2	1	1	+	2	+	1	+	+	1		2
1	1	+	2	2	1	3	4	2	1	3	2	+	2	2	4	4	3	2	1	2	1	2	2	3	2	
1	1	1	1	3	2	1	2	3	1	1	2	2	1	+	1	2	2	2	+	1	1		1	1	1	2

confuse rather than to clarify but he recommends the use of dominants as well as constants to characterize the lowest vegetation units which he terms *noda*. *Dominants,* he defines as those species occurring with the highest cover-abundance ratings using the Domin scale (see above) and *constants* as those species occurring in more than 80% of like relevees. Where, within a nodum, the number of species occurring in Constancy Class V (> 80% of the lists) is greater than the number occurring in Constancy Class IV (60–80% of the lists) the unit is raised to the status of the association. Poore utilizes Sorensen's coefficient (see above) of community similarity as a more objective means of comparing like noda. Poore's approach is very similar to that of Dahl (1956) in his description of the mountain vegetation of Rondane, in Norway, that of McVean & Ratcliffe (1962) and Burnett (1964) for Scottish vegetation, and that of Edgell (1969) for a survey of Welsh upland vegetation. Fundamentally all these methods are attempting to provide a realistic frame of reference within which the vegetation of a relatively large geographical area can be adequately described. Poore thus recognizes that two noda may overlap in their ranges and suggests that any new stand of vegetation described from a similar area could be located within the framework of variation provided by his reference points, the noda.

The advantages which these methods of vegetation classification provide are several. They are primarily reconnaissance methods and are particularly useful at a local to regional scale. They are easy to apply in the field and quick to execute. The arrival of sorting routines which can be computerized will remove the tedious and 'muddled' procedure required to compare hundreds of lists. As computing facilities are extended, they will be capable of handling very large samples so that even if systematic and random sampling is preferred, the enormous numbers of samples generated could be processed. At the moment the subjective selection of the sample stands whether utilizing a divisive or agglomerative survey method presents the most obvious disadvantage of these methods. But, in situations where a rapid inventory is required the Braun-Blanquet method has immediate advantages over more time-consuming objective methods. Certainly it would be extremely isolationist to disregard the vast body of European work already existing on the classification of vegetation types and familiarization with this approach would do much to foster better international understanding among ecologists. On a more academic level too, it would seem desirable to have a frame of reference points to which more local vegetation studies could be referred. Unless a worker has some indication of the representativeness or exclusiveness of his sample stands, the conclusions he can draw from his studies are extremely limited. It seems implausible to invest a great deal of research effort on the detailed analysis of certain vegetation units, for example detailed production studies, unless the results have wider application elsewhere. The methods of vegetational description of Fosberg, Ellenberg, and Braun-Blanquet attempt to provide this frame of reference.

Despite the simplicity of most of the systems for classifying vegetation from a physiognomic or structural basis and the logical approach of many of them (e.g. Fosberg's) there remains the problem of sampling. The ecologist needs to decide how to choose his sample stands and to determine their size and number. Whilst the discussion about sampling design for quadrats (see above) is appropriate here there are special problems encountered with the methods used on an extensive scale. It is generally accepted that these techniques are subjective and many authors advise that 'representative' areas should be described. But, who is to choose what is representative, and at what scale? If mapping is the ultimate objective, it may be acceptable to recognize only areas sufficiently large to be mappable at the scale selected (Kuchler, 1967). If, however, nature conservation is the primary objective, there may be extremely small areas such as vertical cliff faces, to which endemic or rare species are restricted, whose description should be included. It is recommended that the area of study should be subjected to extensive initial reconnaissance prior to sampling to familiarize the ecologist with the range of vegetation represented. The use of aerial photographs may speed-up this stage but extensive direct observation on the ground is nevertheless essential. The range of the main vegetation types should be noted, the approximate extent of each estimated, and the number of samples determined in proportion to the area of each type.

The arrangement of stands within each vegetation type ideally should approximate to randomness, and the size of each should be chosen on the basis of its minimal area (see above). It is thus desirable that several minimal area curves be prepared for each vegetation type in the study area, but if time is limiting, and the worker is familiar with the vegetation, a visual estimate may be sufficient.

In the final analysis however, as the phytosociologists on the Continent of Europe have reported, sampling at a small scale (large-area) must rely heavily upon the experience of the research worker both with respect to his familiarity with the methods and the area under study.

3.3.2 Large-scale (small-area) classifications

The Braun-Blanquet method has been used by a number of workers as a basis for detailed investigations of vegetation/environmental relations, but it is frequently argued that the vegetation units erected by the small-scale classifications are so subjective and generalized that they do not provide useful reference points. This dissatisfaction arises from the growing awareness of the complexity of vegetational variation. But even on a large scale there is still a need for some means of simplifying this complexity, and the simplification must not be at the expense of informativeness. Physiognomic and structural differences do not adequately reflect the complexities of large-scale variation whereas both qualitative and quantitative measures of species occurrence are considered to be relevant since they are likely to reflect more closely local differences in environmental gradients. It is for these reasons that large-scale studies take recourse to some numerical treatment of species occurrence (see below) and to the numerical classification of groups based on differences and/or similarities between floristic samples. These numerical classifications can either proceed by division from above, as for example in Goodall (1953) and Lambert & Williams (1966) or by fusion from below, as in the agglomerative methods such as cluster analysis. Fundamentally, however, these two different strategies work to the same end.

Divisive methods

When qualitative (presence and absence) data only are available the association between pairs of species can be determined using a 2 × 2 contingency table and the appropriate chi-squared test (see above). Goodall (1953) used this approach to subdivide a data set to a level at which all associations between species disappeared and the resultant groups could be designated as homogeneous. In effect, homogeneity can be achieved by either or both of a pair of species being present or absent, but Goodall after experimenting with all these possibilities, recommended that those end groups characterized by the possession of one or two species be adopted. In this case only positive associations were taken into consideration. His method proceeds by using the most frequent species showing association with other species to divide the data into those quadrats containing the species (A), and those from which it is absent (a). Group A is then re-examined to select the most frequent species showing association and Group B established on its presence. Likewise group b is established on its absence. Remaining quadrats $(a + b \ldots + n)$ are pooled and the procedure repeated until all associations have been removed. Williams & Lambert (1959) subsequently modified this method to include both positive and negative associations, since they argued that for a large heterogeneous area, the main divisions of the data are likely to be supported by equally strong negative associations and will thus strengthen the analysis. Williams and Lambert also experimented with several association indices other than the one chosen by Goodall, and while all were based on some measure of chi-squared the index $\sqrt{\chi^2/N}$ is the one most frequently used, where N refers to the number of quadrats. In fact this is a measure of heterogeneity of the data and by plotting the fall in the value of the index at each stage of the division, the relative heterogeneity of each level can be seen. The pathway leading to the final end group is a dichotomizing hierarchy and each group can be characterized by the presence and absence of particular species. Williams and Lambert called this method association-analysis and it is now widely available as a computer library programme. Its execution demands the setting up of a two dimensional matrix in which species are correlated in all possible pairs. That species generating the highest $\Sigma\sqrt{\chi^2/N}$ is used to sub-divide the population into two groups; positive and negative associations are given equal weight throughout the analysis and are summed regardless of sign. The two groups generated by the first division are subsequently re-examined for successive associations in the manner described above. In most programmes a significance level of 5% ($P=0·05$) is established below which no further division occurs either because all associations are indeterminate or because they are not significant. An example of this normal association-analysis hierarchy is given in Fig. 3.16.

NORMAL ANALYSIS OF A HEATHLAND COMMUNITY

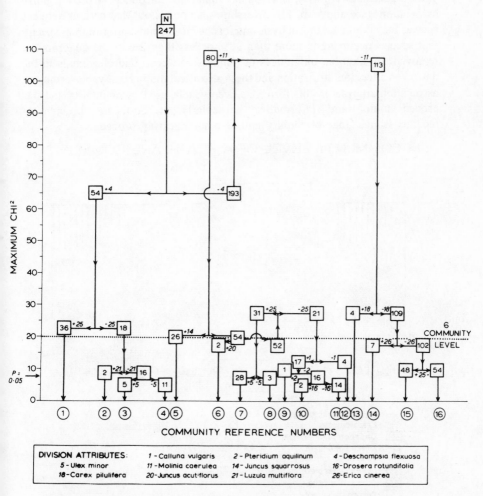

Figure 3.16 An example of a normal association-analysis hierarchy (from Harrison, 1970).

This method has been widely used in many different community types ranging from the tropical savannas (Kershaw, 1968) and tropical rain forest (Austin & Greig-Smith, 1968) to temperate grasslands (Gittins, 1965) and heaths (Harrison, 1970; 1971). Most analyses are performed on data collected from local areas or small regions, although Proctor (1967) used data collected on a national basis to examine the distribution of liverworts in Britain. It is considered to be one of the most useful classificatory procedures for the preliminary analysis and reconnaissance of vegetation and the end groups can be mapped (see Fig. 3.17). The data are easy to collect, although some difficulties arise, notably in the choice of quadrat size and the influence of pattern (Kershaw, 1961), since small quadrats introduce strong negative associations (see above, p. 116). Kershaw also points out that certain frequent species may dominate the analysis and for this reason most programmes specify that species occurring in more than 98% or less than 2% of the quadrats are removed. In addition, differences of richness between quadrats remain in the data even after standardization and these also affect the hierarchy. Nevertheless, association-analysis in this form has been widely used by ecologists and has proved useful and informative particularly at both the preliminary reconnaissance stage of enquiry and as a summarizing procedure .

SIX COMMUNITY LEVEL OF A HEATHLAND COMMUNITY

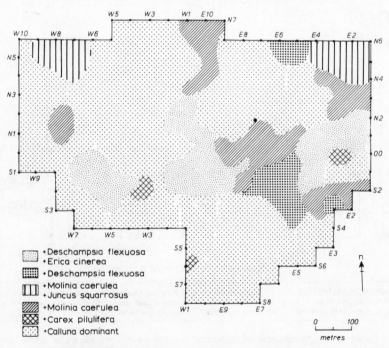

Figure 3.17 Map of normal association-analysis end groups.

Association-analysis also exists in inverse (Williams & Lambert, 1961) and nodal form (Lambert & Williams, 1962). With inverse analysis the inter-relationships between samples are examined by inverting the data matrix so that species become individuals and the quadrats in which they occur, their attributes. The resultant end groups reflect species of similar performance. For example in Fig. 3.18 end group *A* represents a mixed heath community, and group *B* a scrub community and end group *E* a distinctive community of wet hollows. The remainder represents variable grass heath units. Nodal analysis draws together the results of both normal and inverse analysis into a single, two-way table in which the normal end groups are arranged horizontally and the inverse groups, vertically. Each cell of the matrix is then subjected to an approximation of both normal and inverse analysis. The sites and species on which the divisions are made in each cell are termed coincidence parameters. Where a cell is fully defined in both directions by the presence of two coincidence parameters, the cell is termed a nodum (Fig. 3.19). Where the cell is only fully defined in one direction it is termed a sub-nodum (Lambert & Williams, 1962). In this way groups of species which characterize certain site groups can be revealed and vice versa. Technical difficulties in execution, the

INVERSE ANALYSIS OF A HEATHLAND COMMUNITY

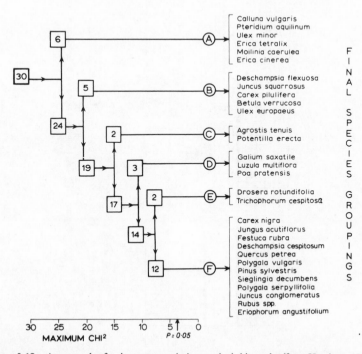

Figure 3.18   An example of an inverse association-analysis hierarchy (from Harrison, 1970).

134

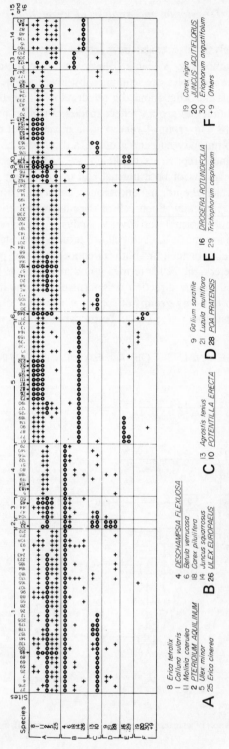

Figure 3.19 A two-way table of a heathland community (nodal analysis) (from Harrison, 1970)

problem of species richness and abundance, together with the emergence of other techniques which offer similar facilities (Lambert & Williams, 1966) have meant that in practice inverse and nodal analysis have only limited use in ecological investigations. However, it is pertinent to point out that both the underlying aims and the mechanics of these numerical methods have much in common with those of more traditional phytosociological methods. For example, Ivimey-Cook & Proctor (1966) suggest that 'association-analysis is of value in examining the consistency of evidence and conclusions in phytosociology and in detecting directions of variation overlooked by traditional methods, and that it provided a powerful tool for the detection of faithful (character) and differential species'.

Agglomerative methods

Where quantitative data on species abundance are thought desirable the investigator has available to him a number of agglomerative classificatory procedures. Numerical agglomerative methods, commonly called cluster analysis or similarity analysis, proceed by scanning the whole data set and by examining the relationships between all possible pairs of individuals $[\frac{1}{2}n(n-1)]$. Subsequently, individuals are united into groups because their similarity exceeds some arbitrary level. The measure of similarity employed varies considerably (Sokal & Sneath, 1963), but includes the correlation coefficient, squared Euclidean distance, standardized squared Euclidean distance, non-metric coefficients such as that of Czekanowski (see below) and used by Curtis (1959) (see below) and the informations statistic. These coefficients are reviewed by Williams *et al.* (1966) together with worked examples. The precise sorting strategy which is used to direct the path taken by the hierarchy can also vary. For example when at some stage there are r clusters, the two which possess the least of the $\frac{1}{2}r(r-1)$ inter-cluster distances are joined, and the distances between the resulting clusters and the other $(r-2)$ clusters are calculated. The calculations require a formula, $D_{k.ij}$ which is a measure of the distance from cluster K to the union of clusters I and J. Several measures of this distance function are currently in use and include nearest neighbour, furthest neighbour, group averaging, centroid sorting and minimum variance clustering. All these sorting strategies are considered by Pritchard & Anderson (1971) and Williams & Dale (1965). Many different kinds of agglomerative methods are available to ecologists and new approaches are constantly being developed. In particular those methods developed for taxonomic studies are frequently of use to ecologists seeking a numerical classificatory approach. One agglomerative method which has proved its usefulness in complex ecological situations is information analysis (Williams *et al.*, 1966). The statistic employed (I) is derived from the concept of entropy and may be regarded as a measure of disorder. When all the members of the group are identical it reduces to zero, so that the most efficient way of erecting a hierarchy using this statistic, is to fuse those two individuals or groups which on fusion produce the smallest increase in I (I-gain or ΛI). This statistic has an advantage over other similarity indices, such as nearest-neighbour indices, for it is completely additive through an analysis. Thus, if the ΛI values are accumulated up the hierarchy, a value of the information-gain (heterogeneity) is obtained which genuinely refers to the groups defined. This allows successive groups of the hierarchy to be plotted along a graduated ordinate such that their position reflects their overall heterogeneity. In this way the observer can select the units which are potentially of most use for his purpose. The clarity of the resulting hierarchy recommends the method, particularly in complex ecological situations (Lambert & Williams, 1966).

In comparison to association-analysis, these agglomerative methods are probably more flexible both in the data which they will accept and in the indices which they employ. However, their execution may be more time consuming and hence expensive. This is because a large number of calculations have to be made at the start of the analysis, and which in terms of their informativeness are of least value to the investigator. By contrast normal association-analysis can be terminated at particular levels during the analysis and well before the final groups which have the least interest are calculated. Chaining effects do occur in association-analysis, when one, or a few individuals (quadrats) are split off at an early stage in the analysis and then the remaining quadrats are progressively split into smaller groups. The diffuse hierarchy resulting often leads to difficulties of interpretation. Similar problems, however, arise with clustering techniques when a given group grows in size by the addition of single individuals rather than by fusion with other groups of comparable size.

To summarize, it is most fruitful to treat these numerical methods of classification as complementary to one another since they all have a similar aim. Association-analysis may lead to mis-classification and over-simplification, because it is highly dependent upon the vagaries of a sampling design using qualitative (presence and absence) information, and because as a monothetic procedure (i.e. using a single character for division) creates groups which are defined by the presence or absence of a *single* species. Polythetic agglomerative methods (i.e. using several characters for fusion) use quantitative data and frequently take into account the total characteristics of the group as a whole and may be preferred. On the other hand association-analysis may usefully resolve ambiguities generated by different agglomerative methods. However, only by adopting a flexible approach can a 'best' method be determined and each problem will have its own 'best' solution. (Pritchard & Anderson, 1971, Moore *et al.*, 1970, Frenkel & Harrison, 1974).

The choice of a classificatory approach to vegetational analysis need not imply that any assumptions have been made about the basic nature of vegetational variation. All classifications are necessarily approximations to reality but they do provide a simplifying procedure for the description of complex vegetational patterns. For this reason a classificatory approach has often proved useful during the preliminary investigations of large heterogeneous areas and before more detailed lines of enquiry are initiated. For example in a study of the vegetation of a tropical island Austin & Greig-Smith (1968) used a classificatory approach before later adopting an ordination approach to examine plant/environment relationships in more detail. On a large scale too the numerical classificatory methods can prove useful as a basis for hypothesis generation about plant/environment relationships. For example the end groups of an association-analysis can provide a guide to the existence of more or less homogenous habitat conditions over those sites. These type locations may then be examined in detail later. A third use of a classificatory approach is as a summarizing procedure after detailed investigations have been completed. It is generally much easier to communicate the characteristics of groups than it is to discuss the relationship of axes between a spread of points as is the case with ordination (see below). Nevertheless, the informativeness of the groups will depend upon the scale of enquiry so that the groups from small-scale classifications can only contribute to a general discussion while the groups of large-scale classifications have relevance on a more local scale.

3.3.3 Vegetation mapping

Many methods of vegetation classification lend themselves very readily to vegetation mapping and other methods have been designed specifically for this purpose. Kuchler (1967) has provided a valuable summary of these methods and of the kinds of decisions which have to be taken during the preparation of such maps. For example, the first consideration is the purpose for which the map is designed and the second is the scale at which the map is to be drawn. A third consideration is the method for characterizing the mappable unit, that is between physiognomic/structural characteristics and community units based on floristics. Small-scale maps with a representative fraction of between 1:1,000,000 and 1:100,000 will only allow generalized features of the vegetation to be shown, while large-scale maps on a scale between 1:10,000 and 1:50,000 allow a greater amount of detail to be shown. None of these decisions, however, can be made without reference to a fourth factor, namely the characteristics of the area to be mapped. For example a large, flat area of uniform vegetation requires different considerations from the same area in a complex mountainous district. In effect the design of a map has therefore to be a compromise between these four considerations.

Kuchler (1967) and Dansereau (1957) have both designed methods of vegetation mapping for use on a small scale. Because each is prepared for the

sole purpose of mapping there is a heavy reliance on the use of symbols and no
implications are made about the universal applicability of the classificatory
units. Kuchler's method relies upon preparatory work being accomplished on
air photographs (see below) before field work commences, and most methods
of small-scale vegetation classification such as those of Fosberg and Ellenberg
(see above) require that air photographs or ground survey maps be available
before mapping can proceed. These photographs and maps need to be on as
large a scale as possible for use in the field even though the final map may be
prepared on a small scale. In tropical countries and the more remote parts of
the globe such basic information may be lacking, in which case vegetation
mapping can only proceed alongside ground survey work, or if special
reconnaissance flights are undertaken (Poore & Robertson, 1964). Where
floristic information is required for mapping on scales between 1:5,000 and
1:50,000, Braun-Blanquet's method may prove suitable. In this case the
method is based on detailed floristic inventories made in the field and reference
to air photographs is only made for ease of field orientation or providing a base
map upon which mapping in the field can proceed. Mapping the vegetation
units generated during the laboratory analysis of selectively chosen samples
(see above) proceeds using cadastral or topographic maps, which are readily
available in most European countries on a range of relatively large scales. The
community units recognized in the laboratory are used to design a key and the
units are checked for their 'real' value in the field and the vegetation within the
area assigned to one of them. Maps prepared using this system of classification
are published at a variety of scales, many of which are beautifully coloured,
and Kuchler's bibliography includes a full inventory of the classic maps. The
community units of this system of classification were originally designed to
embrace the vegetation of the world but current practice among the European
phytosociologists reduces the emphasis on universal applicability and each
map is designed to represent adequately the floristic composition of the
vegetation for the particular area under examination (Ellenberg, 1954).

Mapping on a large scale (small-area) presents more problems to the
ecologist because large scale cadastral or topographic maps may not be
available or are out of date. In this case the student has to prepare his own
base map using aerial photographs or ground survey methods such as the
theodolite traverse, chain surveying and plane tabling. Suitable manuals which
include methods of survey for large scale maps are Curtin & Lane (1955) and
Cain & Castro (1959). All maps should be provided with a scale and a linear
grid for ease of interpretation and for the location of randomly selected or
systematic samples. Such an approach is obviously useful, in the preparation
of maps to illustrate the end groups of the numerical classificatory methods
described above. Preparing maps in open and wooded country can each
present their difficulties. Taylor (1969) offers a method for the rapid
reconnaissance mapping of open country on relatively large scales, but

mapping in woodland can best proceed by careful ground surveying as suggested above. In either case reference should be made to all available reference material such as topographic maps, estate maps and air photographs before embarking on time consuming and precise surveys.

The use of isonomes for the diagrammatic representation of local variability in community composition is a technique similar to the transect in that it permits the diagrammatic representation of vegetation variation in the field (Ashby & Pidgeon, 1942). A grid of sample units is marked out in the field and information about the species and often the environmental factors are collected and transferred to graph paper. Contours (isonomes) are then drawn around species showing the distribution of uniform areas of abundance and so producing a series of maps. This simple technique, however, only works satisfactorily where gradients of variation are pronounced and the study covers a small area.

3.3.4 Use of aerial photography

The use of aerial photography in ecology and especially in the preparation of vegetation maps can be traced back many decades (for example, Stamp, 1925) and has been widely used overseas in resource survey work (Bawden, 1967). The range of methods currently available includes various types of remote sensing, i.e. infra-red imagery, multi-spectral photography and radar imagery, as well as photography *sensu stricto*, and the choice of method depends upon the purpose.

Overlapping, vertical panchromatic prints are usually viewed in pairs, to produce stereoscopic images. Simple mosaics of prints may be useful for initial interpretation and photographs taken at an oblique angle to the ground may be used for special purposes, such as identifying archaeological features. The production of photogrammetric maps is a subject in its own right and usually requires an initial ground survey to correct for vertical and horizontal aberrations in the plane's flight path. However, the transfer of ecological information to such maps can be carried out by ecologists providing they have had some introductory training. Although the preparation of vegetation maps is the most commonly encountered application there are a variety of specialist uses reported in the literature and these include recording the extent of fire, incidence of crop disease (Colwell, 1956), counting of animals (Perkins, 1971), recording of individual tree species (Howard, 1970, p. 235) and mapping of exposure (Goldsmith, 1973a). In theory the ecologist is faced with the choice of type of imagery, scale and time of year for flying and should consult a standard text on aerial photography (Howard, 1970). In practice, however, if he is working on a limited budget, the characteristics of the imagery may be determined by the type of cover available and his problem is principally one of interpretation.

The advantages of using aerial photography are that large areas can be covered rapidly, can be viewed stereoscopically, and vegetation boundaries can be mapped very accurately. These advantages become increasingly important as the scale of the area of study decreases and its degree of inaccessibility increases.

The major disadvantages of vegetation mapping using these techniques are that characterization must be largely based on physiognomy and the problem of producing an initial classification has to be resolved. In practice a ground survey may be conducted initially and then as many vegetation types as possible characterized on the imagery using differences in tone, texture and pattern (spacing).

Examples of the application of aerial photography by the Nature Conservancy in the British Isles are discussed by Goodier (1971) and this includes papers describing the Dartmoor Ecological Survey in which both vegetation maps and maps of the distribution of grazing animals were prepared from air photographs.

Trials to compare four types of multispectral imagery (infra-red colour, infra-red panchromatic, true colour, panchromatic) have recently been conducted and some of these (Goldsmith, 1972) indicate that a combination of types of imagery may provide more ecological information than can be obtained from any one alone.

3.4 VEGETATIONAL AND ENVIRONMENTAL GRADIENTS

Gradients in vegetation have generated considerable debate about their causes, significance, and the most appropriate methods of study. Vegetational gradients usually reflect environmental gradients and may be static or dynamic. An extreme case of a static gradient is a zonation. If dynamic gradients have been demonstrated and are moving in a directional manner they are part of a successional sequence (Clements, 1916). The principal disparity of opinion occurs between workers who emphasize the homogeneity of vegetation, play down the importance of gradients and consider classification the most appropriate way of representing vegetational variation (Poore, 1955, Moore, 1962), and those who consider such variation to be essentially continuous. The proponents of the *continuum concept* (McIntosh, 1967b) favour different methods for representing vegetational variation such as gradient analysis and ordination. The nature of vegetational variation, however, hovers tantalizingly between these two extreme points of view (Webb, 1954). The situation is further complicated by the cyclical changes which take place in vegetation (see above).

In practice, it is important that the research worker be familiar with the alternative concepts of vegetation which can be achieved by reference to the key papers given above and then he must select his method by considering his objectives. If he wants a classification he should choose a classificatory technique and if he wants his results in the form of a gradient analysis or ordination he should choose accordingly. Classes of vegetation can nevertheless be ordinated.

It may be useful to distinguish between direct gradient analysis where vegetation samples are arranged using observed environmental gradients, and indirect gradient analysis, where the samples are arranged along axes generated from the vegetational data (Shimwell, 1971; Whittaker, 1967). The former include the continuum method (Curtis & McIntosh, 1951) and gradient analysis (Whittaker, 1956) and the latter the various types of ordination procedure (Goodall, 1954, Greig-Smith, 1964).

3.4.1 Gradient analysis

Methods of gradient analysis were first proposed by Whittaker (1956) and they offer a means of representing a continuum in a very simple manner. However, they give most satisfactory results where one or two environmental gradients have marked overall control of vegetational variation. Vegetation samples or species are arranged along axes characterized by these controlling environmental factors (Fig. 3.20).

Loucks (1962) has modified and developed this technique by constructing synthetic scalars for the axes. These integrate two or more environmental factors to position sites in the form of a two-dimensional diagram. However, the selection of factors and the construction of the scalars is highly subjective and most likely to be successful where the principal controlling factors are obvious. Elsewhere the construction of ordinations based on more objective methods (see below) is to be recommended.

Gradient analysis may also be used to relate vegetational variation to aspect by constructing polar diagrams. This approach has also been successfully used by Perring (1959) to demonstrate variation on chalk grassland and Goldsmith (1967) to examine the pattern of deposition of salt spray on sea-cliffs (Fig. 3.21).

3.4.2 Ordination

An ordination is a spatial arrangement of samples such that their position reflects their similarity. It is used as a framework upon which to compare species and environmental factors as a basis for erecting hypotheses about cause-and-effect relationships. There are two basic types of ordination:

1. Environmental ordination, where the axes are constructed from environmental data.

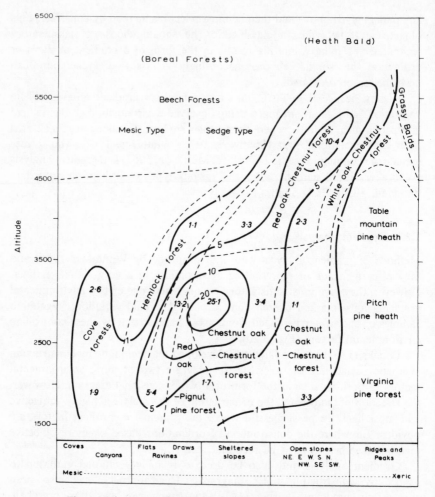

Figure 3.20   An example of gradient analysis (from Whittaker, 1956)

2. Phytosociological ordination, where the axes are constructed from vegetational data.

Methods of gradient analysis may be considered types of environmental ordination *sensu lato* but normally the term is reserved for techniques that use all the variables of the data set. Phytosociological ordinations are considered more flexible since all the vegetational variables are immediately obvious and can be used. Individual environmental factors can then be compared as they become available.

The earlier, simpler ordinations used reference stands to represent the two extremes of variation within the vegetation samples (stands) described. More

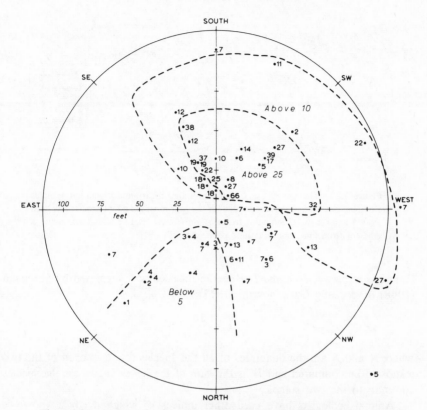

Figure 3.21 An example of a polar gradient analysis. Sea-cliff stands have been arranged using altitude on the radii and aspect on the circumference of the graph. Soil conductivity values have been plotted against the site positions to indicate the distribution of salt spray on sea-cliffs.

recent techniques, usually variants of factor analysis, extract the principal axes of variation directly from the data. The earlier techniques were developed before electronic computers were generally available and are still useful for small projects, work in areas where computers are not available and for teaching purposes.

The procedure of continuum analysis was first advocated by Curtis & McIntosh in 1951. A measure of abundance or importance value (*I.V.*) for each species is multiplied by a 'climax adaptation number' (*C.A.N.*). The bell-shaped curves for each species can then be compared with those for environmental factors (Fig. 3.22). The process is highly subjective and has been largely superseded by the method proposed by Bray & Curtis (1957). These authors replaced the continuum index by a similarity coefficient for

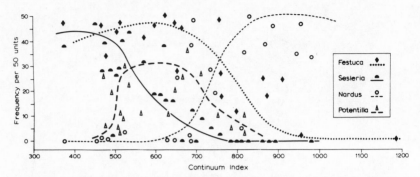

Figure 3.22 The use of the Continuum Index to arrange vegetation samples along an axis. The Importance Value for the four most abundant species has been plotted against the Continuum Index to show their relative abundance along a gradient from limestone grassland on acidic drift.

comparison between stands. The coefficient used was suggested by Sorenson (1948) or possibly Czekanowski (1913):

$$C = \frac{2W}{A+B}$$

where A and B are the quantities of all the species found in each of the two stands to be compared and W is the sum of the lesser values for the species common to the two stands.

Animal ecologists have used other indices of which the best known is probably Mountford's (1962). These are discussed in Southwood (1966). Coefficients of similarity are then inverted ($100 - C$ or $Cmax - C$) to produce a distance coefficient which is then entered in a half-matrix and the two most dissimilar stands chosen as the end points of the first axis. All other stands are located along this axis by intersecting arcs drawn proportional in length to the distances between the stand to be located and the first two reference stands. A second axis can be constructed at right angles to this first axis by using, as reference stands, that pair of stands closest to the mid-point of the original axis but having the largest interstand distance entered in the half-matrix. On the arrangement of stands in two-dimensional space, species performance and the levels of selected environmental factors may be plotted (Fig. 3.23). By visual comparison between species/environmental relationships tentative hypotheses may be erected regarding causal relationships. The same approach can be used to extract a third axis although interpretation usually becomes difficult after the third.

Various improvements on this basic approach have been made. For example, Beals (1960) suggests a formula for positioning stands along each axis with respect to the reference stands and so permits the electronic

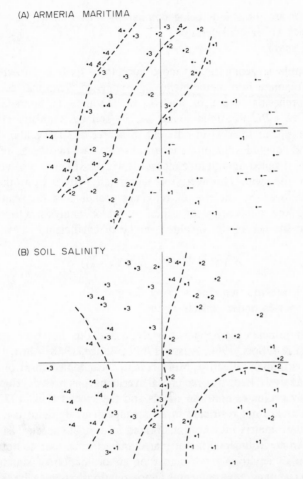

(A) ARMERIA MARITIMA

(B) SOIL SALINITY

Figure 3.23 Examples of ordination diagrams derived from sea-cliff frequency data using the weighted similarity coefficient and principal components analysis, (A) shows the abundance of *Armeria maritima* (as quartiles), and (B) shows the distribution of soil salinity (as quartiles) (from Goldsmith, 1973b).

computation of the ordination diagrams (Orloci, 1966). Coefficients, such as Interstand Distance (Sokal, 1961), which are theoretically more acceptable have also been suggested and are now more widely used than the original one suggested by Sørensen

$$D_{j.h.} = \sqrt{\sum_{n}^{i=1} (X_{ij} - X_{ih})^2}$$

where j and h are the stands being compared,
i is the species under consideration,
and x is the score.

This formula is geometrically more acceptable as it is an extension of Pythagoras' theorem into multidimensional space. It overcomes some of the failings of Sørensen's coefficient, such as the distance AC sometimes being longer than $AB + BC$ for three stands, A, B, and C. Bannister (1968) has discussed the informativeness of different measures of species abundance for ordination and concludes that presence-or-absence data may be as informative as some more detailed quantitative measures which are more time consuming to collect in the field. However, this approach suffers from the serious limitations imposed by the use of reference stands and this can only be overcome by the application of types of factor analysis. Orloci (1966) recommended the use of the weighted similarity coefficient,

$$W.S.C. = \sum_{n}^{i=1} (X_{ij} - \bar{X}_i)(X_{ih} - \bar{X}_i)$$

where j and h are the two stands being compared,
and i is the species under consideration;

together with principal components analysis and this has been extensively tested (Austin & Orloci, 1966; Austin 1968b; Bunce, 1968; Allen, 1971) and found to be extremely effective. More recently a modified form of principal components analysis, known as reciprocal averaging, has been developed which simultaneously ordinates both the stands and the species (Hill, 1973).

Ordinations are not restricted to an arrangement of stands using species but the raw data matrix may be transformed to ordinate species using stands (Fig. 3.24). An environmental factor matrix may also be used to obtain either an ordination of environmental factors or an ordination of stands. In this manner, species/stand/environmental factor relationships may be thoroughly explored.

Ordination has now become one of the most popular techniques for analysing vegetational variation. However, a study of the literature indicates that it is usually used for two, sometimes complementary, purposes. These are (1) to prepare a framework for describing vegetational variation and (2) in the search for causal factors. In both types of application the raw data is invariably floristic and usually quantitative. Nevertheless there are no *a priori* reasons why presence-or-absence data should not be used and some ecologists recommend their use (Bannister, 1966; Greig-Smith, 1969).

Ordination techniques are appropriate to any scale of study and published accounts range from the relatively local scale studies of Gittins (1965) and Austin (1968) on small areas of calcareous grassland, to the more extensive studies of Ashton (1964) in the Dipterocarp forests of Brunei and of Greig-

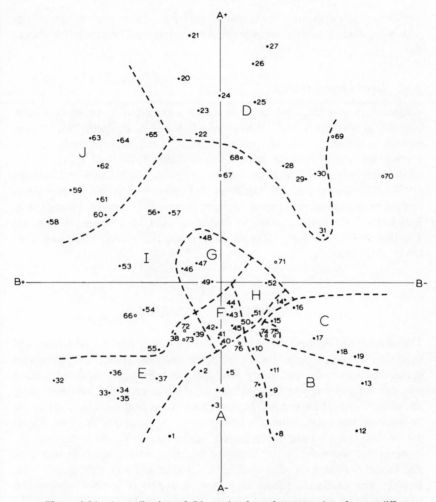

Figure 3.24 An ordination of 76 species from frequency data for sea-cliff vegetation processed using the weighted similarity coefficient and principal components analysis (from Goldsmith, 1973b).

Smith, Austin & Whitmore (1967) in the Solomon Islands. In theory ordination is appropriate in situations where vegetational variation is continuous; in practice it is used as an alternative to classification. The choice between these two major alternative approaches to vegetation is an empirical one, depending on whether the ecologist requires categories of vegetation or ordination diagrams (Anderson, 1965). It is important to note, however, that in the search for causality ordination only serves to erect hypotheses and, as in any correlation-type approach, the testing of such hypotheses must be carried out with other data or by experimentation, for example see Goldsmith (1973b and c).

3.4.3 Trend-surface analysis

Geographers have frequently mapped trends in quantitative variables by fitting polynomial surfaces to their observations (Chorley & Haggett, 1965). Many ecological variables, such as species' abundance and levels of environmental factors, are mappable in the same manner. Gittins (1968) has applied trend-surface analysis to his Anglesey limestone grassland data and to Goodall's (1954) Australian mallee. He concludes that the method should prove a useful tool for vegetation mapping and relating mathematically extracted components back to the field. However, many ecologists consider the technique to represent a further abstraction from reality which is only useful in certain circumstances (Norcliffe, 1969).

4 CHOICE OF METHODS

This chapter has outlined the methods most frequently used for describing and analysing vegetation. In some circumstances the choice of method is immediately obvious, but more frequently the ecologist is faced with a selection from as many as half-a-dozen, all of which appear to have equivalent advantages and disadvantages. In this situation the final decision will have to be made on the basis of the characteristics of the particular problem. Figure 3.1 is designed to assist the research worker select his method. It is not, however, the last word on the choice of an appropriate measure or technique and should only be used after reading this chapter and then with considerable thought and caution. It deals only with that part of a research problem concerning the vegetation and similar decision making processes will be required for studies of fauna and soil.

Stage one must always be to decide the objectives of the study. It often helps to list these on paper and whenever possible discuss them with another ecologist. Six major types of problem frequently confronted by ecologists have been recognized. Every reader will be able to think of others and most people will realize that some studies will include two or more of the categories in Fig. 3.1. If this is the case each category should be examined and more than one approach considered. Each major category of problem is followed by a series of questions and yes/no answers. These should be traced until the recommended method is identified. For example if a zoologist wishes to describe the vegetation on his research site he would start at 'site description'. He would ask himself whether or not the site had an obvious gradient of topographic, edaphic or vegetational variation and if the answer is negative he is asked whether the description is for a zoological, botanical or other type of study. He would end up with the suggestion that he should use a method based on vegetation structure. Reference would then have to be made to the variety of methods described in the text, for example, those of Du Rietz (1931), Raunkiaer (1934) and Elton & Miller (1954). Before making a decision, however, he is advised to consult the references appropriate to that method to obtain a full description of them and examples of their application. Decision-making is a slow process but a period of deliberation spent at this stage in the project is likely to save several days or weeks unnecessary effort during the relatively tedious process of collecting the data.

5 REFERENCES

AGNEW A.D.Q. (1961) The ecology of *Juncus effusus* L in North Wales. *J. Ecol.* **49**, 83–102.

ALLEN T.F.H. (1971) Multivariate approaches to the ecology of algae on terrestrial rock surfaces in North Wales. *J. Ecol.* **59**, 803–826.

ANDERSON D.J. (1961) The structure of some upland plant communities in Caernarvonshire. I. The pattern shown by *Pteridium aquilinum*. *J. Ecol.* **49**, 369–376. II. The pattern shown by by *Vaccinium myrtillus* and *Calluna vulgaris*. *J. Ecol.* **49**, 731–738.

ANDERSON D.J. (1965) Classification and ordination in vegetation science: Controversy over a non-existent problem? *J. Ecol.* **53**, 521–526.

ANDERSON D.J. (1967) Studies on structure in plant communities. III. Data on pattern in colonizing species. *J. Ecol.* **55**, 397–404.

ARCHIBALD E.E.A. (1949) The specific character of plant communities. I. Herbaceous communities. *J. Ecol.* **37**, 260–273.

ASHBY E. & PIDGEON T.M. (1942) A new quantitative method of analysis of plant communities. *Aust. J. Sci.* **5**, 19.

ASHTON P.S. (1964) Ecological studies in the mixed Dipterocarp forests of Brunei State. *Oxford Forestry Memoirs N. 25.* Oxford, Clarendon Press.

AUSTIN M.P. (1968a) Pattern in a *Zerna erecta* dominated community. *J. Ecol.* **56**, 197–218.

AUSTIN M.P. (1968b) An ordination of a chalk grassland community. *J. Ecol.* **56**, 739–758.

AUSTIN M.P. & GREIG-SMITH P. (1968) The application of quantitative methods to vegetation survey. II. Some methodological problems of data from rain forest. *J. Ecol.* **56**, 827–844.

AUSTIN M.P. & ORLOCI L. (1966) Geometric models in ecology. II. An evaluation of some ordination techniques. *J. Ecol.* **54**, 217–229.

BANNISTER P. (1966) The use of subjective estimates of cover-abundance as the basis for ordination. *J. Ecol.* **54**, 665–674.

BANNISTER P. (1968) An evaluation of some procedures used in simple ordinations. *J. Ecol.* **56**, 27–34.

BARCLAY-ESTRUP P. & GIMINGHAM C.H. (1969) The description and interpretation of cyclical processes in a heath community. I. Vegetational change in relation to the *Calluna* cycle. *J. Ecol.* **57**, 737–758.

BAWDEN M.G. (1967) Applications of aerial photography in land system mapping. *Photogrammetric Record, V* **30**, 461–464.

BEALS E. (1960) Forest bird communities in the Apostle Islands of Wisconsin. *The Wilson Journal* **72**, 156–181.

BEARD J.S. (1944) Climax vegetation in tropical America. *Ecology* **25**, 127–158.

BECKING R.W. (1957) The Zurich-Montpellier school of phytosociology. *Bot. Rev.* **23**, 411–488.

BILLINGS W.D. & MOONEY H.A. (1959) An apparent frost hummock-sorted polygon cycle in the alpine tundra of Wyoming. *Ecology* **40**, 16–20.

BIRSE E.L. & ROBERTSON J.S. (1967) Chapter 5 in *The Soils of the Country Around Haddington and Eyemarth* (ed. J.M. Ragg and D.W. Futty). Mem. Soil Survey. Gt. Br.

BISHOP O.N. (1966) *Statistics for Biology.* London, Longmans.

BOALER S.B. & HODGE C.A.H. (1962) Vegetation stripes in Somaliland. *J. Ecol.* **50**, 465–474.

BOATMAN D.J. & ARMSTRONG W.A. (1968) A bog type in north-west Sutherland. *J. Ecol.* **56**, 129–141.

BRAUN-BLANQUET J. (1932) *Plant Sociology: The Study of Plant Communities.* New York, McGraw-Hill.

BRAUN-BLANQUET J. (1964) *Pflanzensoziologie,* 3rd edn, Wien, Springer-Verlag.

BRAY J.R. & CURTIS J.T. (1957) An ordination of the upland forest communities of southern Wisconsin. *Ecol. Monogr.* **27**, 325–349.

BROWN D. (1954) Methods of surveying and measuring vegetation. *Comm. Bur. Pastures and Field Crops. Bull.* **42**, Comm. Agric. Bureau.

BUNCE R.G.H. (1968) An ecological study of Ysgolion Duon, a mountain cliff in Snowdonia. *J. Ecol.* **56**, 59–76.

BURGES A. (1960) Time and size as factors in ecology. *J. Ecol.* **48**, 273–285.

BURNETT J.H. (Ed.) (1964) *The Vegetation of Scotland.* Edinburgh, Oliver and Boyd.

CAIN S.A. (1932) Concerning certain phytosociological concepts. *Ecol. Monogr.* **2**, 475–508.

CAIN S.A. & CASTRO G.M. (1959) *Manual of Vegetation Analysis.* New York, Harper and Bros.

CHORLEY R.J. & HAGGETT P. (1965) Trend-surface mapping in geographical research. *Trans. Inst. Br. Geogr.* **37**, 47–67.

CLAPHAM A.R. (1932) The form of the observational unit in quantitative ecology. *J. Ecol.* **20**, 192–197.

CLAPHAM A.R., TUTIN T.G. & WARBURG E.F. (1962) *Flora of the British Isles* (2nd edition). Cambridge University Press.

CLEMENTS F.E. (1916) Plant succession: an analysis of the development of vegetation. *Carnegie Inst. Washington Publ.* **242**, 1–512.

COLWELL R.N. (1956) Determining the prevalence of certain cereal crop diseases by means of aerial photography. *Hilgardia* **26**, (5), 223–286.

COTTAM G. (1947) A point method for making rapid surveys of woodlands. *Bull. Ecol. Soc. Amer.* **28**, 60.

COTTAM G. & CURTIS J.T. (1949) A method for making rapid surveys of woodlands by means of pairs of randomly selected trees. *Ecology* **30**, 101–104.

COTTAM G. & CURTIS J.T. (1955) Correction for various exclusion angles in the random pairs method. *Ecology* **36**, 767.

COTTAM G. & CURTIS J.T. (1956) The use of distance measures in phytosociological sampling. *Ecology* **37**, 451–460.

COUPLAND R.T. & JOHNSON R.E. (1965) Rooting characteristics of native grassland species in Saskatchewan. *J. Ecol.* **53**, 475–508.

CURTIN W. & LANE R.F. (1955) *Concise Practical Surveying.* London, English Universities Press.

CURTIS J.T. (1959) *The Vegetation of Wisconsin: an Ordination of Plant Communities.* Madison, Wisconsin University Press.

CURTIS J.T. & MCINTOSH R.P. (1951) An upland forest continuum in the prairie-forest border region of Wisconsin. *Ecology* **32**, 476–496.

CZEKANOWSKI J. (1913) *Zarys Metod Statystcznyck.* Warsaw.

DAHL E. (1956) Rondane: Mountain vegetation in south Norway and its relation to the environment. *Skr. norske Vidensk-Akad. Mat. Naturv. Kl.* **3**, 374.

DANSEREAU P. (1957) *Biogeography: An Ecological Perspective.* New York, Ronald.

DANSEREAU P. (1961) Essai de représentation cartographique des éléments structuraux de la végétation, in Gaussen, H. (ed.) *Méthodes de la Cartographie de la Végétation.* 233–255. Paris.

DAVIES T.A.W. & RICHARDS P.W. (1933) The vegetation of Moraballi Creek, British Guiana: an ecological study of a limited area of tropical rain forest. Part I. *J. Ecol.* **21**, 350–372.

DRUDE O. (1913) *Die Okologie der Pflanzen* Braunschweig, Vieweg., (Publ.).

DU RIETZ G.E. (1930) Classification and nomenclature of vegetation. *Svensk. Bot. Tidskr.* **24**, 489–503.

DU RIETZ G.E. (1931) Life-forms of terrestrial flowering plants. *Acta Phytogeographica Suecica* **3**, 1–95.

DUFFIELD B.S. & FORSYTHE J.F. (1972) Assessing the impacts of recreation use on coastal sites in East Lothian in *The Use of Aerial Photography in Countryside Research.* Countryside Commission, London.

EDGELL M.C.R. (1969) Vegetation of an upland ecosystem: Cader Idris, Merionethshire. *J. Ecol.* **57**, 335–359.

ELLENBERG H. (1954) Zur Entwicklung der Vegetationssystematik in Mitteleuropa *Angew, PflSoziol* (Wien) Festschr. Aichinger, **1**, 134–143.

ELLENBERG H. (1956) Aufgaben und Methoden der Vegetationskunde, in H. Walter *Einfuhrung in die Phytologie* Vol. IV, pt I. Stuttgart.

ELLENBERG H. & MUELLER-DOMBOIS D. (1967) Tentative physiognomic-ecological classification of plant formations of the Earth. *Ber. geobot. inst. Stiftg. Rubel* **37**, 21–46.

ELTON C.S. (1966) *The Pattern of Animal Communities.* London, Methuen.

ELTON C.S. & MILLER R.S. (1954) The ecological survey of animal communities with a practical system of classifying habitats by structural characters. *J. Ecol.* **42**, 460–496.

FISHER R.A. (1960) *The Design of Experiments.* Edinburgh, Oliver & Boyd.

FISHER R.A. & YATES F. (1963) *Statistical Tables for Biological, Agricultural and Medical Research.* London, Oliver & Boyd.

FOSBERG F.R. (1961) A classification of vegetation for general purposes. *Trop. Ecol.* **2**, 1–28.

FOSBERG F.R. (1967) A classification of vegetation for general purposes, in G.F. Peterken (ed.) *Guide to the Checksheet for I.B.P. Areas.* I.B.P. Handbook 4. Oxford, Blackwell.

FRENKEL R.E. & HARRISON C.M. (1974) An assessment of the usefulness of phytosociological and numerical classificatory methods for the community biogeographer. *Journal of Biogeography,* **1**, 27–56.

GITTINS R. (1965) Multivariate approaches to a limestone grassland community. I. A stand ordination. II. A direct species ordination. III. A comparative study of ordination and association analysis. *J. Ecol.* **53**, 385–425.

GITTINS R. (1968) Trend-surface analysis of ecological data. *J. Ecol.* **56**, 845–869.

GOLDSMITH F.B. (1967) *Some Aspects of the Vegetation of sea-cliffs* Ph.D. thesis, University of Wales.

GOLDSMITH F.B. (1972) Vegetation mapping in upland areas and the development of conservation management plans, in: *The Use of Aerial Photography in Countryside Research.* Countryside Commission, London.

GOLDSMITH F.B. (1973a) The ecologist's role in development for tourism: a case study in the Caribbean, *Biol. Linn. Soc.* **5**, 265–287.

GOLDSMITH F.B. (1973b) The vegetation of exposed sea cliffs at South Stack, Anglesey, I. The multivariate approach, *J. Ecol.* **61**, 787–818.

GOLDSMITH F.B. (1973c) The vegetation of exposed sea cliffs at South Stack, Anglesey, II. Experimental Studies, *J. Ecol.* **61**, 819–830.

GOLDSMITH F.B. (1974) An assessment of the Fosberg and Ellenberg methods of classifying vegetation for conservation purposes, *Biol. Cons.* **6,** 3–6.

GOODALL D.W. (1953) Objective methods for the classification of vegetation. I. The use of positive interspecific correlation. *Aust. J. Bot.* **1,** 39–63.

GOODALL D.W. (1954) Vegetational classification and vegetational continua. *Angew. PflSoziol.* **1,** 168–182.

GOODIER R. (1971) *The Application of Aerial Photography to the Work of the Nature Conservancy.* Nature Conservancy, Edinburgh.

GREIG-SMITH P. (1952) Ecological observations on degraded and secondary forest in Trinidad, British West Indies. I. General features of the vegetation. *J. Ecol.* **40,** 283–315. II. Structure of the communities. *Ibid.* 316–330.

GREIG-SMITH P. (1961) Data on pattern within plant communities. I. The analysis of pattern. *J. Ecol.* **49,** 695–708.

GREIG-SMITH P. (1964) *Quantitative Plant Ecology,* 2nd edition. London, Butterworths.

GREIG-SMITH P. (1969) Analysis of vegetation data: the user viewpoint. *Proceedings of the International Symposium on Statistical Ecology.* Newhaven.

GREIG-SMITH P., AUSTIN M.P. & WHITMORE T.C. (1967) The application of quantitative methods to vegetation survey. I. Association-analysis and principal component ordination of rain forest. *J. Ecol.* **55,** 483–503.

GROENEWOUD H. VAN (1965) Ordination and classification of Swiss and Canadian forests by various biometric and other methods. *Ber. Geobot. Inst. E.T.A., Stiftg. Rubel* **35,** 28–102.

GRUBB P.J. *et al.* (1969) The ecology of chalk heath: its relevance to the calcicole–calcifuge and soil acidification problems. *J. Ecol.* **57,** 175–212.

HALL J.B. (1971) Pattern in a chalk grassland community. *J. Ecol.* **59,** 749–762.

HARPER J.L. (1967) A Darwinian approach to plant ecology. *J. Ecol.* **55,** 247–270.

HARRISON C.M. (1970) The phytosociology of certain English heathland communities. *J. Ecol.* **58,** 573–589.

HARRISON C.M. (1971) Recent advances in the description and analysis of vegetation. *Trans. Inst. Br. Geogr.* **52,** 113–127.

HARRISON C.M. & WARREN A. (1970) Conservation, stability and management. *Area 2,* 26–32.

HILL M.O. (1973) Reciprocal averaging: an eigenvector method of ordination. *J. Ecol.* **61,** 237–250.

HOPE-SIMPSON J.F. (1940) On the errors in the ordinary use of subjective frequency estimations in grassland. *J. Ecol.* **28,** 193–209.

HOPKINS B. (1955) The species area relations of plant communities. *J. Ecol.* **43,** 409–426.

HOWARD J.A. (1970) *Aerial Photo-Ecology.* London, Faber & Faber.

IVIMEY-COOK R.B. & PROCTOR M.C.F. (1966) The application of association-analysis to phytosociology. *J. Ecol.* **54,** 179–192.

KERSHAW K.A. (1957) The use of cover and frequency in the detection of pattern in plant communities. *Ecology* **38,** 291–299.

KERSHAW K.A. (1958, 1959) An investigation of the structure of a grassland community. I. The pattern of *Agrostis tenuis. J. Ecol.* **46,** 571–592. II. The pattern of *Dactylis glomerata, Lolium perenne* and *Trifolium repens. Ibid.* **47,** 31–43. III. Discussion and conclusions. *Ibid.* **47,** 44–53.

KERSHAW K.A. (1961) Association and co-variance analysis of plant communities. *J. Ecol.* **49,** 643–654.

KERSHAW K.A. (1962) Quantitative ecological studies from Landmannahellir, Iceland. I. *Eriophorum angustifolium,* and II. The rhizome behaviour of *Carex bigelowii* and *Calamagrostis neglecta. J. Ecol.* **50,** 163–179.

KERSHAW K.A. (1968) Classification and ordination of Nigerian savanna vegetation. *J. Ecol.* **56,** 483–495.

KERSHAW K.A. (1973) *Quantitative and Dynamic Ecology,* 2nd edition. London, Edward Arnold.

KOPPEN W. (1928) *Klimakarte der Erde.* Gotha, Perthes.

KUCHLER A.W. (1967) *Vegetation Mapping.* New York, Ronald Press.

LAMBERT J.M. & WILLIAMS W.T. (1962) Multivariate methods in plant ecology. IV. Nodal analysis. *J. Ecol.* **50,** 775–802.

LAMBERT J.M. & WILLIAMS W.T. (1966) Multivariate methods in plant ecology. VI. Comparison of information-analysis and association-analysis. *J. Ecol.* **54,** 635–664.

LEEUWEN C.G. VAN (1966) A relation theoretical approach to pattern and process in vegetation. *Wentia* **15,** 25–46.

LOUCKS O.L. (1962) Ordinating forest communities by means of environmental scalars and phytosociological indices. *Ecol. Monogr.* **32,** 137–166.

MAAREL E. VAN DER (1971) Plant species diversity in relation to management. In Duffey and Watt (eds.). *The Scientific Management of Animal and Plant Communities for Nature Conservation.* Oxford, Blackwell.

MAC ARTHUR R.H. (1965) Patterns of species diversity. *Biol. Rev.* **40,** 510–533.

MARGALEF R. (1968) *Perspectives in Ecological Theory.* Chicago, University Press.

MC INTOSH R.P. (1967a) An index of diversity and the relation of certain concepts to diversity. *Ecology* **48,** 392–404.

MC INTOSH R.P. (1967b) The continuum concept of vegetation. *Bot. Rev.* **33,** 130–187.

MC VEAN D.N. & RATCLIFFE D.A. (1962) *Plant Communities of the Scottish Highlands.* Monographs of the Nature Conservancy, No. 1. London, H.M.S.O.

MOORE J.J. (1962) The Braun-Blanquet system: A reassessment. *J. Ecol.* **50,** 761–769.

MOORE J.J., FITZSIMONS P., LAMBE E. & WHITE J. (1970) A comparison and evaluation of some phytosociological techniques. *Vegetatio* **20,** 1–20.

MOORE N.W. (1962) The heaths of Dorset and their conservation. *J. Ecol.* **50,** 369–391.

MOUNTFORD M.D. (1962) An index of similarity and its application to classificatory problems. In P.W. Murphy (ed.), *Progress in Soil Zoology,* 43–50. London, Butterworth.

NORCLIFFE G.B. (1969) On the use and limitations of trend surface models. *Canadian Geographer* **8,** 338–348.

OBERDORFER E. (1957) *SudDeutsche Pflanzengesellschaften.* Jena, Fisher.

ODUM E.P. (1969) The strategy of ecosystem development. *Science* **164,** 262–270.

ORLOCI L. (1966) Geometric models in ecology. I. The theory and application of some ordination methods. *J. Ecol.* **54,** 193–216.

OSVALD H. (1923) Die vegetation des Hochmoores Komosse. *Akad. Abh. Uppsala.*

PEARSALL W.H. (1924) The statistical analysis of vegetation: a criticism of the concepts and methods of the Uppsala school. *J. Ecol.* **12,** 135–139.

PERKINS D.F. (1971) Counting sheep, cattle and ponies on Dartmoor by aerial photography. In Goodier (ed.) *The Application of Aerial Photography to the work of the Nature Conservancy.* The Nature Conservancy.

PERRING F. (1959) Topographical gradients of chalk grassland. *J. Ecol.* **47,** 447–481.

PETERKEN G.F. (1967) *Guide to the check sheet for I.B.P. Areas.* International Biological Programme, Handbook 4. Oxford, Blackwell.

PHILLIPS M.E. (1954) Studies in the quantitative morphology and ecology of *Eriophorum angustifolium* Roth. Part III. *New Phytol.* **53,** 312–342.

PIELOU E.C. (1959) The use of point-to-plant distances in the study of the pattern of plant populations. *J. Ecol.* **47,** 607–613.

PIELOU E.C. (1962) The use of plant-to-neighbour distances for the detection of competition. *J. Ecol.* **50,** 357–367.

POORE M.E.D. (1955) The use of phytosociological methods in ecological investigations. I. The Braun-Blanquet System. *J. Ecol.* **43,** 226–244. II. Practical issues involved in an attempt to apply the Braun-Blanquet system. *J. Ecol.* **43,** 245–269.

POORE M.E.D. & ROBERTSON V.C. (1964) *An approach to rapid description and mapping of biological habitats based on a survey of the Hashemite Kingdom of Jordan.* Sub-commission on Conservation of Terrestrial Biological Communities. I.B.P.

PRITCHARD N.M. & ANDERSON A.J.B. (1971) Observations on the use of cluster analysis in botany with an ecological example. *J. Ecol.* **59**, 727–748.

PREIS K. (1939) Die *Festuca vallesiaca—Erysimum crepidifolium* Assoziation aux Basalt, Glimmerschiefer und Granitgneis. *Beih Bot. Cbl.* **59B**, 478–530.

PROCTOR M.C.F. (1967) The distribution of British liverworts: a statistical analysis. *J. Ecol.* **55**, 119–136.

RATCLIFFE D.A. (1971) Criteria for the selection of nature reserves. *Adv. of Sci.* **27**, 294–296.

RATCLIFFE D.A. & WALKER D. (1958) The Silver Flowe, Galloway, Scotland. *J. Ecol.* **46**, 407–445.

RAUNKIAER C. (1934) *The Life-Forms of Plants and Statistical Plant Geography.* Oxford, Clarendon Press.

RICHARDS P.W. (1952) *The Tropical Rain Forest.* Cambridge University Press.

RICHARDS P.W., TANSLEY A.G. & WATT A.S. (1940) The recording of structure, life-forms and flora of tropical forest communities as a basis for their classification. *J. Ecol.* **28**, 224–239.

RUBEL E. (1930) *Die Pflanzengesellschaften der Erde.* Bern, Huber Verlag.

RUTTER A.J. (1955) The composition of wet-heath vegetation in relation to the water-table. *J. Ecol.* **43**, 507–543.

SALISBURY E.J. (1916) The Oak-hornbeam woods of Hertfordshire. Parts I and Ii. *J. Ecol.* **4**, 83–120.

SEARLE S.R. (1966) *Matrix Algebra for the Biological Sciences.* New York, Wiley.

SHIMWELL D.W. (1971) *The Description and Classification of Vegetation.* London, Sidgwick & Jackson.

SIMPSON E.H. (1949) Measurement of diversity. *Nature* **163**, 688.

SOCIETY FOR THE PROMOTION OF NATURE RESERVES (1969) *Biological Sites Recording Scheme.* S.P.N.R. Conservation Liaison Committee Technical Publication, **I**.

SOKAL R.R. (1961) Distance as a measure of taxonomic similarity. *Syst. Zoology.* **10**, 70–79.

SOKAL R.R. & SNEATH P.H.A. (1963) *Principles of Numerical Taxonomy.* San Francisco, Freeman.

SORENSON T. (1948) A method of establishing groups of equal amplitude in plant sociology based on similarity of species content. *Kong. Dan Vidensk. Selsk. Biol. Skr.* **5**, 1–34.

SOUTHWOOD T.R.E. (1966) *Ecological Methods.* London, Methuen.

STAMP L.D. (1925) The aerial survey of the Irrawaddy Delta forests (Burma). *J. Ecol.* **13**, 262–276.

SZAFER W. (1966) *The Vegetation of Poland.* International Series of Monographs in Pure and Applied Biology **9**. Oxford, Pergamon.

TANSLEY A.G. (1935) The use and abuse of vegetational concepts and terms. *Ecology* **16**, 284–307.

TANSLEY A.G. (1939) *The British Islands and their Vegetation.* Cambridge University Press.

TANSLEY A.G. & ADAMSON R.S. (1913) Reconnaissance in the Cotteswolds and the Forest of Dean. *J. Ecol.* **1**, 81–89.

TANSLEY A.G. & ADAMSON R.S. (1926) A preliminary survey of the chalk grasslands of the Sussex Downs. *J. Ecol.* **14**, 1–32.

TAYLOR J.A. (1969) Reconnaissance surveys and maps in *Geography at Aberystwyth.* (ed.) E.G. Bowen, H. Carter and J.A. Taylor. University of Wales Press.

THOMAS A.S. (1960) Changes in vegetation since the advent of myxomatosis. *J. Ecol.* **48**, 287–306.

THOMPSON H.R. (1958) The statistical study of plant distribution patterns using a grid of quadrats. *Aust. J. Bot.* **6**, 322–342.

THORNTHWAITE C.W. (1948) An approach towards a rational classification of climate. *Geog. Rev.* **38**, 55–94.

TÜXEN R. (1937) Die Pflanzengesellschaften nordwest-Deutschlands. *Mitt. Flor.-soz. Arbeitsg. in Niedersachsen*, **3**, 1–170.

WADSWORTH R.M. (1970) An invisible marker for experimental plots. *J. Ecol.* **58**, 555–557.

WARMING E. (1923) Okologiens Grundformer, Udkast til en systemastisk Ordning. *K. Danske Vidensk. Selsk. Naturv. Math. Afd. Skr.* **8**, R4, 119–187.

WARREN WILSON J. (1960) Inclined point quadrats. *New Phytol.* **59**, 1–8.

WATT A.S. (1947) Pattern and process in the plant community. *J. Ecol.* **35**, 1–22.

WATT A.S. (1962) The effect of excluding rabbits from Grassland A (Xerobrometum) in Breckland, 1936–60. *J. Ecol.* **50**, 181–198.

WATT A.S. (1964) The community and the individual. *J. Ecol. (Suppl.)* **52**, 203–212.

WEAVER J.E. & CLEMENTS F.E. (1938) *Plant Ecology*. London, McGraw-Hill.

WEBB D.A. (1954) Is the classification of plant communities either possible or desirable? *Bot. Tiddsk.* **51**, 362–370.

WEBB L.J., TRACEY J.G., WILLIAMS W.T. & LANCE G.N. (1970) Studies in the numerical analysis of complex rain-forest communities. V. A comparison of the properties of floristic and physiognomic-structural data. *J. Ecol.* **58**, 203–232.

WHITTAKER R.H. (1956) Vegetation of the Great Smoky Mountains. *Ecol. Monogr.* **26**, 1–80.

WHITTAKER R.H. (1965) Dominance and diversity in land plant communities. *Science* **147**, 250–260.

WHITTAKER R.H. (1967) Gradient analysis of vegetation. *Biol. Rev.* **49**, 207–264.

WHITTAKER R.H. (1969) Evolution of diversity in plant communities. In *Diversity and Stability in Ecological Systems. Brookhaven Symp. Biol.* **22**, 178–196.

WHITTAKER R.H. (1970) *Communities and Ecosystems*. London, Macmillan.

WILLIAMS C.B. (1964) *Patterns in the Balance of Nature and Related Problems in Quantitative Ecology*. London, Academic Press.

WILLIAMS W.T. & DALE M.B. (1965) Fundamental problems in numerical taxonomy. *Adv. Bot. Res.* **2**, 35–68. (ed.) R. D. Preston. New York, Academic Press.

WILLIAMS W.T. & LAMBERT J.M. (1959) Multivariate methods in plant ecology. I. Association analysis in plant communities. *J. Ecol.* **47**, 83–101.

WILLIAMS W.T. & LAMBERT J.M. (1960) II. The use of an electronic digital computer for association analysis. *J. Ecol.* **48**, 689–710.

WILLIAMS W.T. & LAMBERT J.M. (1961) III. Inverse association-analysis. *J. Ecol.* **49**, 717–729.

WILLIAMS W.T., LAMBERT J.M. & LANCE G.N. (1966) Multivariate methods in plant ecology. V. Similarity analyses and information-analysis. *J. Ecol.* **54**, 2; 427–445.

WILLIS A.J. (1973) *Introduction to Plant Ecology*. London, Allen & Unwin.

WINKWORTH R.E. & GOODALL D.W. (1962) A crosswise sighting tube for point quadrat analysis. *Ecology* **43**, 342–343.

YARRANTON G.A. (1966) A plotless method of sampling vegetation. *J. Ecol.* **54**, 229–238.

YARRANTON G.A. (1969) Plant ecology: a unifying model. *J.Ecol.* **57**, 245–250.

YARRANTON G.A. (1971) Mathematical representations and models in plant ecology: response to a note by R. Mead. *J. Ecol.* **59**, 221–224.

CHAPTER 4

PRODUCTION ECOLOGY AND NUTRIENT BUDGETS

S.B. CHAPMAN

1   **Introduction** 158

2   **The ecosystem** 158
2.1   The ecosystem concept 158
2.2   Ecosystem modelling 160

3   **Production, decomposition and accumulation** 160
3.1   Definitions, concepts and units 160
   3.1.1   Production 161
   3.1.2   Decomposition 162
   3.1.3   Accumulation 162
   3.1.4   Units 167
3.2   Gross primary production 168
3.3   The measurement of the above ground standing crop 170
   3.3.1   Direct harvest methods 171
   3.3.2   Indirect estimation of the standing crop 174
   3.3.3   Crop meters 176
   3.3.4   Frequency of sampling 177
   3.3.5   Treatment of samples 177
3.4   Litter production 178
   3.4.1   The direct estimation of litter production 178
   3.4.2   The indirect estimation of litter production 180
3.5   Decomposition and accumulation of litter 181
   3.5.1   Quadrat methods for the estimation of litter disappearance 182
   3.5.2   The measurement of litter disappearance from contained samples 183
   3.5.3   The measurement of litter disappearance from labelled or tagged samples 184

3.5.4   Respirometry and litter disappearance 184
3.5.5   Comparative studies of litter decomposition 185
3.5.6   The measurement of litter accumulation 186
3.6   Roots and soil organic matter 188
   3.6.1   The estimation of root biomass and examination of root systems 189
   3.6.2   The extraction of roots from soil samples 190
   3.6.3   The separation of live and dead roots 193
   3.6.4   Radio-isotopes and root studies 193
   3.6.5   The relationship between root biomass and production 194
   3.6.6   The accumulation of organic matter in seral sites 195
   3.6.7   The measurement of soil respiration and carbon dioxide under field conditions 196
   3.6.8   Temperature and soil respiration 201
3.7   Grazing 202
   3.7.1   Grazing exclosures 202
   3.7.2   The direct estimation of consumption 203
   3.7.3   Faecal analysis 204

4   **Nutrient and energy budgets** 205
4.1   The estimation of nutrient and calorific content 206
4.2   Nutrient inputs 208
   4.2.1   Precipitation and atmospheric fallout 208

4.2.2   Salt spray   209
4.2.3   Nitrogen fixation   211
4.3   Nutrient losses   211
    4.3.1   The loss of nutrients in soil
            solution   212
    4.3.2   The loss of nutrients by
            fire   215
4.4   The measurement of calorific
      values   215

4.4.1   Calorific values and
        calorimetry   215
4.4.2   The preparation of
        samples   216
4.4.3   Calorimeter corrections   216

5   **Summary**   218

6   **References**   219

1 INTRODUCTION

An ecologist can find himself involved in the study of biological production
and nutrient budgets for one or a number of reasons. He may wish to use the
estimates of primary production for a comparison of sites within a particular
type of ecosystem, or to use them as the basis for comparing very different
types of ecosystem (Westlake, 1963). The study of primary production,
nutrient budgets and energy flow are important in attempting to understand
the function of natural communities, but it should be remembered that they
represent only one particular approach to the problem and that other
viewpoints in ecology may be just as important in helping to obtain a more
complete analysis and understanding of ecosystems and ecological processes.

One particularly important feature of production ecology is the way that
it provides a strong and unifying link between a number of different aspects of
the subject. The distinctions between types of ecologists working in this field
tend to breakdown, and an individual engaged in a production study may often
wonder whether he is a botanist, a zoologist or even a pedologist.

This chapter is divided into three main sections; the first provides an
introduction to some of the more important definitions and concepts that relate
to production ecology, the second describes some of the methods that are
available for the estimation of primary production and associated
processes, and the third deals with methods that are relevant to the study of
nutrient budgets.

2 THE ECOSYSTEM

2.1 THE ECOSYSTEM CONCEPT

In a paper presented in 1935, Sir Arthur Tansley dealt with a number of
terminological and conceptual problems that beset ecologists of the time. He
rejected such contemporary terms as 'complex organisms' and 'biotic

community', and introduced the term ECOSYSTEM in the following terms:

'Though the organisms may claim our primary interest when we are trying to think fundamentally we cannot separate them from their special environment, with which they form one physical system.'

The ecosystem has since been defined by many authors as a functional unit that includes the biotic components (organisms including man) and the abiotic components (environmental physico-chemical) of a specified area. While it is generally recognized that the ecosystem includes inter-relationships

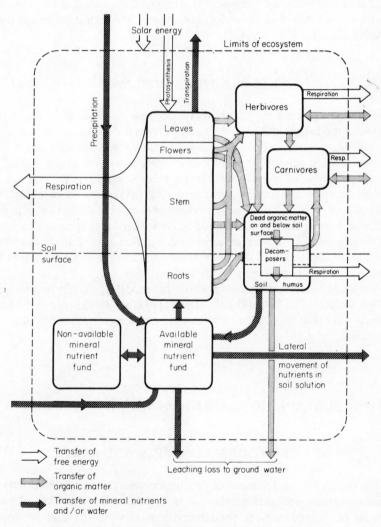

Figure 4.1 General representative diagram of an ecosystem.

between biotic and abiotic factors it is often forgotten that the definition includes 'a specified area'. Tansley stated 'ecosystems are of the most various kinds and sizes. They form one category of the multitudinous physical systems of the universe, which range from the universe as a whole down to the atom'. Unfortunately many people now use the term ecosystem in a general way without defining the limits of the system to which they refer. Whenever the term ecosystem is used in reference to an ecological study the physical and biological limits of the system should be made clear. It should be remembered that ecosystems are inter-related functional units and that any separation is an artificial division for the purposes of simplification and investigation. The historical development of the ecosystem concept has been described in detail by Major (1969).

2.2 ECOSYSTEM MODELLING

In recent years most ecologists have become familiar with the terms 'ecosystem modelling' and 'systems analysis' even if they are not sure of their full meaning or implications. A model is no more than an abstraction that serves to describe or simulate all or part of some process or situation. Models can take a number of forms, mathematical models, word models, box models and flow diagrams. It is inevitable that modelling, the systems approach, and the ability of computers to handle large amounts of complex data, have a great deal to offer ecology in the future. It should be emphasized that systems analysis is not the only way of studying an ecosystem, but that a particularly important feature of the approach is in the way that it emphasizes the need to define and to quantify the basic components of the system.

A detailed account of systems analysis and modelling cannot be attempted here but Smith (1970), Patten (1971), Jeffers (1972) and Reichle *et al.* (1973) are references that will be of interest to anyone requiring an introduction to the subject.

3 PRODUCTION, DECOMPOSITION AND ACCUMULATION

3.1 DEFINITIONS, CONCEPTS AND UNITS

For the successful measurement of production, and the associated processes of decomposition and accumulation, it is necessary for a number of basic terms to be defined and for the relationships between them to be clearly understood.

3.1.1 Production

PRODUCTION is the weight or biomass of organic matter assimilated by an organism or community over a given period of time.

PRIMARY PRODUCTION is the production of organic matter by photosynthesis and SECONDARY PRODUCTION the subsequent conversion of that organic matter by heterotropic organisms.

Primary production can be expressed in two ways:

(a) GROSS PRIMARY PRODUCTION, the total amount of organic matter produced (including that lost in respiration) over a given period of time.

(b) NET PRIMARY PRODUCTION, the amount of organic matter incorporated by a plant or an area of vegetation (gross primary production minus the loss due to respiration) over a given period of time.

It is net primary production that is generally the concern of the plant ecologist and it is often further qualified by reference to some particular part of the plant or vegetation (aerial, root or seed production, etc.).

BIOMASS or STANDING CROP is the weight of organic matter per unit area present in some particular component of the ecosystem at a particular instant of time. Biomass is generally expressed in terms of dry weight and on occasion may be given in terms of ash free dry weight (see section 4.4.1).

The relationship between biomass, time and production, has been given in a standard form in the handbooks produced for the International Biological Programme (Newbould, 1967; Milner & Hughes, 1968).

B_1 = Biomass of a plant community at time t_1.

B_2 = Biomass of a plant community at time $t_2 (= t_1 + \Delta t)$.

ΔB = Change in biomass during the period $t_1 - t_2$

L = Plant losses by death and shedding during $t_1 - t_2$.

G = Plant losses by grazing etc. during $t_1 - t_2$.

P_N = Net production by the community during $t_1 - t_2$.

In terms of these symbols:

$$P_N = \Delta B + L + G,$$

so that if ΔB, L and G can be estimated satisfactorily P_N can be calculated. The use of this relationship requires estimates of biomass to be made at least twice, and generally with at least a year between the determinations.

There is an alternative approach (Newbould, 1967), if the components of production can be recognized at the end of the growing season it should be possible to obtain an estimate of production from a single visit to the site.

$$P_N = P_{\text{flowers}} + P_{\text{green}} + P_{\text{wood}} + P_{\text{roots}}$$

By the end of the growing season some of the current year's production may well have been lost by grazing, or by death and loss as litter, so that the apparent growth increment will be an underestimate of production.

The addition of the total grazing and litter losses to the apparent growth increment will result in overestimation of the net production as part of the losses will have been from previous years' production.

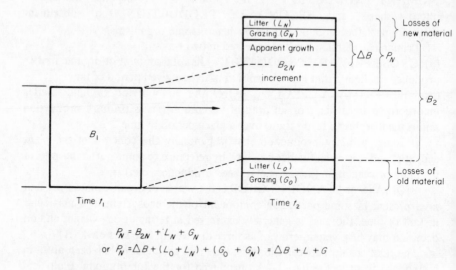

$$P_N = B_{2N} + L_N + G_N$$
$$\text{or } P_N = \Delta B + (L_O + L_N) + (G_O + G_N) = \Delta B + L + G$$

Figure 4.2   The relationship between production, apparent growth increment and change in biomass (based on Newbould, 1967).

3.1.2 Decomposition

Decomposition is the process by which organic matter is physically broken down and converted to simpler chemical substances, resulting in the production of carbon dioxide, water and the liberation of energy.

Decomposition represents a loss of energy and material from the ecosystem as well as the transformation and movement of organic matter within the system. The decomposition of organic matter is of fundamental importance in the release of plant nutrients from the litter and soil organic matter, making them available for uptake and further plant growth.

3.1.3 Accumulation

ACCUMULATION is the rate of change in weight of some part of the ecosystem as a result of production and decomposition.

The accumulation of litter has been described and examined against the background of a simple mathematical model by Jenny *et al.* (1949) and by Olson (1963).

If $B =$ Biomass or weight of organic matter.

$B_{ss} =$ Biomass or weight of organic matter present under steady state conditions.

$P =$ Production (input to the system).

$k =$ Instantaneous fractional loss rate.

The rate of change of biomass in the system over some discrete time interval (t), such as a day or year can be expressed as:

$$\frac{\Delta B}{\Delta t} = \text{input} - \text{losses (for that time interval).}$$

If the input to the system remains constant, the instantaneous rate of change of weight (rate when the limits of ΔB and Δt approach zero) will be:

$$\frac{dB}{dt} = P - kB.$$

Under steady state conditions it follows that inputs to the system must equal losses,

$$P = kB_{ss},$$

so that if these assumptions are valid and two of these parameters can be measured then the third can be calculated. If at all possible production and the rate of decomposition should be measured independently.

If constant rates of production and decomposition are assumed (for some objections to this simple model see Section 3.5.6) a number of further relationships can be derived. Where sources of input to the system are removed, such as in litter bag experiments (Section 3.5.2) or in the case of soil organic matter under fallow conditions it can be shown that:

$$B = B_o\, e^{-kt}$$

where, $B =$ the amount of organic matter remaining after time t.

B_o the initial weight of organic matter present at time t_o.

This expression for exponential decay is the same as that used to calculate the amount of a radio-active isotope remaining after a given period of time.

If the input to the system (production) is now assumed to remain at some value (P) instead of being zero, an expression can be derived that will provide an estimate of the weight of organic matter that will have accumulated after any period of time.

$$B = B_{ss}(1 - e^{-kt})$$

or

$$B = \frac{P}{k}(1 - e^{-kt})$$

This accumulation curve, often referred to as the MONO-MOLECULAR growth curve, shows that with constant production and decomposition the weight of accumulated organic matter increases with time to a steady state value (B_{ss}). It is possible to calculate the time required to approach steady state conditions for any particular combination of rates of production and decomposition.

$$\frac{0 \cdot 6931}{k} = \text{time required to reach 50\% steady state biomass.}$$

$$\frac{3}{k} = \text{time required to reach 95\% steady state biomass.}$$

In these expressions the constant k has been defined as the instantaneous fractional loss rate, but in many ecological investigations it is the fractional loss rate over a definite period of time, such as a month or year, (k') that is measured or estimated. When the loss is expressed as a fraction of the original weight the relationships between the two loss rates are:

$$k' = 1 - e^{-kt}$$
$$k = -\log_e (1 - k')$$

When the time interval is short (e.g. loss per day), or where decomposition is slow there are only small differences between the values of k and k', but when litter decomposition is measured over a whole year the differences between the two decay rates can be very important.

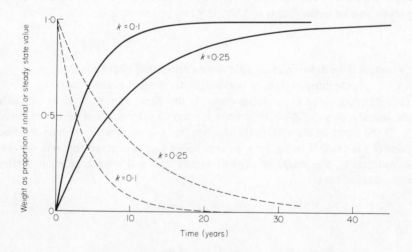

Figure 4.3 Monomolecular decay and accumulation curves with different values of k.

In this discussion it has been assumed that production remains constant with time and in many situations this is obviously not the case. The papers referred to by Jenny *et al.* and Olson discuss the application of these expressions to forest conditions where litter production is seasonal. Despite the assumptions of a simple exponential model it has much to offer in many ecological problems and demonstrates the basic relationships between production, decomposition and accumulation.

The monomolecular growth curve is just one of a series of curves that have been used to describe biological growth, or accumulation. The monomolecular curve, so called because of its relevance to first-order chemical reactions, assumes that the rate of growth at any time will be

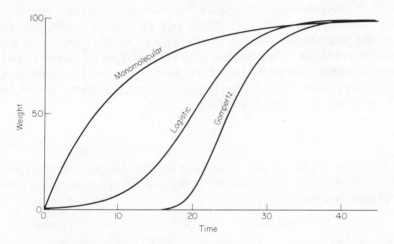

Figure 4.4 Monomolecular, logistic and gompertz growth curves.

proportional to the difference between Biomass (B) and the maximum weight that will be attained or accumulated (B_{ss}) under steady state conditions.

$$\frac{dB}{dt} = k(B_{ss} - B)$$

In the LOGISTIC growth curve it is assumed that the rate of growth at any particular time is proportional to the weight at that time, as well as to a function of that weight in relation to the maximum weight that will be attained.

$$\frac{dB}{dt} = kB\left(\frac{B_{ss} - B}{B_{ss}}\right)$$

from which can be derived:

$$B = \frac{B_{ss}}{1 + e^{a - kt}}$$

where a = a constant.

The GOMPERTZ growth curve, developed by actuaries in the analysis of human mortality, is described by the equation:

$$B = B_{ss}\, e^{(-e^{a-kt})}$$

with the growth rate given by:

$$\frac{dB}{dt} = kB \log_e\left(\frac{B_{ss}}{B}\right)$$

When growth curves such as these are fitted to ecological data they can be useful for comparative and descriptive purposes, but it is important not to attach too great a degree of biological significance to the constants obtained in the process.

To fit these curves to sets of data it is necessary to obtain a 'best fit' for three variables (B_{ss}, a and k). While this presents little difficulty when computer facilities are available there may well be times when it is more convenient to obtain an approximate fit by alternative means. In many cases a reasonable estimate of the final weight (B_{ss}) can be obtained by inspection of a plot of the raw data. When this has been done the weights (B) can be expressed as proportions of the final weight (W).

$$W = \frac{B}{B_{ss}}$$

When the weights are presented in this way the expressions for the growth curves can be written in terms of two variables (a and k) and in a form suitable for plotting in linear form.

	Monomolecular	Logistic	Gompertz
Equation in terms of two variables	$W = 1 - e^{a-kt}$	$W = \dfrac{1}{1 + e^{a-kt}}$	$W = e^{-e^{a-kt}}$
Expression in linear form	$\log_e (1 - W) = a - kt$	$\log_e\left(\dfrac{1-W}{W}\right) = a - kt$	$\log_e\left[\log_e\left(\dfrac{1}{W}\right)\right] = a - kt$

If data that can be described by a logistic curve is taken and the function

$$\log_e\left(\frac{1-W}{W}\right)$$

plotted against time the result will be found to approximate to a straight line. An estimate of the growth constant (k) can be obtained from the slope of the line and of the constant (a) from the intercept of the line on the 'y axis'. Ricklefs (1967) has described a similar method of obtaining an approximate fit for growth curves. He has shown that if a set of data is plotted in a linearized form and the lower weights approximate to a straight line then the particular choice of growth curve was probably appropriate. If the graph is curved with respect to the higher values then replotting after a minor revision of the estimate of B_{ss} will be necessary. The best choice of B_{ss} will be the value that gives the best approximation to a straight line. Unless the data is 'very good' it will not be possible to obtain a good fit for weights greater than about 0·9 of the final value. An example of a Gompertz curve fitted to the growth of heathland vegetation is shown in Fig. 4.5.

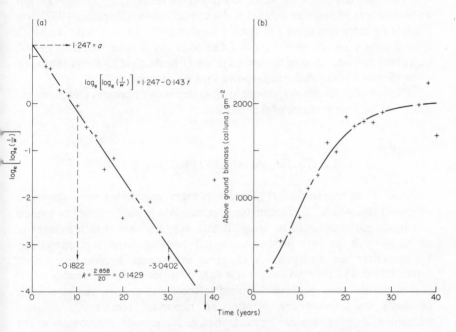

Figure 4.5 A Gompertz growth curve fitted to the above ground standing crop of *Calluna* from heathland of different ages.
(a) Transformed data plotted in linear form.
(b) Resulting growth curve plotted through original data. (Data from Chapman *et al.*, 1975a.)

3.1.4 Units

Various combinations of metric units have been, and are used to indicate the magnitude of dry matter production and associated processes. Whilst there would appear to be a move towards standardization and the use of grammes per square metre (g m^{-2}) as the basic unit for the study of primary production there are several points worth considering before choosing a set of units to express a particular set of results. There is a considerable advantage in using either kilogrammes per hectare (kg ha^{-1}), or tonnes per hectare (tonnes ha^{-1}) when the weights of organic matter involved are large.

$$1 \text{ g m}^{-2} = 10 \text{ kg ha}^{-1} = 0.01 \text{ tonnes ha}^{-1}$$

A further point, apparently not thought to be important by many workers, is the scale at which a particular investigation is being conducted. For example if seral changes within a relatively small area of heathland are being considered (Barclay-Estrup, 1970) it might be appropriate to use the square metre as the basic unit upon which to express the results (i.e. g m^{-2}), but if it is an overall area of heathland that is under consideration (Chapman, 1967) then it might be more appropriate to use the hectare as the basic unit (i.e. kg ha^{-1}). Even if some such convention is not accepted or found to be practical, it is still important that the scale of the investigation is made clear to avoid confusion when the results from different investigations are compared.

The use of other terms and units such as dry weight, calories and joules are dealt with in later sections of this chapter.

3.2 GROSS PRIMARY PRODUCTION

It would be most satisfactory if gross primary production were the basic estimate upon which other estimates of production could be based or against which comparisons could be made. If this were the case then a number of problems such as root production would possibly seem less intractable. Unfortunately the measurement of gross production involves the use of sophisticated and expensive apparatus that does not lend itself readily to the degree of replication generally required in ecological studies, and reasonable proximity to a laboratory is often an important requirement of such techniques. A good case can be made that, as far as some components of the ecosystem are concerned, it is only net primary production that is important. In the case of some herbivores it may only be one particular fraction of the net primary production that is of interest. In a great deal of the ecological literature primary production is taken to be synonymous with net production. It is not intended to deal with methods for the estimation of gross primary production in any detail, but a short account is given in the hope that developments will make it possible to apply some of them more widely in the future.

The estimation of carbon dioxide assimilation by the plant or vegetation is the most direct approach to the measurement of gross production. This has been achieved in a number of different ways, absorptiometric methods have been used (Stocker & Vieweg, 1960), but in recent years recording infra-red gas analysers have been used (e.g. Bliss & Hadley, 1964; Bourdeau & Woodwell, 1965; Billings *et al.*, 1966; Botkin *et al.*, 1970; Lange & Schulze,

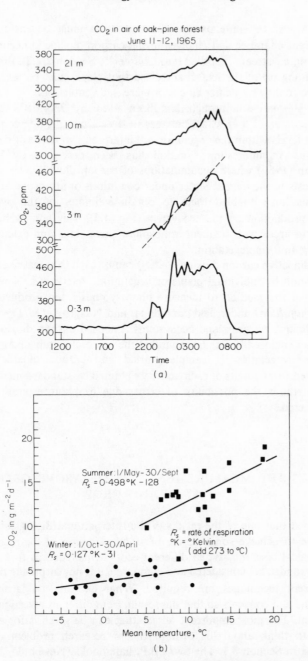

Figure 4.6 CO_2 concentrations at different heights in the Brookhaven Forest (USA) during a temperature inversion (a), and derived respiration rates plotted against mean temperature (b). (From Woodwell & Dykeman, 1966.)

1971). A leaf or some part of the vegetation must be enclosed within a transparent chamber and changes in the carbon dioxide concentration of the ventilating air stream recorded (see Chapter 5, Section 2.2.2). Problems occur with heating up of the leaf chamber so that steps must be taken to provide adequate cooling by Peltier blocks or increased ventilation.

An alternative approach has been used by Montieth (1962, 1963), Montieth *et al.* (1964) where changes in the carbon dioxide concentration at different levels within the vegetation are interpreted in terms of carbon dioxide exchange. An interesting situation has been described by Woodwell & Dykeman (1966) where accumulations of carbon dioxide over areas of oak-pine forests in the United States under conditions of atmospheric temperature inversions have enabled them to obtain estimates of the overall carbon dioxide production by the ecosystem (Fig. 4.6). This approach can only be applied to areas with frequent atmospheric inversions and extensive areas of relatively uniform vegetation.

Radioactive carbon dioxide ($^{14}CO_2$) can be used to measure carbon dioxide assimilation by individual plants or vegetation. Peterken & Newbould (1966) have used the method to measure photosynthesis and production by Holly *(Ilex aquifolium)* under field conditions; and Nasyrov *et al.* (1962) have used the technique on grassland ecosystems. Labelled carbon dioxide is liberated within an enclosed leaf chamber and at the end of a suitable period of time the plant or vegetation is sampled, dried and carbon dioxide assimilation calculated from counts of radioactivity obtained by standard methods. In all of these methods the possibility of errors due to photorespiration should be remembered.

3.3 THE MEASUREMENT OF THE ABOVE GROUND STANDING CROP

It is hoped that one of the results of the International Biological Programme would be to achieve some degree of standardization of the techniques used in ecological research. When different ecosystems, such as forest, grassland, bog and tundra are considered and compared, it is not surprising that a number of different techniques are required. What can be standardized are the definitions and concepts so that the results from different investigations can be compared. The most important advice that can be given is for an individual worker to think about his own particular research problem in the terms described in Section 3.1. The two I.B.P. handbooks (Newbould, 1967; Milner & Hughes, 1968) describe and review methods applicable to forest and grassland ecosystems while many individual research papers give details of methods in relation to other particular types of vegetation.

3.3.1 Direct harvest methods

1. The individual plant method

This approach is most suitable where separate plants occur at low densities and where only a few different species are present within the sampling area. Estimates of biomass per individual plant, are combined with estimates of species density (see Chapter 3, Section 2.2.2) to obtain overall figures for the standing crop.

2. The harvested quadrat method

In vegetation such as grassland or heathland it is not possible to differentiate between individual plants and the vegetation must be sampled by means of random quadrats. The location and numbers of quadrats required for a particular degree of accuracy are discussed in Chapter 3, Section 2.2.3. It has been suggested by Milner & Hughes that a standard error in the order of 10% of the mean is an acceptable level of accuracy for I.B.P. studies.

The size and shape of quadrats are discussed in Chapter 2, but an important factor in the measurement of standing crop is the amount of plant material that a particular size of quadrat will provide. Chapman (1967) found that a quadrat 50×50 cm provided sufficient sample from younger stands of heathland but not so much from older stands that would cause practical difficulties in handling the samples for analysis.

Any technique used for sampling the above ground standing crop must provide accurate and reproducible results, preferably with the minimum of effort on the part of the observer. If possible the results obtained should be independent of the actual person involved in the sampling process. This is important especially when a sampling programme extends over a long period and a number of different people will be involved in the work.

The particular type of cutting device that is chosen will depend upon the particular study, it may be hand shears, secateurs, scissors or some form of mechanical device. Where the vegetation sample needs to be sorted into separate species it may be most convenient to do this at the cutting stage. Although this will increase the length of time spent in the field it will often save time, especially where species of grass are involved. Milner and Hughes suggest that point quadrats (Warren Wilson, 1960, 1963) can be used to determine the relative proportions of the different species in the vegetation, and that by calibrating the results against cut samples they can be applied to other harvested material. Heady & Van Dyne (1964) have developed a laboratory point sampling technique where proportions of the species presented are determined by placing chopped vegetation on a tray and using a binocular microscope. Neither of these methods provide samples of individual plant species for chemical analysis, this can only be done by means of direct sorting. Any standing dead material must also be sorted out and separated from the harvested material.

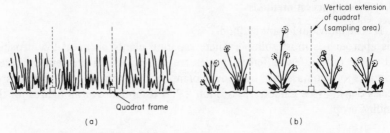

Figure 4.7 Diagrams showing the relationship between standing crop and vegetation rooted within a quadrat.

(a) Where rooted vegetation might approximate to standing crop per unit area.

(b) Where rooted vegetation cannot be equated with standing crop per unit area.

When the vegetation enclosed by a quadrat is to be sampled for the weight of standing crop it is important to minimize the edge effects and to collect only the vegetation contained within vertical extensions of the quadrat boundaries. The weight of the standing crop is not necessarily the same as the weight of vegetation rooted within the quadrat (Fig. 4.7) although in some cases they may be almost the same.

Satoo (1970) has listed three different ways in which harvesting techniques can be applied to woodland and forest conditions.

1. Clear felling, normally used for the standardization of other indirect methods. An estimate of the standing crop (W) of an area is given by:

$$W = \Sigma w \tag{1}$$

where w = weight of each individual tree.

2. Trees of mean cross-sectional area are harvested and the mean value of the sample trees ($\bar{w}$) are either multiplied by the number of trees per unit area (N), or the sum of the sample trees Σw is multiplied by the ratio of the basal area of all the trees in the area (G) to the sum of the cross-sectional area of the sample trees (g).

$$W = N.\bar{W} \tag{2}$$

or

$$W = \Sigma w, \left(\frac{G}{\Sigma g}\right) \tag{3}$$

Satoo points out that it is often difficult to obtain sufficient 'average trees', and that a possible alternative is to stratify the trees contained within the sample area into different size classes.

3. Another variation of this last method is to sample trees of different size classes at random and to calculate the total standing crop by means of equation (3).

In the majority of cases where changes in the standing crop are being measured it will be possible to distinguish and use the same limits for sampling, such as the soil surface, at the beginning and the end of the observational period; but situations do occur where this can be difficult unless special methods are adopted. An example is the case of a sand dune where sand may be added to or lost from the soil surface during an experimental period. A similar situation exists upon a growing peat surface (Forrest, 1971), and the problem is most relevant where it is the *Sphagnum* upon the bog surface whose growth is being measured. Clymo (1970) has reviewed and tested a number of methods that are suitable for estimating the growth and production of *Sphagnum*. The growth of *Sphagnum* is predominantly apical and the difficulty is to establish some reference point against which the increase in standing crop can be measured.

Clymo has divided the methods that can be used into four groups:

1. The use of innate time markers. These time markers include cyclic fluctuations in the arrangement of branches and the use of ^{14}C dated peat profiles.

2. The use of reference markers outside the plant. These markers include wires placed upon the *Sphagnum* carpet ('the cranked wire method') enabling relocation and measurement of samples with minimum disturbance, and the use of thread tied around the stems.

3. The use of plants cut to known lengths. A method based upon an increase in length is probably more reliable than one based directly upon an increase in weight as material is likely to be lost from the stems during the observational period.

4. Direct estimates of change in weight. Clymo has developed a method that is based upon weighing the plant under water at the beginning and at the end of the growth period. After the second underwater weighing the plant is dried and re-weighed in air so that the specific gravity (d) can be calculated.

$$d = \frac{D_h}{D_h - W_h}$$

where D_h = dry weight at harvest,

W_h = weight under water at harvest.

If the specific gravity remains constant, and measurements suggest that this is a reasonable assumption, then:

$$D_s = \frac{W_s d}{(d-1)}$$
$$G = D_h - D_{ss}$$
$$= D_h(1 - W_s/W_h)$$

where D_s = dry weight at start,

W_s = weight under water at start,

G = growth (increase in weight).

As long as care is taken to remove gas bubbles, by evacuating the plants, and corrections made to allow for the effects of surface tension on the balance at the air—water interface it is claimed that increases in weight in the order of 2 mg can be measured.

3.3.2 Indirect estimation of the standing crop

In situations such as a forest or woodland the complete harvesting of a series of sample areas, or even a single area, will not be possible in the majority of cases. Under such circumstances alternative methods are needed to estimate the above ground standing crop. If some readily measured parameter such as stem diameter or tree height can be correlated with biomass of harvested samples then the relationship obtained can be used to obtain estimates of the standing crop in other similar areas of vegetation. Such correlations have been used extensively in the study of woodlands and forests and have been reviewed by Newbould (1967) and Satoo (1970). Statistical relationships that are of use in estimating standing crop are based either upon an assumption as to the shape or form of the plant, or upon a direct correlation with weight.

The simplest assumption is that a log or stem is cylindrical in form with a mean cross-sectional area that can be based upon measurements taken at the mid-point or from the two ends of the log. Newbould (1967) suggests that the square root mean (D_m) of the end diameters $(D_1$ and $D_2)$ should be taken.

$$D_m = \sqrt{\frac{D_1^2 + D_2^2}{2}}$$

It has been assumed that the volume of a tree trunk approximates to a parabaloid of rotation, the volume of which is given by:

$$V_p = \frac{\pi r^2 h}{2}$$

where r = radius at breast height,
   h = tree height.

Whittaker & Woodwell (1968) found that the parabolic volume (V_p) was often an over-estimate of the true volume for shrubs (where the radius is measured at the base), while in the case of small trees it was an under-estimate because the radius at breast height was small in relation to the true basal radius. For larger trees they found that the estimates of volume were close to the true values.

Where slices of the stem are available or increment cores can be taken with a Pressler type borer (see Chapter 2, Section 3.2.5) the radial wood increment can be measured. The mean annual increment should be estimated from the previous 5–10 years' growth and a series of measurements made from different points around the circumference of the stem.

Measurements of radial increment can be combined with estimates of basal area to provide basal area increments (A_i).

$$A_i = \pi \left[r^2 - (r-i)^2 \right]$$

where r = radius at base,
    i = mean annual increment.

An estimate of the annual volume increment (V_i) can be obtained from half the basal increment times the height of the tree,

$$V_i = 0.5(A_i h)$$

According to Newbould this estimate is often an under-estimate, the true value is often somewhere between 1·0 and 1·5 times the estimated value. Whittaker & Woodwell found close correlations between the estimated volume increment and wood growth in their studies.

Estimates of rate of wood volume increment can be based upon assumptions of linear or exponential growth.

$$\Delta V = \frac{(v - v')}{n} \qquad\qquad \text{Linear growth}$$

$$\Delta V = V(1 - e^{-r}) \qquad\qquad \text{Exponential growth}$$

where V = volume at time of felling,
    V' = volume n years ago,

$$r = \frac{1}{n} \log_e (V/V')$$

To convert estimates of volume to weight it is necessary to determine the specific gravity. This can be done from samples obtained by destructive sampling, but in many cases it will be necessary to make use of increment cores. When this is done precautions must be taken to avoid compression errors (Stage, 1963; Walters & Bruckmann, 1964).

The choice of a suitable regression model for predictive purposes should be based upon known or reasonably assumed relationships. A model that has been widely employed is based upon the law of allometric growth:

$$\log_e Y = a + b \log_e X$$

where Y = weight of the standing crop, or some component of the standing
    crop or production,
    X = some readily measured parameter of the standing crop,
    a & b are constants.

This type of expression has been used to relate measurements such as basal diameter, or diameter at breast height (DBH) to weight, volume or production (Kittredge, 1944; Ovington & Madgwick, 1959; Whittaker & Woodwell, 1968; Satoo, 1970; Andersson, 1970, and others).

In some cases a more satisfactory estimate can be obtained by including measurements of both diameter and height in the regression:

$$\log_e Y = a + b \log_e (d^2 h)$$

Once a series of suitable relationships have been established it is possible to calculate the weight of the standing crop from the relevant dimensions of the individuals present in the sample area. If the estimate of the total standing crop (W) is obtained from an allometric regression, by taking the sum of the antilogs of the predicted values for the individuals, it will be biassed and an underestimate of the true value. Mountford & Bunce (1972) suggest multiplication by a factor $e^{s^2/2}$ to correct for this bias:

$$W = e^{s^2/2} \sum_{i=1}^{N} e^{a+bx_i}$$

where s^2 = estimated variance about the regression line.

Mountford and Bunce also discuss the calculation of confidence limits for estimates derived from allometric relationships. Corrections for bias in regression estimates after logarithmic transformation have also been discussed by Beauchamp & Olson (1973).

A logarithmic regression cannot be used in situations where the weight of some component, such as fruiting bodies or dead wood, may be absent. In such cases Whittaker & Woodwell (1968) have suggested the use of regressions of the type:

$$W_d = a + bD^3$$

where W_d = weight of dead wood or fruit,

D = branch basal diameter.

Regression techniques can also be of value in non-woodland types of vegetation. Leaf length or tussock diameter might be correlated with standing crop in grassland (Scott, 1961; Mark, 1965) and used to obtain estimates of standing crop between harvests. Bliss (1966) has used such a combination of regression and clipping techniques in the study of an alpine ecosystem in the United States.

3.3.3 Crop meters

Electrical capacitance is a function of the surface area of the capacitor plates, their arrangement and the nature of the di-electric material between them. If a suitable apparatus is placed on the ground so that the vegetation lies between an arrangement of electrodes the resulting capacitance will depend upon the weight and moisture content of the standing crop. Once calibrated such an instrument can provide rapid and non-destructive estimates of the weight of standing crop. The method was first used by Fletcher & Robinson (1956) and other workers have developed the method (Alcock, 1964; Hyde & Lawrence,

1964; Johns *et al.*, 1965). The use of such an instrument for grassland is described by Alcock & Lovett (1967). In practice sub-samples of the vegetation must be taken to correct for moisture content, difficulties may arise if used on very wet ground and in some cases the results obtained may not be sufficiently accurate to justify the use of the apparatus. An alternative approach to the non-destructive estimation of standing crop is the use of Beta-ray attenuation (Mott *et al.*, 1965).

3.3.4 Frequency of sampling

The frequency of sampling that will be required in a production study will depend very much upon the particular investigation. Measurements of the standing crop of an area of woodland made over a period of from 3–10 years will allow the rate of change to be calculated. When this estimate of the rate of change of the standing crop is combined with litter production and grazing losses it will provide an estimate of net aerial production.

On lowland heathlands, where the weight of standing crop varies with age, the sampling programme must cover the whole range of different aged stands of vegetation, and these must be sampled at suitable intervals to build up a growth curve for the standing crop (Fig. 4.5). To do this it is necessary to establish the age of a number of stands of vegetation. This can be done by ring counts from cut stems or increment borings, or from historical records (see Chapter 2). In all cases it is advisable to obtain a number of estimates of age from as many independent sources as possible. In some investigations a single harvest at the end of the growing season will not be sufficient, for example a study of the ground flora in a woodland will require a very different sampling programme than that for the tree species. The phenology of the individual plant species must be considered when designing any sampling programme (Lieth, 1970).

3.3.5 Treatment of samples

All primary production data are expressed in terms of dry weight so that harvested samples must be dried in an oven at temperatures somewhere between 80°C and 105°C. The exact temperature will depend upon circumstances but it is important to dry the material quickly to minimize the loss of weight of organic matter by decomposition. Forced draught ovens are available that allow temperature to be reached quickly and to be maintained throughout the oven (Grassland Research Institute Staff, 1961). When for some reason the plant material is not dried at 105°C it is suggested that sub-samples should be dried at this temperature and a factor used to convert the weight obtained to equivalent dry weights at 105°C.

Having dried and weighed the samples they must then be milled or ground for any subsequent analysis (see Chapter 8, Section 3.4).

3.4 LITTER PRODUCTION

The production of litter by the above ground vegetation represents a major component of the net primary production, and its measurement is important whether it be in relation to primary production, or for consideration of other relationships within the ecosystem. Bray & Gorham (1964) have examined the production of litter within forest ecosystems and their review should be consulted by anyone concerned with litter production. The mineralization and release of plant nutrients within the litter layer are processes in which the soil fauna play an important part, and it is easy to see that the soil zoologist and the production botanist have many problems in common. It has been shown that production, decomposition and accumulation are inter-related (Section 3.1.3) but for the sake of convenience the methods involved in the estimation of litter production will be dealt with in this section and those for litter decomposition and accumulation in Section 3.5.

Litter production can be defined as the weight of dead material (of both plant and animal origin) that reaches unit area of the soil surface within a standard period of time. All material that dies does not immediately fall to the ground, and in some types of vegetation, such as grassland and other tussocky plant communities, the litter layer may contain only a small proportion of the dead material. It was to distinguish between the different locations of dead organic matter in the ecosystem that the term standing dead was introduced (Odum, 1960; Gore & Olson, 1967; Forrest, 1971).

3.4.1 The direct estimation of litter production

Different types of litter trap have been described in the literature, but as has been stated previously the design of a piece of apparatus for a particular project must depend upon the special requirements of the individual problem. Whatever the local conditions a litter trap must fulfil a number of basic requirements:

1. The trap must intercept the litter fall before it reaches the ground with as little aerodynamic disturbance as possible.

2. The trap must retain material once it has been trapped.

3. Trap should be designed or placed so that litter already on the soil surface cannot enter the trap.

4. Water must be allowed to drain from the trap without loss of litter (especially fine litter material).

5. The size and number of traps must provide an estimate of the required degree of accuracy.

A selection of litter traps is shown in Fig. 4.8.

One difficulty that arises in the use of litter traps especially in open or exposed situations is the failure to trap, or the subsequent loss of trapped material from the traps. Chapman *et al.* (1975a) have found a good correlation between the numbers of *Calluna* seed capsules in litter traps and the numbers produced per unit area of older and denser heathland vegetation, but in younger and more open stands the numbers of capsules trapped were significantly lower than the numbers produced by the vegetation. Wherever possible some similar check upon the efficiency of a litter trap should be made.

In some cases the amount of litter being added to the soil surface each year will include a 'blow in' and a 'blow off' component; and it is important to consider whether any such lateral movement of material within the ecosystem

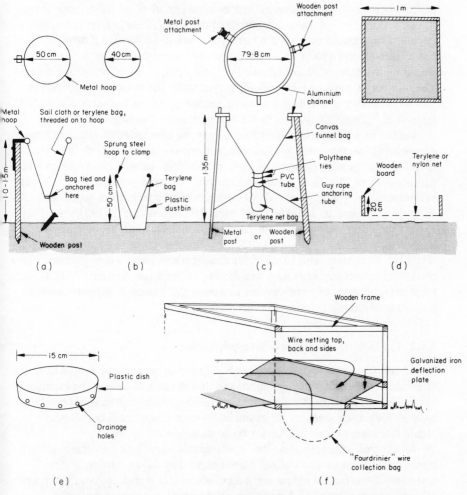

Figure 4.8 A selection of litter traps suitable for use in different types of vegetation (a to d from Newbould, 1967, e after Chapman, 1975a, f after Woodell (pers. comm.))

is likely to be important. S.R.J. Woodell (pers. comm.) has used suitable traps (Fig. 4.8f) to study the movement of litter in a narrow band of old beechwood in Oxfordshire, and has been able to measure the amounts of litter entering and leaving the wood as a result of wind action.

The length of time between emptying traps should be kept as short as possible. Weekly collections are desirable, especially in damp and humid areas or where the breakdown and leaching of litter may be appreciable between trapping and collection. The phenology of the particular plant species present will be important when deciding upon the actual timing and frequency of emptying the traps. It is suggested that estimates of litter production should be based upon observations carried out over a period of at least three years. Sykes & Bunce (1970) have demonstrated the fluctuations in litter fall within a deciduous woodland over such a three-year period. Medwecka-Kornás (1971) has suggested that a suitable level of accuracy for estimates of litter production would be to obtain 95% confidence intervals equal to about 5% of the mean. The number of traps be required will vary, and an initial sampling period may well be necessary to establish the actual number required, but in most cases 20 traps at each sampling site will be the sort of number required.

Under some types of vegetation it may be necessary to estimate some components of the litter by independent means and combine the results to obtain an overall estimate of litter production. In woodland ecosystems components of the litter such as branches, or even whole trees, will not be measured by trapping; and it will be necessary to record the appearance of these larger items in suitably sized plots (Healey & Swift, 1971). As they appear they must be removed, or labelled so that they can be recognized at a later date.

When litter material is required for chemical analysis it will be necessary to prevent chemical contamination and fouling by birds (see Section 4.2.1). If the traps are not made of suitably inert materials the frames or supports must be painted with bitumastic or some other suitable paint.

3.4.2 The indirect estimation of litter production

It will soon become apparent that it is difficult or almost impossible to trap litter under some types of vegetation, and that in some situations the estimates obtained from litter traps will not be very reliable, younger stands of heathland have already been given as an example. In such cases it will be necessary to obtain estimates of litter production by some indirect method.

As leaves senesce and die considerable amounts of material are translocated to other parts of the plant before they fall as litter. It is therefore extremely important to assess the magnitude of this translocation if estimates of the standing crop of leaves are to be used to predict potential litter production. An estimate of the weight of material lost from a leaf before it falls as litter can be obtained from a comparison of the dry weight/surface area ratios of live and dead leaves. Such comparisons assume that no changes in surface area take place but this assumption can be checked against a series of measurements on individual marked leaves. In some cases it may be more satisfactory to compare dry weight/leaf length ratios to obtain an estimate of the weight lost by reabsorption.

In cases where the green matter produced in a single season is not all shed before the next growing season it is necessary to develop a model that takes into account the life expectancy of the green material, and to include this in the estimation of potential litter. Estimates of litter production by *Calluna* heathland can be checked by such an indirect method (Chapman *et al.,* 1975a). Much of the green matter produced by *Calluna* is in the form of lateral short shoots, these short shoots either drop as litter at the end of the growing season or persist to grow and increase in length during the next season. The growth increments can be recognized, so that the age of an individual shoot can be determined, and a partial estimate of the potential litter production obtained from the difference between the weight of green shoot material present in one year and the weight of the 'old short shoots' remaining at the end of the following year. This difference in weight of green material must be corrected for loss of weight due to translocation (approximately 20%) and added to the weight of the flower and woody components of the litter production. Bray & Gorham (1964) and Burrows (1972) report similar losses of about 20% in the dry weight of leaves from a variety of plant species before litter fall. If it can be assumed that the production of green matter is constant from year to year then the annual loss of green matter can be estimated from the difference between the total weight of green shoots and the weight of old green shoots present at the end of the growing season. If measurements of the amount of shoot material that have been consumed by grazing animals have been made they should be incorporated in the calculation of litter production.

3.5 DECOMPOSITION AND ACCUMULATION OF LITTER

The rate of accumulation of litter upon the surface of the soil is the result of the interaction between litter production and the rate of litter disappearance.

Litter disappears from the ecosystem as a result of combined losses due to decomposition, mineralization, leaching, animal consumption, wind transport and in some cases harvest by man (Medwecka-Kornás, 1971, Anderson, 1973b). In some cases the disappearance of litter can be equated with decomposition, but this is not necessarily so and it may be necessary to apply a correction to the rate of disappearance to obtain an estimate of the rate of litter decomposition. As the accumulated litter is the result of a series of annual inputs combined with the progressive decomposition of the litter layer it can be divided into a number of horizons depending upon the degree of decomposition or humification.

3.5.1 Quadrat methods for estimating litter disappearance

A direct approach to the measurement of litter disappearance has been used by Wiegert & Evans (1964) in their study of old field systems in Michigan. They developed a technique where a series of 'paired plots' were selected, and in which individual pairs of plots were assumed to be identical in terms of weight and composition of the litter present. At the start of the experiment (t_0) all dead vegetation present on one of the plots was removed and weighed (W_0). To prevent any more dead material being added to the plots during the observation period all the green matter was removed from both the plots. After a suitable period of time (t_1) the litter on the second plot was sampled and weighed (W_1). The rate of disappearance of litter (r) was calculated assuming an exponential rate:

$$r = \frac{\log_e (W_0/W_1)}{t_1 - t_0}$$

This basic method has been modified by Lomnicki *et al.* (1968) where they have avoided the assumption that decay remains constant despite the fact that the live vegetation has been removed from the plots. At the beginning of the experiment (t_0) they sampled the dead material (W_0) from one of the plots but did not remove any of the live vegetation. At the end of their observational period (t_1) they sampled the dead material from both of the plots. The weight of dead material collected from the previously sampled plot they designated (h) and that from the unsampled plot (g) it follows that:

$$g = W_0 + h - (W_0 - W_1),$$

and

$$W_1 = g - h.$$

This modification of the basic method of Wiegert and Evans assumes that the litter produced during the observational period $(t_1 - t_0)$ is not affected by the removal of the dead material from the plot. Lomnicki and his associates compared their techniques with the original method and found good agreement between the two procedures, but claim that their modification of the method is more convenient to use. The combination of their parameter (h) representing the weight of litter produced during the observation period can be combined with the change in standing crop (b) to provide an estimate of net aerial production (P_n).

$$P_n = h + (b_1 - b_0)$$

To obtain this estimate of net production (h) should be measured over suitably short periods and the results summed to provide an annual estimate. This method can be applied to a variety of vegetation types especially those where standing dead is an important component of the ecosystem.

3.5.2 The measurement of disappearance from contained litter samples

If samples of litter, either freshly produced or collected from the litter layer, are contained in some way that makes it possible to retrieve them at a later date then the rate of disappearance of the litter can be measured directly. Commercially available nylon hairnets have been employed for this purpose by Bocock & Gilbert (1957), Bocock *et al.* (1960) and Bocock (1964). Nylon net bags, made from ballroom dresses, were used by Shanks & Olson (1961) to demonstrate that the rate of loss of weight by leaves on the soil surface was dependent upon the plant species involved, their chemical composition and the prevailing climatic conditions.

Edwards & Heath (1963) have shown that the size of the mesh used in nylon bag experiments is important because it controls the types of organism that can enter the bags and participate in the decomposition process. In their experiments they assessed the rate of decomposition by estimating the area of leaf discs (2·5 cm in diameter) that had been lost after a given period of time. Results of the effect of mesh size upon the rate of breakdown in litter bag experiments have also been reported by Bocock (1964) and Anderson (1973b).

The choice of a particular size of mesh for litter bag experiments must take into account the type of litter under investigation. In an experiment on the breakdown of *Calluna* litter by Chapman (1967) it was found that a mesh larger than 2 mm could not be used as it was unable to retain the relatively small sized heather shoots, but that this size of mesh probably did not exclude any important component of the heathland soil fauna.

Where the object of the investigation is to study the loss of weight of even aged or 'fresh litter' then plant material derived from current year's growth is used to fill the litter bags. An alternative approach is to fill the bags with material collected from the accumulated litter on the soil surface, when the sample will be composed of organic matter of different ages. The results will provide an integrated estimate of the rate of loss of organic matter from the whole litter layer, but as stated this estimate will only represent disappearance of litter and will be greater than the equivalent loss of carbon from the ecosystem as carbon dioxide.

In some cases significant weights of animal material may enter the litter bag and unless removed will affect the results obtained. Nylon litter bags have in fact been used by a number of workers to study the invasion of plant litter by soil organisms (Crossley & Hoglund, 1962).

The rate of disappearance of litter from mesh bags is estimated from random samples taken at intervals throughout the experiment and dried in an oven. It is important to make sure that all the bags contain the same dry weight of litter at the start of the experiment and as the litter bags need to be sampled several times throughout the experiment it means that sufficiently large numbers of the bags must be set out at the beginning of the exercise. It is possible that the method used by Clymo (1970) to estimate the initial dry

weights of shoots of *Sphagnum* in growth experiments (Section 3.3.1) might be
used to obtain estimates of the initial dry weights of litter bags.

3.5.3 The measurement of litter disappearance from labelled or tagged samples

An alternative to the use of litter bags is to mark or label individual fragments
or items of litter so that they can be retrieved at a later date. An example is the
work of Frankland (1966) where she studied the decay of bracken *(Pteridium
aquilinum)* by labelling individual petioles with plastic labels. Latter & Cragg
(1967) marked 200 individual leaves of *Juncus squarrosus* with paint in a
study of decomposition that also used litter bags. Hayes (1965a,b) attached
coniferous leaf litter to lengths of nylon thread to enable him to relocate
individual leaves, and compared results with those obtained from litter
bags. He found lower rates of disappearance from the litter bags as fragments
were not so easily lost from the experimental system.

Murphy (1962) used a paint containing the radioisotope Tantalum-182 to
label leaves while they were still on trees, and was able to relocate individual
leaves for up to 12 months after they had fallen as litter. He measured
decomposition directly from the loss of weight.

Olson & Crossley (1963) combined the use of nylon litter bags with the
application of radioactive isotopes in a study of forest litter decomposition.
They introduced radioactive tracers into the trees, producing radioactive
leaves and subsequently radioactive litter. The litter was placed in nylon mesh
bags 10 cms square that were then contained in plastic sandwich boxes with
holes in the sides and glass fibre mesh at the base. The boxes were placed on
the forest floor and counts of the radioactivity of the bags enabled the amount
of radioactive material that had been transferred to the forest floor to be
calculated.

3.5.4 Respirometry and litter decomposition

It has already been shown that the loss of weight from a litter bag, or from the
litter layer itself cannot necessarily be equated with the mineralization and loss
of carbon from the ecosystem in the form of carbon dioxide. It would seem
that the obvious approach to this problem would be to measure the uptake of
oxygen, or the evolution of carbon dioxide by respirometry. Parkinson &
Coups (1963) measured the respiration of soil and litter that had been placed
in specially designed respirometer flasks. Howard (1967) has described an 'experimental tube' (Fig. 4.9) that can be placed upright in a box in the field and
protected from rain, but kept moist by watering with distilled water when
necessary (about 5 ml every 2 weeks). These tubes can be returned to the
laboratory at intervals and connected to a Gilson respirometer in a flask of the

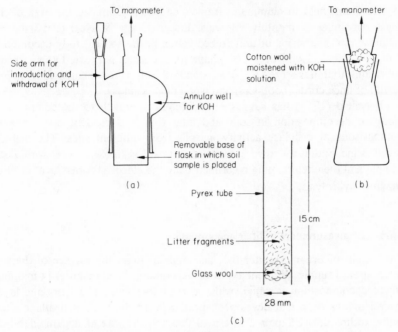

Figure 4.9 Flasks and container enabling litter samples to be placed in respirometer for measurement of respiration.

type illustrated, and at the end of the experiment the contents of the tubes can be dried, weighed, ashed and the weight of organic matter lost during the experimental period calculated.

If sufficient attention is paid to the moisture content and aeration of the experimental material (Ross & Boyd, 1970) it would seem that the main disadvantages of respirometry as a method for the study of litter decomposition are disturbance and the isolation of the litter from the natural litter layer. If these limitations are recognized, respirometry can provide useful estimates of carbon losses from the litter layer. The direct estimation of the amount of carbon dioxide evolved from the soil and litter layer can be measured by other methods but these are described in Section 3.6.7.

3.5.5 Comparative studies of litter decomposition

In many cases the object of a decomposition study is to investigate litter decomposition at one particular type of site, but as the rate of decomposition is dependent upon both the type of organic matter and upon other site factors, an alternative approach is to investigate the decomposition of a standard substrate at a number of different sites. Golley (1960) buried cellulose sheet

(5 cm × 5 cm) held in aluminium frames, and then estimated the area of the sheet remaining at monthly intervals. Latter, Cragg & Heal (1967) buried cellulose film and strips of unbleached calico in the soil to study decomposition, and Went (1959) used cellophane for the same purpose. Other organic substrates that have been used as standard substrates include filter paper, cotton or linen cords (bootlaces) and wood pulp cellulose.

Benefield (1971) has suggested a different approach to the comparative study of decomposition in soils and litter that is dependent upon making comparisons of cellulase activity in soils from different sites. The method involves incubating a known amount of standard cellulose powder with a soil sample and then determining colorimetrically the glucose formed as a result of enzymic hydrolysis.

3.5.6 The measurement of litter accumulation

The weight of organic matter that accumulates upon the surface of the soil varies greatly under different types of vegetation. Accumulation is minimal where decomposition is rapid, while at the other extreme an organic layer several metres deep can develop on peatlands. In the case of peatland sites dating techniques (Chapter 2) are available and the rate of accumulation of organic matter with time can be investigated (Durno, 1961; Clymo, 1965).

Direct estimation of the weight of accumulated litter can be difficult when there is no well defined boundary between the soil and litter layers, but in most cases it is possible to define practical limits that make reproducible sampling possible. It is often most convenient to combine sampling of the litter layer with that of the standing crop, especially where the litter is to be sampled by quadrats.

The accumulation of litter on the soil surface can also be sampled with a corer or tubular sampling tool. Capstick (1962) used samplers that had different cross sectional areas (1·0, 2·63, 8·5 and 41·5 cm²) to sample the litter layer in a forest ecosystem, and found that a 1·0 cm² sampler cut through the litter layer easily and with little disturbance of the experimental area. The results from this size of sampler showed only a low degree of scatter and agreed well with estimates derived from larger samples. He found inconsistencies when trying to follow changes in the weight of litter after litter fall, but these were just as likely to occur with samplers of a larger size. Frankland *et al.* (1963) used a corer 81 cm² in cross sectional area, to sample the litter layer from woodlands in the Lake District, but found it difficult to demonstrate any significant differences between their sampling sites. Before using a corer to sample the litter layer it is important to carry out a trial sampling programme to establish the size and number of samples that may be required; or whether such a method is even suitable for the investigation about to be undertaken.

The litter layer will often contain considerable amounts of mineral soil and to make the estimates of litter accumulation comparable with those obtained for litter production and above ground standing crop it is advisable to determine the loss on ignition (L.O.I.) of the litter samples and to express the results on an ash free, or a standard (e.g. 5%) ash weight basis.

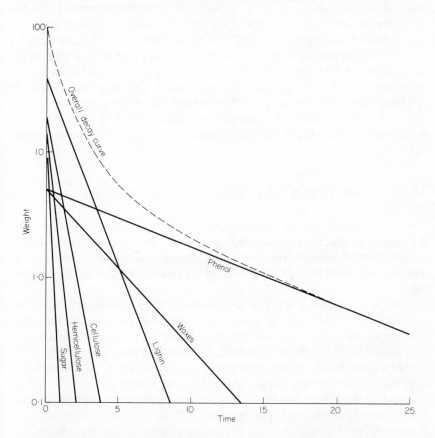

Figure 4.10   The relationship between the overall decomposition curve of organic matter and the exponential decay curves of the individual constituents. (Based on data from Minderman, 1968.)

Root growth within the litter layer can be a problem when one is trying to estimate the weight of litter that has accumulated at some sites, Chapman *et al.* (1975b) found it to be an important factor when dealing with the litter layer from older stands of *Calluna* heathland. In such cases some estimate of the root content of the litter layer and its contribution to the accumulation of litter must be made.

Jenny *et al.* (1949) and Olson (1963) have described and discussed litter accumulation within forest ecosystems by means of a simple exponential model (see Section 3.1.3). While the assumptions assumed in this model make it convenient for the purposes of computation, and may be reasonable in some situations, the processes of decomposition and accumulation cannot always be described adequately by such a simple model. Minderman (1968) has stated that 'it is difficult to give a quantitative gravimetric picture of this phenomenon', and has shown that if the different chemical constitutents of the litter decay exponentially, but at different rates then the overall decay curve for the litter will be curved when plotted on a semi-logarithmic basis (Fig. 4.10). This departure from the predictions of the simple exponential model is most important where decomposition is rapid, but is also dependent upon the chemical composition of the litter being studied. The effect of the assumptions, that are implied when the simple exponential model is used to make predictions, is discussed in relation to roots and soil organic matter in Section 3.6.6.

3.6 ROOTS AND SOIL ORGANIC MATTER

The importance of root production need not be emphasized, comparisons of the relative productivity of different ecosystems or studies of their nutrient budgets that are based solely upon above-ground data are so incomplete that one may well doubt the validity of many of the conclusions that are drawn from them.

The same principles apply to the estimation of root production as to the measurement of above-ground production except that the practical problems are very much greater. The information required for the direct estimation of root production is the change of weight of root and the amount of root that has died in a given period of time. When some part of the above-ground standing crop dies it generally falls to the surface of the soil as litter or can be identified as standing dead, in either case it can be measured. When part of the root system dies it is retained and is often difficult or almost impossible to distinguish from the living roots. This factor combined with the difficulty of sampling root systems will help to explain the relative lack of data available about root production. In 1964 Olson stated that 'the kinetics of underground production and loss rates remain as one of the most challenging ecological and agricultural problems during the half century to come'. Despite the fact that about one fifth of that half century has now passed it is probably almost as true a statement as it was in 1964.

The problems involved in the estimation of root production have been described by Newbould (1967), Milner & Hughes (1968), Ghilarov *et al.* (1968) and Head (1971). The relationship between root production and net-productivity has been reviewed by Bray (1963). Kononova (1961) has dealt with many aspects of soil organic matter in his book entitled *Soil Organic Matter*. As the problems involved in the estimation of root production are so closely linked to those of soil organic matter they are considered jointly in this section.

3.6.1 Estimation of root biomass and examination of root systems

The estimation of root biomass is entirely dependent upon the sampling and extraction of roots from soil samples or from the soil profile, both of which are reviewed by Schuurman & Goedewaagen (1971) and Lieth (1968). The methods of sampling fall into three main categories, the exposed soil face, soil cores and blocks, and the excavation or exposure of the root system *in situ*.

The use of soil pits involves considerable effort and labour and is only of use where few replicates are required or plenty of labour is available. The exposed soil face can be sampled as an intact monolith by using a specially made container (Chapter 6), or use can be made of a 'pinboard' (Schuurman & Goedewaagen, 1971; Ashby, 1962; Schuster, 1964; Sheikh & Rutter, 1969). The pinboard consists of a baseboard of suitable size with wires or nails protruding from one side at closely and regularly spaced intervals. The board is pressed against the face of the soil pit so that the nails are pushed into the soil and hold the root system in place when the soil particles are washed away. If necessary a car jack can be used to press the board into place. At times it may be possible to use jets of water from a hose and motor driven pump to expose the root system of an area of vegetation or an individual plant by direct washing in the field.

In the majority of investigations it will probably be found that some form of soil corer or auger will be the most convenient way of obtaining soil and root samples. A great variety of corers and augers are described in the literature, and as is usually the case when a whole range of variations in a particular technique exists, it means that no single corer will be suitable for all situations. Corers vary from relatively simple tubular devices to complex power operated augers (Schuurman & Goedewaagen, 1971; Kelly *et al*. 1948; Welbank & Williams, 1968). The particular type of corer chosen for a particular project will depend upon the type of soil, the diameter and depth of the core required.

Root growth has been studied by the installation of inspection windows against the sides of soil pits (Ovington & Murray, 1968; Rogers & Head, 1968). Periodic observation or the use of time lapse photography has enabled the periodicity of root growth to be studied and for the life history of individual roots to be followed. While a great deal of information has, and can be obtained in this way it is very difficult to express the results in terms of root production.

Roots can also be studied *in situ* by the preparation of thin sections of soil, and a number of techniques have been described for the preparation of soil sections (see Chapter 6). Not many of the methods described were intended, or have been used, for the examination of roots, but Sheikh & Rutter (1969) have used soil sections to investigate the distribution of roots in relation to the size of the pore spaces in the soil The methods of Haarlov & Weis-Fogh (1953), Minderman (1956), and that of Anderson & Healey (1970) have much to recommend them for the purpose of examining roots. These methods use gelatine or agar for embedding so that the soil does not have to be dried, as when being impregnated with resin, and means that the roots and soil animals can be seen with great clarity. Some difficulty may be encountered when cutting sections of soil embedded in gelatine if mineral grains are present in any quantity, but this problem was overcome by Minderman where he used hydrofluoric acid to dissolve away sand grains between the stages of embedding and sectioning his soil samples.

The length of larger roots can be measured from photographs or shadowgrams prepared from the extracted roots. The measurements can be made with a wheel-type measurer. Newman (1966) has described a method of measuring the length of fine roots. The roots contained in a transparent flat bottomed dish are placed over a base marked with a series of randomly placed straight lines. It can be shown that whatever the shape of the root, an estimate of its length is given by:

$$R = \frac{\pi N A}{2H}$$

where R = total length of root sample,

   N = number of intersections between roots and straight lines,

   A = area of rectangle inside which lines are drawn,

   H = total length of the straight lines.

3.6.2 The extraction of roots from soil samples

Once the soil samples have been collected it remains to separate and extract the roots. Pinboard samples are immersed and allowed to soak in water. The water level is then maintained just below the surface of the sample and sprinkling is carried out either by hand or by mechanical means to wash the soil material away, leaving the roots in place between the nails or pins.

Samples obtained by corer or auger are generally divided into sub-samples representing different depths or soil horizons. The relative ease or difficulty that will be encountered in extracting the roots will depend very much upon the particular type of soil. Pre-soaking the sample in aqueous sodium pyrophosphate (270 g per 100 litres) or 1% sodium hexametaphosphate may help to disperse some soils. Dahlman & Kucera (1965) followed such a treatment by soaking in 0·8% sodium hypochlorite to assist in separating the clay and humus fractions from the roots. The effects of any such pre-treatment, as well as soaking and washing in water should be considered if root samples are required for any subsequent chemical analysis.

The separation of roots from the remainder of the soil can be done by one of two main methods. The sample can be placed over a nest of sieves of different mesh sizes and washed, or it can be subjected to some sort of flotation or elutriation technique. In the first case the coarser soil material will be retained and have to be sorted by hand while the roots will be collected upon a number of different sized sieves. With flotation methods the soil organic matter, roots and finer mineral fractions will be separated from and leave behind the larger and denser mineral fractions.

Small samples can be placed in a litre beaker or graduated cylinder, and allowed to soak. After stirring and dispersion the roots, organic matter

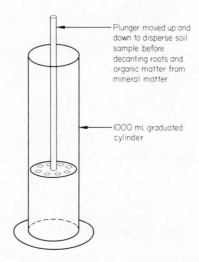

Figure 4.11   Simple apparatus for separating samples of soil and roots.

and finer soil material can be separated by decantation. The roots can then be separated by washing over a nest of sieves. A simple piece of apparatus for separating roots by this method is illustrated in Fig. 4.11.

The flotation method of root separation can be mechanized to make it suitable for use with larger numbers of samples, and make the process more reproducible by giving the samples a more uniform treatment. Apparatus has been described by McKell *et al.* (1961) and Cahoon & Morton (1961), both of which are shown in Fig. 4.12, and operate by jets of water entering the bottom of the apparatus, creating a vortex that carries the roots, organic matter and less dense fractions upwards and out into a nest of sieves. It has been suggested by Chapman (1970) that ultrasonic cleaners might be of use in 'cleaning' the root fractions collected on the nest of sieves.

Other flotation methods such as the one described by Salt & Hollick (1944) for the extraction of soil fauna may sometimes be useful for the extraction of root material. The flotation is carried out in concentrated solution of magnesium sulphate (specific gravity *c.* 1·2), sodium chloride, potassium bromide or zinc chloride (see Southwood, 1966) and as air is bubbled through the solution from the bottom of the container (the Ladell can) the surplus liquid and organic matter pass over the lip and are filtered through bolting silk. The animals can then be separated from the other organic matter by differential wetting in xylene or benzene.

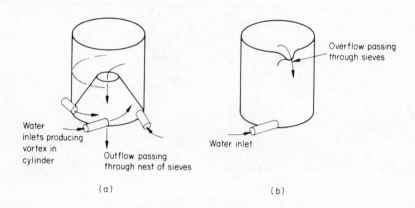

Figure 4.12 Apparatus for extracting roots from soil samples by elutriation. (After (a) Cahoon & Morton, 1961, and (b) McKell *et al.*, 1961.)

Froth flotation has been used to extract sub-fossil seeds from soils in archaeological studies (Charles, 1972). A suitable chemical (the collector) is chosen for which the surface of the material to be extracted has a preferential adsorption so rendering it water repellent and aerophilic. When air is passed through a suspension of mixed solids those particles with aerophilic surfaces will be raised into a stabilized froth from which they can be separated. In the case of the sub-fossil seeds cresol or kerosene was used as the collector and pine oil (terpineol) as the frothing agent.

Having separated the roots by one means or another they can be sorted into different size classes, dried and weighed. However carefully the roots have been washed they will still contain a certain amount of mineral soil contamination and it will be necessary to ash the dried root samples and to express the results on an ash-free dry weight basis, or to correct them to some standard ash weight content (i.e. 5%, Newbould, 1967).

3.6.3 Separation of live and dead roots

One of the greatest problems in calculating root production from root biomass data is that of distinguishing live from dead roots. Jacques & Schwass (1956), and Aimi & Fujimaki (1958) have tried to use tetrazolium salts to distinguish live roots but the success of such a method depends upon the individual root system and must by its nature be tedious to apply, especially to very fine roots. A number of workers have based their separations of live from dead roots upon visual assessments, but this is very difficult especially with the fine roots. Konova (1961) describes a flotation method (ascribed to Pankova) for the separation of roots in which it is claimed that living roots can be separated from dead roots by repeated elutriation. Sator & Bommer (1971) tried both chemical and enzymatic methods to differentiate living from dead but reported little success. They were of the opinion that radioactive labelling techniques were the most promising.

3.6.4 Radioisotopes and root studies

Radioisotopes have been used in root studies to label live roots, to estimate rates of decay and turnover times, and to follow the transfer of materials from one part of the plant to another.

The root system can be labelled by the incorporation of ^{14}C into the plant through the leaves, either by the foliar application of a substance such as urea (Yamaguchi & Crafts, 1958; Nielson, 1964) or by the photosynthetic assimilation of $^{14}CO_2$ (Ueno *et al.*, 1967; Dahlman & Kucera, 1968; Ellern *et al.*, 1970). The assimilation takes place within some form of enclosure covering the vegetation or an individual leaf, and after a suitable period the roots can be sampled and their activity measured either by counting or by autoradiography.

An alternative technique is the injection of an isotope into the soil at selected depths, and to establish the pattern of root activity by sampling the above ground vegetation for radioactivity. Boggie *et al.* (1958) used ^{32}P as a tracer to investigate the rooting depths of plants growing on blanket peat by this method. They describe an improved technique for placing the tracer at different depths in a later paper (Boggie & Knight, 1962).

The root extension of forest trees (Hough *et al.* 1965; Ferrill & Woods, 1966) and the rooting pattern of forest understory species (Nimlos *et al.*, 1968) have been studied using iodine-131 as the tracer. When using labelled iodine it has been reported that prior sterilization of the soil with methylbromide around the point of application increases the rate of absorption of the isotope (Woods *et al.*, 1962).

Phosphorus-32 and Calcium-45 have been injected into the stumps of tree species (Woods & Brock, 1964; Woods, 1970) to demonstrate the transfer of labelled material from the roots of one species to another in the ecosystem. It was concluded that root exudates or mutually shared mycorrhizal fungi were probably the important factors in the transfer rather than root grafts.

The problem of root exudates and their contribution to the total net primary production is generally recognized but in most cases it is then conveniently forgotten. The reasons for this are the practical difficulties that remain to be solved. The use of radioactive tracers will undoubtedly figure prominently in the solution of the problem when it is achieved. To date most of the work on root exudates would appear to have been confined to studies carried out under sterile conditions in the laboratory and the extension of the work to field conditions is a major step.

Dahlman & Kucera (1968) have used $^{14}CO_2$ to label the root system of grassland vegetation, and by following the reduction in the radioactivity of the roots over a period of about two years they have obtained estimates of a root system turnover time of just over four years. Caldwell & Camp (1974) have followed the dilution of C^{14}/C^{12} ratios in structural carbon of root systems to estimate turnover times and root production.

Radioactive tracers will be found to be of greater use in the study of roots of some types of vegetation than in others and they do not provide a universal panacea to all root production problems. It must also be remembered that radioactive materials can only be used in accordance with local regulations.

3.6.5 The relationship between root biomass and production

The relationship between biomass and production that applies to the above-ground component of the vegetation applies equally to the root system but it is more difficult to apply.

$$P = \Delta B + L + G$$

Other expressions relating root production to biomass have been used or suggested, such as:

$$\text{root production} = \frac{\text{max. root biomass}}{\text{turnover time}}$$

The turnover time is the time required for the decomposition of a weight of organic matter equal to the weight of the root biomass (the reciprocal of the decomposition rate K) so that the expression is the same as:

$$L = kX_{ss}$$

and therefore assumes steady state conditions.

Remezov *et al.* (1963) have suggested a slight modification of this type of expression to calculate the annual loss of roots.

$$R = W_a + \frac{w - w_a}{n}$$

where R = annual loss of roots,

W_a = weight of roots of annual species,

W = total weight of roots,

n = average length of life of roots of perennial species.

Newbould (1967) suggests that:

$$\frac{\text{above ground production}}{\text{above ground biomass}} = K \times \frac{\text{below ground production}}{\text{below ground biomass}}$$

but points out that as few accurate estimates are available for root production it is difficult to estimate the value of K, but he suggests that the further assumption that K is equal to unity might be made until better estimates are available.

Dahlman & Kucera (1965) measured the annual increment of the root system of prairie vegetation by taking the differences between the maximum and minimum values that were obtained for the root biomass during the year. The periods of greatest difference were from April to July in the A_1 horizon, from July to January in the A_2 horizon and from July to October in the B_2 horizon. By making the assumption that these increments could be equated with root production they calculated that approximately 25% of the root system would be replaced each year. As production and decomposition are simultaneous processes their estimates of root production must therefore be minimum values. In some other ecosystems, such as *Calluna* heathland (Chapman, unpublished), it has not proved possible to demonstrate significant seasonal variations in the root biomass. The differences are either small compared with the variance of the mean or production and decomposition proceed in such a way that the net root biomass remains relatively constant.

3.6.6 The accumulation of organic matter in seral sites

There are obvious advantages, to the ecologist, where an ecosystem can be assumed to exist in a steady state condition, but in many cases this is clearly not true, and although the investigation may be complicated because the system is changing with time it can be an advantage when studying some aspects of the ecosystem.

Where distinct and dateable stages in a succession exist, such as the series of glacial moraines studied by Crocker & Major (1955) and Crocker & Dickson (1957), or the sand dune systems described by Salisbury (1922, 1925), Burges & Drover (1953), Olson (1958) and Wilson (1960), it should be possible to measure the rate of certain parameters and to obtain estimates of the rates of input and loss from the system. One way that this might be achieved would be to fit accumulation curves of the type described by Olson (1963), but estimates obtained in this way would be only approximate and subject to the assumptions of constant rates of input and decomposition. Objections to the assumption of constant decay rates have been discussed in Section 3.5.6, and if a monomolecular accumulation curve is fitted to the hypothetical data given by Minderman (1968) it will be found to produce an under estimate of the true input to the system.

3.6.7 The measurement of soil respiration and carbon dioxide under field conditions

Root and litter production are the main sources of organic matter input to the soil. The total metabolism of the soil has been reviewed by Macfadyen (1971), and the evolution of carbon dioxide is one method that has been suggested, by a number of workers, as a measure of this total metabolism. Where it has been measured it is generally found that the production of carbon dioxide is in excess of the amount that can be attributed to litter production and the decomposition of organic matter (e.g. Wanner, 1970).

The sources of carbon dioxide evolution from the soil are shown in Fig. 4.13 and are from decomposition processes and from respiration by live roots. The carbon sources for the decomposer cycle are mainly litter and root production. If the soil ecosystem can be assumed to approximate to a steady

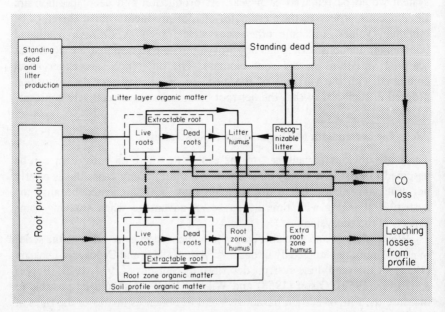

Figure 4.13 Flow diagram to show the inputs of organic matter to the soil and the sources of CO_2 evolved from the soil.

state condition then the annual carbon dioxide production will be proportional to litter and root production, plus root respiration. In such an argument it is important to consider soil respiration on an annual basis so that differences due to the translocation of organic matter to and from the roots are largely cancelled out. If it were possible to subtract the amount of carbon dioxide due to root respiration from the total amount of carbon dioxide produced annually in the soil it would provide some form of estimate of the input of organic carbon to the soil. Estimates of root respiration derived from excised roots (Crapo & Coleman, 1972) have obvious disadvantages, but might be used to make such calculations in the first instance. Kucera & Kirkham (1971) have suggested an alternative approach, if estimates of total soil respiration can be obtained from a series of soils that contain different combinations of weight of root and organic matter then it should be possible to apportion the total carbon dioxide to its various sources.

A number of methods have been used to measure soil respiration under field conditions (Witkamp, 1966; Witkamp & Frank, 1969; Howard, 1966; Kosonen, 1968; Brown & Macfadyen, 1969; Wanner, 1970; Kucera & Kirkham, 1971; Chapman, 1971, and others).

The actual methods used can be classified according to the methods used to collect the gas samples, and to estimate the carbon dioxide produced (Bowman, 1968). In the majority of cases some form of container ('open box' Witkamp, 1966) is placed over the soil to collect the carbon dioxide, and it is preferable that it is set into the soil some time prior to the measurements being made. The box or cylinder is then left in place so that the effects of disturbance are minimized, and is covered with an airtight lid while actual measurements are being made.

The carbon dioxide evolved has been measured by means of infra-red gas analysers (Witkamp, 1966; Witkamp & Frank, 1969; Reiners, 1968), by absorption in soda-lime (Howard, 1966), in barium hydroxide (Witkamp, 1969), or by absorption in potassium or sodium hydroxide (Witkamp, 1966; Brown & Macfadyen, 1969, and Chapman, 1971). Where the carbon dioxide has been absorbed in potassium hydroxide it has generally been by diffusion (Conway, 1950), and determined by titration.

Titration of the carbon dioxide evolved has the disadvantage of providing only one determination at the end of the observation period, and the rate is therefore calculated from this one single reading. Chapman (1971) has described a soil respirometer operating on the same principles as those just described but in which the carbon dioxide absorbed can be determined from readings of the conductivity of the electrolyte taken at intervals throughout the period of measurement.

A modification of this conductiometric method is to build the electrode system into a syringe (Fig. 4.14) so that simpler containers for the electrolyte can be used. If the lids of commercially obtainable plastic pots are used to cover the respirometer cylinders a great number of units can be used at any

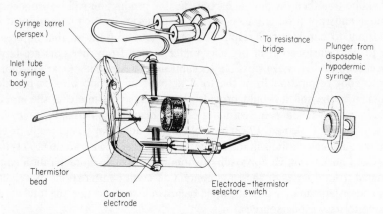

Figure 4.14   A syringe conductivity cell for use with a simple soil respirometer.

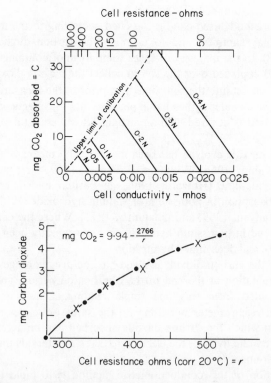

Figure 4.15   Calibration curves for conductimetric soil respirometer showing the relationship between CO_2 absorbed, conductivity and strength of (KOH). (From Chapman, 1971.)

one time. The electrolyte is drawn up into the syringe and the cell resistance and temperature of the solution measured at the end of the experimental period.

There are situations other than those already described where the ecologist will need to measure the concentration of carbon dioxide in the field. A selection of methods that are available will be described, but it is up to the individual to assess their suitability for his own particular use.

Macfadyen (1970), in a paper describing a number of field techniques, has described an electrolytic syringe based upon the method of Köpf (1953). Details of the apparatus are shown in Fig. 4.16. When a direct electric current is passed through a solution of sodium chloride, from a silver anode to a platinum cathode, sodium hydroxide will be produced.

$$NaCl \rightarrow Na^+ + Cl^-$$
$$Na^+ + H_2O \rightarrow NaOH + \tfrac{1}{2}H_2$$
$$Cl^- + Ag^+ \rightarrow AgCl$$

If a gas sample is shaken up and dissolved in the syringe an equivalent amount of sodium hydroxide can then be generated. Equivalence is indicated by the production of a pink colour from phenolphthalein contained in the electrolyte. The amount of sodium hydroxide generated can be calculated from the product of the current passed and the time.

$$1 \text{ mA/min} \equiv 0.01445 \text{ ml } CO_2 \text{ at } 10°C$$

If the end-point, compared against a standard made from dilute eosin, is passed the reaction can be reversed and the time subtracted from that in the forward direction. Periodically it will be necessary to pass a higher reverse current through the apparatus to remove deposits of silver chloride that make it difficult to maintain a constant current. A current of 1 mA passed for 1 minute is equivalent to 0.3% CO_2 with a 5 ml air sample, and as the end point can only be timed to within about 1.5 seconds the expected accuracy will not be greater than about 0.01% CO_2, but this error is independent of concentration.

Macfadyen has used this syringe in combination with small thin walled polythene bags (*c.* 15 mm × 120 mm) buried in the soil. The bags were provided with sealed tubes leading to the surface of the soil allowing gas samples to be withdrawn and analysed at suitable intervals. Other methods of estimating the carbon dioxide concentration in the soil have been described by Rutter & Webster (1962), and Martin & Pigott (1965). Rutter and Webster used a probe incorporating a ceramic cup filled with distilled water, that was allowed to equilibrate with the soil solution. One problem was to find the time that was required to reach equilibrium, they employed four probes and measured the rate of change of pH and used it to indicate equilibration. Martin and Pigott inserted sterilized pyrex tubes, with small 'windows' covered with

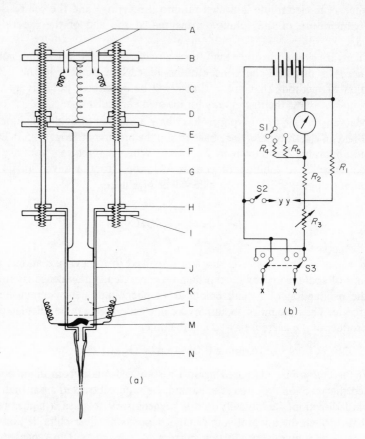

(a)

(b)

Figure 4.16   An electrolytic syringe analyser for carbon dioxide. (Macfadyen, 1970.)

(a) Syringe assembly made from a 10 ml disposable plastic syringe. A, sockets of non-reversible connector; B, upper plastic sheet bolted to brass rods; C, tension spring—or rubber band; D, locked nuts constituting a stop for the open position; E, middle plastic sheet cemented to plunger of syringe; free to slide over brass rods; F, syringe plunger; G, brass rods, threaded at ends; H, nut, locked on thread serves as stop for closed position; I, lower plastic sheet cemented to syringe barrel and rods; J, syringe barrel; K, position of plunger when depressed; L, silver anode, cemented through barrel wall; M, platinum cathode cemented through barrel wall; N, jet made from rigid polythene tubing.

(b) Circuit of current supply system. The 6 V dry battery normally supplies the syringe (at XX) through the 1 mA F.S.D. milliammeter and control resistances R2 (1 kΩ) and R3 (10 kΩ) and the reversing switch S3. A separate plug YY is available for rapid removal of silver chloride through resistance R1 (100 Ω). Shunts are provided (R4 and R5) to extend the scale of the meter if required.

polytetrafluorethylene membrane, into the soil. The tubes contained 0.01M sodium bicarbonate and after 10 days the pH of the solution was measured (to within 0.05 pH units) and the carbon dioxide concentration estimated from the relationship with pH (Table 4.1).

This electrometric method for determining the concentration of carbon dioxide in the soil has been examined further by Lee & Woolhouse (1966), and they emphasize the necessity for calibrating individual experimental procedures. They also point out that a particular weakness of the method is the long equilibration time that is required with the range of CO_2 concentrations that might be found in the soil. The equilibration time might even be expected to be up to 3 weeks in some cases.

Table 4.1   pH of 0.01M sodium bicarbonate solution at different concentrations of CO_2. (From Martin & Pigott, 1965.)

$\%CO_2$	pH
0	9.53
0.03	9.31
0.25	8.65
1.00	8.05
4.00	7.45
100.0	6.05

3.6.8 Temperature and soil respiration

The intensity of soil respiration is strongly correlated with soil temperature and shows both an annual and a diurnal cycle (Witkamp, 1966b, 1969; Witkamp & Frank, 1969). For comparative purposes it is normal to convert observed levels of respiration to equivalent values at some standard temperature. To do this and to be able to use recordings of soil temperature to obtain estimates of the annual evolution of carbon dioxide from the soil it is necessary to establish a relationship between soil respiration and temperature for a particular ecosystem.

The most generally used relationship between temperature and respiration is:

$$\log R = a + bT$$

where R = rate of respiration,
     T = temperature ($°C$),
     a = constant,
     b = temperature coefficient.

The temperature coefficient can be expressed as the ratio of the rate of respiration at a given temperature and that at a second temperature 10°C lower. The symbol Q_{10} is used to represent this coefficient.

$$Q_{10} = \text{antilog } (10.\text{b})$$

In practice temperature coefficients with values in the order of 2·0 are obtained. Objections can be made regarding the use of this relationship over wide ranges of temperatures, but within the ranges occurring in the soil it has been found satisfactory (Witkamp, 1969; Wiant, 1967b; Reiners, 1968; Anderson, 1973a). Other relationships between temperature and respiration have been proposed (Krogh, 1944) and some workers have obtained better correlations from log-log regressions (Kucera & Kirkham, 1971).

Soil respiration is also a function of soil moisture (Wiant, 1967a, Macfadyen, 1971; Froment, 1972; Anderson, 1973a) and other factors such as the carbon dioxide concentration of the soil atmosphere (Macfadyen, 1971).

As a result of the diurnal cycle of carbon dioxide evolution from the soil it is preferable to base estimates of the daily respiration upon measurements taken over 24 hour periods.

3.7 GRAZING

The importance of grazing must be considered in any investigation into the primary production of an area. It can be seen that the basic equation for primary production given in Section 3.1.1 includes a term for 'plant losses by grazing'. Milner & Hughes (1968) have pointed out that the grazing of grassland by large herbivores and the measurement of net primary production present two inter-related and mutually interfering problems. The magnitude and importance of grazing is not always so obvious as in grassland and many plant ecologists are only too willing to accept the absence of cattle, deer, ponies or sheep as evidence of an ungrazed ecosystem. It is hoped that with present day teaching of ecology that this attitude will disappear and that a more balanced and integrated approach will emerge. Details of methods available for the study of secondary production and its associated problems can be found in works such as Southwood (1966), and Petrosewicz & Macfadyen (1970). A number of techniques relating to the study of grazing that would seem particularly relevant in the study of net primary production are described in this section.

3.7.1 Grazing exclosures

The most straight forward method of estimating the amount of the standing crop that is being consumed by grazing animals is to exclude them from the system and to measure the difference. The removal of grazing pressure may have a considerable effect upon the primary production; Vickery (1972) has shown that the production of an area of grassland can vary under different grazing pressures and that primary production with 20 sheep per hectare was higher than with either 10 or 30 sheep per hectare. Pearson (1965) found at least 12% higher net primary production upon grazed than ungrazed desert in America. The effect of an exclosure can be kept to a minimum if the period without grazing is kept short compared to the life cycle of the more important plant species (Green, 1949).

Different types of cage and exclosure have been described and they vary widely in size and shape. Lock (1972) has described a 30 m square exclosure surrounded by a ditch 2·4 m wide to prevent Hippopotamus grazing in an area of grassland. Welch & Rawes (1965) employed small cages 1·5 × 1 m in size to study the production of Pennine grassland, moving the position of the cages every four to five weeks. Shaw (1968) has described a series of cages used to study the losses and germination of acorns under field conditions. These cages were constructed of resin coated wire mesh to overcome the problems of zinc toxicity to plants that were reported by Harris (1946). The design of a cage for a particular purpose must be suited to the local requirements and the size of the mesh suitable for the size of animal to be excluded from the experimental area. Where burrowing animals are present the walls of the cage or exclosure must be buried to a suitable depth in the soil.

3.7.2 The direct estimation of consumption

In some circumstances it may be possible to measure the amount of plant material that has been consumed by direct measurement. Bray (1961, 1964) collected both freshly fallen and attached leaves from trees to make outline drawings of the leaves on graph paper. The drawings included holes within the leaf and areas where leaf-mining insects had removed all the tissue except for the veins, where necessary an estimate of the original leaf margin was made and drawn in. The percentage leaf consumption ranged from 3·2 for *Acer rubrum* to 15·0 for *Prunus virginiana* with a mean value for all species of 8·3% (Bray, 1961). Bray estimated that these values represented something between 0·5 and 1·4% of the total primary production. Petrusewicz & Macfadyen (1970) point out that it may be necessary to make control holes to correct for 'growth' of insect produced holes after they have been made.

The effect of phytophagous insects upon areas of vegetation extend beyond the simple consumption of foliar material, insects can damage individual plants in a number of ways so reducing their photosynthetic capacity and net primary production. Rafes (1970) has reviewed the effects of phytophagous insects upon the production of forest ecosystems under four main headings, foliar consumption, sap consumption, consumption and damage to woody tissue, and damage to the reproductive organs. It is clear that the overall effects of invertebrate grazing can be considerably greater than the amount of plant material actually consumed.

Radio-isotopes have been used to measure consumption of plant material by invertebrates (Crossley, 1966; Paris & Sikora, 1967).

3.7.3 Faecal analysis

It may be necessary to establish which particular plant species are being selected and taken as food by animals grazing within a particular area. In some types of vegetation direct observation may provide the information but in others it may be difficult to determine exactly which species are being taken.

The technique of faecal analysis depends upon the fact that plant cuticles bear sufficient characteristic features to allow identification, and in many cases down to the level of an individual species. The cuticle, consisting of a waxy material secreted over the epidermis, is impermeable to water and resistant to many types of chemical treatment with the result that cuticular fragments can be retrieved and identified from faeces produced by grazing animals. Techniques have been developed for the preparation of reference material, and for the examination of cuticles extracted from faeces collected from a variety of different animals (Baumgartner & Martin, 1939; Martin, 1955; Croker, 1959; Brunsven & Mulkern, 1960; Storr, 1961; Stewart, 1967; Williams, 1969; Zyznar & Urness, 1969). Their methods differ in detail but can be summarized as follows:

Reference material

Segments of the reference plant material are cut around the edges to allow easier separation of the cuticles, macerated in 10% nitric acid (more concentrated if necessary) and washed in water. The samples are then transferred to 70% alcohol, stained in safranin, acid fuchsin or gentian violet and mounted in Euparol.

Preparation and examination of faecal samples

The faecal samples are collected and stored in a formalin–acetic acid–alcohol mixture. The simplest method of preparing the samples for examination is that described by Croker, who dispersed the faeces in water, and after sub-sampling counted the cuticular fragments under a microscope. The method developed by Williams includes dispersing the samples in 70% alcohol, to extract and remove chlorophyll, and placing them in 500 ml of boiling water. The samples are then allowed to stand overnight and then the water treatment repeated but with the addition of 10–15 ml of 5% sodium hypochlorite after a further 3–4 hours. The samples are then washed, passed through 70% alcohol, stained in safranin and mounted in Euparol. Zynar and Urness have recommended soaking the faecal pellets of some animals in 10% sodium hydroxide to remove any mucous coating that would interfere with the separation and dispersal of the sample.

When the faecal material has been prepared, stained and mounted the slides are examined under the microscope and at least 200 cuticular fragments identified from each sample.

4 NUTRIENT AND ENERGY BUDGETS

The cycling of nutrients, the sources of additional nutrients and the pathways by which they are lost are of great importance in attempting to analyse and obtain an understanding of the working of an ecosystem. The production of organic matter and the flow of energy are important processes but are both influenced by the availability of nutrients and water. The relative importance of the nutrient budget (the nutrient income and loss account) varies from one type of ecosystem to another. Where the source of nutrients is restricted to rainfall, as in the case of ombrotrophic peatland, the overall nutrient budget is obviously of great interest. In more nutrient rich situations the nutrient budget may assume relatively less importance than the internal cycling of some particular element, but it can be very unwise to try and rank such factors in order of importance when they are so intimately related.

To compile a nutrient or energy budget for any ecosystem it is essential to have a reliable framework of organic matter production and turnover upon which to base the estimates. The calculations of the nutrient or energy contents of the difference components of the ecosystem are based upon the combination of biomass and estimates of composition or calorific content.

While there is no hard and fast division between a nutrient budget and a nutrient cycle, the study of the budget often presents fewer practical problems and may be the initial approach to any such investigation. As the individual worker can, and must, define the limits of the ecosystem under investigation, he can also sub-divide the overall system into smaller units and it is the recombination of the budgets from these sub-systems that produces the overall picture of nutrient or energy flow through the complete system. This is illustrated in Fig. 4.17 where the complete ecosystem (X) is enclosed by a dotted line, and has been divided into component systems A,B,C & D. The transfers of material in and out of these sub-systems are represented by arrows numbered from 1–10. It can be seen that the budgets of the individual sub-systems are:

$$1 = 2 + 3 + 9 \pm \Delta A$$
$$3 = 4 + 7 + 8 \pm \Delta B$$
$$4 = 6 + 5 \pm \Delta C$$
$$9 + 8 + 6 = 10 \pm \Delta D$$

and that overall,

$$1 = 2 + 5 + 7 + 10 \pm \Delta X$$

where ΔA etc. = change in weight or biomass of a particular component of the ecosystem.

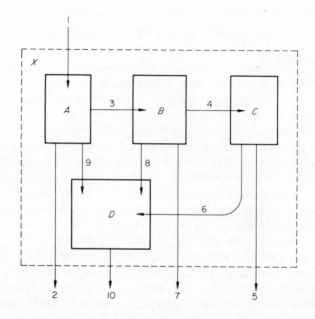

Figure 4.17 A simple hypothetical ecosystem to illustrate the relationship between the budgets of components of the system, and transfer and flow of material through the complete system.

When some of the inputs to an ecosystem have been measured the question will often arise as to whether they represent a total or true input, or whether part merely represents some degree of recycling within the limits of the ecosystem. An example is found with the estimation of the nutrient input by rainfall. If the rain samples are filtered they will be found to contain mineral particles, pollen grains and other organic debris. Some of this material will have originated within the ecosystem and should not therefore be included in any estimate of nutrient input, and presents a very real practical problem.

References to relevant published work are given in the sections dealing with particular aspects of nutrient and energy flow, but general references providing valuable background information include Rodin & Bazilovich (1967), and papers included in Young (1968) and Reichle (1970).

4.1 THE ESTIMATION OF NUTRIENT AND CALORIFIC CONTENT

Where independent estimates of weight and chemical composition (or calorific content) are combined to calculate the nutrient (or calorific content) of some particular component of the ecosystem, the relationship between the means and the standard deviations of the means (standard errors) are:

$$\bar{N} = \bar{k} \cdot \bar{w}$$

where
$$\bar{N} = \quad \text{mean nutrient or energy content,}$$
$$\bar{k} = \quad \text{mean composition of calorific value,}$$
$$\bar{w} = \quad \text{mean weight or biomass.}$$

$$S_{\bar{N}}^2 = \bar{k}^2 S_{\bar{w}}^2 + \bar{w}^2 S_{\bar{k}}^2 + S_{\bar{w}}^2 S_{\bar{k}}^2$$

where
$$S_{\bar{N}}^2 = \text{variance of the mean nutrient content } (\bar{n}),$$
$$S_{\bar{k}}^2 = \text{variance of the mean composition } (\bar{k}),$$
$$S_{\bar{w}}^2 = \text{variance of the mean biomass } (\bar{w}).$$

and where:

$$S_x^2 = \frac{\Sigma x^2 - \dfrac{(\Sigma x)^2}{n}}{n(n-1)}$$

when
$$x = \text{single observation,}$$
$$n = \text{number of observations.}$$

The derivation and proof of this relationship between the means and variances of combined estimates are given by Colquhoun (1971) ar.d have been used by Rawes & Welch (1969) and Gyllenberg (1969).

To obtain an estimate of the nutrient content of the soil it is necessary to combine measurements of composition (i.e. %N), bulk density and soil volume. Methods of measuring the density of soil are described in Chapter 6. A relationship between soil density and the loss-on-ignition for a wide range of soils has been shown by Jeffrey (1970) (see Table 4.2) and in the absence of actual determinations of soil density this relationship can be used to estimate the nutrient content of soil horizons. The expression for predicting the bulk density of the soil (y) from the percentage loss-on-ignition data (x) is:

$$y = 1 \cdot 482 - 0 \cdot 6786 \log_{10} x$$

Table 4.2 Predictive table for estimating bulk density from ignition loss, utilizing the regression of bulk density on log ignition loss. (From Jeffrey, 1970.)

Ignition loss (%)	Bulk density (g/ml)	Confidence limits ± 0.01	Confidence limits ± 0.05
1·0	1·4823	0·1365	0·1029
2·5	1·2123	0·1018	0·0767
5·0	1·0080	0·0781	0·0588
7·5	0·8885	0·0665	0·0501
10·0	0·8037	0·0597	0·0450
15·0	0·6842	0·0537	0·0404
20·0	0·5994	0·0527	0·0397
25·0	0·5337	0·0535	0·0403
30·0	0·4800	0·0555	0·0418
35·0	0·4345	0·0581	0·0438
40·0	0·3952	0·0607	0·0457
45·0	0·3605	0·0632	0·0476
50·0	0·3294	0·0658	0·0496
55·0	0·3013	0·0685	0·0516
60·0	0·2757	0·0708	0·0533
65·0	0·2521	0·0731	0·0551
70·0	0·2242	0·0760	0·0573
75·0	0·2097	0·0774	0·0583
80·0	0·1909	0·0794	0·0598
85·0	0·1731	0·0813	0·0613
90·0	0·1564	0·0832	0·0627
95·0	0·1413	0·0851	0·0641
100·0	0·1251	0·0868	0·0654

4.2 NUTRIENT INPUTS

There are a number of different sources of input of nutrients to an ecosystem, these include precipitation and atmospheric fallout, the lateral movement of nutrients through the soil from adjacent areas, nitrogen fixation, faunal migration, the weathering of soil minerals and in some cases the application of fertilizers (see Fig. 4.1).

4.2.1 Precipitation and atmospheric fallout

A survey of the chemical content of precipitation in north-western Europe has been carried out by Egnér *et al.* (1955–1960). The chemical composition of rainfall on a daily basis has been related to the amount of precipitation, to wind direction, to wind velocity and to temperature by Gorham (1958) working in the Lake District of England.

Where it is intended to collect rain samples for chemical analysis it is recommended that the amount of rainfall be measured with a standard raingauge (Chapter 7), and that a separate gauge made of polythene or some other chemically inert material be used to collect samples for analysis. The plastic funnels should be fitted with gauze filters to prevent contamination by insects or other debris, and the collecting bottle kept within a dark container to prevent algal growth. Microbiological growth can affect the composition of the samples by nutrient uptake and assimilation, but this can be reduced by impregnating the walls of the collecting bottle with iodine (Heron, 1962). This can be done by placing a few crystals of iodine in the bottle and putting the stoppered bottle into a warm oven (60°C) for several hours. Fouling of the raingauges by birds is a problem that is generally encountered, and can sometimes be reduced by fitting the gauge with a crown of spikes (Egnér *et al.,* 1956), or by a ring around the funnel to act as a 'decoy' perch. Some people seem to have found these precautions useful while others have reported only little benefit (Allen *et al.,* 1968). The gauges should be emptied as frequently as possible and the samples stored in a deep freeze until the analyses are carried out.

Where rain samples are collected after passing through the tree canopy or after running down the trunks or stems of trees (see Chapter 7) they are found to contain higher concentrations of dissolved nutrients than the incident rainfall (Madgwick & Ovington, 1959; Carlisle *et al.,* 1966). While some of this enrichment is due to the leaching of materials out of the plant tissue part is due to the capture of airborne particles (aerosols) by the vegetation. These aerosol components arise from smoke, reactions between gases in the atmosphere, sea spray and mineral dust.

The capture of airborne particles by a woodland canopy has been studied by White & Turner (1970). Their sampling apparatus was mounted on a tower in and above the tree canopy, but they suggest that their results overestimate the nutrient income in the form of airborne particles, especially during the windier months of the year.

A different method has been used by Nihlgård (1970) in an attempt to assess the importance of aerosols in the enrichment of the precipitation passing through the woodland canopy. He arranged 10 layers of plastic netting, one above another, in a woodland clearing with raingauges underneath. The concentration of nutrients in the water collected underneath the netting was found to be higher than in gauges collecting direct rain samples.

4.2.2 Salt spray

When working in coastal habitats it is important to consider both the direct effects of sea-spray upon the vegetation and also its importance as a source of plant nutrients.

Edwards & Claxton (1964) devised an instrument where a square of filter paper is held at right angles to the wind. Etherington (1967) in a study of potassium cycling and the production of a dune heath ecosystem used a set of plastic baffles over a funnel to intercept the salt spray. The drainage from the funnel was passed through a cation exchange column to absorb sodium and potassium. A 30 cm square of towelling material, weighted at the base and suspended beneath a plastic shelter was used by Randall (1970) to collect salt

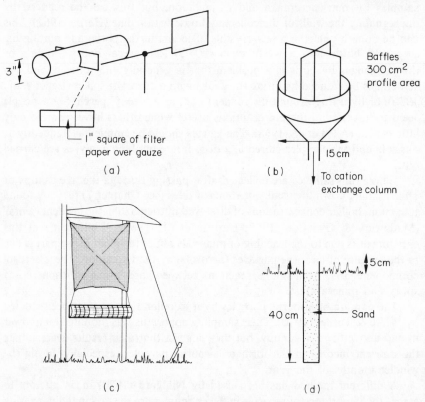

Figure 4.18 Types of apparatus used to measure salt deposition in coastal habitats: (a) Edwards & Claxton (1964); (b) Etherington (1967); (c) Randall (1970); (d) Malloch (1972).

spray. The salt was extracted from the towelling by placing it in 300 ml of water and then estimated conductimetrically. Malloch (1970) buried sand-filled tubes vertically in the soil to estimate salt deposition on the Lizard Peninsula in Cornwall. The salt was extracted in 500 ml of deionized water and sodium determined by flame photometry after periods of 2–3 months. An advantage of this apparatus is that it is capable of being 'hidden from predation by tourists' a point worth considering when any apparatus must be left in a relatively public place.

4.2.3 Nitrogen fixation

The biological fixation of nitrogen has been studied by a number of different methods, kjeldahl analyses, ^{15}N-enrichment assayed by mass spectrometer, and by means of the radio-isotope ^{13}N, even with its great disadvantage of a 10-minute half-life. These methods are all far from ideal in that they are either insensitive, time consuming or require complex apparatus such as a mass spectrometer.

With the discovery of the acetylene reduction technique by Dilworth (1966) and Schöllhorn & Burris (1967) these problems were largely overcome. The method has the advantage of being sensitive, capable of being used in the field and relatively inexpensive. The technique depends upon the fact that the enzyme nitrogenase is capable of reducing acetylene to ethylene and therefore uses the rate of acetylene reduction as an index of nitrogen fixation.

The use of the technique in the field has been described by Stewart *et al.* (1967) and by Hardy *et al.* (1968). Waugham (1971) has described a modified technique that simplifies the field procedure and uses apparatus that can be easily transported in the field. He uses incubation chambers consisting of 28 ml serum bottles fitted with rubber caps so that gas samples can be injected and withdrawn by the use of hypodermic syringes. The gas mixture (4:1 Argon–Oxygen mixture) is carried in a football bladder with a valve, and acetylene is carried in a second bladder fitted with a tap and a serum cap.

The incubation bottle containing the sample material is flushed through with the argon–oxygen mixture and acetylene is injected to give a final mixture of 70% argon, 20% oxygen and 10% acetylene. Other workers have reported using gas mixtures that include carbon dioxide in addition to these three gases.

After incubation a sample of the gas phase is transferred to a serum bottle by displacement and the use of syringe. The resulting gas sample is returned to the laboratory for analysis by gas chromatography.

The removal of the gas sample at the end of the incubation period is an alternative to fixing the sample with a metabolic poison such as trichloroacetic acid (T.C.A.). It has been shown (Thake & Rawle, 1972) that ethylene can be produced within a few hours by the action of T.C.A. upon rubber serum caps, and that unless the samples are to be analysed within a very short time the technique of gas withdrawal is to be recommended.

Although the acetylene reduction technique has provided a convenient method for the assay of nitrogen fixation under field conditions a great deal of work is still required before the overall importance of nitrogen fixation in many ecosystems can be assessed.

4.3 NUTRIENT LOSSES

The principal losses of nutrients from an ecosystem are through leaching and the lateral movement of nutrients in the soil solution, by grazing, by the removal of harvested plant material, by the removal of nutrients in animal bodies, and by other processes such as fire, soil erosion and solifluction. It can be seen that a number of these losses of nutrients from one particular area represent an input to some other adjacent area or ecosystem.

4.3.1 The loss of nutrients in the soil solution

The methods available for the measurement of nutrient losses by leaching include direct sampling of the soil solution, the use of lysimeters and the analyses of drainage waters from catchment areas. In part these methods depend upon the combination of hydrological and analytical methods described in Chapters 7 and 8 respectively.

When water is present in the soil in excess of field capacity it is free to drain through the soil and be lost from the ecosystem. Under these conditions the soil solution can be sampled by interception or by suction sampling.

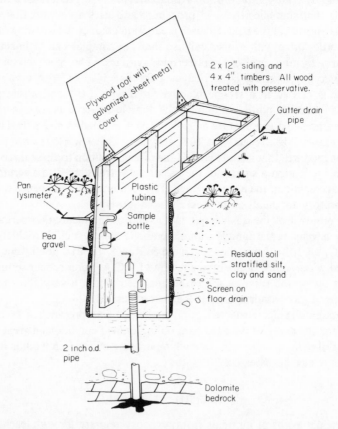

Figure 4.19 Trench lysimeter for collection of soil water samples. (From Parizek & Lane, 1970.)

Parizek & Lane (1970) have reviewed a number of these techniques and describe a trench lysimeter (Fig. 4.19) where the drainage water is intercepted by metal pans inserted at intervals down the soil profile. The trench lysimeter that they described was used in conjunction with an irrigation experiment, and in the absence of such an experimental treatment samples of water for analysis would only have been obtained in small quantities and at infrequent intervals. They considered the constructional problems involved and concluded that

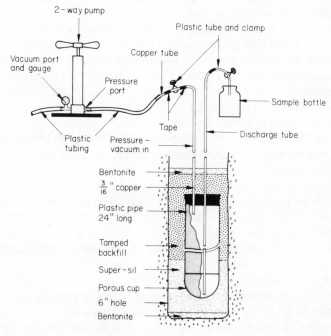

Figure 4.20 Cross section of a typical pressure-vacuum lysimeter installation with the pressure pump attached to the lysimeter, the apparatus is ready to recover a water sample. (From Parizek & Lane, 1970.)

other procedures were preferable especially where sampling was required at a number of different sites. In the same paper they describe the use of vacuum or suction lysimeters of the type previously used by Wagner (1962). These samplers consist of unglazed ceramic cups cemented to the end of plastic pipes (Fig. 4.20) that are connected to tubes to which a vacuum can be applied and any soil solution available collected and withdrawn for chemical analysis. This type of sampler will only extract water from the soil when it is present at tensions less than the suction that can be supplied by a vacuum pump. The amount of water sample that will be obtained is dependent upon both the water content and the 'water supplying power' of the soil. The results of any chemical analyses carried out on samples obtained from suction lysimeters must be used in combination with soil moisture determinations to obtain estimates of nutrient flow through the soil profile (see Chapter 6).

Grover & Lamborn (1970) found that some types of ceramic cup used for sampling soil solution added calcium, sodium and potassium to solutions drawn through them and that they absorbed significant amounts of phosphorus from the soil solution. They found that leaching the cups in 1 N HCl reduced the contamination from all but calcium and reduced the loss of phosphorus to an acceptable level.

An investigation into the water and mineral budget of an area of Pennine moorland by Crisp (1966) provides an example of the estimation of nutrient losses from an ecosystem by the analysis of stream flow. Relationships between the concentrations of nutrients and the rate of run-off were established enabling estimates of the quantities of nutrients lost in solution to be calculated from the volume of water leaving the catchment area.

It is important that the variation of nutrient concentration with stream flow be measured directly, or alternatively some device that will collect water samples that are proportional in volume to the rate of run-off be used. If such a proportional sampler is used then the water samples collected at regular time intervals can be used to provide an estimate of nutrients lost in solution. A simple proportional sampler of this type has been described by Eggink & Duvigneau (1963) consisting of a sampling chamber that is placed upstream from a weir and so shaped that the volume of water that it contains at any particular time is proportional to a function of the head of water, and therefore to the rate of flow over the weir. At intervals a valve in the base is opened and the sampling chamber fills to a level governed by the head of water flowing over the weir. The sample is transferred to a collecting chamber by means of compressed air so providing a composite water sample whose composition is representative of the total volume of run-off that has left the area during the sampling period.

In addition to the loss of nutrients in solution the movement of solid material in suspension may be important. Crisp obtained estimates of the amount of peat lost from his catchment area by using a siphon sampling system that incorporated a self-cleansing filter. Periodic samples of the filtrate were taken to allow a correction to be made for the loss of fine material through the filters. The relationship between the rate of flow through the siphons and the total flow in the stream was used to calculate the weight of solid material lost in suspension.

Perrin (1965) has examined the possibility of using drainage water analyses in the study of soil development, and in particular to the development of a chalk soil and to the genesis of shallow brown soils on shales in central Wales.

4.3.2 The loss of nutrients by fire

In a number of habitats, such as savanna and heathland the above ground vegetation is burnt off at periodic intervals. Such fires are important for a number of reasons, the maintenance of a particular type of vegetation, the encouragement of new growth as potential grazing material or the mineralization and return of plant nutrients to the soil. After a fire the soil surface will be covered by a nutrient rich layer of ash, but significant amounts of the nutrients that were contained in the above ground vegetation and litter layer will have been lost from the ecosystem. Where the particular ecosystem is extensive many of the nutrients lost from one particular area will be distributed over a surrounding area of the same type of vegetation, but when the ecosystem is smaller or has become fragmented, such as heathland in lowland Britain (Moore, 1962), the loss of nutrients in a fire may represent a complete loss from the ecosystem.

The magnitude of these losses from heathland have been measured by placing cut vegetation over sheets of steel, burning and collecting the resulting ash (Chapman, 1967; Evans & Allen, 1971). Samples of plant material have been burnt under laboratory conditions (Allen, 1964; Evans & Allen, 1967) in attempts to control the conditions of the experiment rather more than is possible in the field procedure just described.

The actual losses of nutrients under field conditions depend very much upon local conditions at the time of the burn, such as wind, the weight and structure of the standing crop, and the temperature of the burn. The temperatures reached in natural fires have been measured by means of chemicals with different melting points (Beadle, 1940), by the use of heat sensitive paints (Whittaker, 1961) and by the use of thermocouples (Kenworthy, 1963). The distribution and the changes in the nutrient content of heathland soils after fires have been examined by Allen *et al.* (1969) and by Hansen (1969).

4.4 THE MEASUREMENT OF CALORIFIC VALUES

4.4.1 Calorific values and calorimetry

In the establishment of a nutrient budget the results of chemical analyses are combined with biomass data to obtain estimates of nutrient content, so in the construction of an energy budget are calorific values combined with weights to provide estimates of calorific content.

The GROSS CALORIFIC VALUE is the number of heat units that are liberated when unit weight of material is burnt in oxygen; and the residual materials are oxygen, carbon dioxide, sulphur dioxide, nitrogen, water and ash.

The basic unit of energy is the joule (J) but in the past the unit used to express the energy content of biological material has often been the calorie (c) or the kilocalorie (C) and results expressed in terms of kilocalories per gramme dry weight. In many cases it is more meaningful to calculate and express the results in terms of ash free dry weight. Ecologists have often overlooked the fact that there are a number of differently defined types of calorie (15° calorie, steam calorie, international table calorie) each having a slightly different value and it is now universally accepted that the SI unit the joule (4·1840 joules = 1 Thermochemical calorie) should be used to express all results in the study of ecological energetics (Phillipson, 1971; Leith, 1968).

Wet combustion techniques (Ivlev, 1934), that depend upon conversion factors being applied to carbon, nitrogen and sometimes sulphur contents, have been used to obtain estimates of the calorific values of biological material. These methods are not generally as satisfactory or as convenient as the use of calorimetry.

In bomb calorimetry a sample of material is ignited and burnt inside a thick walled stainless steel container (the bomb) that contains oxygen under pressure of up to 30 atmospheres. The increase in temperature of the bomb, as a result of the combustion of the sample, is measured and by comparison with the results obtained from the combustion of a standard material the calorific value of the sample can be calculated. The standard generally used is Benzoic acid having a calorific value of 26,447 joules per gramme.

4.4.2 The preparation of samples

The collection and preparation of samples for the determination of calorific values may require greater care and precautions than are required when sampling for the estimation of the weight of standing crop. The sample once collected must be dried in such a way that losses of volatile constituents are kept to a minimum. Freeze drying and the use of vacuum ovens are recommended by some authors. When dried the material must be milled, homogenized and compressed into pellets. Lieth (1968) describes methods by which the sample material can be melted into wax or contained in gelatine capsules as alternatives to the use of pellets. In some cases known amounts of benzoic acid have been added to the sample to aid combustion (Malone & Swartout, 1969), or to enable calorific values to be determined on small amounts of material (Richman, 1958).

4.4.3 Calorimeter corrections

Bomb calorimeters are of two main types, adiabatic and non-adiabatic. In the adiabatic type of calorimeter the bomb is surrounded by a water jacket whose temperature is controlled and remains the same as that of the bomb at all times. This means that no heat is transferred to or from the bomb and the calorific value (V) of a sample can be calculated from:

$$V = \frac{W . \Delta t - \Sigma c}{G}$$

where W=the number of joules necessary to raise the temperature of the
water bath (and the bomb) by one degree centigrade,
Δt=the rise in temperature of the water bath,
Σc=the sum of necessary corrections (see below),
G=dry weight of the sample.

The micro-bomb calorimeter described by Phillipson (1964), suitable for small samples (5–100 mg dry weight) and marketed commercially, is an example of a non-adiabatic calorimeter. In this type of calorimeter heat will flow either in or out of the apparatus depending upon its temperature relative to that of the surroundings. This means that suitable corrections must be applied to the results obtained from such a calorimeter to allow for any heat exchange. The most convenient cooling correction (C) is the application of Dickinson's formula:

$$C = r_1 (T_a - T_0) + r_2 (T_n - T_a)$$

where r_1 = the pre-firing rate of temperature change, if the temperature is
rising r is negative,
r_2 = the post-firing cooling rate,
T_0 = time at temperature t_0,
T_n = time at temperature t_n,
T_a = time at temperature $t_0 + 0 \cdot 60 (t_n - t_0)$.

Where a chart recorder is used to record the temperature changes within the calorimeter the cooling correction can be applied by a graphical method direct on the chart (Fig. 4.21).

An alternative type of non-adiabatic calorimeter is the ballistic bomb calorimeter where larger samples of material are combusted. The time taken for the bomb to reach its maximum temperature is short (45 to 60 seconds) so that heat exchange is reduced and cooling corrections are unnecessary. This type of calorimeter is capable of producing results quickly and is convenient to use where large numbers of determinations are required and there is no shortage of sample material.

In addition to the cooling corrections described there are a number of additional corrections that may be necessary to apply in the calculation of calorific values. When electrical energy is used to ignite the sample there is an input of energy, this can be measured by means of blank determinations and subtracted from actual determinations of calorific value. If any of the firing wire is burnt during combustion heat will be released, but when platinum wire is used

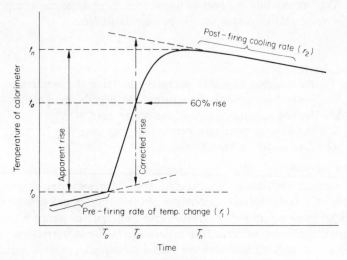

Figure 4.21 The graphical solution of Dickinson's cooling correction for non-adiabatic bomb calorimeters (for details see text).

this weight is small and the correction insignificant ($418 \, \text{J/g}^{-1}$ platinum, $1,402 \, \text{J/g}^{-1}$ nichrome). In the definition of calorific value it is assumed that the products of combustion will include nitrogen and sulphur. In practice these elements are oxidized to nitric and sulphuric acid that results in the evolution of extra heat. The washings obtained from the bomb can be titrated and used to estimate the quantities of these acids that have been formed, and the necessary corrections applied. With biological materials these corrections are small, generally less than $0\cdot1\%$ (Leith, 1968), and are often omitted in ecological research.

5 SUMMARY

This chapter has attempted to provide an introduction and background to the methods and concepts that are required by anyone attempting to investigate the primary production and nutrient circulation within a terrestrial ecosystem. It will be apparent that anyone hoping to find a single set of methods that can be applied to any set of ecological circumstances will be disappointed. It is, or should be, one of the attractions of research that the means of obtaining the relevant information may be one of the most difficult problems that has to be solved.

While it has not been possible to give much of an account of the results obtained from the methods described, the references provided in the text should provide starting points that will enable anyone who is interested to examine his own results against a more general background.

The International Biological Programme has encouraged and stimulated the study of complete ecosystems in different parts of the world and as the results of these studies are published over the next few years they will add greatly to our knowledge of production, energy and nutrient flow within natural communities. These studies will also highlight many of the difficulties that exist in the measurement of certain production parameters, and emphasize many of the practical difficulties that remain to be solved.

It is hoped that this chapter has already drawn attention to some of these problems and that it will encourage the co-operation and liaison between botanists and zoologists in the study of natural communities.

6 REFERENCES

AIMI R. & FUJIMAKI K. (1958) Cell-physiological studies on the function of root. 1. Chemical and enzymatic constitution associated with the structural differentiation of root in rice plant. *Proc. Crop Sci. Soc. Japan* **27**, 21–24.

ALCOCK M.B. (1964) An improved electronic instrument for estimation of pasture yield. *Nature, Lond.* **203**, 1309.

ALCOCK M.B. & LOVETT J.V. (1967) The electronic measurement of the yield of growing pasture. 1. A statistical assessment. *J. agric. Sci., Camb.* **68**, 27–38.

ALLEN S.E. (1964) Chemical aspects of heather burning. *J. appl. Ecol.* **1**, 347–368.

ALLEN S.E., CARLISLE A., WHITE E.J. & EVANS C.C. (1968) The plant nutrient content of rainwater. *J. Ecol.* **56**, 497–504.

ALLEN S.E., EVANS C.C. & GRIMSHAW H.M. (1969) The distribution of mineral nutrients in soil after heather burning. *Oikos* **20**, 16–25.

ANDERSON J.M. (1973a) Carbon dioxide evolution from two temperate, deciduous woodland soils. *J. appl. Ecol.* **10**, 361–378.

ANDERSON J.M. (1973b) The breakdown and decomposition of Sweet Chestnut (*Castanea sativa* Mill.) and Beech (*Fagus sylvatica* L.) leaf litter in two deciduous woodland soils. 1. Breakdown, leaching and decomposition. *Oecologia* **12**, 251–274.

ANDERSON J.M. & HEALEY I.N. (1970) Improvements in the gelatine embedding technique for woodland soil and litter samples. *Pedobiologia* **10**, 108–120.

ANDERSSON F. (1970) Ecological studies in a Scanian woodland and meadow area, Southern Sweden. II. Plant biomass, primary production and turnover of organic matter. *Bot. Notiser* **123**, 8–51.

ASHBY W.C. (1962) Root growth in American Basswood. *Ecology* **43**, 336–339.

BARCLAY-ESTRUP P. (1970) The description and interpretation of cyclical processes in a heath community. II. Changes in biomass and shoot production during the *Calluna* cycle. *J. Ecol.* **58**, 243–249.

BAUMGARTNER L.L. & MARTIN A.C. (1939) Plant histology as an aid in squirrel food-habitat studies. *J. Wildl. Mgmt* **3**, 266–268.

BAZILEVICH N.I. & RODIN L.E. (1966) The biological cycle of nitrogen and ash elements in plant communities of the tropical and sub-tropical zones. *For. Abstr.* **27**, 357–368.

BEADLE N.C.W. (1940) Soil temperatures and their effect on the survival of vegetation. *J. Ecol.* **28**, 180–192.

BEAUCHAMP J.J. & OLSON J.S. (1973) Corrections for bias in regression estimates after logarithmic transformation. *Ecology* **54**, 1403–1407.

BENEFIELD C.B. (1971) A rapid method for measuring cellulase activity in soils. *Soil Biol. Biochem.* **3**, 325–329.

BILLINGS W.D., CLEBSCH E.E.C. & MOONEY H.A. (1966) Photosynthesis and respiration rates of Rocky Mountain alpine plants under field conditions. *Am. Midl. Nat.* **75**, 34–44.

BLISS L.C. (1966) Plant productivity in alpine micro-environments on Mt. Washington, New Hampshire. *Ecol. Monogr.* **36**, 125–155.

BLISS L.C. & HADLEY E.B. (1964) Photosynthesis and respiration of alpine lichens. *Am. J. Bot.* **51**, 870–874.

BOCOCK K.L. (1964) Changes in the amounts of dry matter, nitrogen, carbon and energy in decomposing woodland leaf litter in relation to the activities of the soil fauna. *J. Ecol.* **52**, 273–284.

BOCOCK K.L. & GILBERT O.J.B. (1957) The disappearance of leaf litter under different woodland conditions. *Pl. Soil* **9**, 179–185.

BOCOCK K.L., GILBERT O.J.B., CAPSTICK C.K., TWINN D.C., WAID J.S. & WOODMAN M.J. (1960) Changes in leaf litter when placed on the surface of soils with contrasting humus types. *J. Soil Sci.* **11**, 1–9.

BOGGIE R., HUNTER R.F. & KNIGHT A.H. (1958) Studies of the root development of plants in the field using radioactive tracers. *J. Ecol.* **46**, 621–640.

BOGGIE R. & KNIGHT A.H. (1962) An improved method for the placement of radioactive isotopes in the study of root systems of plants growing in deep peat. *J. Ecol.* **50**, 461–464.

BOTKIN D.B., WOODWELL G.M. & TEMPEL N. (1970) Forest productivity estimated from carbon dioxide uptake. *Ecology* **51**, 1057–1060.

BOURDEAU P.F. & WOODWELL G.M. (1965) Measurements of plant carbon dioxide exchange by infra-red absorption under controlled conditions and in the field. In *Methodology of Plant Eco-physiology—Proceedings of the Montpellier Symposium*, pp. 283–290.

BOWMAN G.E. (1968) The measurement of carbon dioxide concentration in the atmosphere. In R.M. Wadsworth (ed.) *The Measurement of Environmental Factors in Terrestrial Ecology, British Ecol. Soc. Symp.* **8**, pp. 131–139.

BRAY J.R. (1961) Measurement of leaf utilization as an index of minimum level of primary consumption. *Oikos* **12**, 70–74.

BRAY J.R. (1963) Root production and the estimation of net productivity. *Can. J. Bot.* **41**, 65–72.

BRAY J.R. (1964) Primary consumption in three forest canopies. *Ecology* **45**, 165–167.

BRAY J.R. & GORHAM E. (1964) Litter production in forests of the world. *Adv. Ecol. Res.* **2**, 101–157.

BROWN A. & MACFADYEN A. (1969) Soil carbon dioxide output and small-scale vegetation pattern in a *Calluna* heath. *Oikos* **20**, 8–15.

BRUNSVEN M.A. & MULKERN G.B. (1960) The use of epidermal characteristics for the identification of plants recovered in fragmentary condition from the crops of grasshoppers. *N. Dak. agr. Exp. Sta. Res. Rep.* **3**, 11.

BURGES A. & DROVER D.P. (1953) The rate of podsol development in the sands of Woy Woy district. *Aust. J. Bot.* **1**, 83–94.

BURROWS W.H. (1972) Productivity of an arid zone shrub (*Eremophila gilesii*) community in south-western Queensland. *Aust. J. Bot.*, **20**, 317–329.

CAHOON G.A. & MORTON E.S. (1961) An apparatus for the quantitative separation of plant roots from soil. *Proc. Am. Soc. hort. Sci.* **78**, 593–596.

CALDWELL M.M. & CAMP L.B. (1974) Below ground productivity of two cool desert communities. *Oecologia* **17**, 123–130.

CAPSTICK C.K. (1962) The use of small cylindrical samplers for estimating the weight of forest litter. In P.W. Murphy (ed.) *Progress in Soil Zoology*, pp. 353–356. London, Butterworths.

CARLISLE A., BROWN A.H.F. & WHITE E.J. (1966) The organic matter and nutrient elements in the precipitation beneath a sessile oak (*Quercus petraea*) canopy. *J. Ecol.* **54,** 87–98.

CHAPMAN S.B. (1967) Nutrient budgets for a dry heath ecosystem in the south of England. *J. Ecol.* **55,** 677–689.

CHAPMAN S.B. (1970) The nutrient content of the soil and root system of a dry heath ecosystem. *J. Ecol.* **58,** 445–452.

CHAPMAN S.B. (1971) A simple conductimetric soil respirometer for field use. *Oikos* **22,** 348–353.

CHAPMAN S.B., HIBBLE J. & RAFAREL C.R. (1975a) Net aerial production by *Calluna vulgaris* on lowland heath in Britain. *J. Ecol.* **63,** 233–258.

CHAPMAN S.B., HIBBLE J. & RAFAREL C.R. (1975b) Litter accumulation under *Calluna vulgaris* on a lowland heathland in Britain. *J.Ecol.* **63,** 259–271.

CHARLES J.A. (1972) Physical science and archaeology. *Antiquity* **46,** 134–138.

CLYMO R.S. (1965) Experiments on breakdown of *Sphagnum* in two bogs. *J. Ecol.* **53,** 747–758.

CLYMO R.S. (1970) The growth of Sphagnum: methods of measurement. *J. Ecol.* **58,** 13–49.

COLQUOHOUN D. (1971) *Lectures on Biostatistics.* Oxford.

CONWAY E.J. (1950) *Microdiffusion Analysis and Volumetric Error,* 3rd edition. London, Crosby Lockwood.

CORMACK E. & GIMINGHAM C.H. (1964) Litter production by *Calluna vulgaris* (L.) Hull. *J. Ecol.* **52,** 285–298.

CRAPO N.L. & COLEMAN D.C. (1972) Root distribution and respiration in a Carolina old field. *Oikos* **23,** 137–139.

CRISP D.T. (1966) Input and output of minerals for an area of Pennine moorland: the importance of precipitation, drainage, peat erosion and animals. *J. appl. Ecol.* **3,** 327–348.

CROCKER R.L. & DICKSON B.A. (1957) Soil development on the recessional moraines of the Herbert and Mendenhall glaciers, south-eastern Alaska. *J. Ecol.* **45,** 169–185.

CROCKER R.L. & MAJOR J. (1955) Soil development in relation to vegetation and surface age at Glacier Bay, Alaska. *J. Ecol.* **43,** 427–448.

CROKER B.H. (1959) A method of estimating the botanical composition of the diet of sheep. *N.Z. J. agric. Res.* **2,** 72–85.

CROSSLEY D.A. JR. (1966) Radioisotope measurement of food consumption by a leaf beetle species, *Chrysomela knabi* Brown. *Ecology* **47,** 1–9.

CROSSLEY D.A. JR. & HOGLUND M.P. (1962) A litter bag method for the study of micro-arthropods inhabiting leaf litter. *Ecology* **43,** 571–573.

DAHLMAN R.C. (1968) Root production and turnover of carbon in the root-soil matrix of a grassland ecosystem. In *Methods of Productivity Studies in Root Systems and Rhizosphere Organisms.* Leningrad, USSR Acad. Sciences.

DAHLMAN R.C. & KUCERA C.L. (1965) Root productivity and turnover in native prairie. *Ecology* **46,** 84–89.

DAHLMAN R.C. & KUCERA C.L. (1968) Tagging native grassland with carbon-14. *Ecology* **49,** 1199–1203.

DILWORTH M.J. (1966) Acetylene reduction by nitrogen fixing preparations from *Clostridium pasteurianum. Biochim. biophys. Acta* **127,** 285–294.

DURNO S.E. (1961) Evidence regarding the rate of peat growth. *J. Ecol.* **49,** 347–351.

EDWARDS C.A. & HEATH G.W. (1963) The role of soil organisms in breakdown of leafy material. In J. Doeksen & J. van der Drift (eds.) *Soil Organisms,* pp. 76–84. Amsterdam, North Holland Publishing Co.

EDWARDS R.S. & CLAXTON S.M. (1964) The distribution of air-borne salt of marine origin in the Aberystwyth area. *J. appl. Ecol.* **1,** 253–263.

EGGINK H.J. & DUVIGNEAU H. (1963) Bemonstering van afvalwater naar hoeveelheid (sampling proportional to rate of sewage flow). *Water* **47,** 69–71.

EGNÉR H., BRODIN G. & JOHANSSON O. (1956) Sampling technique and chemical examination of air and precipitation. *K. LantbrHögsk. Annlr.* **22**, 369–382.

EGNÉR H., ERIKSSON E. & BRODIN G. (1955–60) Current data on the composition of air and precipitation. *Tellus* 7–12.

ELLERN S.J., HARPER J.L. & SAGAR G.R. (1970) A comparative study of the distribution of the roots of *Avena fatua* and *A. strigosa* in mixed stands using a carbon-14 labelling technique. *J. Ecol.* **58**, 865–868.

ETHERINGTON J.R. (1967) Studies of nutrient cycling and productivity in oligotrophic ecosystems. 1. Soil potassium and wind blown sea-spray in a south Wales dune grassland. *J. Ecol.* **55**, 743–752.

EVANS C.C. & ALLEN S.E. (1971) Nutrient losses in smoke produced during heather burning. *Oikos* **22**, 149–154.

FERRILL M.D. & WOODS F.W. (1966) Root extension in a long-leaf pine plantation. *Ecology* **47**, 97–102.

FLETCHER J.E. & ROBINSON M.E. (1956) A capacitance meter for estimating forage weight. *J. Range Mgmt.* **9**, 96–97.

FORD E.D. & NEWBOULD P.J. (1970) Stand structure and dry weight production through the Sweet Chestnut (*Castanea sativa* Mill.) coppice cycle. *J. Ecol.* **58**, 275–296.

FORREST G.I. (1971) Structure and production of north Pennine blanket bog vegetation. *J. Ecol.* **59**, 453–479.

FRANKLAND J.C. (1966) Succession of fungi on decaying petioles of *Pteridium aquilinum*. *J. Ecol.* **54**, 41–63.

FRANKLAND J.C., OVINGTON J.D. & MACRAE C. (1963) Spatial and seasonal variations in soil litter and ground vegetation in some Lake District woodlands. *J. Ecol.* **51**, 97–112.

FROMENT A. (1972) Soil respiration in a mixed Oak forest. *Oikos* **23**, 273–277.

GHILAROV M.S., KORDA V.A., NOVICHKOVA-IVANOVA L.N., RODIN L.E. & SVESHNIKOVA V.M. (Eds.) (1968) *Methods of Productivity Studies in Root Systems and Rhizosphere Organisms.* USSR, Acad. Sciences, Leningrad.

GILBERT O.J.B. & BOCOCK K.L. (1962) Some methods of studying the disappearing and decomposition of leaf litter. In P.W. Murphy (ed.) *Progress in Soil Zoology*, pp. 348–352. London, Butterworths.

GOLLEY F.B. (1960) An index to the rate of cellulose decomposition in the soil. *Ecology* **41**, 551–552.

GOLLEY F.B. & GENTRY J.B. (1965) A comparison of variety and standing crop of vegetation on a one-year and a twelve-year abandoned field. *Oikos* **15**, 185–199.

GORE A.J.P. & OLSON J.S. (1967) Preliminary models for accumulation of organic matter in an *Eriophorum/Calluna* ecosystem. *Aquilo Ser. Botanica* **6**, 297–313.

GORHAM E. (1958) The influence and importance of daily weather conditions in the supply of chloride, sulphate and other ions to freshwaters from atmospheric precipitation. *Phil. Trans. R. Soc. Ser. B.* **241**, 147–178.

GRASSLAND RESEARCH INSTITUTE STAFF (1961) *Research Techniques in use at the Grassland Research Institute, Hurley.* Bulletin 45. Commonwealth Bureau of Pastures and Field Crops, Hurley, Berks., England.

GREEN J.O. (1949) Herbage sampling errors and grazing trials. *J. Br. Grassld. Soc.* **4**, 11–16.

GROVER B.L. & LAMBORN R.E. (1970) Preparation of porous ceramic cups to be used for extraction of soil water having low solute concentrations. *Proc. Soil Sci. Soc. Am.* **34**, 706–708.

GYLLENBERG G. (1969) The energy flow through a *Chorthippus parallelus* (Zett.) (Orthoptera) population on a meadow in Tvärminne, Finland. *Acta Zool. fenn.* **123**, 1–74.

HAARLØV M. & WEIS-FOGH T. (1953) A microscopical technique studying the undisturbed texture of soils. *Oikos* **4**, 44–57.

HANSEN K. (1969) Edaphic conditions of Danish heath vegetation and the response to burning-off. *Bot. Tidsskr.* **64**, 121–140.

HARDY R.W.F., HOLSTEIN R.D., JACKSON E.K. & BURNS R.C. (1968) The acetylene ethylene assay for N$_2$ fixation. Laboratory and field evaluation. *Plant Physiol.* **43**, 1185–1207.

HARRIS T.M. (1946) Zinc poisoning of wild plants from wire netting. *New Phytol.* **45**, 50–55.

HAYES A.J. (1965a) Studies on the decomposition of coniferous leaf litter. 1. Physical and chemical changes. *J. Soil Sci.* **16**, 121–140.

HAYES A.J. (1965b) Studies on the decomposition of coniferous leaf litter. 2. Changes in external features and succession of micro fungi. *J. Soil Sci.* **16**, 242–257.

HEAD G.C. (1971) Plant roots. In J. Phillipson (ed.) *Methods of Study in Quantitative Soil Ecology: Population, Production and Energy Flow,* IBP Handbook No. 18, pp. 14–23. Oxford, Blackwell.

HEADY H.F. & VAN DYNE G.M. (1965) Prediction of weight composition from point samples of clipped herbage. *J. Range Mgmt.* **18**, 144–148.

HEALEY I.N. & SWIFT M.J. (1971) Aspects of the accumulation and decomposition of wood in the litter layer of a coppiced beech-oak woodland. In *VI Colloquium pedobiologiae,* pp. 417–430. Dijon 1970. Paris. Inst. National de la Recherche Agronomique.

HERON J. (1962) Determination of PO$_4$ in water after storage in polyethylene. *Limnol. Oceanogr.* **7**, 316–321.

HOUGH W.A., WOODS F.W. & MCCORMACK M.L. (1965) Root extension of individual trees in surface soils of a natural Longleaf Pine–Turkey Oak stand. *For. Sci.* **11**, 223–242.

HOWARD P.J.A. (1966) A method for the estimation of carbon dioxide evolved from the surface of soil in the field. *Oikos* **17**, 267–271.

HOWARD P.J.A. (1967) A method for studying the respiration and decomposition of litter. In O. Graff & J.E. Satchell (eds.) *Progress in Soil Biology,* pp. 464–472. Braunschweig, Vieweg.

HYDE F.J. & LAWRENCE J.T. (1964) Electronic assessment of pasture growth. *Electron. Engng.* **36**, 666–670.

IVLEV V.G. (1934) Eine mikromethode zur bestimmung des kaloriengehalts von nahrstoffen. *Biochem. Z.* **275**, 49–55.

JACQUES W.H. & SCHWASS R.H. (1956) Root development in some common New Zealand pasture plants. *N.Z. J. Sci. Technol.* **37**, 569–583.

JEFFERS J.N.R. (1972) *Mathematical Models in Ecology.* British Ecol. Soc. Symp. 12. Oxford, Blackwell.

JEFFREY D.W. (1970) A note on the use of ignition loss as a means for the approximate estimation of soil bulk density. *J. Ecol.* **58**, 297–299.

JENNY H., GESSEL S.P. & BINGHAM F.T. (1949) Comparative study of decomposition rates of organic matter in temperate and tropical regions. *Soil Sci.* **68**, 419–432.

JOHNS G.G., NICOL G.R. & WATKIN B.R. (1965) A modified capacitance probe for estimating pasture yield. *J. Br. Grassld. Soc.* **20**, 212–217.

KELLY O.J., HARDMAN J.A. & JENNINGS D.S. (1948) A soil sampling machine for obtaining two-, three-, and four-inch diameter cores of undisturbed soil to a depth of six feet. *Proc. Soil Sci. Soc. Am.* **12**, 85–87.

KENWORTHY J.B. (1963) Temperatures in heather burning. *Nature, Lond.* **200**, 1226.

KITTREDGE J. (1944) Estimation of the amount of foliage of trees and stands. *J. For.* **42**, 905–912.

KONONOVA M.M. (1961) *Soil Organic Matter—Its Nature, its Role in Soil Formation and in Soil Fertility.* Oxford, Pergamon.

KÖPF H. (1952) Laufende Messungen der Bodenatmung im Freiland. *Landw. Forsch.* **4**, 186–194.

KOSONEN M. (1968) The relation between the carbon dioxide production in the soil and the vegetation of a dry meadow. *Oikos* **19**, 242–249.

KOSONEN M. (1969) Carbon dioxide production in relation to plant mass. *Oikos* **20**, 335–343.

KROGH A. (1914) The quantitative relation between temperature and standard metabolism in animals. *Int. Z. phys.-chem. Biol.* **1**, 491–508.

KUCERA C.L. & KIRKHAM D.R. (1971) Soil respiration studies in tallgrass prairie in Missouri. *Ecology* **52,** 912–915.

LANGE O.L. & SCHULZE E.D. (1971) Measurement of CO_2 gas-exchange and transpiration in the Beech (*Fagus sylvaticus* L.) In H. Ellenberg (ed.) *Integrated Experimental Ecology,* pp. 16–28. Berlin, Springer-Verlag.

LATTER P.M. & CRAGG J.B. (1967) The decomposition of *Juncus squarrosus* leaves and microbiological changes in the profile of *Juncus* moor. *J. Ecol.* **55,** 465–482.

LATTER P.M., CRAGG J.B. & HEAL O.W. (1967) Comparative studies on the microbiology of four moorland soils in the northern Pennines. *J. Ecol.* **55,** 445–464.

LEE J.A. & WOOLHOUSE H.W. (1966) A reappraisal of the electrometric method for the determination of the concentration of carbon dioxide in soil atmospheres. *New Phytol.* **65,** 325–330.

LIETH H. (1968) The measurement of calorific values of biological material and the determination of ecological efficiency. In F.E. Eckardt (ed.) *Functioning of Terrestrial Ecosystems at the Primary Production Level,* pp. 233–242. Proc. Copenhagen Symposium, UNESCO.

LIETH H. (1968) Methods for the determination of the productivity of the underground organs. *Ibid.*

LIETH H. (1970) Phenology in productivity studies. In D.E. Reichle (ed.) *Analysis of Temperate Forest Ecosystems,* pp. 29–46. Berlin, Springer-Verlag.

LOCK J.M. (1972) The effects of hippopotamus grazing on grasslands. *J. Ecol.* **60,** 445–467.

LOMNICKI A., BANDOLA E. & JANKOWSKA K. (1968) Modification of the Wiegert-Evans method for estimation of net primary production. *Ecology* **49,** 147–149.

MACFADYEN A. (1970) Simple methods for measuring and maintaining the proportion of carbon dioxide in air, for use in ecological studies of soil respiration. *Soil Biol. Biochem.* **2,** 9–18.

MACFADYEN A. (1971) The soil and its total metabolism. In J. Phillipson (ed.) *Methods of Study in Quantitative Soil Ecology,* pp. 1–13. IBP Handbook No. 18. Oxford, Blackwell.

MADGWICK H.A.I. & OVINGTON J.D. (1959) The chemical composition of precipitation in adjacent forest and open plots. *Forestry* **32,** 14–22.

MAJOR J. (1969) Historical development of the ecosystem concept. In G.M. Van Dyne (ed.) *The Ecosystem Concept in Natural Resource Management,* pp. 9–22. New York, Academic Press.

MALLOCH A.J.C. (1972) Salt-spray deposition on the maritime cliffs of the Lizard peninsula. *J. Ecol.* **60,** 103–112.

MALONE C.R. & SWARTOUT M.B. (1969) Size, mass and caloric content of particulate organic matter in old-field and forest soils. *Ecology* **50,** 395–399.

MARK A.F. (1965) The environment and growth rate of narrow-leaved snow tussock, *Chionochloa rigida,* in Otago. *N.Z. J. Bot.* **3,** 73–103.

MARTIN D.J. (1955) Features of plant cuticle: an aid to the analysis of the natural diet of grazing animals, with special reference to Scottish hill sheep. *Trans. Proc. bot. Soc. Edinb.* **36,** 278–288.

MARTIN M.H. & PIGOTT C.D. (1965) A simple method for measuring carbon dioxide in soils. *J. Ecol.* **53,** 153–155.

MCKELL C.M., WILSON A.M. & JONES B.M. (1961) A flotation method for easy separation of roots from soil samples. *Agron. J.* **53,** 56–57.

MEDWECKA-KORNÁS A. (1971) Plant litter. In J. Phillipson (ed.) *Methods of Study in Quantitative Soil Ecology,* pp. 24–33. IBP Handbook No. 18. Oxford, Blackwell.

MILNER C. & HUGHES R.E. (1968) *Methods for the measurement of the primary production of grassland.* IBP Handbook No. 6. Oxford, Blackwell.

MINDERMAN G. (1956) The preparation of microtome section of unaltered soil for the study of soil organsims *in situ. Pl. Soil* **8,** 42–48.

MINDERMAN G. (1968) Addition, decomposition and accumulation of organic matter in forests. *J. Ecol.* **56**, 355–363.

MOTT G.O., BARNES R.F. & RHYKERD C.L. (1965) Estimating pasture yield *in situ* by beta-ray attenuation techniques. *Agron. J.* **57**, 512–513.

MOUNTFORD M.D. & BUNCE R.G.H. (1973) Regression sampling with allometrically related variables, with particular reference to production studies. *Forestry* **46**, 203–212.

MONTEITH J.L. (1962) Measurement and interpretation of CO_2 fluxes in the field. *J. Agron. Sci.* **10**, 334–346.

MONTEITH J.L. (1963) Gas exchange in plant communities. In L.T. Evans (ed.) *Environmental Control of Plant Growth*, pp. 95–112. New York, Academic Press.

MONTEITH J.L., SZEICZ G. & YABUKI K. (1964) Crop photosynthesis and the flux of carbon dioxide below the canopy. *J. appl. Ecol.* **1**, 321–337.

MOORE N.W. (1962) The heaths of Dorset and their conservation. *J. Ecol.* **50**, 369–391.

MURPHY P.W. (1962) A radioisotope method for determination of rate of disappearance of leaf littter in woodland. In P.W. Murphy (ed.) *Progress in Soil Zoology*, pp. 357–363. London, Butterworths.

NASYROV J.S., GILLIER J.E., LOGINOV M.A. & LEBEDOV V.N. (1962) The use of ^{14}C for studying the photosynthetic balance of plant communities. *Bot. Zh. S.S.S.R.* **47**, 96–99.

NEWBOULD P.J. (1967) *Methods for Estimating the Primary Production of Forests.* IBP Handbook No. 2. Oxford, Blackwell.

NEWMAN E.I. (1966) A method of estimating the total length of root in a sample. *J. appl. Ecol.* **3**, 139–146.

NEWTON J.D. (1923) Measurement of carbon dioxide evolved from the roots of various crop plants. *Scient. Agric.* **4**, 268–274.

NIELSON J.A. JR. (1964) Autoradiography for studying individual root systems in mixed herbaceous stands. *Ecology* **45**, 644–646.

NIHLGÅRD B. (1970) Precipitation, its chemical composition and effect on soil water in a beech and a spruce forest in south Sweden. *Oikos* **21**, 208–217.

NIMLOS T.J., VAN METER W.P. & DANIELS L.A. (1968) Rooting patterns of forest understory species as determined by radioiodine absorption. *Ecology* **49**, 1146–1151.

ODUM E.P. (1960) Organic production and turnover in old-field succession. *Ecology* **41**, 34–49.

OLSON J.S. (1958) Rates of succession and soil changes on southern Lake Michigan sand dunes. *Bot. Gaz.* **119**, 125–170.

OLSON J.S. (1963) Energy storage and the balance of producers and decomposers in ecological systems. *Ecology* **44**, 322–331.

OLSON J.S. (1964) Gross and net production of terrestrial vegetation. *J. Ecol.* **52** (Suppl.), 99–118.

OLSON J.S. & CROSSLEY D.A. JR. (1963) Tracer studies of the breakdown of forest litter. In V. Schultz & A.W. Klements (eds.) *Radioecology*, pp. 411–416. New York, Reinhold.

OVINGTON J.D. & MADGWICK H.A.I. (1959) The growth and composition of natural stands of birch. 1. Dry matter production. *Pl. Soil.* **10**, 271–283.

OVINGTON J.D. & MURRAY G. (1968) Seasonal periodicity of root growth of Birch trees. In M.S. Ghilarov *et al.* (eds.) *Methods of Productivity studies in Root Systems and Rhizosphere Organisms*, pp. 146–161. Leningrad, USSR Acad. Sciences.

PARIZEK R.P. & LANE B.E. (1970) Soil-water sampling using pan and deep pressure-vacuum lysimeters. *J. Hydrol.* **11**, 1–21.

PARIS O.H. & SIKORA A. (1967) Radiotracer analysis of the trophic dynamics of natural isopod populations. In K. Petrusewicz (ed.) *Secondary Productivity of Terrestrial Ecosystems (Principles and Methods)*, pp. 741–771. Warsawa-Kraków. Institute of Ecology, Polish Academy of Sciences.

PARKINSON D. & COUPS E. (1963) Microbial activity in a podzol. In J. Doeksen & J. van der Drift (eds.) *Soil Organisms*, pp. 167–175. Amsterdam, North Holland Publishing Co.

PATTEN B.C. (ed.) (1971) *Systems Analysis and Simulation in Ecology*. Vol. 1. New York, Academic Press.

PEARSON L.C. (1965) Primary production in grazed and ungrazed desert communities of eastern Idaho. *Ecology* **46**, 278–285.

PERRIN R.M.S. (1965) The use of drainage water analyses in soil studies. In E.G. Hallsworth & D.V. Crawford (eds.) *Experimental Pedology*, pp. 73–96. London, Butterworths.

PETERKEN G.F. & NEWBOULD P.J. (1966) Dry matter production by *Ilex aquifolium* L. in the New Forest. *J. Ecol.* **54**, 143–150.

PETRUSEWICZ K. & MACFADYEN A. (1970) *Productivity of Terrestrial Animals: Principles and Methods*. IBP Handbook No. 13. Oxford, Blackwell.

PHILLIPSON J. (1964) A miniature bomb calorimeter for small biological samples. *Oikos* **15**, 130–139.

PHILLIPSON J. (1971) Other Arthropods. In J. Phillipson (ed.) *Methods of Study in Quantitative Soil Ecology*, pp. 262–287. IBP Handbook No. 18. Oxford, Blackwell.

RAFES P.M. (1970) Estimation of the effect of phytophagous insects on forest production. In D.E. Reichle (ed.) *Analysis of Temperate Forest Ecosystems*, pp. 100–106. Berlin, Springer-Verlag.

RANDALL R.E. (1970) Salt measurement on the coasts of Barbados, West Indies. *Oikos* **21**, 65–70.

RAWES M. & WELCH D. (1969) Upland productivity of vegetation and sheep at Moor House National Nature Reserve, Westmorland, England. *Oikos Suppl.* **11**, 7–72.

REICHLE D.E. (ed.) (1970) *Analysis of Temperate Forest Ecosystems*. Berlin, Springer-Verlag.

REICHLE D.E., O'NEILL R.V., KAYE K.V., SOLLINS P. & BOOTH R.S. (1973) Systems analysis as applied to modeling ecological processes. *Oikos* **24**, 337–343.

REINERS W.A. (1968) Carbon dioxide evolution from the floor of three Minnesota forests. *Ecology* **49**, 471–483.

REMEZOV H.P., RODIN L.E. & BAZILEVICH N.I. (1963) Instructions on methods of studying the biological cycle of ash elements and nitrogen in the above ground parts of plants in the main natural zones of the temperate belt [Russian]. *Bot. Zh. S.S.S.R.* **48**, 869–877.

RICHMAN S. (1968) The transformation of energy by *Daphnia pulex*. *Ecol. Monogr.* **28**, 273–291.

RICKLEFS R.E. (1967) A graphical method of fitting equations to growth curves. *Ecology* **48**, 978–983.

RODIN L.E. & BAZILEVICH N.I. (1967) *Production and Mineral Cycling in Terrestrial Vegetation*. Edinburgh, Oliver & Boyd.

ROGERS W.S. & HEAD G.C. (1968) Studies of roots of fruit plants by observation panels and time-lapse photography. In M.S. Ghilarov *et al.* (eds.) *Methods of Productivity Studies in Root Systems and Rhizosphere Organisms*, pp. 176–185. Leningrad, USSR Acad. Sciences.

ROSS D.J. & BOYD I.W. (1970) Influence of moisture and aeration on oxygen uptakes in Warburg respiratory experiments with litter and soil. *Pl. Soil* **33**, 251–256.

RUTTER A.J. & WEBSTER J.R. (1962) Probes for sampling ground water for gas analysis. *J. Ecol.* **50**, 615–618.

SALISBURY E.J. (1922) The soils of Blakeney Point: A study of soil reaction and succession in relation to plant covering. *Ann. Bot.* **36**, 391–431.

SALISBURY E.J. (1925) Note on the edaphic succession in some sand dune soils with special reference to the time factor. *J. Ecol.* **13**, 322–328.

SALT G. & HOLLICK F.S.J. (1944) Studies of wireworm populations. 1. A census of wireworms in pasture. *Ann. appl. Biol.* **31**, 52–64.

SATOO T. (1970) A synthesis of studies by the harvest method: primary production relations in the temperate deciduous forests of Japan. In D.E. Reichle (ed.) *Analysis of Temperate Forest Ecosystems*, pp. 55–72. Berlin, Springer-Verlag.

SATOR CH. & BOMMER D. (1971) Methodological studies to distinguish functional from non-functional roots of grassland plants. In H. Ellenberg (ed.) *Integrated Experimental Ecology*, pp. 72–78. Berlin, Springer-Verlag.

SCHÖLLHORN R. & BURRIS R.H. (1967) Acetylene as a competitive inhibitor of nitrogen fixation. *Proc. natn. Acad. Sci. U.S.A.* **58,** 213–216.

SCHUSTER J.L. (1964) Root development of native plants under three grazing intensities. *Ecology* **45,** 63–70.

SCHUURMAN J.J. & GOEDEWAAGEN M.A.J. (1971) *Methods for the Examination of Root Systems and Roots* (2nd edition). Wageningen, Centre for Agricultural Publications and Documentation.

SCOTT D. (1961) Methods of measuring growth in short tussocks. *N.Z. J. agric. Res.* **4,** 282–285.

SHANKS R.E. & OLSON J.S. (1961) First year breakdown of leaf litter in southern Appalachian forests. *Science, N.Y.* **134,** 194–195.

SHAW M.W. (1968) Factor affecting the natural regeneration of Sessile Oak (*Quercus petraea*) in north Wales. 2. Acorn losses and germination under field conditions. *J. Ecol.* **56,** 647–660.

SHEIKH K.H. & RUTTER A.J. (1969) The responses of *Molinia caerulea* and *Erica tetralix* to soil aeration and related factors. 1. Root distribution in relation to soil porosity. *J. Ecol.* **57,** 713–726.

SMITH F.E. (1970) Analysis of ecosystems. In D.E. Reichle (ed.) *Analysis of Temperate Forest Ecosystems*, pp. 7–18. Berlin, Springer-Verlag.

SOUTHWOOD T.R.E. (1966) *Ecological Methods: with particular reference to the study of insect populations*. London, Methuen.

STAGE A.R. (1963) Specific gravity and tree weight of single-tree samples of Grand Fir. *US For. Serv. Res. Pap. Intermt. For. Range Exp. Stn. No.* INT-4, pp. 11.

STEWART D.R.M. (1967) Analysis of plant epidermis in faeces: a technique for studying the food preferences of grazing herbivores. *J. appl. Ecol.* **4,** 83–111.

STEWART W.D.P., FITZGERALD G.P. & BURRIS R.H. (1967) *In situ* studies on N_2 fixation using the acetylene reduction technique. *Proc. Acad. Sci. USA* **58,** 2071–2078.

STOCKER O. & VIEWEG F.H. (1960) Die Darmstädter apparatur zur momentanmessung der photosynthese unter ökologischen bedingungen. *Ber. dt. bot. Ges.* **73,** 198–208.

STORR G.M. (1961) Microscopic analysis of faeces, a technique for ascertaining the diet of herbivorous mammals. *Aust. J. biol. Sci.* **14,** 157–164.

SYKES J.M. & BUNCE R.G.H. (1970) Fluctuations in litter fall in a mixed deciduous woodland over a three-year period 1966–68. *Oikos* **21,** 326–329.

TANSLEY A.G. (1935) The use and abuse of vegetational concepts and terms. *Ecology* **16,** 284–307.

THAKE B. & RAWLE P.R. (1972) Non-biological production of ethylene in the acetylene reduction assay for Nitrogenase. *Arch. Mikrobiol.* **85,** 39–43.

UENO M., YASHIHARA K. & OKADA T. (1967) Living root system distinguished by the use of Carbon-14. *Nature, Lond.* **213,** 530–532.

VAN DYNE G.M. (1966) The use of a vacuum clipper for harvesting herbage. *Ecology* **47,** 624–623.

VICKERY P.J. (1972) Grazing and net primary production of a temperate grassland. *J. appl. Ecol.* **9,** 307–314.

WAGNER G.H. (1962) Use of porous ceramic pots to sample soil water within the profile. *Soil Sci.* **94,** 379–386.

WALTERS C.S. & BRUCKMANN G. (1964) A comparison of methods for determining volume of increment cores. *J. For.* **62,** 172–177.

WANNER H. (1970) Soil respiration, litter fall and productivity of tropical rain forest. *J. Ecol.* **58,** 543–547.

WARREN WILSON J. (1960) Inclined point quadrats. *New Phytol.* **59,** 1–8.

WARREN WILSON J. (1963) Estimation of foliage denseness and foliage angle by inclined point quadrats. *Aust. J. Bot.* **11,** 95–105.

WAUGHAM G.J. (1971) Field use of the acetylene reduction assay for nitrogen fixation. *Oikos* **22,** 111–113.

WAUGHAM G.J. (1972) Acetylene reduction assay for nitrogen fixation in sand dunes. *Oikos* **23,** 206–212.

WELBANK P.J. & WILLIAMS E.D. (1968) Root growth of a Barley crop estimated by sampling with portable powered soil-coring equipment. *J. appl. Ecol.* **5,** 477–482.

WELCH D. & RAWES M. (1965) The herbage production of some Pennine grasslands. *Oikos* **16,** 39–47.

WENT J.C. (1959) Cellophane as a medium to study cellulose decomposition in forest soils. *Acta bot. neerl.* **8,** 490–491.

WESTLAKE D.F. (1963) Comparisons of plant productivity. *Biol. Rev.* **38,** 385–425.

WHITE E.J. & TURNER F. (1970) A method of estimating income of nutrients in a catch of airborne particles by a woodland canopy. *J. appl. Ecol.* **7,** 441–461.

WHITTAKER E. (1961) Temperatures in heath fires. *J. Ecol.* **49,** 709–716.

WHITTAKER R.H. & WOODWELL G.M. (1968) Dimension and production relations of trees and shrubs in the Brookhaven Forest, New York. *J. Ecol.* **56,** 1–25.

WIANT H.V. (1967a) Influence of moisture content on soil respiration. *J. For.* **65,** 902–903.

WIANT H.V. (1967b) Influence of temperature on the rate of soil respiration. *J. For.* **65,** 489–490.

WIEGERT R.G. & EVANS F.C. (1964) Primary production and the disappearance of dead vegetation on an old field. *Ecology* **45,** 49–63.

WILLIAMS O.B. (1969) An improved technique for identification of plant fragments in herbivore faeces. *J. Range Mgmt.* **22,** 51–52.

WILSON K. (1960) The time factor in the development of dune soils at South Haven peninsula, Dorset. *J. Ecol.* **48,** 341–360.

WITKAMP M. (1966) Rates of carbon dioxide evolution from the forest floor. *Ecology* **47,** 492–494.

WITKAMP M. (1969) Cycles of temperature and carbon dioxide evolution from litter and soil. *Ecology* **50,** 922–924.

WITKAMP M. & FRANK M.L. (1969) Evolution of carbon dioxide from litter, humus and subsoil of a Pine stand. *Pedobiologia* **9,** 358–366.

WITKAMP M. & OLSON J.S. (1963) Breakdown of confined and non-confined Oak litter. *Oikos* **14,** 138–147.

WOOD S.F.W. (1970) Interspecific transfer of inorganic materials by root systems of woody plants. *J. appl. Ecol.* **7,** 481–486.

WOODS F.W. & BROCK K. (1964) Interspecific transfer of Ca-45 and P-32 by root systems. *Ecology* **45,** 886–889.

WOODS F.W., FERRILL M.D. & MCCORMACK M.L. (1962) Methyl bromide for increasing Iodine-131 uptake by Pine trees. *Radiat. Bot.* **2,** 273–277.

WOODWELL G.M. & DYKEMAN W.R. (1966) Respiration of a forest measured by carbon dioxide accumulation during temperature inversions. *Science, N.Y.* **154,** 1031–1034.

YAMAGUCHI S. & CRAFTS A.S. (1958) Autoradiographic method for studying absorption and translocation of herbicides using Carbon-14 labelled compounds. *Hilgardia* **28,** 161–191.

YOUNG H.E. (ed.) (1968) *Symposium on Primary Production and Mineral Cycling in Natural Ecosystems.* University of Maine Press.

ZYZNAR E. & URNESS P.J. (1969) Qualitative identification of forage remnants in deer faeces. *J. Wildl. Mgmt.* **33,** 506–510.

CHAPTER 5

PHYSIOLOGICAL ECOLOGY AND PLANT NUTRITION

P. BANNISTER

1   **Introduction**   230

2   **Photosynthesis and respiration**   233
2.1   The response of plants to radiant energy   233
   2.1.1   Growth as a response to radiant energy   234
2.2   Measures of photosynthesis   236
   2.2.1   Carbohydrate production   237
   2.2.2   The uptake of carbon dioxide (CO_2)   238
   (a) Closed systems   240
   (b) Open systems   240
   (c) Combined systems   241
   (d) The measurement of CO_2 241
   2.2.3   The evolution of oxygen   242
   2.2.4   The measurement of photosynthesis in the field   242
2.3   Respiration   244
2.4   Compensation points and photosynthetic optima   244
   2.4.1   Light and temperature compensation points   245
   2.4.2   Carbon dioxide 'compensation point'   246

3   **Temperature relations**   247
3.1   The measurement of plant temperature   247
3.2   The resistance of plants to extremes of temperature   248
   3.2.1   High temperatures   248
   3.2.2   Low temperatures   249
   3.2.3   Measurement of resistance to extremes of temperature   249

4   **Water relations**   251
4.1   Introduction   251
4.2   The water status of plant tissues   252
   4.2.1   Water content   252

4.2.2   The measurement of relative water content   253
4.2.3   Water potential   255
(a) The use of tissue characteristics in water potential determinations   256
(b) Techniques involving the measurement of a solution characteristic   258
(c) Techniques involving the determination of a vapour or gaseous characteristic   259
4.2.4   Measurement of osmotic potential   260
4.2.5   The relationship between water potential and relative water content   262
4.3   The loss of water from plants   263
4.3.1   The measurement of transpiration   264
4.4   Stomatal aperture   266
   4.4.1   Direction observation   266
   4.4.2   Indirect methods   267
   4.4.3   Porometers   267
   (a) Viscous flow porometers   268
   (b) Diffusive flow porometers   268
   4.4.4   The relationship between stomatal closure and water deficit   268
4.5   The diffusive resistance of leaves   271
4.6   Drought resistance   272

5   **The mineral nutrition of plants**   274
5.1   Introduction   274
5.2   Field experiments   275
5.3   Laboratory experiments   277
5.4   Nutrient Deficiency   278
5.5   Excess of nutrient and other ions   280
5.6   Competition and mineral nutrition   283

6   **References**   285

1 INTRODUCTION

It is appropriate to start this chapter by considering a possible definition of 'physiological plant ecology'. Physiology is the study of the functioning of organisms; ecology is the study of organisms in relation to their environment and thus physiological plant ecology may be defined as the study of the functioning of organisms in relation to their environment. This chapter considers 'whole plant physiology' in relation to ecology although physiological plant ecology can also be pursued at a cellular or biochemical level (e.g. Crawford, 1966; Woolhouse, 1969).

The physiological approach to plant ecology is as old as the subject of plant ecology itself. Classical ecological texts such as Warming (1895, 1909) and Schimper (1898, 1903) were well aware of the relevance of physiological studies. Latterly, two Presidential Addresses to the British Ecological Society (Clapham, 1956; Harper, 1967) have encouraged the development of the physiological approach to plant ecology in the English-speaking world.

Field studies offer the raw material for physiological investigations as they allow species abundance to be correlated with environmental variability. Cause and effect may be inferred from field observation but can only be verified through critical experimentation; this is where the physiological approach comes into its own.

However, experimental studies are not without their own problems. Growth, usually as dry matter production, is often measured with the inference that the plants showing the best growth will also show an optimal ecological response. Rorison (1969) has discussed the problems of interpreting dry weight gain in physiological experiments. A major problem is whether the biggest is necessarily the best; in extreme habitats slow-growing species may be the most successful. A partial solution is to compare the responses of different species to a range of conditions. In an extreme situation, a slow-growing species may actually grow better than a rapidly-growing species. Some measurement of relative response may be useful, for a species growing at 90% of its optimum rate would appear to be better adapted to a situation than a species growing at 50% of its optimum rate, even if the latter species showed the greater amount of growth. If all species showed an identical response to environmental factors then the field of the physiological plant ecologist would be identical with that of the plant physiologist. Species' responses differ and thus the contribution of the physiological approach to the study of ecology includes the comparison of species' responses.

There are various attributes of species which determine success or failure in the field. Each species has a range of absolute requirements (e.g. a maximum and minimum temperature at which it can survive). The physiological plant ecologist can determine these limits for different species. Within the absolute requirements there is a range of tolerance which is likely to differ between species. The tolerance range of a species in the field (ecological amplitude) differs from the range found in laboratory experiments (physiological amplitude) (Fig. 5.1). Wide amplitudes may result from either a very plastic (phenotypic) response or the evolution of distinct genetic races (genotypic response). Physiological investigations are needed to investigate the mechanism of a plastic response and to compare the behaviour of ecological races (ecotypes).

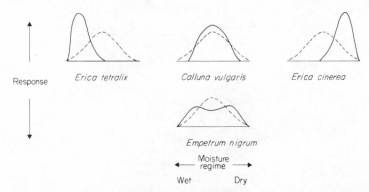

Figure 5.1 A diagrammatic comparison of ecological (—) and physiological (---) amplitude of four heathland dwarf shrubs with respect to soil moisture (cf. Ellenberg (1958) and Rorison (1969)).

Some of the possible relationships between physiological and ecological amplitudes are given in Fig. 5.1; these are well illustrated by reference to heath plants. In laboratory conditions most dwarf shrubs show their best growth on intermediate, rather than wet or dry moisture régimes (Bannister, 1964b) but the ecological response is often different. *Calluna vulgaris* alone shows a similarity between laboratory and field response while *Erica tetralix* is typical of wet régimes in the field, *E. cinerea* of dry régimes and *Empetrum nigrum* of either wet or dry régimes (Gimingham, 1972). Competition causes the difference between ecological and physiological responses. A vigorous competitor is likely to have similar physiological and ecological amplitudes whereas feeble competitors are limited to a part of their tolerance range where they have some competitive advantage. *Erica tetralix* is more tolerant of water-logging and *E. cinera* more drought resistant than *Calluna* (Bannister, 1964d) thus they can compete successfully on wet and dry sites respectively (Fig. 5.1). The response of *Empetrum* is probably due to distinct ecotypes (Gimingham, 1972).

Competition may explain the difference between laboratory and field responses and may be inferred from measurements such as relative growth rates and growth form. Its influence can be confirmed only by appropriate experiments; these are comparatively rare.

It is also important to consider the phase of the life cycle that is to be studied. An investigation should commence with a consideration of conditions necessary for the germination and establishment of seeds and seedlings. If these requirements are found to be crucial, then there is less point in studying

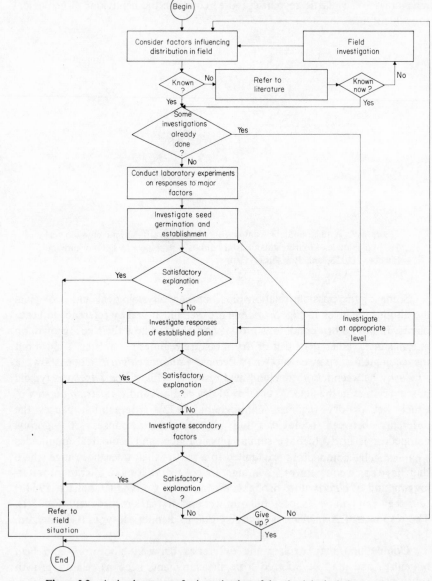

Figure 5.2 A simple strategy for investigation of the physiological plant ecology of a species or group of species.

the physiology of the mature plant. However, a study of the causes of failures in seedling establishment would include physiological investigations similar to those considered in the rest of this chapter. A possible strategy of investigation is outlined in Fig. 5.2; this emphasizes the laboratory and physiological approaches but recognizes the need of field studies both to initiate and interpret experiments.

Some of the approaches to physiological experimentation and problems are considered under three main headings: the plant's response to the aerial environment, water relations, and, finally, mineral nutrition. The treatment is necessarily superficial but should provide not only some indication of the techniques, potentialities and difficulties of the physiological approach to plant ecology but also adequate reference to more specialized reading.

2 PHOTOSYNTHESIS AND RESPIRATION

2.1 THE RESPONSE OF PLANTS TO RADIANT ENERGY

Radiant energy is usually separated into light and heat by most plant ecologists, but, because of interactions (Fig. 5.3), the influence of these two components is best considered jointly. Various plant responses may be examined and given an ecological interpretation (Fig. 5.3). Some of these, such as stomatal aperture, transpiration and tissue water balance, are considered later (4.1–4.6).

Plant growth integrates a variety of physiological events and is readily

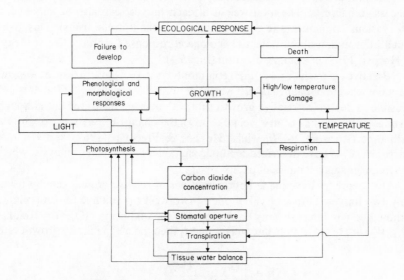

Figure 5.3 Plant response to light and temperature.

related to ecological responses (see Introduction). Various aspects of growth can be quantified by growth analysis (2.1.1) and are relevant to the study of the effects of radiant energy on plant responses. Direct measurements of photosynthesis and respiration (2.2, 2.3) may be used to make models of growth and production (e.g. Grace & Woolhouse, 1970) or to characterize the responses of species. Light also has a photoperiodic effect on growth and development. If the requirement is not met then the plant will fail to develop or some modification of growth or development may change the response and affect the ecology of the species. Extremes of temperature (3.2) may cause the death of the plant or alter the pattern or amount of growth.

2.1.1 Growth as a response to radiant energy

Growth integrates a variety of responses but is not a good measure of the rate of photosynthesis. As a plant grows, it produces more photosynthetic tissue which then allows more photosynthesis. This leads to an exponential increase and, although the exponent varies both with time and environment the results of experiments which measure growth (e.g. as height, dry weight) should therefore be analysed in logarithmic rather than arithmetic units.

More sophisticated measurements of growth have led to the science of 'growth analysis' which was developed for crop plants (Watson, 1952). A comprehensive work on this subject has recently been produced (Evans, 1972) and it would be both impossible and impertinent to attempt to cover the whole subject here.

The first increase in sophistication is to analyse growth over a number of successive harvests. The total weight at each harvest can also be subdivided into various components (e.g. roots, leaves, stems, flowers, fruits). This gives much information about potential ecological response ('reproductive strategy' of Harper, 1967) in a simple diagram (Fig. 5.4).

'Relative growth rates' are commonly used. They measure the instantaneous rate of change of dry weight per unit dry weight with time. It is not possible to determine relative growth rate with a single measurement although a series of logarithms of dry weights shows the change of relative growth rate with time (Rorison, 1969), whilst Hughes & Freeman (1967) and Hunt & Parsons (1974) have described methods, suitable for computer analysis, which use frequent small samples of plants.

Most workers have made approximations of relative growth rate by using only two harvests. An interval of one or two weeks is suitable for crop plants although longer intervals may be appropriate for wild plants. If the dry weights (W_1, W_2) are taken at time intervals t_1 and t_2 then the mean relative growth rate $(\bar{R})$ is given by:

$$\bar{R} = \frac{\ln (W_2) - \ln (W_1)}{t_2 - t_1}$$

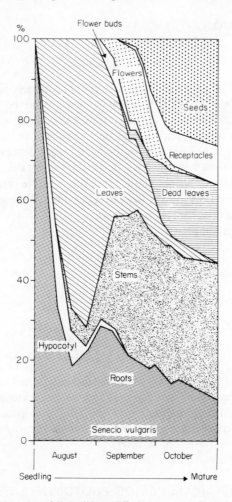

Figure 5.4   Allocation of dry weight to different structures through the life cycle of *Senecio vulgaris*. (Harper & Ogden, 1970.)

The instantaneous relative growth rate (**R**) is given by

$$\bar{R} = \frac{1}{W} \cdot \frac{dW}{dt} \quad \text{(where } W \text{ is dry weight).}$$

The dry weight increment of the plant is a function of leaf area. An estimate of leaf area (A_2, A_1) allows the calculation of the mean unit leaf rate, $\bar{E}$ (Evans, 1972) also known as the net assimilation rate (NAR).

$$\bar{E} = \frac{W_2 - W_1}{t_2 - t_1} \cdot \frac{\ln(A_2) - \ln(A_1)}{A_2 - A_1}$$

Or instantaneously

$$E = \frac{dW}{dt} \cdot \frac{1}{A}$$

The measures of weight and leaf area must be linearly related, this is valid only for short time intervals. **E** is not a measure of the photosynthesis of individual leaves as it includes the respiration losses of the whole plant. Thus **E** can fall, without any changes in the photosynthetic or respiration rates of leaves, when the ratio of leaf area to total dry weight decreases, e.g. by loss of leaves or by an increase in non-leafy parts. Leaf area ratio (LAR) is the ratio of leaf area to total dry weight and is the product of the ratio of total leaf dry weight (W_L) to plant dry weight (leaf weight ratio, LWR) and the ratio of leaf area to leaf weight (specific leaf area SLA).

Thus, at an instant in time,

$$R = E \times LAR = E \times LWR \times SLA.$$

i.e.

$$\frac{1}{W} \cdot \frac{dW}{dt} = \frac{dW}{dt} \cdot \frac{1}{A} \times \frac{W_L}{W} \times \frac{A}{W_L}$$

$$\quad\ \ \mathbf{R} \qquad\quad \mathbf{E} \qquad \mathbf{LWR} \ \ \mathbf{SLA}$$

$$\underbrace{\qquad\qquad\qquad}_{\textstyle LAR}$$

The results of growth analysis should be interpreted with care. Relative growth rates change with time (Fig. 5.5) and the rate measured at one stage of growth may not be representative of the plant's response at other times. Moreover, high relative growth rates are not necessarily a pre-requisite of ecological success (cf. Bradshaw, 1969).

2.2 MEASURES OF PHOTOSYNTHESIS

The process of photosynthesis uses carbon dioxide, water and radiant energy to produce carbohydrate, water and oxygen. Theoretically any one component could estimate photosynthetic activity but some must be dismissed as impractical. Photosynthesis both consumes and produces water, but photosynthetic water would be swamped and rendered undetectable by the bulk of other water in the plant. Similarly, small errors in the measurement of fluxes of radiant energy could produce misleading estimates of photosynthesis as this uses only a small fraction of the total energy. Such methods (e.g. Montieth, 1968) are suitable for measuring the photosynthesis of large homogenous stands of vegetation.

The production of carbohydrate, the evolution of oxygen and the utilization of carbon dioxide are all affected by respiration which utilizes

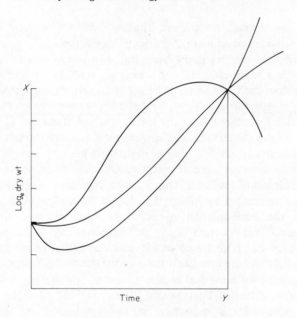

Figure 5.5 A diagrammatic representation of a range of growth patterns which may achieve the same yield Y at time X. Relative growth rate at any instant in time is given by the slope of the graph. (After Rorison, 1969.)

carbohydrate and oxygen and produces carbon dioxide. Apparent (net) photosynthesis is the difference between gross (real) photosynthesis and respiration. Dark respiration is often used as an estimate of the respiration component, but respiration in the light (photorespiration) is probably higher than in the dark (Heath, 1969).

A detailed consideration of methods of measuring photosynthesis are given by Heath (1969) and Šesták *et al.* (1971).

2.2.1 Carbohydrate production

Carbohydrate analyses do not measure absolute rates of photosynthesis although they may indicate qualitative changes. However, longer-term analyses (e.g. over a growing season) of changes in carbohydrate content can be used to examine and compare the responses of individuals to the environment (e.g. Stewart & Bannister, 1973). There are many methods of carbohydrate analysis; the anthrone technique (e.g. Yemm & Willis, 1954) is quite convenient and allows the estimation of carbohydrate as ethanol-soluble and perchloric acid-soluble fractions (viz. sugars and starch).

Photosynthetic carbohydrate may be rapidly translocated and readily converted into other compounds (e.g. amino-acids, proteins and organic acids), but is only lost through respiration. Thus, the dry weight increment of the whole plant provides an estimate of the net photosynthesis.

Sach's half-leaf method is the basis for most procedures which use dry weight, provide a rapid estimate, and which are suitable for field use. Half a leaf is taken from each of a series of leaves of known area at the beginning of the period of measurement. Alternatively discs of known area are cut from one half of the leaf. The other half of the leaf is sampled at the end of the period. Translocation from the leaves and changes in leaf area are sources of error. A turgid leaf is stretched and has a lower dry weight per unit area than a flaccid leaf and thus an apparent increase in dry weight could be caused by water loss and the contraction of leaf area. Translocation represents a net loss from the leaf and may be limited by killing the phloem by applying hot wax to the petiole, but the accumulation of products may influence the rate of photosynthesis. Translocation may be estimated by measurements on darkened leaves but both translocation and respiration rates are probably higher in the light and are thus likely to be underestimated. Changes in leaf area may be accounted for if the leaf area is measured both initially and when the half leaf is finally detached (Portsmouth, 1949).

If it is assumed that translocation and respiration are the same in illuminated and darkened leaves, then gross photosynthesis can be estimated

viz. $$P = w_2 - w_1 + r + t$$

where P is the dry weight increase due to photosynthesis, w_2 and w_1 are the final and initial weight of leaf samples, r and t are losses due to respiration and translocation.

As weight changes (w_1', w_2') in the dark are attributed solely to translocation and respiration,

$$w_1' - w_2' = r + t$$

Thus $$P = w_2 - w_1 + w_1' - w_2'$$

with adequate sampling $$w_1 = w_1'$$

therefore $$P = w_2 - w_2'$$

2.2.2 The uptake of carbon dioxide (CO_2)

Photosynthesis reduces the atmospheric concentration of CO_2 (0·03%) to lower levels which must be measurable if photosynthesis is to be detected. Alterations of CO_2 concentration affect both the rate of photosynthesis and stomatal aperture. A lowering of CO_2 concentration in an assimilation chamber could limit the rate of photosynthesis, cause an increase in stomatal aperture (Heath & Williams, 1948) and increase transpiration leading to an increased water vapour pressure and a greater retention of the heat (Fig. 5.6).

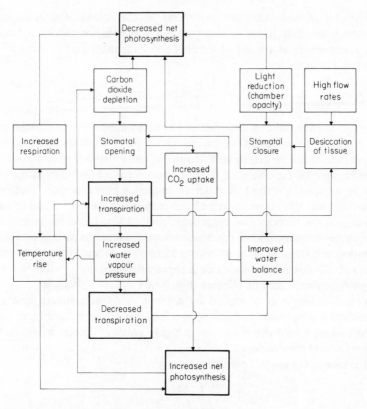

Figure 5.6 Some possible outcomes of enclosing a plant part in a chamber for the measurement of either photosynthesis or transpiration.

The material of the chamber may alter the short-wave radiation balance and reduce or even increase the photosynthetic rate (Jarvis, 1970). The retention of long-wave radiation may result in acute heating. The problems of chamber climate are more acute under field conditions. Lange *et al.* (1969) have employed a sophisticated (and expensive) chamber which reproduces the conditions of the external environment. Such chambers can also provide experimental environments but it can still be contended that they do not reproduce the environment of the leaf (Jarvis, 1969). An alternative is to use a chamber for only a very short period (Lange, 1962). In laboratory measurements, where the object is to relate various factors to photosynthetic rate, the problem may be partially overcome by monitoring the components of the microenvironment with the chamber. It is still difficult to relate laboratory performance to the field as pot-grown plants have developed differently and are isolated from the competition and shading which occurs under natural conditions.

Methods of measuring photosynthesis by the uptake of CO_2 require an instrument, sensitive to small changes, and a system for containing the gas. Such systems involve the use of a chamber for the plant part in either an open or a closed system.

(a) *Closed systems*

If a plant is enclosed in a chamber of known volume and the change in concentration of carbon dioxide measured during photosynthesis, then the rate of photosynthesis can be determined. However, most methods of analysis require the exhaustion of the chamber. Where direct measurement is possible the air must be adequately mixed in order to minimize concentration gradients of carbon dioxide. Enclosure must be brief because of the exhaustion of carbon dioxide and the problems of chamber climate. Closed systems consequently incorporate some method of continuous flow. Changes in carbon dioxide concentration are small in systems with a large volume and must be accurately measured. However, carbon dioxide is rapidly depleted in a small volume of air and photosynthesis may be affected (Fig. 5.6).

Volatile substances produced by a plant could accumulate in a closed system and inhibit photosynthesis or even register on the measuring system. A closed system should therefore have a highly sensitive detection system and a large volume of circulating air.

The rate of uptake (U) per unit time is given by:

$$U = \frac{v(c_1 - c_2)}{t_2 - t_1}$$

where v is the volume of the system, c_1, c_2 are the initial and final concentrations, t_1, t_2 are the initial and final times.

Carbon dioxide may be injected at a very slow rate into an otherwise closed system; when a steady state is obtained the rate of injection is equal to the photosynthetic rate (Meidner, 1967).

(b) *Open systems*

In open systems there is usually an airflow from an external source which is buffered from external fluctuations in CO_2 concentration by using a large storage vessel. The concentration of carbon dioxide is measured both before and after passage over the plant. The rate of flow must also be measured. The uptake is then given by

$$U = f(c_1 - c_2)$$

where the symbols are as before and f is the rate of flow.

A slow rate of flow allows a large change in concentration which is easy to measure, but it becomes difficult to specify the concentration over the leaf (Heath, 1969) and the slow rate allows complications to develop (Fig. 5.6). A higher rate of flow requires more sensitive methods of detection. Too high a flow rate may lead to desiccation, stomatal closure and a decrease in photosynthesis.

Ungerson & Scherdin (1968) have described a method for field use in which the carbon dioxide passing over a plant is absorbed and compared with a similar sample without a plant.

(c) *Combined systems*

In semi-closed circuits, a slow flow of CO_2 enriched air at a known concentration (c) is fed into a continuous circuit with a more rapid rate of flow. Air is allowed to escape at the same rate as it enters. When a steady state of photosynthesis occurs, the concentration in the closed part of the system is constant. This use of a null-point makes the method sensitive.

The amount of CO_2 entering the chamber is the sum of the amount entering (f_1c_1) and the amount circulating in the system (f_2c_2). The amount leaving is the sum of the amount circulating (f_2c_2) and the amount leaving the system (f_1c_2). Thus

$$U = (f_1c_1 + f_2c_2) - (f_2c_2 + f_1c_2) = f_1(c_1 - c_2)$$

The rate of flow around the closed part of the system (f_2c_2) is not needed in the calculation of the assimilation rate.

(d) *The measurement of* CO_2

The infra-red gas analyser is probably the most sensitive instrument available for measuring CO_2 concentration (Bowman, 1968). The principle of the instrument is the absorption of infra-red radiation by carbon dioxide. Water vapour also absorbs radiation of the same spectral range and thus air streams must be dried before measurement or filters (interference filters or filters filled with water vapour) used to eliminate the effects of water vapour. The instrument may also be used for the estimation of water vapour (Decker & Wien, 1960). In modern differential instruments, the infra-red source is passed into a sample cell containing a known standard (a mixture of air and CO_2 or CO_2 free air) and another containing the unknown. The infra-red radiation which is not absorbed by the cells enters two chambers of a detector filled with CO_2. Different samples produce differential heating which moves a diaphragm within the detector. Any movement is translated into an electrical signal. The differential instrument can be used in an open circuit system with the incoming air as a standard. Sensitivity is enhanced but calibration requires considerable care (cf. Ludlow & Jarvis, 1971).

2.2.3 The evolution of oxygen

Most methods use solutions and are more suitable for aquatic plants. The bubbling method of Sachs is, despite modifications by Wilmott (1921) and Audus (1953), unsuitable for quantitative estimation except when the oxygen concentration is measured. Oxygen concentration may be determined chemically. Winkler's method is widely used (e.g. Carritt & Carpenter, 1966), Linossier's method as modified by Miller (1914) is recommended by Heath (1969). Chemical methods have been used in conjunction with flow systems (cf. James, 1928) but there is no instrument as sensitive as the infra-red gas analyser) to make continuous measurements. However, polarographic techniques with suitable electrodes (oxygen electrodes) can detect changes in oxygen saturation (cf. Šesták *et al.*, 1971). As the diffusion of carbon dioxide in water is much slower than in air, bicarbonate buffers (which give an instant supply of CO_2) are often added during the determination of the photosynthesis of aquatic plants. The buffers produce unnaturally high pH levels. The Warburg technique may be used to measure photosynthesis in aquatic plants (e.g. Spence & Chrystal, 1970). Pressure changes are produced by the different solubilities of oxygen and carbon dioxide in the liquid medium. Assumptions have to be made about the relationships within one flask, or a parallel series of flasks of different volume must be used or buffers used to maintain a constant carbon dioxide concentration (Heath, 1969).

2.2.4 The measurement of photosynthesis in the field

The chief problem when photosynthesis is estimated in the field is the maintenance of 'natural' conditions during measurement (particularly when assimilation chambers are used). This is discussed elsewhere (2.2.2).

One solution to the problem of chamber climate is to enclose the plant part only for a short time (e.g. Lange, 1962; Ungerson & Scherdin, 1968) another is to dispense with a chamber and analyse the total flux of energy and CO_2. This requires sophisticated measurement and, if the appropriate calculations are to be valid, an idealized stand of uniform vegetation (Montieth, 1968).

The half-leaf method has the virtue that the environment of the plant is relatively undisturbed, although a considerable physiological disturbance may result from cutting the leaf.

The most commonly used methods in the field are those using assimilation chambers. Both infra-red gas analysis and conductivity measurements have been used successfully to measure the CO_2 content. An adequate power supply is needed for field use: this may involve a mobile laboratory (e.g. Eckardt, 1968) but the conductivity method as modified by Ungerson & Scherdin (1968) is hand-portable. The external climate in the field cannot be controlled and in temperate climates humid air is drawn through tubes which run over the cold ground leading to condensation within the tubes and measuring apparatus.

The short-term measurement of photosynthesis in water plants may be possible with polarographic techniques (cf. Šesták *et al.,* 1971). Radioactive carbon dioxide (from a bicarbonate source in a closed assimilation chamber) has been used to estimate photosynthetic rates in free-living freshwater macrophytes (cf. Campbell, 1972; Vosnesenskii & Zalenskii, 1971). (The use of labelled carbon dioxide [$^{14}CO_2$] in the measurement of gross production is discussed in Chapter 4.) With algal populations measurement of total oxygen production or even of changes in biomass provide a more readily usable technique.

Field measurements are likely to differ from laboratory measurements of photosynthesis. Laboratory plants are usually well grown and supplied with near optimum amounts of all the essentials for growth. Thus, even when subjected to stress, they may be influenced by their favourable prehistory. In contrast plants in the field are in competition for light, space, nutrients and are commonly under conditions of water stress. Consequently some workers (e.g. Hesketh & Baker, 1967) have distinguished physiological and ecological optima for photosynthesis (Fig. 5.7).

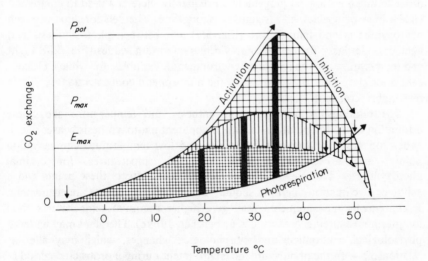

Figure 5.7 Relationships between photosynthesis, respiration, temperature and the physiological and ecological responses of a plant. P_{pot}, potential physiological maximum of gross photosynthesis (all requirements, including CO_2, at optimum levels). P_{max}, physiological maximum (CO_2 limiting). E_{max}, ecological maximum (most factors limiting). Vertical bars represent optima for net photosynthesis. Vertical arrows indicate temperature compensation points. (After Hesketh & Baker, 1967: and Larcher, 1969.)

2.3 RESPIRATION

Most of the techniques in Section 2.2 can be used to estimate respiration. Darkened vessels prevent photosynthesis but it must be appreciated that conditions inside them may vary considerably from those inside transparent chambers. The evolution of CO_2 is readily measured, but it is also possible to monitor respiration by dry weight loss (e.g. Grime, 1966) or the utilization of oxygen. Dark respiration is used in most ecological studies because there is no simple and reliable technique for measuring photorespiration which may differ from dark respiration (Jackson & Volk, 1970).

Additional methods of estimating respiratory CO_2 are described in Chapter 4 (Section 3.2) in association with problems of decomposition and soil respiration.

2.4 COMPENSATION POINTS AND PHOTOSYNTHETIC OPTIMA

The physiological plant ecologist usually wishes to compare different species in order to make ecological judgements. Frequently there is a need to compare a wide range of species in a relatively short time. The relationships between photosynthesis and temperature (Fig. 5.7) and between photosynthesis and light provide bases for determining various important values (Fig. 5.8). Light and temperature are the two main environmental variables, for while CO_2 concentration may limit photosynthesis, the atmospheric concentration of CO_2 is reasonably constant.

The relationship between photosynthesis and temperature allows the estimation of the temperatures at which apparent photosynthesis ceases due to either too much or too little heat (high and low temperature compensation points) and a temperature, or range of temperatures, for optimal photosynthesis. The previous history of the plant affects these points and a suitable acclimation period should be employed before measurement. Instantaneous measurements may give higher optimum temperatures than long-term measurements (Jost, 1906; Larcher, 1969b). The plant may undergo physiological, anatomical and morphological changes, which may alter its relationship with the photosynthetic environment during a protracted period of measurement (Mooney & Shropshire, 1968).

The relationship with light has two important values (Fig. 5.8). The low light intensity at which apparent photosynthesis ceases (light compensation point) and the intensity above which no increase in photosynthesis occurs (light saturation). These values are influenced by the previous history of the plant as can be seen in comparisons of sun and shade plants (e.g. Larcher, 1969a). The resistance to the uptake of CO_2 probably determines light saturation, thus an abrupt cut-off at saturation occurs when the plant has closed stomata but is

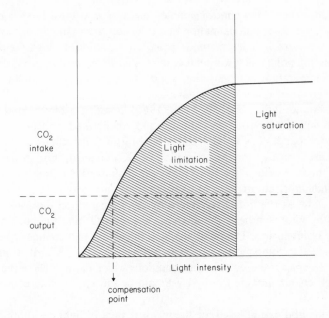

Figure 5.8 Diagrammatic representation of the relationship between CO_2 exchange and light intensity.

likely to be gradual when the diffusive resistance is low (cf. Heath, 1969). It is also difficult to determine light saturation precisely when only a few values are available for the curve.

There is an interaction between the effects of light and temperature on photosynthesis. At high light intensities the optimum is shifted towards higher temperatures and the low and high temperature limits for photosynthesis may also be extended (e.g. Pisek *et al.,* 1969).

Accurate measurements of CO_2 assimilation are needed to determine optimal rates of photosynthesis and the light intensity at saturation. However, compensation points, which are points of no net change of CO_2 concentration, may be easily measured as only the presence or absence of photosynthesis needs to be determined.

2.4.1 Light and temperature compensation points

At the point at which net photosynthesis just ceases, dry weight is constant (as there is no photosynthetic gain or respiratory loss) and there is also no apparent change in CO_2 concentration. Both these facts simplify the estimation of compensation points.

The light compensation point has been determined by comparing the dry weight of leaf discs floating on water under a range of light intensities, with the dry weight of untreated control samples (e.g. Grime, 1966). The waterlogging and sinking of discs and exudation and diffusion from cut edges may provide sources of error. The method could be used for high temperature compensation points as water temperature is easily controlled but is unsuitable for low temperature compensation points as most harmless solutions would freeze.

Bicarbonate indicators (e.g. Lieth, 1960) in which a bicarbonate solution made up with an appropriate indicator and equilibrated with normal air, are sensitive to changes in CO_2 concentration. Martin & Pigott (1965) have used a bicarbonate indicator for the estimation of CO_2 concentration in the soil and this has been modified for use in schools (e.g. Nuffield Biology, 1966). In Lieth's indicator, an increase in CO_2 concentration results in a change from red to yellow as the cresol-red indicator takes up carbon dioxide. If CO_2 is depleted by photosynthesis the mixture releases CO_2 and the colour remains the same or intensifies. The point at which the colour just changes can be used to estimate the compensation point (e.g. Pisek et al., 1967). If a number of replicate samples are used then the conditions which cause 50% of the samples to change colour provide the best estimate of the appropriate compensation point.

This method can be used for the measurement of light and both high and low temperature compensation points.

All the compensation points are integrated by the relationship between light compensation and temperature. Light compensation occurs at higher intensities as temperature increases. The relationship (e.g. Meidner, 1970) predicts whether a plant part will be photosynthesizing or respiring at given light and temperature and could be useful in interspecific comparisons and even in simple models of the production of stands of vegetation.

2.4.2 Carbon dioxide 'compensation point'

This value is hardly used in ecological studies. It is a compensation point as there is no apparent influx or efflux of CO_2. However, its relationship to respiration and photosynthesis is not straightforward and some workers (e.g. Heath, 1969) prefer to use the neutral symbol, Γ. The carbon dioxide compensation point can be determined by circulating air over an illuminated leaf in a closed system with an infra-red gas analyser until the carbon dioxide level remains constant. It can also be obtained by equilibrating a leaf in a closed vessel and estimating the CO_2 by a suitable method (e.g. by titration).

As the carbon dioxide compensation point is the lowest concentration that can be produced by photosynthesis, it has been used to estimate the CO_2 concentration at the chloroplasts. This concentration is required for models of the photosynthesis of plants and stands of vegetation (e.g. Ludlow & Jarvis, 1971).

Plants which have alternative photosynthetic (C_4) pathways involving dark carboxylation (Hatch et al., 1967) are capable of reducing the carbon dioxide concentration to zero (Meidner, 1967). A steep gradient of CO_2 concentration can be maintained and when light, temperature or water are not limiting, can provide for an increased assimilation of carbon dioxide. Thus the study of CO_2 compensation points may help to identify species with potentially high assimilation and growth rates.

3 TEMPERATURE RELATIONS

3.1 THE MEASUREMENT OF PLANT TEMPERATURE

Air temperatures and plant temperatures are rarely indentical. Consequently it is advisable to measure the plant temperatures directly. This is simple for massive organs but difficult for organs, such as leaves, with a small heat capacity.

Short-wave radiation may be variously reflected and transmitted by a leaf, leaving a variable amount which is absorbed, whilst long-wave radiation is almost completely absorbed. The leaf loses some energy by re-radiation but most is lost either through the evaporation of water from transpiring leaves or through convective transfer from leaves with closed stomata. The net gain of energy during the day usually results in leaf temperatures exceeding air temperatures whilst the net radiation loss during the night makes for leaf temperatures lower than in the air. Large leaves, dark green leaves, and low transpiration rates all produce high leaf temperatures while smallness, light colouration and high transpiration rates will dissipate heat more readily (cf. Lange & Lange, 1963).

The thermometer must not change the temperature of the organ being measured. The insertion of a mercury thermometer into a tree-trunk is unlikely to affect its temperature, but the same thermometer will have a major effect on the temperature of a leaf. The thermometer can affect the leaf temperature by conduction, shading, by evaporation and transpiration, or by mechanical disturbance. Nevertheless investigators have measured leaf temperature by wrapping leaves around the bulb of a mercury-in-glass thermometer (e.g. von Guttenberg, 1927). More satisfactory methods involve the use of thermocouples and thermistors (see Chapter 7). The heat capacity of small thermo-electric sensors is small in relation to that of plant organs, but a proper contact must be maintained between the organ and the sensor, if correct estimates of temperature are to be made. Clamps are commonly used (Fig. 5.9) but they must not upset the thermal relations of the organ; a bulky clamp holding a minute sensor could be as deleterious as a large sensor.

The temperature of large fleshy leaves may be measured by the injection of needles carrying a thermocouple (e.g. Lange, 1953) or a thermistor (Lange, 1965a).

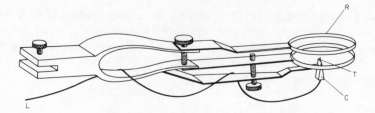

Figure 5.9 Clamp for securing a thermocouple to the underside of a leaf. T, thermoelement; C, cork edge; R, clamping rings; L, connecting leads.

Individual sensors and even complete measuring systems must be precisely calibrated (Perrier, 1971) as there is no absolute relationship with temperature.

Radiometers, which are expensive, appear to be ideal for the measurement of leaf temperatures as the leaf need not be touched. However, leaves are scarcely ever still and even a remote radiometer will alter the radiation background of the leaf. Various types of remote sensor are discussed by Perrier (1971).

3.2 THE RESISTANCE OF PLANTS TO EXTREMES
OF TEMPERATURE

3.2.1 High temperatures

Leaf temperatures commonly exceed air temperatures under sunny conditions. There is a real possibility of thermal damage in hot climates; temperatures of 45–55°C usually cause the death of plant organs. Lange & Lange (1963) recorded a leaf temperature of 47·7°C (18·4°C above air temperature in *Lonicera*. In cold climates plant organs may be raised to a temperature more suitable for metabolic activity. The leaves of *Rubus chamaemorus*, a plant of boreal distribution, have been described as 'warm to the touch' under natural conditions and leaf temperatures of *c.* 40°C have been predicted from theoretical considerations (Idle, 1970).

The measurement of the heat resistance of plant organs gives a temperature beyond which damage is likely to occur, but can also be used to study the adaptation of different species to temperature régime. High heat resistance has been found in plants from low altitude, in leaves which attain high temperatures and during hot summers (Lange, 1961); with lower resistances in plants from high altitude (Lange & Lange, 1962), in leaves which maintain low temperatures (Lange & Lange, 1963) and during cool summers (Bannister, 1970). High heat resistances are often associated with high resistance to cold and drought (cf. Kappen, 1964) and this may be associated with a common physiological mechanism.

3.2.2 Low temperatures

In temperate, polar and alpine regions, plants may be subjected to intense winter cold which could cause damage to plant tissues. Sudden reversals in temperature in autumn and winter may kill susceptible species or cause damage sufficient to alter their competitive ability. Under natural conditions most plants are well adapted to prevailing temperatures and are likely to be killed only in exceptionally cold years. Polwart (1970) has shown that *Vaccinium uliginosum* in Argyllshire is resistant to winter temperatures below $-30°C$, but during two years' investigation in the area the minimum temperature recorded was not much below $-10°C$.

Tissue damage is more likely in autumn and spring when resistance is being acquired or lost. The maximum resistance that is developed to low temperatures often seems to be a function of geographical origin, with plants from colder climates showing greater resistance to cold than those from warmer climates, even when they are growing in the same habitat (e.g. Till, 1956; Flint, 1972).

Extreme temperatures do not only cause physical damage, they may produce physiological disturbance. Exposure to low temperatures which does not cause visible damage may result in reductions of photosynthetic activity for several days (e.g. Pisek & Kemnitzer, 1968). Ecologists have tended to neglect the study of such effects.

3.2.3 Measurement of resistance to temperature extremes

Resistance is measured by exposing the plant, or plant part, to the appropriate temperature for a given length of time. Ideally, whole plants should be used, but this is limited by the size of the plants. Such methods would be obviously unsuitable for studies of mature trees although valuable information can be obtained from the investigation of plant deaths during exceptional periods of heat or cold. Unfortunately such periods are uncommon and the number of species affected small (Bannister, 1973).

If large numbers of plants can be exposed to extremes of temperature then the temperature at which 50% of the plants are killed provides a standard for comparative purposes. Such a method may be unsuitable for the measurement of heat resistance as it may be difficult to separate the effects of drought and heat unless the plants can be presented with very humid air at the correct temperature.

In most cases temperature resistances of detached plant parts may be determined, as they do not appear to differ significantly from those of whole plants (Lange, 1965b). The simplest method of measuring heat resistance of plants or parts of plants is to submerge them in a water bath at an appropriate temperature, for a period of 30 minutes (Lange, 1961). The samples may be immersed directly in the water or enclosed in a suitable container, such as a close-fitting envelope made from polythene film. The physiological state of the exposed tissue is important and it may be best to use tissue which is fully turgid as the presence of water deficits usually leads to an increase in resistance (Kappen, 1966; Bannister, 1970).

The rates of cooling and thawing have been found to be critical in the determination of frost resistance (Levitt, 1966). Arrangements are often made for freezing and thawing rates which are comparable to those found in nature. The cooling apparatus used may vary from ice-salt mixtures in Dewar vessels (Ulmer, 1937) to deep freezes (Irving & Lanphear, 1967), alcohol evaporating machines (Till, 1956) or a refrigerated bath using mixtures of water and alcohol (or water and glycol) (Polwart, 1970). Thermoelectric cooling is possible, but the power needed to attain sufficiently low temperatures may be quite considerable.

The major problem in estimating frost and heat resistance is not the exposure of samples but the estimation of the damage produced. The simplest method is to place samples in a humid atmosphere (or in a closed container with their cut ends in water) and to monitor the appearance of symptoms of damage. These symptoms may involve browning or blackening, a change to a glossy or translucent appearance, and often, before decay sets in, a characteristic sweet or sickly odour. After the symptoms have developed the samples can be assessed and the temperature at which half the samples are killed taken as a standard. If individual leaves or shoots are used then the temperature at which 50% of the leaf area or shoot is damaged may be calculated (Lange, 1961). The damage often takes several days to develop but occasionally weeks or even months are needed. More rapid methods include the use of the cessation of protoplasmic streaming as a criterion for tissue damage (Alexandrov, 1964), the measurement of the electrical conductivity of leachates of the samples or the use of vital staining techniques. The damage of tissues results in a loss of semipermeability and the release of electrolytes into the surrounding solution, hence the higher the conductivity of the leachate the greater the damage. The amount of vital staining (often with a tetrazolium salt) decreases with increased tissue damage (e.g. Steponkus & Lanphear, 1967). Tetrazolium salts work best at a neutral pH and in the dark: they may not be effective in detecting death due to heat or toxins and care must be taken to differentiate staining due to the presence of micro-organisms (Mackay, 1972). The use of either technique requires the standardization of samples and the establishment of upper and lower limits of damage. Healthy tissue will release some electrolyte and completely damaged material may produce some vital staining, so that the introduction of both untreated and completely killed controls is essential. The simplest way of expressing results is to calculate the damage to the treated sample (C) as a proportion of the difference between the undamaged control (C_o) and a completely killed sample (C_d), i.e.

$$I = \frac{C - C_o}{C_d - C_o}$$

A more complicated index, which takes differences in sample size into account, is given by Flint *et al.* (1967). This index may be unreliable if a completely damaged sample produces a different conductivity from heat killed control (Bannister, 1970).

Figure 5.10 compares the heat resistances obtained by direct observation of damage and by relative conductivity. There is close agreement in the temperature at which 50% damage occurs, although the temperature differences would be somewhat greater at other levels of the index.

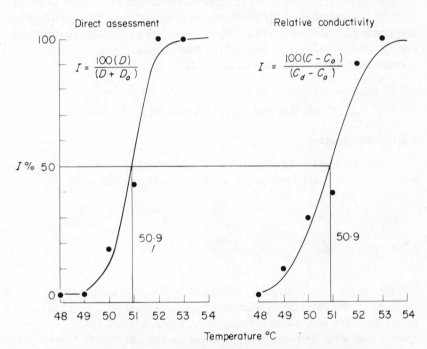

Figure 5.10   Heat resistance of *Calluna vulgaris,* January 1968. Comparison of two different methods of estimating the temperature causing 50% damage. *D, D_O,* amount of damaged and undamaged material (visually assessed). *C, C_d, C_o,* conductivity of damaged, completely damaged and undamaged samples. Curves fitted by probit analysis.

4 WATER RELATIONS

4.1 INTRODUCTION

The emphasis of this section is upon the water balance of tissues, the control of water loss and resistance to desiccation, all of which are important in relating water relations of a plant to its ecological amplitude. Water uptake is not considered, although plants from different habitats have been compared in the past (e.g. Huber, 1928; Haines, 1928).

Studies of the control of transpiration appear to be more relevant to the ecologist (cf. Maximov, 1929, 1932) than the uncritical measurement of transpiration rates and considerations of the resistance of leaves to the diffusion of water vapour and carbon dioxide have led to the formulation of a number of mathematical models which can predict both physiological and ecological responses (cf. Lewis, 1972).

The water relations of plants cannot be considered in isolation as changes in water balance affect other physiological processes such as stomatal closure and hence photosynthetic rates. Other inter-relations between environmental factors, water relations and other physiological processes have been considered in earlier sections (2.1 and 2.2).

4.2 THE WATER STATUS OF PLANT TISSUES

4.2.1 Water content

The water content (w_d) of plant tissues is often derived from the fresh or field weight of the sample (f) and its dry weight (d) and expressed as follows

$$w_d = \frac{f-d}{d}$$

or, as a percentage $\qquad w_d = \frac{100(f-d)}{d}$

This expression is strongly influenced by fluctuations in the dry weight component. The ratio of water content to dry weight at full saturation (i.e. s/d) is not constant and varies between populations, species, organs of a single species, and even between different leaves on the same plant. Woody plants and tissues have a lower s/d than fleshy plants or tissues. The use of w_d might be considered acceptable for a series of measurements on the same plant, but dry weights can show considerable diurnal and seasonal fluctuations (Weatherley, 1950). Leaves, frequently used for measurements of water balance, show marked diurnal changes which are related to the production, utilization and translocation of carbohydrate. Such changes will cause fluctuations in w_d, even if the absolute water content of the tissue remains constant.

Young tissues have high ratios of s/d so that w_d is high early in the growing season and declines to a minimum during the dormant season (Fig. 5.11).

Relative water content (w_r) which is the relative turgidity of Weatherley (1950) renamed to take account of the criticisms of Walter (1963), is, at least

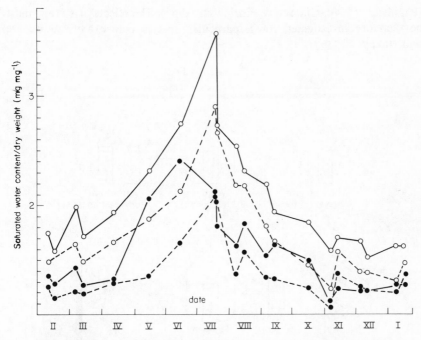

Figure 5.11 The annual course of saturated water content/dry weight in four heath species. —O— *Calluna vulgaris*; --O-- *Erica cinerea*; —●— *Vaccinium myrtillus*; --●-- *Erica tetralix*. (Redrawn from Bannister, 1970.)

theoretically, independent of dry weight. It is complementary to the water deficit (w_s) of Stocker (1929). Both are usually expressed as percentages.

$$w_r = \frac{100(f-d)}{(s-d)} \qquad w_s = \frac{100(s-f)}{(s-d)}$$

4.2.2 The measurement of relative water content

It is usually more convenient to determine water content in the laboratory than to attempt to make a direct measurement in the field. A major problem is the loss of water during transportation, this is least acute with cut shoots and more serious with detached leaves and discs. Shoots may be sampled directly into stoppered containers and weighed, allowed to take up water to saturation, reweighed and then dried and weighed again (Fig. 5.12). Water lost during transportation is then retained within the container and included in the field water content. Errors may arise if the interval between sampling and placing in water impairs uptake. This occurs in some species after a brief period of storage (Clausen & Kozlowski, 1965) but may not be apparent in others

(Bannister, 1964c, Clausen & Kozlowski, 1965). The effects of storage must be therefore investigated for a particular species before any studies are undertaken.

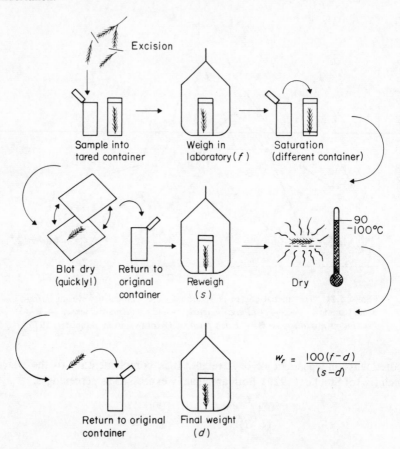

Figure 5.12 Procedure for the determination of relative water content.

The use of leaf discs requires additional precautions as water loss is very rapid over short periods, storage allows cut edges to dry out and become sealed, and the disc may gain or lose weight, or even grow during the period of measurement. Discs may become waterlogged by the injection of water between the cells. The problems of water loss and storage may be overcome by rapid collection into containers, followed by rapid weighing and immediate placement in water. This may be achieved by taking suitable equipment into the field or by working on plants that grow conveniently near the laboratory.

Changes in weight due to respiration and photosynthesis may be minimized by resaturating the discs in light intensities at or near their compensation point (Slatyer, 1967) but are best controlled by shortening the period of water uptake (e.g. Jarvis & Jarvis, 1963b).

Discs or cut sections of leaves often show greater amounts of uptake than intact leaves (Hewlett & Kramer, 1963; Clausen & Kozlowski, 1965) whilst the size of disc may influence the water uptake (Barrs & Weatherley, 1962). Čatský (1960) has used a technique where the discs are placed in holes punched in saturated plastic foam as floating discs often show an indefinite period of water uptake which has been attributed to growth.

4.2.3 Water potential

Crude measures of water content give no indication of water tension in a plant tissue. Water potential (ψ) provides a more appropriate thermodynamic measure of water tension and is related to the difference between the chemical potential of water at some point in the system (μ_w) and the chemical potential of pure, free, water at the same temperature (μ_w°). It is customary to measure plant water potentials in pressure units (e.g. atmospheres or, latterly, bars) which are derived from the energy units of chemical potential by dividing by the partial molal volume of water (V_w) (Table 5.1).

$$\psi = \mu_w - \mu_w^\circ = RT \ln \left(\frac{e}{e_o}\right)$$

Table 5.1 Conversion table for various expressions of water potential.

Atmospheres (STP)	Bar	N m^{-2}	J g^{-1}	m of water
1	1·013	1·013 × 10^5	0·1013	10·33
0·987	1	10^5	0·1	10·17
9·87 × 10^{-6}	10^{-5}	1	10^{-6}	1·017 × 10^{-4}
9·87	10	10^6	1	101·7
9·70 × 10^{-2}	9·833 × 10^{-2}	9·833	9·833 × 10^{-3}	1

The conversions assume that 1 cm^3 of water weighs 1 g.

where e, e_o are the vapour pressure of water in the system and of pure water at the same temperature (T), R is the gas content and T the absolute temperature (°K).

The water potential of plant cells and tissues is generally less than zero (i.e. negative) as the chemical potential of pure, free, water is zero. Water in most plant systems is under some form of constraint which lowers its chemical potential. Previous measures of water stress such as the suction force ('Saugkraft') of Ursprung & Blum (1916), and the diffusion pressure deficit (DPD) of Meyer (1945) used positive units of pressure and are similar in concept to water potential. Some workers (e.g. Tinklin & Weatherley, 1968) have used a water potential deficit, $\Delta\psi$, which is given a postitive value.

Plant physiologists usually identify the water potential of a cell as the resultant of two forces: the depression of water potential due to solutes in the cell sap (ψ_s) and a pressure component identified with turgor pressure (ψ_p).

Thus,

$$\psi_{cell} = \psi_s + \psi_p$$

where ψ_p is usually positive and ψ_s negative. The statement is often given in DPD terminology as

$$DPD = OP - TP$$

Direct measurements of turgor pressure (e.g. Arens, 1939) cannot normally be used by the plant ecologist and turgor pressure is usually estimated from the difference between ψ_{cell} and ψ_s. ψ_s is sometimes lower (i.e. more negative) than ψ_{cell} and this has been taken as evidence for the development of negative pressures within cells and tissues (e.g. Grieve, 1961). The relationship given by the equation is a simplification and forces other than ψ_p and ψ_s may determine the water potential of cells and negative turgor pressures, although theoretically possible, may not occur (Slatyer, 1967).

(a) *The use of tissue characteristics in water potential determinations*

Changes in the weight, length, diameter and volume of suitable samples of tissue equilibrated in or above solutions of known osmotic potential are frequently used in the determination of water potential.

Any loss of water from the tissue in between collection and equilibration will result in an inaccurate estimate of water potential and a number of the precautions mentioned in Section 4.2.1 are appropriate.

Tissue should be equilibrated at constant temperatures (as water potential is influenced by temperature) in or over a range of solutions of different concentration. The solute must not be taken up to any great extent by the equilibrating tissue (cf. Slatyer, 1967) and solutions of sucrose and mannitol, although not ideal, are often employed (Barrs, 1968). Polyethylene glycol (carbowax) has been used to adjust the osmotic potential of culture solutions and could be used for tissue equilibration (Jarvis & Jarvis, 1963c). Exudation from the tissue can also cause inaccuracies, particularly when small volumes of bathing fluid are used (Gaff & Carr, 1964).

A relatively short equilibration (about two hours) is often sufficient for tissues submerged in a solution. The weight, or some other characteristic of the tissue, must be measured both initially and after equilibration and particular care must be taken when excess solution is removed by blotting as inadequate blotting will mask changes in weight while rough handling may cause damage.

The solution which matches the water potential of the tissue is found by interpolation in a graph of changes of the tissue against the characteristics of the bathing solution (Fig. 5.13). Solute equilibration is unsuitable for the determination of water potentials which cause plasmolysis (Slatyer, 1958) as the retention of the external solution within the plasmolysed tissue and the rigidity of the cell wall prevents further changes in the weight or dimensions of plasmolysed tissue (Fig. 5.13). These defects may be overcome either by using equilibration in vapour of known water potential or by measuring a solution rather than a tissue characteristic.

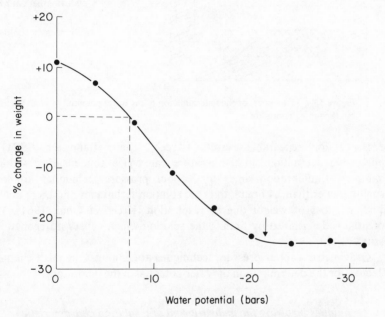

Figure 5.13 Determination of tissue water potential by solute equilibration. A hypothetical example. Dashed lines indicate the point of no change which estimates the tissue water potential. The vertical arrow indicates the point at which the method becomes invalid, because of plasmolysis (it could be used to estimate the osmotic potential of the cell sap).

In vapour pressure equilibration the tissue is not in contact with the solution, thus a variety of solutes can be used, e.g. sodium chloride (Owen, 1952; Kreeb, 1960). Plant material is placed on a grid within a chamber made of plastic foam and containing the osmoticum (Fig. 5.14). A precise temperature control is essential as the water potential of vapour changes dramatically with small changes in temperature. Accordingly, some authors have designed and used temperature baths in which they claim control to ±0·001°C (Owen, 1952). Such a precise control does not seem to be

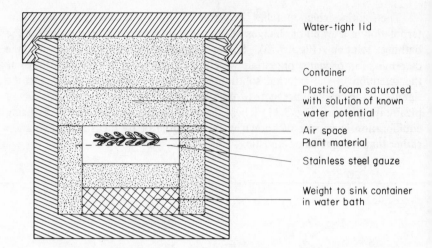

Water-tight lid

Container

Plastic foam saturated
with solution of known
water potential

Air space
Plant material

Stainless steel gauze

Weight to sink container
in water bath

Figure 5.14   Chamber for the determination of the water potential of plant tissue
by vapour equilibration.

necessary for repeatable results (Kreeb, 1960; Bannister, 1971) but
appreciable fluctuations in temperature can cause condensation within the
chambers. Equilibration times for vapour pressure techniques are longer,
usually greater than 24 hours, than for solution techniques and there may be a
significant loss of weight due to respiration (Kreeb & Önal, 1961). Water
potential is determined by finding the solution which causes no change in the
tissue.

Gravimetric vapour pressure techniques are amongst the most simple and
effective for the determination of water potential in the laboratory.

(b) *Techniques involving the measurement of a solution characteristic*

The influx and efflux of water are equal in tissue in equilibrium with the
external solution and thus the characteristics of the solution should be constant
and allow the water potential of the tissue to be determined.

However, the amount of water taken up or released by equilibrating tissue
is often so small that it has little effect on the properties of the bathing solution,
therefore minimal quantities of solution must be used. The gain of small
amounts of solute from dust and the exudation of cut surfaces, or loss by
absorption and adsorption now become critical (Gaff & Carr, 1964; Hellmuth
& Grieve, 1969). The material can be rinsed in the equilibration solution
which is discarded and replaced with fresh solution of the same concentration
and exudation and absorption can be controlled by selecting samples with an
appropriate ratio of cut edge to surface.

Solution characteristics that have been used include the refractive index of solutions such as sucrose (e.g. Gaff & Carr, 1964) and changes in the density of solutions. This latter characteristic is the basis of the technique attributed to Shardakov, although Barrs (1968) traces it back further, in which tissue is equilibrated with small amounts of solution tinted with a non-toxic dye (e.g. congo red, bromothymol blue). A drop of the coloured solution is released very gently under the surface of a colourless control solution at the original concentration; drops which rise have abstracted water from the tissue, those which fall have given up water. The water potential of the tissue is estimated from the real or interpolated concentration where the drop stands still. Control and test solutions must be at the same temperature as their density is influenced by temperature. The technique is intended for field use and has been critically evaluated by Hellmuth & Grieve (1969).

(c) *Techniques involving the determination of a vapour or gaseous characteristic*

Water vapour in equilibrium with plant tissues is usually more than 95% saturated. The measurement of small changes in vapour pressure in nearly saturated air is not easy and sophisticated techniques have to be used.

The thermo-electric technique of Spanner (1951) relies on the Peltier effect to cool a thermocouple junction (see also Chapter 6, Section 6.3.5) inserted in a small chamber containing the plant material. In the humid atmosphere of the chamber a bead of moisture is formed on the cooled junction. The evaporation of this drop generates a minute electric current which is proportional to the depression of wet-bulb temperature. The method of Richards & Ogata (1958) depends upon the temperature difference between wet and dry thermocouples and is determined at a steady rate of evaporation at constant temperature depression.

Macklon & Weatherley (1965a) have evolved a simple method which relies on the rate of evaporation of a drop extruded from a micropipette. All these instruments are calibrated by reference to the water potential of vapour above solutions of known characteristics.

Some workers demand an extremely precise control of temperature (e.g. Slatyer, 1958) whilst others (e.g. Macklon & Weatherley, 1965a) have found this not to be necessary.

The 'pressure bomb' (Fig. 5.15) of Scholander *et al.* (1965) is a radically different technique which is included here as it employs air pressure. A detached leaf or shoot, is inverted in the bomb with the cut end protruding through a pressure-tight seal of soft plastic or rubber. Air pressure is applied to the bomb until a drop of cell sap exudes from the cut end. The pressure is taken as a direct measure of water potential. The method is suitable for hard tissues as it is difficult to effect a pressure-tight seal for soft tissues without damaging them. Measurements of electrical conductivity have been used to

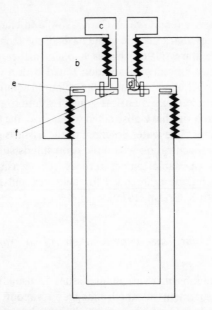

Figure 5.15 Pressure bomb (not showing tap or pressure gauge). a, main body of pressure bomb; b, cap screwed onto main body; c, hollow screw, various sizes; d, soft rubber gland; e, rubber washer; f, round plate with hole, various sizes. (Rainford, 1972.)

determine the end point as the water extruded from cells contains many solutes and can therefore be distinguished from loosely held water which is low in solutes and has a poor conductivity. Suitably refined electrodes have been described by Richter & Rottenburg (1971).

4.2.4 Measurements of osmotic potential

The osmotic potential of cell sap gives an indication of water stress. Walter (1931, 1963) contends that the hydration of the protoplast is more important to the plant than water balance and his concept of 'hydrature' is a sort of 'protoplasmic relative humidity' which is directly related to osmotic potential. Slatyer (1967) criticizes the hydrature concept whilst Walter (1963) provides a criticism of water potential.

The cell sap must be extracted in order to determine its osmotic potential and the extraction may influence its composition. Extraction by pressure may dilute the cell sap with water from the cell wall; freezing may extract extracellular solutes; whilst boiling may hydrolyse both the cell contents and extracellular materials. The osmotic potential of extracted sap may be altered by microbial degradation unless it is refrigerated.

Both plant and animal physiologists usually determine the osmotic potential of small samples by the depression of their freezing points (e.g. Walter, 1931, 1963; Ramsay & Brown, 1955). The depression of freezing point of an ideal molal solution with an osmotic potential of 22·4 atm is 1·86°C. A direct proportionality exists, thus

$$\frac{\text{unknown OP}}{-22\cdot4} = \frac{\text{unknown freezing point depression}}{1\cdot86}$$

This simplifies to

$$\text{unknown OP} = -12\cdot04 \times \text{unknown freezing point depression}$$

where the osmotic potential is in atmospheres. Conversions are readily made to other units (cf. Table 5.1).

Solutions are usually supercooled and sudden freezing is followed by a rapid rise in temperature to the true freezing point. Harris & Gortner (1914) give corrections for supercooling. Refractrometric methods (Gaff & Carr, 1964) and thermo-electric and other techniques (e.g. Weatherley, 1960) described in 4.2.3 may be used to determine osmotic potential. Field instruments, employing refractrometry and electrical conductivity have been described by Kreeb (1963) and Shimshe & Livne (1967).

In the 'method of limiting plasmolysis', strips of tissue are mounted in suitable solutions (4.2.3a) of known osmotic potential and examined for plasmolysis. The solution which causes 50% of the cells to plasmolyse (Fig. 5.16) is used to estimate the water potential at the point where the protoplast just loses contact with the wall and turgor pressure is effectively zero (i.e. $\psi_p = 0$).

Thus, at equilibrium, $\qquad \psi_{\text{external solution}} = \psi_{\text{cell}} = \psi_s + \psi_p, \qquad \psi_p = 0,$

therefore $\qquad \psi_{\text{external solution}} = \psi_{\text{cell}} = \psi_s$

The method is well suited to tissues containing cells with coloured contents but can be used for colourless tissues, although some are unsuitable. Crafts *et al.* (1949) consider that the defects of the method are insufficiently serious to discourage its use.

Osmotic potential can be estimated from changes in the volume ($V_1 V_2$) of the protoplast or vacuole of individual cells, as at constant temperature, the osmotic potential is inversely proportional to volume and at equilibrium, the osmotic potential of a plasmolysed cell is equal to that of the external solution ($\pi_2 = \pi_{\text{ext}}$).

Thus, $\qquad \pi_1 V_1 = \pi_2 V_2$

therefore $\qquad \pi_1 = \pi_2 \dfrac{V_2}{V_1} = \pi_{\text{ext}} \dfrac{V_2}{V_1}$

This is essentially the plasmometric method of Hofler (1917).

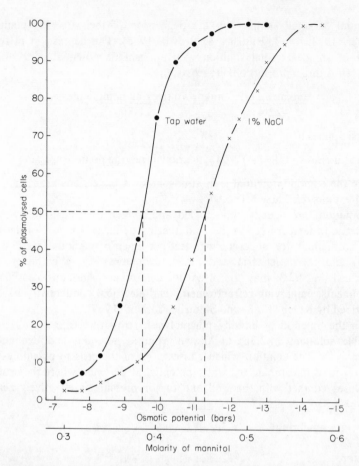

Figure 5.16 Method of limiting plasmolysis. Estimation of the osmotic potential of the cell sap of epidermal strips of *Saxifraga sarmentosa*, pretreated with either tap water or 1% NaCl. Dashed lines indicate the mannitol solution causing 50% plasmolysis, i.e. the estimate of the osmotic potential of the cell sap. (Data of H. Meidner.)

4.2.5 The relationship between water potential and relative water content

There is no unique relationship between water content and water potential of plant tissues. This is similar to the situation in soils and their water relations are often characterized by relating soil moisture tensions to measures of water content (Chapter 6).

Weatherley & Slatyer (1957) established an analogous relationship between water potential and water content for tomato and privet. These, and most subsequent curves, are 'desorption' curves in which turgid tissue is equilibrated with surroundings of known water potential and, as the water content after equilibration is used in the calibration, they have the advantage that water losses during previous transfers are unlikely to be critical (cf. 4.2.3a). The relationship was thought to be constant and characteristic of a particular species (Slatyer, 1960) but it can be altered experimentally (Jarvis & Jarvis, 1963b,c) and varies with habitat (Bannister, 1971) and season (Knipling, 1967) within plants of the same species.

Xerophytic plants often show shallow curves with relatively small reductions in water content for large differences in water potential (Fig. 5.17).

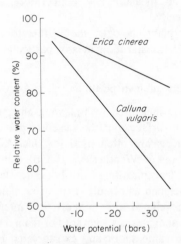

Figure 5.17 Relationship between relative water content and water potential in two species of different ecological amplitudes growing together in the same site. *Erica cinerea* is characteristic of drier habitats than *Calluna*. (Simplified from data of Bannister, 1971.)

As stomatal closure and desiccation damage are often more readily related to water content than to water potential (Jarvis & Jarvis, 1963c), these plants can maintain open stomata and a favourable water balance despite water stresses in the environment. Mesophytes show steeper relationships and are more likely to close their stomata and thus limit both water loss and the uptake of carbon dioxide.

4.3 THE LOSS OF WATER FROM PLANTS

The physiological ecologist is concerned with the individual plant and the relationship of its water economy to its distribution. Early workers (cf. Seybold, 1929) investigated a wide variety of plants in the hope of establishing general relationships between water loss (transpiration) and habitat, but they found few consistent trends (e.g. Schratz, 1932). Maximov (1929, 1932) realized that the control of water loss was more important than the rate of transpiration. Accordingly, the physiological plant ecologist ought to direct his attention to the water balance of tissues and the control of transpiration, although ecologists interested in the water relations of ecosystems may require to measure and integrate the water loss from individual plants.

Transpiration cools leaves (Lange & Lange, 1963) and fluctuations in transpiration rate may indicate the degree of stomatal control and the potential for the assimilation of carbon dioxide. However, these are indirect applications and it is preferable to measure leaf temperature, stomatal aperture and assimilation rates directly.

The physiological plant ecologist should therefore consider carefully the reasons for measuring transpiration in any particular circumstance.

4.3.1 The measurement of transpiration

Transpiration is readily measured by following changes in weight. In potted plants the pot and soil surface must be sealed to prevent losses of water and polythene bags or sheets are often used for this. Automatic weighing and recording (e.g. Macklon & Weatherley, 1965b) can be used to provide a continuous record of transpiration. Cut shoots (or leaves) can be removed from the plant and weighed as rapidly as possible. They are then replaced in the vegetation and re-weighed after a short time interval. Transpiration can be estimated from the short-term weight loss or from a more extended series of readings taken on the same shoot and extrapolated back to zero time (e.g. Willis & Jefferies, 1963). The transpiration of cut shoots may increase (Weinemann & Le Roux, 1946), or decrease (Decker & Wien, 1960) or do both (Balasubramaniam & Willis, 1969) after excision.

Meidner (1965) found a pattern of stomatal movements in excised bean leaves which would account for both increases and decreases in transpiration. Sudden changes in transpiration on cutting are most likely in turgid plants and less likely as water deficits develop (Balasubramanian & Willis, 1969). Most plants in natural situations are under water stress, even when well-watered, so that the effects of excision may not be too serious. Stocker (1956) and Eckardt (1960) reviewed different techniques and concluded that the cut shoot method is as reliable as any other and can certainly give indications of the relative, if not the absolute, amounts of transpiration (Fig. 5.18).

Other methods of measuring transpiration usually require part of the plant to be contained in a chamber or cuvette. This usually leads to a rapid alteration in the microclimate within the cuvette (cf. 2.2.2, Fig. 5.6). Water vapour will accumulate in a closed system, causing a decrease in the vapour pressure gradient between leaf and air and consequently in transpiration rate. In open systems the transpiration rate will be influenced by the rates of air flow and by changes in stomatal aperture resulting from alterations in water vapour

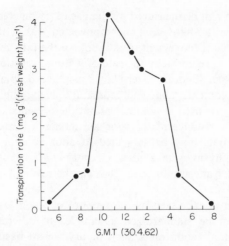

Figure 5.18 Daily course of transpiration in *Calluna vulgaris,* estimated by the quick-weighing method. (Data of Bannister, 1964c.)

and CO_2 content. It is, therefore, difficult to predict what will happen to the transpiration rate once a plant part is enclosed in a cuvettte (Fig. 5.6).

There are many possible ways of measuring the water content of the air that has passed over a suitably enclosed plant. The moisture content of the incoming air must be monitored as the passage of completely dry air over the plant would affect its water balance. Gravimetric determinations of the absorption of water by phosphorus pentoxide (Gregory *et al.,* 1950), calcium chloride (Freeman, 1908) or magnesium perchlorate have been used, whilst Huber & Miller (1954) measured the increase of temperature when water is absorbed by sulphuric acid. Hygrometers such as a suitably sized psychrometer to measure wet and dry bulb temperatures (Glover, 1941), the corona hygrometer of Andersson *et al.* (1954), the bimetallic hygrometer of Eckardt (1957), the microwave hygrometer of Falk (1966) or the use of infra-red absorption (Decker & Wien, 1960) have all been used to measure transpiration. tion.

Cobalt chloride or cobalt thiocyanate papers have been used to measure transpiration. Dry papers are strapped to the relevant plant part and the time taken for them to change to a standard colour is inversely proportional to the rate of water loss. The initially dry paper and the exclusion of light and moving air from the covered part create atypical conditions and Bailey *et al.* (1952) could not show any consistent relationship between the cobalt chloride method and gravimetric determinations. If they are made rapidly such measurements probably give a better indication of stomatal resistance and aperture than of transpiration rate. The same is true of the hydro-photographic method of Sivadjian (1952).

Transpiration can be measured by the use of potometers where cut shoots are sealed in a water-filled container connected with a narrow tube and a reservoir. The rate of movement of a bubble in the tube measures the rate of transpiration, or more correctly the rate of water absorption by the cut shoot. The placing of a cut shoot in water allows unimpeded uptake which produces abnormal water contents, abnormal stomatal opening and high transpiration. Intact plants may be used in potometers but the presence of free water means that there is no resistance at the root/soil interface (Tinklin & Weatherley, 1966) and the roots may be subjected to poor aeration. Potometers have proved useful in physiological studies (Gregory *et al.*, 1950) but are unlikely to give measures of transpiration rates that can be related to those in the intact plant.

Transpiration is expressed as the amount of water lost over a period of time, but it must be standardized if samples of different sizes are to be compared. Standardization on a fresh or dry weight basis is often used but expression upon a surface area basis is preferable as transpiration is a surface phenomenon. Leaf areas may be determined simply by tracing onto graph paper or from silhouettes made with 'daylight' photographic papers although elaborate instruments such as photometric digitizers are available. The area of minute leaves can be estimated if precise measurements are made on small samples and related to some easily measured parameter such as dry weight or volume.

All comparisons must be interpreted with care as the morphology and anatomy of the transpiring surfaces may have quite different relationships to leaf area, even within the same species (Decker, 1955).

4.4 STOMATAL APERTURE

There are many techniques available for the measurement of stomatal aperture (cf. Meidner & Mansfield, 1968), but no method is suitable for all plants; those with hairy leaves (e.g. *Verbascum*) and with stomata on a permanently inrolled surface (e.g. *Erica*) may defy direct measurement of any sort.

4.4.1 Direct observation

It is occasionally possible to observe the stomata on the intact plant, but the observer is normally restricted to the microscopic examination of mounted material. Stalfelt (1932) used small strips of leaves mounted in liquid paraffin which could be observed directly using an oil-immersion objective. Lloyd (1908) fixed epidermal strips in absolute alcohol; but Heath (1969) considers that this drastic method can only indicate broad qualitative trends.

The advent of non-toxic silicone rubbers (Sampson, 1961) revitalized techniques that involve making impresssions of the leaf surface (Buscaloni & Pollacci, 1901). The negative impressions are painted with nail varnish (cellulose acetate) which is peeled off to produce a positive replica. Nail varnish may be applied directly to the leaf but the tissues are immediately damaged.

Direct observation is limited to leaves in which the narrowest portion of the stomatal pore is visible. The manipulation of material can alter stomatal reactions, and, as stomatal behaviour is not uniform over the surface of a leaf, it is necessary to make a large number of measurements.

4.4.2 Indirect methods

Stomatal aperture can be estimated by the infiltration of liquids of low surface tension. The technique is often suitable for leaves that are not amenable to other methods. The penetrated area appears dark or translucent (depending on the angle of view) and the time taken for penetration, or the area of penetration, can be used to estimate stomatal aperture.

Heath (1950) has used gentian violet in absolute alcohol which produces a stain that persists even in dried leaves and Michael (1969) has described a similar method for use with coniferous needles. Solutions of xylol and benzol have also been used (Molisch, 1912). Liquids of different viscosities can be used (Alvim & Havis, 1954) and a point is found where one solution just penetrates and the next remains on the surface of the leaf. Injection under pressure has been used for conifer needles (Fry & Walker, 1967; Lopushinsky, 1969a), the pressure needed to cause injection being a function of stomatal aperture.

The cobalt chloride paper method (Stahl, 1894) and the hydro-photographic method (Sivadjian, 1952) can be used as indirect measures of stomatal aperture. The loss of moisture from a saturated mesophyll to a desiccated paper is related to stomatal diffusive resistance, but long periods of equilibration may be necessary and the exclusion of light may cause stomatal closure.

These qualitative methods can often be made quantitative by calibration with more precise measurements of stomatal aperture.

4.4.3 Porometers

In porometers the leaf is connected to a measuring apparatus by a cup or chamber. Viscous flow porometers draw air through the leaf, whilst diffusion porometers measure the diffusion of gases from the enclosed surface.

Porometer cups must not be positioned permanently as lowered carbon dioxide concentrations can cause stomatal opening (Heath & Williams, 1948). An ideal porometer cup is one that is airtight but can be readily removed between readings. Seals were originally effected with beeswax (Darwin & Pertz, 1911) but subsequently gelatine washers and various soft plastics have been used.

(a) *Viscous flow porometers*

The disadvantage of viscous flow porometers is that air must be dragged through the leaf which means that not only the stomatal resistance is being measured but the resistance of the mesophyll and additional stomata outside the area covered by the cup. Closed stomata offer a resistance to air flow which is often too large to be measured by a viscous flow porometer even though water is still being lost from the leaf surface.

A number of instruments are suitable for field studies, one of the most simple and effective of these is the hand porometer of Meidner (1965) (Fig. 5.19). The clip is attached to the leaf and the deflated bulb connected with the apparatus. The time taken for the bulb to reinflate is a measure of the resistance to air flow. In the field porometers of Alvim (1965) (Fig. 5.19) and Bierhuizen *et al.* (1965), the time taken for an applied positive pressure to decrease to a predetermined level is taken as a measure of resistance. Weatherley (1966) has designed a miniaturized version of Knight's porometer (Knight, 1917).

Resistance porometers are more sophisticated instruments and can be calibrated in absolute units (Heath, 1939). The Gregory and Pearse porometer (1934) which has been modified by Spanner & Heath (1952) and Heath & Mansfield (1962), uses two manometers to compare the resistance of the leaf with that of a fixed capillary resistance, and in the Wheatstone Bridge porometer of Heath & Russell (1951), the resistance of the leaf is balanced with calibrated capillary resistances by a sensitive manometer.

(b) *Diffusive flow porometers*

Most diffusive flow porometers are complex and measure the diffusion of exotic gases such as hydrogen (Gregory & Armstrong, 1936) or nitrous oxide (Slatyer & Jarvis, 1966) through leaves. A simpler diffusion porometer suitable for field use (van Bavel *et al.*, 1965) is described in a modified form by Meidner & Mansfield (1968). In this apparatus the change in electrical resistance of a moisture sensitive element, placed at a standard distance from the leaf, is related to the diffusion which occurs. Such an instrument has been used to calibrate viscous flow porometers in terms of diffusive flow (Meidner & Mansfield, 1968).

4.4.4 The relationship between stomatal closure and water deficit

The relationship between the control of water loss and the water balance of the plant is useful in comparative investigations and can be determined by

The measurement of stomatal aperture

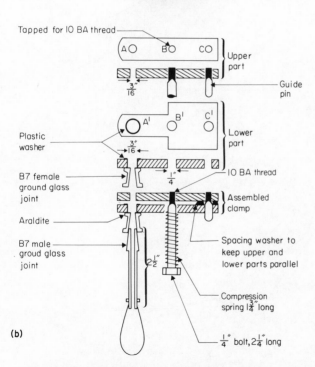

Figure 5.19   (a) Diagram of an Alvim field porometer. (b) Diagram of a simple hand porometer. Copyright © McGraw-Hill Book Co. (U.K.) Ltd. 1968, Meidner & Mansfield. Physiology of Stomata. Reproduced by permission.

following the decline in weight of cut shoots, initially fully saturated, under normal illumination. The assumption that stomata are fully open under such conditions is not always valid and may be checked, where possible, by direct examination of stomatal aperture or by inspection of the form of decline curve. The decline in water content can be separated, by plotting rates of loss (Fig.

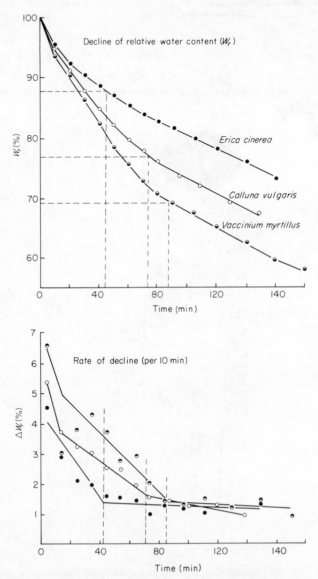

Figure 5.20 Relative water content at stomatal closure, estimated from the water loss of cut shoots. Dotted lines indicate the estimated point of closure.

5.20), into a rapid (stomatal) phase and a less rapid (cuticular) phase, and the effective point of stomatal closure has been shown to coincide with the intercept of the two rates of loss (Bannister, 1964a). This method can be used to estimate the point of stomatal closure in plants (e.g. heathers) which are not amenable to other methods. Hygen (1953) has followed the decline in water content of a wide variety of species although he did not estimate the point of stomatal closure. Relative water contents at stomatal closure vary both between species (Jarvis & Jarvis, 1963b; Bannister, 1964a; Lopushinsky, 1969b) and within species (Bannister, 1964a; Hutchinson, 1970a) and can often be related to the distribution and field behaviour of the individual plants.

The time taken to stomatal closure under standard conditions may prove significant, Bannister (1971) found that plants from open and shaded habitats showed similar water contents at stomatal closure but that the stomata closed more rapidly in the shade plants.

4.5 THE DIFFUSIVE RESISTANCE OF LEAVES

An estimation of the resistance of leaves to the diffusion of water vapour and carbon dioxide is useful in the theoretical analysis of the response of plants to environmental factors.

The resistances to diffusion may be divided into boundary layer resistance (r_a), stomatal resistance (r_s), cuticular resistance (r_c) and mesophyll resistance (r_m). By analogy with electrical resistances, these are usually combined partly in series and partly in parallel (Fig. 5.21). A fundamental relationship for the calculation of the various resistance to water vapour transfer is

$$E = \frac{e_1 - e_a}{R}$$

where E = the evaporation rate (g m^{-2} s^{-1})
e_1 = water vapour concentration at the evaporating surface of the leaf (g m^{-3}) water
e_a = vapour concentration of the air (g m^{-2})
and R = total resistance to diffusion (s m^{-1}).

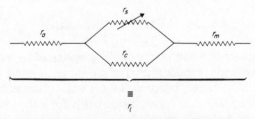

Figure 5.21 Diagrammatic representation of the various resistances to gaseous diffusion in a leaf. r_1, total leaf resistance; r_a, leaf/air resistance; r_s, stomatal resistance (variable); r_c, cuticular resistance; r_m, mesophyll resistance.

The boundary layer resistance can be estimated by the use of model leaves made of blotting paper saturated with water (Cowan & Milthorpe, 1968; Thom, 1968), although more rigid models, particularly of awkward leaf shapes, can be made from plaster of paris. The resistance to other components can be obtained from measuring transpiration under defined environmental conditions (Holmgren *et al.*, 1965). The mesophyll resistance is difficult to measure and is often taken as some function of leaf thickness, such as the ratio of internal to external surface area, although the carbon dioxide compensation point (2.4.2) is often used in estimates of the mesophyll resistance to the diffusion of carbon dioxide.

Workers such as Heath (1969) have taken a purely anatomical approach and have derived formulae such as

$$R = \frac{1}{D}\left[l_i + \frac{\pi d_1}{8} + \frac{a_1}{na_s}\left(l_s + \frac{\pi d_s}{4}\right)\right]$$

where
$D =$ the diffusion coefficient,
$l_i =$ length of internal pathway,
$\pi d_1/8 =$ depth of boundary layer in still air,
$a_1 =$ leaf area,
$n =$ stomatal density,
$a_s =$ stomatal area,
$l_s + \dfrac{\pi d_s}{4} =$ length of the diffusion pathway of a single stomatal pore of depth l_s and diameter d_s.

The relationships can be corrected for moving air (Penman & Schofield, 1951; Milthorpe, 1961) and used to compute heat transfer, vapour transfer and the diffusion of carbon dioxide (e.g. Thom, 1968; Cowan & Milthorpe, 1968). The various responses can be integrated by mathematical models which can be used to investigate and compare the behaviour of leaves (Gates, 1965, 1968a,b; Idle, 1970; Lewis, 1972).

4.6 DROUGHT RESISTANCE

The drought resistance of whole plants can be investigated by monitoring their responses to water stress. Measurements of growth allow some assessment of resistance but are more readily interpreted if the water status of both plant and soil is determined (e.g. Jarvis & Jarvis, 1963b). It is not always possible (3.2.3) to use whole plants and considerable information can be obtained from the desiccation of detached plant parts.

A range of water contents can be produced by allowing turgid shoots to lose water for different periods of time (e.g. Bannister, 1970) or by subjecting them to different degrees of desiccation for the same period of time (e.g. Jarvis & Jarvis, 1963b). Different evaporating conditions can be maintained above saturated solutions of various salts (Nuffield Biology, 1966) or by sulphuric acid/water mixtures (Sutcliffe, 1968).

The water status of samples after desiccation gives a measure of their potential for avoiding drought, while the potential for drought tolerance can be estimated by monitoring their recovery. Damage to tissues can be observed directly or estimated from conductivity measurements or by vital staining (3.2.3) but other methods have been used for desiccated tissue. Michael (1968) has used a bicarbonate indicator to detect abnormal respiration in damaged samples, and the ability of the tissue to resaturate after desiccation has been often used (e.g. Bornkamm, 1958; Rychnovská-Soudková, 1963; Bannister, 1970, 1971).

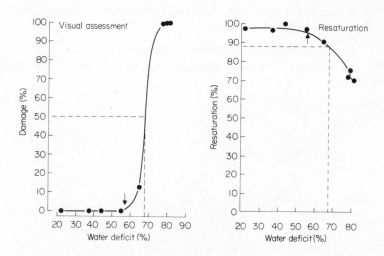

Figure 5.22 Drought resistance of *Erica cinerea*, January 1968. Dotted lines indicate the point causing 50% damage (water deficit = 68%). Arrows indicate point of 95% resaturation (water deficit = 57%).

Undamaged material may fail to show complete (100%) resaturation but material showing 95% resaturation is usually undamaged (Bannister, 1970; Fig. 5.22) while 85% resaturation indicates some damage (Polwart, 1970). Occasionally, damaged material may show complete resaturation or even supersaturation (Rychnovská, 1965) and it is advisable to investigate the relationships between damage and resaturation before using the latter as a measure of recovery.

5 THE MINERAL NUTRITION OF PLANTS

5.1 INTRODUCTION

The plant physiologist has investigated the basic requirements of plants for mineral elements and knows much about their uptake and role in the physiology of the plant. There is still a need to compare species or ecotypes as they may respond differently to the same soil as a result of dissimilar uptake or by having different requirements for mineral nutrients. In extreme soils, the deficiency or toxicity of certain minerals may determine the responses of plants and tolerant species will be better equipped for survival. Competition between species will accentuate their differences and, although high-yielding species would normally be expected to survive, they can occasionally be eliminated by competition from lower-yielding species (the 'Montgomery Effect', Montgomery, 1912).

The opportunities that exist for the physiological-ecological study of mineral nutrition (Fig. 5.23) include:

(a) The growth response of species in relation to nutrient regime.
(b) The tolerance of nutrient deficiency and of excesses of nutrient and other ions.

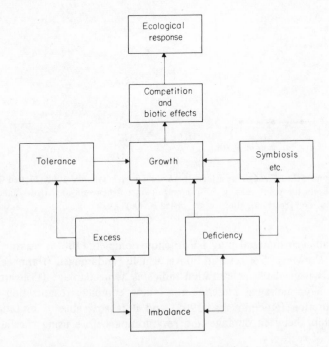

Figure 5.23 Mineral nutrition and ecological response.

(c) The competitive effects of other species and individuals upon the mineral nutrition of plants (symbiotic and other microbial effects can be considered here).

Critical experimentation must be carried out under carefully controlled conditions and laboratory experimentation in growth cabinets may be imperative (Rorison, 1969) but preliminary investigations can be carried out in the field. The results of laboratory experiments must also be related to field conditions if they are to be of any practical value to the ecologist.

5.2 FIELD EXPERIMENTS

It is easier to add nutrients to the soil than to remove them and the application of nutrients will almost always produce some responses in the vegetation as it may counteract deficiency, alleviate toxicity, alter pH and nutrient availability or create imbalances, excesses and toxicities. Analyses of plants and soil, both before and after the addition of nutrients, will help to interpret the responses.

Natural populations of plants are usually well adapted to the soils on which they grow and while individual plants may not show symptoms of toxicity or deficiency, they will often respond to nutrient addition. Visual symptoms of deficiency have been documented for crop plants (Wallace, 1961) and potential deficiencies and toxicities can be investigated by growing appropriate crop plants in particular soils, for example, Proctor (1971a) has used oats for the bioassay of serpentine soils (Fig. 5.24). Crop plants make greater demands upon the pool of soil nutrients than many wild species and a number of workers have used wild plants in order to obtain more ecologically meaningful information. An example is the use of *Rumex acetosa* by Atkinson (1973) in a study of the phosphorus nutrition of plants on sand dune soils. Visible symptoms of deficiency can easily be confused with various toxicities and even with the effects of atmospheric pollutants (e.g. Treshow, 1970).

Soil analyses (Chapter 8), although informative, cannot account for differences in the physiology and rooting pattern of species and for seasonal variations in availability and requirement. Analyses of plant material can give some indication of deficiencies and excesses of mineral nutrients, but only if the 'normal' nutrient content is known and the ecologist may well have to determine the 'normal' content of the species in his investigation. The mineral contents of plants are subject to considerable seasonal fluctuations which may be real or related to growth in a manner analogous to seasonal variations in water content (Guha & Mitchell, 1966).

The chemical analysis of soil and plant material is therefore no substitute for a well designed experiment although it does provide important information.

The design of field experiments owes much to agricultural experience, and

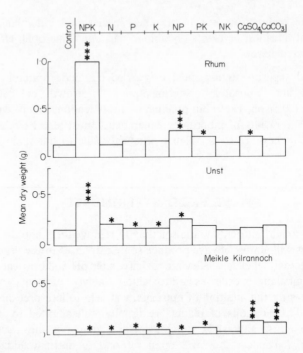

Figure 5.24 The response of oats to different nutrient additions to three serpentine soils. * Responses significantly different from unfertilized soil at $P=0.05$, *** at 0.001. (Proctor, 1971a.)

a long-term study, such as the Park Grass Experiment at Rothamstead (Thurston, 1969) provides a dramatic example of this approach.

The greatest amount of information will be obtained from carefully planned experiments. A factorial design (Bailey, 1959) where the major nutrients, nitrogen, phosphorus and potassium (N, P, K) are added both individually and in various combinations increases the value of an experiment in which the limiting factors are unknown. The possible treatment combinations would be no nutrients, N, P, K, NP, NK, PK, NPK.

These treatments are randomized in sub-plots within a defined area of vegetation (a 'block'). Blocks are replicated at different localities within the experimental area in order to separate the effects of environmental variation from the effects of treatment. The statistical analysis gives information on the variation between the blocks and the significance of any responses to the major nutrients or to interactions between nutrients. Chemical analyses of plants give information on nutrient availability and limitation, elements which increase in concentration and are associated with increased growth are likely to be limiting.

Examples of nutrient addition in field experiments are given by Willis

(1963) and Atkinson (1973) on dune soils which are generally infertile and by Ferreira & Wormell (1971) on serpentine soils which may combine infertility with toxicity.

The principles outlined are not confined to experiments with major nutrients; they may also be applied to problems involving minor nutrients such as copper deficiency in Irish peats (Mulqueen *et al.*, 1961) or the effects of liming on serpentine soils (Proctor, 1971a).

5.3 LABORATORY EXPERIMENTS

Laboratory experiments will often resemble field experiments in that they use field soils with or without added nutrients and examine the responses of individual species or mixtures (e.g. Proctor, 1971a). The advantage of controlled studies is that that number and mixture of individuals can be manipulated and factors other than soil nutrient status can be held relatively constant. A difficulty in experimental design is that different soils have different ion exchange properties and the soil rather than the experimenter may well ultimately control the supply of nutrients to the plant. Often a more satisfactory approach is to grow the plants in an entirely artificial medium. Culture solutions may be used by themselves or added to an artificial supporting medium with virtually no nutrient-supplying capacity of its own (e.g. acid-washed quartz sand). The use of solutions is often preferred as it can be difficult to ensure a homogenous supply of nutrients within a sand culture and the sand offers a large surface for the adsorption of ions from weak solutions. The culture solutions of Knop (1860) and Hoagland (1920) are the forerunners of modern solutions (Table 5.2). A comprehensive account of culture solutions and relevant techniques is given by Hewitt (1966).

Culture solutions must be well aerated and this can be ensured by employing a large volume of liquid in relation to the size of the plant, wide-

Table 5.2 A comparison of Knop's and Hoagland's solutions (all concentrations in g l^{-1}).

Chemical	Knop	Hoagland (I)	Hoagland (II)
$Ca(NO_3)_2.4H_2O$	0·8	1·18	0·95
KNO_3	0·2	0·51	0·61
KH_2PO_4	0·2	0·14	—
$NH_4H_2PO_4$	—	—	0·12
$MgSO_4.7H_2O$	0·2	0·49	0·49
$FePO_4$	trace	—	—
Ferric tartrate	—	0·005	0·005

Trace amounts of B, Mn, Cu, Zn, Mo may be added to the solutions at a final concentration of between 0·02 and 0·2 ppm.

necked vessels which present a large liquid surface to the air, frequent changes of solution and forced aeration or mixing (Clymo, 1962) using pumps.

The absorption of nutrients lowers the concentration of ions present, whilst differential absorption can lead to changes in availability and pH which in extreme cases can be toxic. These are additional reasons for frequent changes of solutions. The composition of the medium may determine the direction of change. Where nitrogen is supplied as NH_4^+, absorption will lead in an increase in acidity as NH_4^+ is likely to be exchanged for H^+ and conversely the use of NO_3^- may result in a higher pH. This means that a continual adjustment of pH is necessary in experiments designed to compare different sources of nitrogen (e.g. Gigon & Rorison, 1972). The substitution of ions in culture solutions will often lead to differences in the availability of ions other than the one in question. When $CaSO_4$ is substituted for $Ca(NO_3)_2$ in studies of nitrate absorption the calcium availability is held constant but the availability of both sulphate and nitrate are altered. Hewitt (1966) gives recommendations for substitutions.

The concentrations of culture solutions are often more suited to crop than wild plants and the normal concentrations of standard solutions may inhibit the growth of some species (e.g. Sheikh, 1969). Most plants can tolerate a very dilute culture solution without a reduction in performance as long as the overall ionic balance is maintained (Wallace *et al.*, 1967).

In extreme dilutions, changes in concentration due to the adsorption of ions onto the surface culture vessesls and contamination by dust and micro-organisms become significant, but may be partly overcome by using a constant flow of large volumes of liquid (Asher *et al.* 1965, 1966). The purity of the reagents used and the mineral reserves of seed, seedling or cutting used in the experiment must be considered when the effects of low concentrations of specific ions are being investigated. The mineral content of the plant may be sufficient to delay any observable responses for several weeks (Rorison, 1969).

5.4 NUTRIENT DEFICIENCY

The adaptation of plants to the mineral status of their habitat (Bradshaw, 1969) means that nutrient deficiencies are not always apparent. Symptoms may include stunting, chlorosis and poor vegetative cover, but frequently the deficiency is apparent only after the addition of nutrients. The characteristic species of the site may respond to the addition of nutrients, but more often other species become dominant and alter the whole character and composition of the community.

The production of wild plants, like that of crop plants, is limited by deficiencies, but the survival of some wild species may depend upon the deficiencies of other plants. Phosphorus is commonly in short supply in British soils and may be almost absent from some acid soils whilst in basic soils it may be present only as unavailable insoluble phosphates. The phosphorus deficiencies of sand dune communities have been demonstrated by Willis (1963) and Atkinson (1973) but the survival of *Kobresia simpliciuscula,* one of the Teesdale rarities, has been shown to depend upon the inhibition of competing species by phosphorus deficiencies (Jeffrey & Pigott, 1973).

While it is useful and important to know that some species respond to nutrient addition and that others depend upon deficiency for survival it is the strategy of individual species that is of interest. Rorison (1969) has discussed the possible effects of a spasmodic, rather than a continuous supply of nutrients. Atkinson & Davison (1971) have shown that *Epilobium montanum* responds differently to a period of phosphorus availability followed by a period of deficiency than to a period of deficiency followed by an adequate supply. Investigations of this type open up promising lines of research as nutrient supply is unlikely to remain constant under field conditions.

The comparison of different species usually involves an examination of their responses over a range of nutrient régimes (e.g. Clarkson, 1967; Rorison, 1968). Some fast-growing species such as *Urtica dioica,* appear to require high concentrations of phosphorus for maximum growth whereas others such as *Deschampsia flexuosa* (Rorison, 1968) and *Agrostis setacea* (Clarkson, 1967) with low growth rates grow quite well at low concentrations of phosphorus. If low growth rates can be generally associated with tolerance to low levels of nutrients it would explain the elimination of characteristic species when nutrients are added to plant communities on nutrient-deficient soils. However, many more species will have to be investigated before such a general conclusion can be confirmed. The ability of plants to accumulate phosphorus in their roots (Jeffery, 1964, 1969; Nassery, 1970) or in mycorrhizas (Harley, 1969) provides a mechanism for ensuring adequate supplies when the external source is poor. Few studies show when and how stored nutrients are made available to the rest of the plant, this may be investigated by determining the root and shoot concentrations in plants which can accumulate nutrients and comparing them with plants that do not have this facility (e.g. Nassery, 1970; Bannister & Norton, 1974) or by the use of radioactive tracers (e.g. Pearson, 1971).

Radio-isotopes are of great value in the study of plant nutrition but some nutrients have the disadvantage that their absorption may be inefficient, that they may have too short or too long a half-life or that they may be readily incorporated into a variety of compounds and disseminated through the tissues. Radio-isotopes are particularly useful in short-term studies such as the investigation of the pattern of nutrient uptake by different root systems (Boggie *et al.,* 1958), the uptake of nutrients by mycorrhizas (Harley, 1969) and the transfer of nutrients from fungus to host (Smith, 1966), or from root to shoot (Pearson, 1971). The availability and convenience of ^{32}P may well have contributed to the amount of work which has been concerned with phosphorus nutrition.

Other limiting nutrients such as nitrogen have been neglected in ecological studies (Bradshaw, 1969) but offer many opportunities for research. The value of leguminous crops in supplementing nitrogen supplies is well known, but the ecological role of plants with root nodules or insectivorous habits is very underinvestigated. Stewart & Pearson (1967) have demonstrated the effect of nitrogen fixation by *Hippophae rhamnoides* on the nitrogen economy of sand-dunes, and Oudman (1936) showed that nitrogen deficient sundew plants benefited from being fed insects, but the effect of leguminous plants on the nutrition of heathland communities or the effects of insectivorous plants on the nitrogen economy of an oligotrophic mire have never been seriously considered.

Studies of symbiotic nitrogen fixation were once hampered by the lack of suitable techniques. The use of ^{15}N and its separation by mass spectrometry (e.g. Stewart & Pearson, 1967) was beyond the means of most ecology laboratories, but the recent acetylene reduction technique (Dilworth, 1966; Hardy *et al.*, 1968) has removed many of the barriers to the study of nitrogen fixation (see Chapter 4, Section 4.2.3).

Deficiencies of trace elements may only become apparent when crop plants are grown on deficient soils, but the ecologist can investigate the mechanisms which enable naturally occurring species to be tolerant of deficiencies while crop plants are susceptible.

The presence and effects of nutrient deficiency can be assessed by nutrient addition in the field, followed by laboratory experiments with controlled nutrient supply, and expanded by a consideration of the strategy of uptake, storage and translocation within the plant. The ecologist must then relate the results of such experiments to field conditions and to the role of individual plants within the community.

5.5 EXCESS OF NUTRIENT AND OTHER IONS

Excesses may be of major or minor nutrients, calcareous soils have an excess of calcium, serpentinic soils may have an excess of magnesium, soils associated with bird colonies may have an excess of nitrogen, and soils derived from metalliferous rocks may have excesses of ions normally needed in only minute amounts (e.g. copper and zinc) or of elements not normally required by the plant (e.g. lead). Acid soils may have excess aluminium, saline soils an excess of sodium and chlorine, whilst exotic elements such as fluorine may accumulate to excess as a result of various industrial processes.

The problem of excess is one of imbalance as well as toxicity, the abnormal abundance of one element may alter soil conditions or patterns of uptake so that other elements become deficient. In calcareous soils, iron may be deficient and magnesium rich soils may effectively be calcium deficient. The interaction between excess and deficiency is well illustrated by studies of lime chlorosis in relation to calcium excess and iron deficiency (Grime & Hodgson, 1969; Hutchinson, 1967).

Experiments with nutrient excesses often involve the comparison of the growth of species or ecotypes on suspect and normal soils (Fig. 5.25) or on contrasting soil types such as acidic and calcareous soils. The addition of

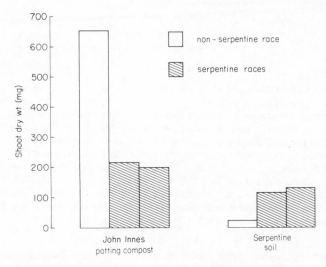

Figure 5.25 A comparison of the growth of serpentine and non-serpentine races of *Rumex acetosa* on 'normal' and serpentine soil. (From data of Proctor, 1971a.)

nutrients and calcium salts may help to counteract excesses, imbalances and toxicities (Proctor, 1971a).

Plants may also be grown on artificial media and supplied with nutrient solutions, but it may be difficult to add an excess of a particular ion without completely altering the properties of the nutrient solution. Plants can, however, be alternated between solutions containing only the ion in question and complete nutrient solutions (Fig. 5.26).

The technique of 'tolerance testing' examines the effect of the suspect ion in water culture upon the rate of root extension. Calcium nitrate is usually added to the solutions as this lessens any toxic effects and allows the suspect ion to be presented at higher concentration (e.g. Gregory & Bradshaw, 1965). Tolerance is usually expressed as the ratio of root growth in the test solution to that in a control solution without the suspect ion. The concentration used for tolerance testing is empirical and derived from an investigation of root growth in a series of concentrations (Fig. 5.27). Tolerance has also been assessed in terms of the solution concentration that causes a critical (e.g. 50%) reduction in root growth (Sparling, 1967).

Indices of tolerance are expected to lie between zero with no growth in the test solution and unity with equal growth in test and control solutions, but

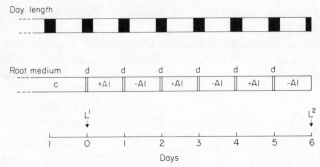

Figure 5.26 Scheme describing the experimental procedure and the solutions used to investigate the response of iron-deficient roots to aluminium treatment. Day length: shaded areas represent dark periods. Root medium: − Al solution, a complete nutrient solution designed by Rorison (1958). The pH was adjusted to 4·0. Iron at 2 ppm (0·5 ppm for *Scabiosa columbaria*) as ferrous sulphate was dissolved in the solution immediately before use and the solution was discarded after 24 hours. + Al solution, a solution of aluminium sulphate adjusted to pH 4·0. c solution, as − Al but with Fe E.D.T.A. as iron source. d solution, distilled water washing between + Al and − Al treatments. L^1, initial measurement of total root length of seedling taken approximately 1 week after germination. L^2, final measurement of total root length. (Grime & Hodgson, 1969.)

occasionally roots may grow better in the test solution than in the control (Proctor, 1971b) and this can be interpreted as extreme tolerance, a requirement for the ion, or inhibition by the control solution. Interpretation is easier when many comparisons are made as species may then be ranked with

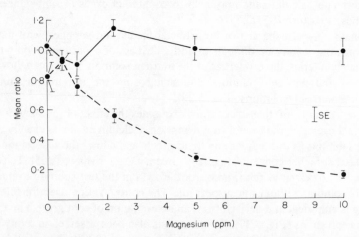

Figure 5.27 Mean ratio of growth in length of longest adventitious roots in magnesium solutions/growth in distilled water. Solid line, *Agrostis canina* from Meikle Kilrannoch. Broken line, *A. stolonifera* from a non-serpentine soil near Oxford. (Proctor, 1971b.)

regard to the index, irrespective of whether they show stimulation or depression.

The degree of physiological tolerance which is measured by tolerance testing can often be related to the concentration of the suspect ion in the soil from which the plant was taken (e.g. Gregory & Bradshaw, 1965), but the ecological tolerance may differ, especially in soils where the particular element is complexed and relatively unavailable to the plant. The presence of abnormal amounts of an element in a soil does not necessarily indicate that the plants are being subjected to an excess.

Co-tolerance occurs when the tolerance to one ion confers tolerance to others and is considered to be rare and anomalous (Turner, 1969). Co-tolerances can be investigated by testing the tolerance of a species or ecotype to a range of elements and may be expected to occur where the size and physico-chemical properties of the different ions are similar. Multiple tolerances may be found where several elements occur in the same soil and may not be as rare as true co-tolerance. For example, an examination of Proctor's (1971b) data for various populations of *Agrostis stolonifera* reveals a significant association between tolerances to magnesium and nickel, both of which occur in serpentine soils. The demonstration of a multiple tolerance does not necessarily indicate that similar mechanisms are involved although the demonstration of co-tolerance may suggest this more strongly.

A tolerance of one ion may be associated with a susceptibility to another. Salt marsh ecotypes of *Festuca rubra* are highly tolerant of NaCl but intolerant of aluminium salts whereas the converse is shown by ecotypes from acid pasture (Hunter, 1971). This combination of tolerances can be used to explain both the presence of the appropriate ecotype on one particular site and its absence from a contrasting site.

Further investigations of the tolerance of plants to excesses of nutrients and other ions can examine the mechanisms involved. The excess may be truly tolerated when the ion is found within the cell in abnormal amounts, or avoided when the element is excluded, precipitated at the root surface or bound in the cell wall. Such mechanisms can be investigated by tissue fractionation prior to analysis and comparing the chemical compositions of the various fractions. Nutrient excess (or deficiency) is not necessarily a direct cause of failure in the field. Hutchinson (1970a,b) has shown that species with lime-chloroses are unable to control their water loss and that this could be the most important reason for their failure under field conditions.

5.6 COMPETITION AND MINERAL NUTRITION

Many experiments on mineral nutrition have concluded that the differences that have been found between species may be of advantage in a competitive situation but relatively few experiments deal with competition itself. The replacement series of de Wit (1960) is commonly used to investigate the interaction between two competing species by comparing their growth in pure culture with their growth in mixtures. If a standard density (m) is used in an experiment with species A & B, a simple design could include the following three treatments:

$$m(A), \quad m(B), \quad \tfrac{1}{2}m(A) + \tfrac{1}{2}m(B).$$

An example of this type of experiment is the comparison of two ecotypes of *Galinsoga ciliata* by Shontz & Shontz (1972). In mixed culture and under an intermediate nutrient régime, a eutrophic ecotype showed a depression of growth when compared with a more oligotrophic ecotype, although both showed comparable growth in pure culture (Fig. 5.28). More complex experimental designs are possible, a sophisticated analysis of competition with regard to mineral nutrition is given by van den Bergh (1969). He showed that under sub-optimal conditions the low-yielding *Agrostis tenuis* and *Alopecurus pratensis* replaced the high-yielding *Dactylis glomerata* and *Lolium perenne* respectively. This is an example of the 'Montgomery Effect'.

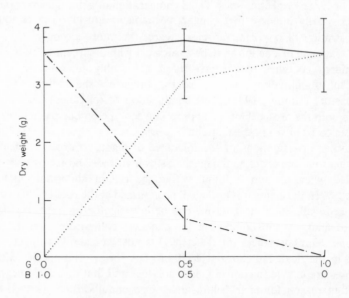

Figure 5.28 Competition between ecotypes of *Galinsoga ciliata*. Growth of plants for 8 weeks in pure and in mixed cultures when supplied with quarter-strength soluble fertilizer. Dotted line represents the glasshouse plants (G). Pecked line indicates the barnyard plants (B). Continuous line represents the summed growth for each pot. Each point represents the mean weight per pot for three replications; standard deviation is indicated for each point. (Shontz & Shontz, 1972.)

Where mineral nutrition gives a species an advantage in a competitive situation, its effect usually operates through a growth form which allows it to exploit some other aspect of the physical environment (cf. van den Bergh, 1969). Improved root growth allows more successful competition for both water and nutrients while improved shoot growth allows better competition for light. In both these situations alterations in light and moisture régime are likely to have just as dramatic effect on the outcome as changes in the nutrient régime and suitable factorial experiments must be designed to separate the various environmental effects.

It may be even more difficult to separate the effects of mineral nutrition from biotic factors especially the effects of micro-organisms (e.g. Barber, 1969) in the rhizosphere and in symbiotic associations, and the possibility of chemical anatagonisms between competing plants or their associated micro-organisms. Grazing animals may select the more nutritious plants (Gimingham, 1972) and their droppings may alter the nutrient supply of a site. 'Heather check' of spruce provides a good example of the complexity of such interactions and can be variously interpreted as competition between plants, competition between mycorrhizas, or chemical inhibition of plant or mycorrhizal growth (Gimingham, 1972).

The inhibition of plant growth by extracts of living and dead plants or of soil and litter is readily demonstrated in the laboratory (e.g. Went, 1970; Lange & Kanzow, 1965; Jarvis, 1964) but it is difficult to show that it is of any significance in nature. A possible approach is to enclose the roots of the test species in containers that would prevent the penetration of roots but allow the diffusion of inhibitors (e.g. Stevenson, 1972; cf. Grümmer & Beyer, 1959).

The many interactions between mineral nutrition, competition and biological factors represent an important, although negelected, area of study.

6 REFERENCES

ALEXANDROV V.Y. (1964) Cytophysiological and cytoecological investigations of heat resistance of plant cells towards the action of high and low temperatures. *Q. Rev. Biol.* **39,** 35–77.

ALVIM P. DE T. (1965) A new type of porometer for measuring stomatal opening and its use in irrigation studies. *Arid Zone Res.* **25,** 325–329.

ALVIM P. DE T. & HAVIS J.R. (1954) An improved infiltration series for studying stomatal opening as illustrated with coffee. *Pl. Physiol., Lancaster* **29,** 97–98.

ANDERSSON N.E., HERTZ C.H. & RUFELT H. (1954) A new fast recording hygrometer for plant transpiration measurements. *Physiol. Pl.* **7,** 753–767.

ARENS K. (1939) Bestimmung des Turgordruckes an einer Einzelzelle mit dem Manometer. *Planta.* **30,** 113–117.

ASHER C.J., OZANNE P.G. & LONERAGAN J.F. (1965) A method for controlling the ionic environment of plant roots. *Soil Sci.* **100**, 149–156.

ASHER C.J.& OZANNE P.G. (1966) Growth and potassium content of plants in solution cultures maintained at constant potassium concentrations. *Soil Sci.* **103**, 155–161.

ATKINSON D. (1973) Observations on the phosphorus nutrition of two sand dune communities at Ross Links. *J. Ecol.* **61**, 117–133.

ATKINSON D. & DAVISON A.W. (1971) Effect of phosphorus deficiency on growth of *Epilobium montanum* L. *New Phytol.* **70**, 789–797.

AUDUS L.J. (1953) A simplified version of an apparatus for the measurement of oxygen evolution in the photosynthesis of *Elodea*. *School Sci. Rev.* **25**, 120.

BAILEY L.F., ROTHACHER J.S. & CUMMINGS W.H. (1952) A critical study of the cobalt chloride method of measuring transpiration. *Pl. Physiol., Lancaster,* **27**, 563–574.

BAILEY N.T.J. (1959) *Statistical Methods in Biology*. London, English Universities Press.

BALASUBRAMANIAM S. & WILLIS A.J. (1969) Stomatal movement and rates of gaseous exchange in excised leaves of *Vicia faba*. *New Phytol.* **68**, 663–674.

BANNISTER P. (1964a) Stomatal responses of heath plants to water deficits. *J. Ecol.* **52**, 151–158.

BANNISTER P. (1964b) The water relations of certain heath plants with reference to their ecological amplitude. I. Introduction: germination and establishment. *J. Ecol.* **52**, 423–432.

BANNISTER P. (1964c) The water relations of certain heath plants with reference to their ecological amplitude. II. Field studies. *J. Ecol.* **52**, 481–497.

BANNISTER P. (1964d) The water relations of certain heath plants with reference to their ecological amplitude. III. Experimental studies: general conclusions. *J. Ecol.* **52**, 499–509.

BANNISTER P. (1970) The annual course of drought and heat resistance in heath plants from an oceanic environment. *Flora, Jena.* **159**, 105–123.

BANNISTER P. (1971) The water relations of heath plants from open and shaded habitats. *J. Ecol.* **59**, 51–64.

BANNISTER P. (1973) A note on some observation on frost damage in the field, with particular reference to various ferns. *Trans. Bot. Soc. Edinb.* **42**, 111–113.

BANNISTER P. & NORTON W.M. (1974) The response of mycorrhizal and non-mycorrhizal rooted cuttings of heather (*Calluna vulgaris* (L.) Hull) to variations in nutrient and water regime. *New Phytol.* **73**, 81–90.

BARBER D.A. (1969) The influence of the microflora on the accumulation of ions by plants. In I.H. Rorison (ed.) *Ecological Aspects of the Mineral Nutrition of Plants*, pp. 191–200. Oxford, Blackwell.

BARRS H.D. (1965a) Heat of respiration as a possible cause of error in the estimation by psychometric methods of water potential in plant tissues. *Nature, London* **203**, 1136–1137.

BARRS H.D. (1965b) Comparison of water potentials in leaves as measured by two types of thermocouple psychrometer. *Aust. J. biol. Sci.* **18**, 36–52.

BARRS H.D. (1968) The determination of water deficits in plant tissues. In T.T. Kozlowski (ed.) *Water Deficits and Plant Growth*, Vol. I, pp. 235–368. New York, Academic Press.

BARRS H.D. & WEATHERLEY P.E. (1962) A re-examination of the relative turgidity technique for estimating water deficits in leaves. *Aust. J. biol. Sci.* **15**, 413–428.

BAVEL C.H.M. VAN, NAKAYAMA F.S. & EHRLER W.L. (1965) Measuring transpiration resistance in leaves. *Pl. Physiol., Lancaster* **40**, 535–540.

BERGH J.P. VAN DEN (1969) Distribution of pasture plants in relation to chemical properties of the soil. In I.H. Rorison (ed.) *Ecological Aspects of Mineral Nutrition of Plants*, pp. 11–23. Oxford, Blackwell.

BIERHUIZEN J.F., SLATYER R.O. & ROSE C.W. (1965) A porometer for laboratory and field operation. *J. exp. Bot.* **16**, 182–191.

BOGGIE R., HUNTER R.F. & KNIGHT A.H. (1958) Studies of root development in the field using radioactive tracers. *J. Ecol.* **46**, 621–639.

BORNKAMM R. (1958) Standortsbedingungen und Wasserhaushalt von Trespen–Halptrockenrasen (Mesobromion) in oberen Leinegebiet. *Flora, Jena.* **146,** 23–67.

BOWMAN G.E. (1968) The measurement of carbon dioxide concentration in the atmosphere. In R.M. Wadsworth (ed.) *The Measurement of Enviromental Factors in Terrestrial Ecology,* pp. 131–140. Oxford, Blackwell.

BRADSHAW A.D. (1969) The ecologist's view point. In I.H. Rorison (ed.) *Ecological Aspects of the Mineral Nutrition of Plants,* pp. 415–427. Oxford, Blackwell.

BUSCALONI L. & POLLACCI G. (1901) L'applicazione delle pellicole di collodio allo studio di alcuni processi fisiologici delle plante ad in particolar modo pella transpirazione. *Atti. Ist. bot. eco. Pavia.* **2,** 44–49.

CAMPBELL R.M. (1972) *Studies on the distribution and productivity of freshwater macrophytes.* Ph.D. Thesis, University of St. Andrews.

CARRITT D.E. & CARPENTER J.H. (1966) Recommended procedure for Winkler analysis of sea water for dissolved oxygen. *J. Mar. Res.,* **24,** 286–318.

ČATSKÝ J. (1960) Determination of water deficit in disks cut out from leaf blades. *Biologia Pl.* **2,** 201–215.

CLAPHAM A.R. (1956) Autoecological studies and the 'Biological flora of the British Isles'. *J. Ecol.* **44,** 1–11.

CLARKSON D.T. (1967) Phosphorus supply and growth rates in species of *Agrostis* L. *J. Ecol.* **55,** 707–731.

CLAUSEN J.J. & KOZLOWSKI T.T. (1965) Use of the relative turgidity technique for measurement of water stresses in gymnosperm leaves. *Can. J. Bot.* **43,** 305–316.

CLYMO R.S. (1962) An experimental approach to part of the calcicole problem. *J. Ecol.* **50,** 707–731.

COOMBE D.E. (1966) The seasonal light climate and plant growth in a Cambridgeshire wood. In R. Bainbridge, G.C. Evans & O. Rackham (eds.) *Light as an Ecological Factor,* pp. 148–166. Oxford, Blackwell.

COWAN I.R. & MILTHORPE F.L. (1968) Plant factors influencing the water status of plant tissues. In T.T. Kozlowski (ed.) *Water Deficits and Plant Growth,* Vol. I, pp. 137–193. New York, Academic Press.

CRAFTS A.S., CURRIER H.B. & STOCKING C.R. (1949) *Water in the Physiology of Plants.* Waltham, Mass., Chronica Botanica Co.

CRAWFORD R.M.M. (1966) The control of anaerobic respiration as a determining factor in the distribution of the genus *Senecio. J. Ecol.* **54,** 403–413.

DARWIN F. & PERTZ D.F.M. (1911). On a new method of estimating the aperture of stomata. *Proc. R. Soc. B* **84,** 136–154.

DECKER J.P. (1955) The uncommon denominator in photosynthesis as related to tolerance. *Forest Sci.* **1,** 88–89.

DECKER J.P. & WIEN J.D. (1960) Transpiration surges in Tamarix and Eucalyptus as measured with an infra-red gas analyser. *Pl. Physiol., Lancaster* **35,** 340–343.

DILWORTH M.J. (1966) Acetylene reduction by nitrogen-fixing preparations from *Clostridium pasterianum. Biochim biophys. Acta* **127,** 285–294.

ECKARDT F.E. (1957) *Biol. Rep.* (1956–1957). Pasadena, Calif. Inst. Tech.

ECKARDT F.E. (1960) Ecophysiological measuring techniques applied to research on the water relations of plants. UNESCO, *Arid Zone Res.* **15,** 139–154.

ECKARDT F.E. (1968) Techniques de mesure la photosynthese sur le terrain basées sur l'emploi d'enceintes climatisées. UNESCO *Nat. Res. Res.* **5,** 289–320.

ELLENBERG H. (1958) Mineralstoffe für die pflanzliche Besiedlung des Bodens. A. Bodenreaktion (einschliesslich Kalkfrage). *Handb. Pfl Physiol.* **IV,** 638–709.

EVANS G.C. (1972) *The Quantitative Analysis of Plant Growth.* Oxford, Blackwell.

FALK F.O. (1966) A microwave hygrometer for measuring plant transpiration. *Z. Pfl Physiol.* **55,** 31–57.

FERREIRA R.E.C. & WORMELL P. (1971) Fertiliser response of vegetation on ultrabasic terraces on Rhum. *Trans. Bot. Soc. Edinb.* **41,** 149–154.

FLINT H.L. (1972) Cold hardiness of twigs of *Quercus rubra* L. as a function of geographic origin. *Ecology,* **53,** 1163–1170.

FLINT H.L., BOYCE B.R. & BEATTIE D.J. (1967) Index of injury—useful expression of freezing injury to plant tissues as determined by the electrolytic method. *Can. J. Pl. Sci.* **47,** 229–230.

FREEMAN G.F. (1908) A method for the quantitative determination of transpiration in plants. *Bot. Gaz.* **46,** 118–129.

FRY K.E. & WALKER R.B. (1967) A pressure infiltration method for estimating stomatal opening in conifers. *Ecology* **48,** 155–157.

GAFF D.F. & CARR D.J. (1964) An examination of the refractometric method for determining the water potential of plant tissues. *Ann. Bot., Lond.* **28,** 352–368.

GATES D.M. (1962) *Energy Exchange in the Biosphere.* New York, Harper & Row.

GATES D.M. (1965) Energy, plants and ecology. *Ecology* **46,** 1–13.

GATES D.M. (1968a) Transpiration and leaf temperature. *Ann. rev. Pl. Physiol.* **19,** 211–238.

GATES D.M. (1968b) Energy exchange in the biosphere. UNESCO, *Nat. Res. Res.* **5,** 33–43.

GIGON A. & RORISON I.H. (1972) The response of some ecologically distinct plant species to nitrate and ammonium nitrogen. *J. Ecol.* **60,** 93–102.

GIMINGHAM C.H. (1972) *The Ecology of Heathlands.* London, Chapman & Hall.

GLOVER J. (1941) A method for the continuous measurement of transpiration of single leaves under natural conditions. *Ann. Bot.* **5,** 25–34.

GRACE J. & WOOLHOUSE H.W. (1970) A physiological and mathematical study of the growth and productivity of a *Calluna-Sphagnum* community. I. Net photosynthesis of *Calluna vulgaris* (L) Hull. *J. appl. Ecol.* **7,** 363–381.

GREGORY F.G. & ARMSTRONG J.I. (1936) The diffusion porometer. *Proc. R. Soc. B* **121,** 27–42.

GREGORY F.G. & PEARSE H.L. (1934) The resistance porometer and its application to the study of stomatal movement. *Proc. R. Soc., B* **114,** 477–493.

GREGORY F.G., MILTHORPE F.L., PEARSE H.L. & SPENCER H.L. (1950). Experimental studies of the factors controlling transpiration. I. Apparatus and experimental technique. *J. exp. Bot.* **1,** 1–14.

GREGORY R.P.G. & BRADSHAW A.D. (1965) Heavy metal tolerance in populations of *Agrostis tenuis* Sibth. and other grasses. *New Phytol.* **64,** 131–143.

GRIEVE B.J. (1961) Negative turgor pressure in sclerophyllous plants. *Aust. J. Sci.* **23,** 376–377.

GRIME J.P. (1966) Shade avoidance and shade tolerance in flowering plants. In R. Bainbridge, G.C. Evans & O. Rackham (eds.) *Light as an Ecological Factor,* pp. 187–207. Oxford, Blackwell.

GRIME J.P. & HODGSON J.G. (1969) An investigation of the ecological significance of lime chlorosis by the means of large-scale comparative experiments. In I.H. Rorison (ed.) *Ecological Aspects of Mineral Nutrition of Plants,* pp. 67–99. Oxford, Blackwell.

GRUMMER G. & BEYER H. (1959) The influence exerted by species of *Camelina* on flax by means of toxic substances. In J.L. Harper (ed.) *The Biology of Weeds.* Oxford, Blackwell.

GUHA M.M. & MITCHELL R.L. (1966) The trace and major element composition of the leaves of some deciduous trees. *Pl. Soil* **24,** 90–112.

GUTTENBERG H. VON (1927) Studien über das Verhalten des immergrünen Laubblattes der Mediterranflora zu verschiedenen Jahreszeiten. *Planta* **4,** 726–779.

HAINES F.M. (1928) A method of investigating and evaluating drought resistivity and the effect of drought conditions upon water economy. *Ann. Bot.* **42,** 667–705.

HANDLEY W.R.C. (1963) Mycorrhizal associations and *Calluna* heathland afforestation. *Bull. For. Comm., Lond.* **36,** 1–70.

HARDY R.W.F., HOLSTEN R.D., JACKSON E.K. & BURNS R.C. (1968) The acetylene-ethylene assay for N_2 fixation. Laboratory and field evaluation. *Pl. Physiol., Lancaster* **43**, 1185–1207.

HARLEY J.L. (1969) *The Biology of Mycorrhiza* (2nd edn). London, Leonard Hill.

HARPER J.L. (1967) A Darwinian approach to plant ecology. *J. Ecol.* **55**, 247–271.

HARPER J.L. & OGDEN J. (1970) The reproductive strategy of higher plants. I. The concept of strategy with special reference to *Senecio vulgaris* L. *J. Ecol.* **58**, 681–698.

HARRIS J.A. & GORTNER R.A. (1914) Notes on the calculation of the osmotic pressure of expressed vegetable saps from the depression of the freezing point, with a table for the values of P for $\Delta = 0 \cdot 001°C$ to $\Delta = 2 \cdot 999°C$. *Am. J. Bot.* **1**, 75–78.

HATCH M.D., SLACK C.R. & JOHNSON H.S. (1967) Further studies on a new pathway of photosynthetic carbon dioxide fixation in sugar-cane and its occurrence in other plant species. *Biochem. J.* **102**, 417–422.

HEATH O.V.S. (1939) Experimental studies of the relation between carbon assimilation and stomatal movement. *Ann. Bot.* **3**, 469–495.

HEATH O.V.S. (1950) Studies in stomatal behaviour. V. The role of carbon dioxide in the light response of stomata. *J. Exp. Bot.* **1**, 29–62.

HEATH O.V.S. (1969) *The Physiological Aspects of Photosynthesis*. London, Heinemann.

HEATH O.V.S. & MANSFIELD T.A. (1962) A recording porometer with detachable cups operating on four separate leaves. *Proc. R. Soc., B* **156**, 1–13.

HEATH O.V.S. & RUSSELL J. (1951) The Wheatstone bridge porometer. *J. exp. Bot.* **2**, 111–116.

HEATH O.V.S. & WILLIAMS W.T. (1948) Studies in stomatal action. *Nature, Lond.* **161**, 178–179.

HELLMUTH E.O. & GRIEVE B.J. (1969) Measurement of water potential of leaves with particular reference to the Schardakow method. *Flora, Jena* **159**, 147–167.

HESKETH J.D. & BAKER D. (1967) Light and carbon assimilation by plant communites. *Crop Sci.* **7**, 285–293.

HEWITT E.J. (1966) Sand and water culture methods used in the study of plant nutrition. *Comm. Ag. Bur. Commn. Bur. Hort.* **22.**

HEWLETT J.D. & KRAMER P.J. (1963) The measurement of water deficits in broadleaf plants. *Protoplasma* **57**, 381–391.

HOAGLAND D.R. (1920) Optimum nutrient solutions for plants. *Science* **52**, 562–564.

HOFLER K. (1917) Die plasmolytisch-volumetrische Methode und ihre Andwendbarkeit zur Messung des osmotischen Wertes lebender Pflanzenzellen. *Ber. bot. Ges.* **35**, 706–726.

HOLMGREN P., JARVIS P.G. & JARVIS M.S. (1965) Resistances to carbon dioxide and water vapour transfer in leaves of different plant species. *Physiol. Pl.* **18**, 557–573.

HUBER B. (1928) Weitere quantative Untersuchungen über das Wasserleitungssystem der Pflanzen. *Jb. wiss. Bot.* **67**, 877–959.

HUBER B. & MILLER R. (1954) Methoden zur Wasserdampf- und Transpirations-H_2SO_4-registrierung im laufenden Luftstrom. *Ber. dt. Bot., Ges.* **67**, 223–233.

HUGHES A.P. & FREEMAN P.R. (1967) Growth analysis using frequent small harvests. *J. appl. Ecol.* **4**, 553–560.

HUNT R. & PARSON I.T. (1974) A computer program for deriving growth functions in plant growth analysis. *J. appl. Ecol.* **11**, 297–307.

HUNTER A.C.F. (1971) Salt tolerance in *Festuca rubra* Hons. Thesis, University of Stirling.

HUTCHINSON T.C. (1967) Lime chlorosis as a factor in seedling establishment on calcareous soil. I. A comparative study of species from acidic and calcareous soils in their susceptibility to lime chlorisis. *New Phytol.* **66**, 697–705.

HUTCHINSON T.C. (1970a) Lime chlorosis as a factor in seedling establishment on calcareous soils II. The development of leaf water deficits in plants showing lime chlorosis. *New Phytol.* **69**, 143–157.

HUTCHINSON T.C. (1970b) Lime chlorosis as a factor in seedling establishment on calcareous soils. III. The ability of green and chlorotic plants fully to reverse large water deficits. *New Phytol.* **69**, 261–268.

HYGEN G. (1953) On the transpiration decline of excised plant samples. *Norske Vid. Akad. Skr. 1 math. nat. Kl.* **1**, 1–84.

IDLE D.B. (1970) The calculation of transpiration rate and diffusion resistance of a single leaf from micro-meteorological information subject to errors of measurement. *Ann. Bot.* **34**, 159–176.

IRVING R.M. & LANPHEAR F.O. (1967) Environmental control of cold hardiness in woody plants. *Pl. Physiol., Lancaster* **42**, 1191–1196.

JACKSON W.A. & VOLK R.J. (1970) Photorespiration. *Ann. Rev. Pl. Physiol.* **21**, 385–432.

JAMES W.O. (1928) Experimental researches on vegetable assimilation and respiration. XIX. The effect of variations of carbon dioxide supply upon the rate of assimilation of submerged water plants. *Proc. R. Soc. B* **103**, 1–42.

JARVIS P.G. (1964) Interference by *Deschampsia flexuosa* (L.) Trin. *Oikos* **15**, 56–78.

JARVIS P.G. (1970) Characteristic of the photosynthetic apparatus derived from its response to natural complexes of environmental factors. In *Prediction and Measurement of Photosynthetic Productivity.* Wageningen, Pudoc.

JARVIS P.G. & JARVIS M.S. (1963a) The water relations of tree seedlings. I. Growth and water use in relation to soil water potential. *Physiol. Pl.* **16**, 215–235.

JARVIS P.G. & JARVIS M.S. (1963b) The water relations of tree seedlings. IV. Some aspects of the tissue water relations and drought resistance *Physiol. Pl.* **16**, 501–516.

JARVIS P.G. & JARVIS M.S. (1963c). Effects of several osmotic substrates on the growth of *Lupinus albus* seedlings. *Physiol. Pl.* **16**, 485–500.

JEFFERY D.W. (1964) The formation of polyphosphates in *Banksia ornata* an Australian heath plant. *Aust. J. biol. Sci.* **17**, 845–854.

JEFFERY D.W. (1969) Phosphate nutrition of five Australian heath plants. II. The function of polyphosphates in five heath species. *Aust. J. Bot.* **16**, 603–613.

JEFFERY D.W. & PIGOTT C.D. (1973) The response of grasslands on sugar limestone to applications of phosphorus and nitrogen. *J. Ecol.* **61**, 85–92.

JOST L. (1906) Uber die Reaktionsgeschwindigkeit im Organismus. *Biol. Zbl.* **26**, 225–244.

KAPPEN L. (1964) Untersuchungen über den Jahreslauf der Frost-, Hitze und Austrocknungsresistenz von Sporophyten einheimischer Polypodiaceen (*Filicinae*). *Flora, Jena* **155**, 123–166.

KAPPEN L. (1966) Der Einfluss des Wassergehaltes auf die Widerstandsfähigkeit von Pflanzen gegenüber hohen und tiefen Temperaturen, untersucht an Blättern einiger Farne und von *Ramonda myconi*. *Flora, Jena* 427–445.

KNIGHT R.C. (1917) The interrelations of stomatal aperture, leaf water content and transpiration rate. *Ann. Bot.* **31**, 221–240.

KNIPLING E.B. (1967) Effects of ageing on water deficit—water potential relationships of dogwood leaves growing in two environments. *Physiol.* (1860), **20**, 65–72.

KNOP W. (1860) Uber die Ernährung der Pflanzen durch wässerige Lösungen bei Ausschluss des Bodens. *Landw. VersStat.* **2**, 65.

KREEB K. (1960) Uber die gravimetrische Methode zur Bestimmung der Saugspannung und das Problem des negativen Turgors. I. Mitteilung. *Planta* **55**, 274–282.

KREEB K. (1963) Hydrature and plant production. In A.J. Rutter & F.H. Whitehead (eds.) *The Water Relations of Plants*, pp. 272–288. Oxford, Blackwell.

KREEB K. & ÖNAL M. (1961) Uber die gravimetrische Methode zur Bestimmung der Saugspannung und das Problem des negativen Turgors. II. Mitteilung Die Berucksichtigung von Atmungsverlusten während der Messung. *Planta* **56**, 409–415.

LANGE O.L. (1953) Hitze und Trockenresistenz der Flechten in Beziehung zu ihrer Verbreitung. *Flora, Jena* **140**, 39–97.

LANGE O.L. (1961) Die Hitzeresistenz einheimischer immer- und wintergrüner Pflanzen im Jahreslauf. *Planta* **56**, 666–683.

LANGE O.L. (1962) Eine 'Klapp-Küvette' zur CO_2- Gaswechsel-registrierung an Blättern von Freilandpflanzen mit dem URAS. *Ber. dt. bot. Ges.* **75**, 41–50.

LANGE O.L. (1965a) Leaf temperatures and methods of measurement. *Arid Zone Res.* **25**, 203–209.

LANGE O.L. (1965b) The heat resistance of plants, its determination and variability. *Ibid.* 399–405.

LANGE O.L. & LANGE R. (1962) Die Hitzeresistenz einiger mediterraner Pflanzen in Abhängigkeit von der Höhenlage ihrer Standorte. *Flora, Jena* **152**, 707–710.

LANGE O.L. & LANGE R. (1963) Untersuchungen über Blattemperaturen, Transpiration und Hitzeresistenz an Pflanzen mediterraner Standorte (Costa Brava, Spanien). *Flora, Jena* **153**, 387–425.

LANGE O.L. & KANZOW H. (1965) Wachstumshemmung an höhren Pflanzen durch abgetötete Blätter und Zwiebeln von *Allium ursinum*. *Flora, Jena* **B156**, 94–101.

LANGE O.L., KOCH W. & SCHULZE E.D. (1969) CO_2-gas exchanges and water relationships of plants in the Negev Desert at the end of the dry period. *Ber. dt. bot. Ges.* **82**, 39–61.

LARCHER W. (1969a) The effect of environmental and physiological variables on the carbon dioxide gas exchange of trees. *Photosynthetica*, **3**, 167–198.

LARCHER W. (1969b) Die Bedeutung des Faktors 'Zeit' für die photosynthetische Stoffproduktion. *Ber. dt. bot. Ges.* **82**, 71–80.

LEWIS M.C. (1972) The physiological significance of variation in leaf structure. *Sci. Prog.* **60**, 25–51.

LEVITT J. (1966) Winter hardiness in plants. In H.T. Meryman (ed.) *Cryobiology*, pp. 495–563. New York, Academic Press.

LIETH H. (1960) Uber den Lichtkompensationspunkt der Landpflanzen. *Planta* **54**, 530–576.

LLOYD F.C. (1908) The physiology of stomata. *Publ. Carnegie Inst. Wash.* **82**, 1–142.

LOPUSHINSKY W. (1969a) A portable apparatus for estimating stomatal aperture in conifers. Pacif. N.W. For. & Range exptl. Stn. (US Dep. Agric., For. Serv.) 1–7.

LOPUSHINSKY W. (1969b) Stomatal closure in conifer seedlings in response to leaf moisture stress. *Bot. Gaz.* **130**, 258–263.

LUDLOW M.M. & JARVIS P.G. (1971) Photosynthesis in sitka spruce (*Picea sitchensis* (Bong.) Carr). I. General characteristics. *J. appl. Ecol.* **8**, 925–953.

MACKAY D.B. (1972) The measurement of viability. In E.H. Roberts (ed.) *The Viability of Seeds*, pp. 172–208. London, Chapman & Hall.

MACKLON A.E.S. & WEATHERLEY P.E. (1965a) A vapour pressure instrument for the measurement of leaf and soil water potential. *J. exp. Bot.* **16**, 261–270.

MACKLON A.E.S. & WEATHERLEY P.E. (1965b) Controlled environment studies of the nature and origins of water deficits in plants. *New Phytol.* **64**, 414–427.

MARTIN M.H. & PIGOTT C.D. (1965) A simple method for measuring carbon dioxide in soils. *J. Ecol.* **53**, 153–156.

MAXIMOV N.A. (1929) *The Plant in Relation to Water*. London, Allen & Unwin.

MAXIMOV N.A. (1932) The physiological significance of the xeromorphic structure of plants. *J. Ecol.* **19**, 273–282.

MEIDNER H. (1965) A simple porometer for measuring the resistance to air flow offered by stomata. *School Sci. Rev.* **47**, 149–151.

MEIDNER H. (1965a) Stomatal control of transpirational water loss. In *The State of Water in Living Organisms*. Symp. Soc. Exptl. Biol. **19**, 185–204.

MEIDNER H. (1967) Further observations on the minimum intercellular space carbon dioxide concentration (Γ) of maize leaves and the postulated role of 'photorespiration' and glycollate metabolism. *J. exp. Bot.* **18**, 177–185.

MEIDNER H. (1970) Light compensation points and photorespiration. *Nature, London* **228**, 1349.

MEIDNER H. & MANSFIELD T.A. (1968) *Physiology of Stomata.* London, McGraw-Hill.

MEYER B.S. (1945) A critical evaluation of the terminology of diffusion phenomena. *Pl. Physiol., Lancaster* **20**, 142–164.

MICHAEL G. (1968) Prüfung der Lebensfähigkeit geschadigter Pflanzen mit Hilfe der kolormetrischen Methode nach Kauko/Alvik. In *Vorträge und Disk. i. Internat. Baumphysiol.*—Symp. 108–111.

MICHAEL G. (1969) Eine Methode zur Bestimmung der Spaltöffnungsweite von Koniferen. *Flora, A.* **159**, 559–561.

MILLER J. (1914) A field method for determining dissolved oxygen in water. *J. Soc. Chem. Ind., Lond.* **33**, 185–186.

MILTHORPE F.L. (1961) Plant factors involved in transpiration. *Arid Zone Res.* **16**, 107–115.

MOLISCH H. (1912) Das Offen- und Geschlossensein der Spaltöffnungen, veranschaulicht durch eine neue Methode (Infilrationsmethode). *Z. Bot.* **4**, 106–122.

MONTEITH J.L. (1918) Analysis of the photosynthesis and respiration of field crops from vertical fluxes of CO_2. *UNESCO, Nat. Res. Res.* **5**, 349–358.

MONTGOMERY E.G. (1912) Competition in Cereals. *Bull. Nebr. Agric. Exp. Sta.* **26**, art. V, 1–22.

MOONEY H.A. & SHROPSHIRE F. (1968) Population variability in temperature related photosynthetic acclimation. *Oecol. Plant.* **2**, 1–13.

MULQUEEN J., WALSHE M.J. & FLEMING G.A. (1961) Copper deficiency on Irish blanket peats. *Scient. Proc. R. Dubl. Soc., B* **1**, 25–35.

NASSERY H. (1970) Phosphate absorption by plants from habitats of different phosphate status. II. Absorption and incorporation of phosphate by intact plants. *New Phytol.* **69**, 197–203.

NUFFIELD BIOLOGY (1966) *Teacher guide III. The maintenance of life.*

OUDMAN J. (1936) Uber Aufnahme und Transport N hältiger Verbindungen durch die Blätter von *Drosera capensis. Ext. Rec. Trav. Bot., Neerl.* **33**, 351–433.

OWEN P.C. (1952) The relation of germination of wheat to water potential. *J. exp. Bot.* **3**, 188–203.

PEARSON V. (1971) *The biology of the mycorrhiza in Ericaceae.* Ph.D. Thesis, University of Sheffield.

PENMAN H.L. & SCHOFIELD R.K. (1951) Some physical aspects of assimilation and transpiration. *Symp. Soc. exptl. Biol.* **4**, 115–129.

PERRIER A. (1971) Leaf temperature and measurement. In Z. Šesták, J. Čatský & P.G. Jarvis (eds.) *Plant Photosynthetic Production, Manual of Methods.* The Hague, Dr. W. Junk, N.V.

PISEK A. (1960) The nature of the temperature optimum and minimum of photosynthesis. *Bull. Res. Coun. Israel., D.* 285–289.

PISEK A. & KEMNITZER R. (1968) Der Einfluss von Frost auf die Photosynthese der Weisstanne (*Abies alba* Mill) *Flora, Jena* **B**, 314–376.

PISEK A., LARCHER W. & UNTERHOLZNER R. (1967) Kardinale Temperaturbereiche der Photosynthese und Grenztemperaturen des Lebens der Blätter verschiedener Spermatophyten. I. Temperaturminimum der Nettoassimilation, Gefrier- und Frostschadensbereiche der Blätter. *Flora, Jena* **157**, 239–264.

PISEK A., LARCHER W., MOSER W. & PACK I. (1969) Kardinale temperaturbereiche und Grenztemperaturen des Lebens der Blätter verschiedener Spermatophyten. III. Temperatureabhängigkeit und optimaler Temperaturbereich der Netto-Photosynthese. *Flora, Jena* **158**, 608–630.

POLWART A. (1970) *Ecological aspects of the resistance of plants to environmental factors.* Ph.D. Thesis, University of Glasgow.

PORTSMOUTH G.B. (1949) The effect of manganese on carbon assimilation in the potato plant as determined by a modified half-leaf method. *Ann. Bot.* **13**, 113–133.

PROCTOR J. (1971a) The plant ecology of serpentine. II. Plant response to serpentine soils. *J. Ecol.* **59**, 397–410.

PROCTOR J. (1971b) The plant ecology of serpentine. III. The influence of a high magnesium/calcium ratio and high nickel and chromium levels in some British and Swedish serpentine soils. *J. Ecol.* **59**, 827–842.

RAINFORD A.E.D. (1972) Guttation and recovery from wilting in cucumber plants, *Cucumis sativas* L. Honours Thesis, University of Stirling.

RAMSAY J.A. & BROWN R.H.J. (1955) Simplified apparatus and procedure for freezing point determination upon small volumes of fluid. *J. Sci. Instr.* **32**, 372–375.

RICHARDS L.A. & OGATA G. (1958) Thermocouple for vapour pressure measurements in biological and soil systems at high humidity. *Science, N.Y.* **128**, 1089–1090.

RICHTER H. & ROTTENBURG W. (1971) Leitfähigkeitmessung zur Endpunktanzeige bei der Saugspannungsbestimmung nach Scholander. *Flora, Jena* **160**, 440–443.

ROBINSON R.K. (1972) The production by roots of *Calluna vulgaris* of a factor inhibitory to growth of some mycorrhizal fungi. *J. Ecol.* **60**, 219–224.

RORISON I.H. (1968) The response to phosphorus of some ecologically distinct species. I. Growth rate and phosphorus absorption. *New Phytol.* **67**, 913–923.

RORISON I.H. (1969) Ecological inferences from laboratory experiments in mineral nutrition. In I.H. Rorison (ed.) *Ecological Aspects of the Mineral Nutrition of Plants*, pp. 155–175. Oxford, Blackwell.

RYCHNOVSKÁ-SOUDKOVÁ M. (1963) Study of the reversibility of the water saturation deficit as one of the methods of casual phytogeography. *Biologia Pl.* **5**, 175–180.

RYCHNOVSKÁ M. (1965) Water relations of some steppe plants investigated by means of the reversibility of the water saturation deficit. In B. Slavik (ed.) *Water Stress in Plants*, pp. 108–116. Prague, Czech. Acad. Sci.

SAMPSON I. (1961) A method of replicating dry or moist surfaces for examination by light microscopy. *Nature, Lond.* **191**, 932.

SCHIMPER A.F.W. (1898) *Pflanzengeographie auf Physiologische Grundlage.* Jena, Fischer.

SCHIMPER A.F.W. (1903) *Plantgeography on a Physiological Basis.* Oxford, Clarendon Press.

SCHOLANDER P.F., HAMMEL H.T., BRADSTREET E.D. & HEMMINGSEN E.A. (1965) Sap pressure in vascular plants. *Science, N.Y.* **148**, 339–346.

SCHRATZ E. (1932) Untersuchungen über die Beziehung zwischen Transpiration und Blattstruktur. *Planta* **16**, 17–69.

ŠESTÁK Z., ČATSKÝ J. & JARVIS P.G. (1971) *Plant Photosynthetic Production, Manual of methods.* The Hague, Dr. W. Junk, N.V.

SEYBOLD A. (1929) *Die Physikalische Komponente der Pflanzlichen Transpiration.* Berlin, Julius Springer.

SHARDAKOV V.S. (1948) A new field method for the determination of the suction pressure of plants [in Russian]. *Dokl. Akad. Nauk. SSSR* **60**, 169–172.

SHEIKH K.H. (1969) The effects of competition and nutrition on the water-relations of some wet-heath plants. *J. Ecol.* **57**, 87–99.

SHIMSHE D. & LIVNE A. (1967) The estimation of the osmotic potential of plant sap by refractrometry and conductivity: A field method. *Ann. Bot.* **31**, 506–511.

SHONTZ N.N. & SHONTZ J.P. (1972) Competition for nutrients between ecotypes of *Galinsoga ciliata*. *J. Ecol.* **60**, 89–72.

SIVADJIAN J. (1952) Recherches sur la transpiration des plantes par la méthode hygrophotographique. *J. Bull. Soc. Bot. Fr.* **99**, 138–141.

SLATYER R.O. (1958) The measurement of DPD in plants by a method of vapour equilibration. *Aust. J. biol. Sci.* **11**, 349–365.

SLATYER R.O. (1960) Internal water balance of *Acacia aneura* F. Muell in relation to environmental conditions. *Arid Zone Res.* **16**, 137–146.

SLATYER R.O. (1967) *Plant Water Relationships.* New York, Academic Press.

SLATYER R.O. & JARVIS P.G. (1966) Gaseous diffusion porometer for continuous measurement of diffusive resistance of leaves. *Science, N.Y.* **151,** 574–576.

SMITH S.E. (1966) Physiology and ecology of *Orchis* mycorrhizal fungi with reference to seedling nutrition. *New Phytol.* **65,** 488–499.

SPANNER D.C. (1951) The Peltier effect and its use in the measurement of suction pressure. *J. exp. Bot.* **2,** 145–168.

SPANNER D.G. & HEATH O.V.S. (1952) Experimental studies of the relation between carbon assimilation and stomatal movement. II. The use of the resistance porometer in estimating stomatal aperture and diffusive resistance. *Ann. Bot.* **5,** 319–331.

SPARLING J.H. (1967) The occurrence of *Schoenus nigricans* L. in blanket bogs. II. Experiments on the growth of *Schoenus nigricans* L. under controlled conditions. *J.Ecol.* **55,** 15–31.

SPENCE D.H.N. & CHRYSTAL J. (1970) Photosynthesis and zonation of freshwater macrophytes. I. Depth distribution and shade tolerance. *New Phytol.* **69,** 205–215.

STAHL E. (1894) Einige Versuche über Transpiration und Assimilation. *Bot. Ztg.* **52,** 117.

STALFELT M.G. (1932) Die stomatäre Regulation in der pflanzlichen Transpiration. *Planta* **17,** 22–32.

STEPONKUS P.L. & LANPHEAR F.O. (1967) Refinement of the triphenyl tetrazolium chloride method of determining cold injury. *Pl. Physiol., Lancaster* **42,** 1423–1426.

STEVENSON A.G. (1972) Interference by *Mercurialis perennis* on other species. Hons. Thesis, University of Stirling.

STEWART W.D.P. & PEARSON M.C. (1967) Nodulation and nitrogen fixation by *Hippophaë rhamnoides* in the field. *Pl. Soil* **26,** 348–360.

STEWART W.S. & BANNISTER P. (1973) Seasonal changes in carbohydrate content of three *Vaccinium* spp. with particular reference to *V. uliginosum* L. and its distribution in the British Isles. *Flora, Jena* **162,** 134–155.

STOCKER O. (1929) Das Wasserdefizit von Gefässpflanzen in verschiedenen Klimazonen. *Planta* **7,** 382–387.

STOCKER O. (1956) Messmethoden der Transpiration. *Hand. Pfl. Physiol.* **III,** 293–311.

SUTCLIFFE J. (1968) *Plants and Water.* London, Arnold.

THOM A.S. (1968) The exchange of momentum, mass and heat between an artificial leaf and the air flow in a wind tunnel. *Q. J. Roy. met. Soc.* **94,** 44–55.

THURSTON J.M. (1969) The effect of liming and fertilisers on the botanical composition of permanent grassland and on the yield of hay. In I.H. Rorison (ed.) *Ecological Aspects of the Mineral Nutrition of Plants,* pp. 3–10. London, Blackwell.

TILL O. (1956) Uber die Frosthärte von Pflanzen sommer-grüner Laubwälder. *Flora, Jena* **143,** 498–542.

TINKLIN R. & WEATHERLEY P.E. (1966) The role of root resistance in the control of leaf water potential. *New Phytol.* **65,** 509–517.

TINKLIN R. & WEATHERLEY P.E. (1968) The effect of transpiration rate on the leaf water potential of sand and soil rooted plants. *New Phytol.* **67,** 605–615.

TRESHOW M. (1970) *Environment and Plant Response.* New York, McGraw-Hill.

TURNER R.G. (1969) Heavy metal tolerance in plants. In I.H. Rorison (ed.) *Ecological Aspects of Mineral Nutrition of Plants,* pp. 399–410. Oxford, Blackwell.

ULMER W. (1937) Uber den Jahresgang der Frosthärte einiger immergrüner Arten der alpinen Stufe, sowie der Zirbe und Fichte. *Jb. wiss. Bot.* **84,** 553–592.

UNGERSON J. & SCHERDIN G. (1968) Jahresgang von Photosynthese und Atmung unter naturlichen Bedingungen bei *Pinus sylvestris* L. an ihrer Nordgrenze in der Subarktis. *Flora, Jena* **157,** 391–434.

URSPRUNG A. & BLÜM G. (1916) Zur Kenntnis der Saugkraft. *Ber. drsch. bot. Ges.* **34,** 525–539.

VOZNESENSKII V.L., ZALENSKII O.V. & AUSTIN R.P. (1971) Methods of measuring rates of photosynthesis using carbon-14 dioxide. In Z. Šesták, J. Čatsky & P.G. Jarvis (eds.) *Plant Photosynthetic Production. Manual of Methods.* The Hague, Dr. W. Junk, N.V.

WALLACE A., FROHLICH E. & LUNT O.R. (1967) Calcium requirements of higher plants. *Nature, Lond.* **209,** 634.

WALLACE T. (1961) The diagnosis of mineral deficiencies in plants by visual symptoms. London, HMSO.

WALTER H. (1931) *Die Hydratur der Pflanzen.* Jena, Fischer.

WALTER H. (1963) Zur Klärung des spezifischen Wasserzustandes im Plasma. II. Methodisches. *Ber. dt. bot., Ges.* **76,** 54–71.

WARMING E. (1895) *Plantesamfund.* Copenhagen.

WARMING E. (1909) *Ecology of Plants.* Oxford, Clarendon Press.

WATSON D.J. (1952) The physiological basis of variation in yield. *Adv. Argon.* **4,** 101–145.

WEATHERLEY P.E. (1950) Studies in the water relations of the cotton plant. I. The field measurement of water deficits in leaves. *New Phytol.* **48,** 81–97.

WEATHERLEY P.E. (1960) A new micro-osmometer. *J. exp. Bot.* **11,** 250–260.

WEATHERLEY P.E. (1966) A porometer for use in the field. *New Phytol.* **65,** 376–387.

WEATHERLEY P.E. & SLATYER R.O. (1957) Relationship between relative turgidity and diffusion pressure deficit in leaves. *Nature, Lond.* **179,** 1085–1086.

WEINEMANN H. & LE ROUX M. (1946) A critical study of the torsion balance of measuring transpiration. *S. Afr. J. Sci.* **42,** 147–163.

WENT F.W. (1970) Plants and the chemical environment. In E. Sondheimer & J.B. Simeone (eds.) *Chemical Ecology,* pp. 71–82. New York, Academic Press.

WILLIS A.J. (1963) Braunton Burrows: the effects on the vegetation of the addition of mineral nutrients to the dune soils. *J. Ecol.* **51,** 353–374.

WILLIS A.J. & JEFFERIES E.L. (1963) Investigations on the water relations of sand dune plants under natural conditions. In A.J. Rutter & F.H. Whitehead (eds.) *Water Relations of Plants,* pp. 168–189. Oxford, Blackwell.

WILMOTT A.J. (1921) Experimental researches on vegetable assimilation and respiration. XIV. Assimilation by submerged plants in dilute solutions of bicarbonate and of acids; an improved bubble counting technique. *Proc. R. Soc. B* **92,** 304–327.

WIT C.T. DE (1960) On competition. *Versl. landbouwk Onderz. Ned.* **66**(8), 1–82.

WOOLHOUSE H.W. (1969) Differences in the properties of the acid phosphatases of plant roots and their significance in the evolution of edaphic ecotypes. In I.H. Rorison (ed.) *Ecological Aspects of the Mineral Nutrition of Plants,* pp. 357–380. Oxford, Blackwell.

YEMM E.W. & WILLIS A.J. (1954) The estimation of carbohydrates in plants by anthrone. *Biochem. J.* **57,** 508–514.

CHAPTER 6

SITE AND SOILS

D. F. BALL

1 **Introduction** 297

2 **Site physical characteristics** 298
2.1 Location 298
2.2 Physiography 299
  2.2.1 Altitude, slope, aspect and exposure 300
  2.2.2 Landform 301
2.3 Geology 301
  2.3.1 Stratigraphy 302
  2.3.2 Mode of occurrence 302
  2.3.3 Lithology 303
2.4 Climate 303

3 **Field description of soils** 306
3.1 Preparation of a soil for description 306
3.2 The soil profile and its recording on data sheets 307
  3.2.1 Features to be observed and recorded in profile descriptions 310
  3.2.2 Horizon symbols 317

4 **Soil classification** 318
4.1 General considerations of soil classification 318
4.2 Soil classification systems 319
  4.2.1 International soil classification 319
  4.2.2 National soil classification 323
  4.2.3 Local soil classification 326

5 **Soil mapping** 327
5.1 General considerations of survey procedure 327
5.2 Survey techniques 329
  5.2.1 'Free' survey 330
  5.2.2 'Grid' survey 331

6 **Soil analyses** 332
6.1 Sampling 332
  6.1.1 Sampling techniques 332
  6.1.2 Sampling programmes 334
6.2 Soil chemistry 338
6.3 Physical analyses 340
  6.3.1 Particle size distribution 340
  6.3.2 Bulk density 343
  6.3.3 Pore size distribution 346
  6.3.4 Structural determinations 347
  6.3.5 Soil moisture 348
  6.3.6 Soil accretion and erosion 355
  6.3.7 Mechanical strength 356
6.4 Mineralogical and fabric analyses 356
  6.4.1 Thin-section techniques 356
  6.4.2 Sand-grain mineralogy 357
  6.4.3 Silt and clay mineralogy 358

7 **Conclusion** 358

8 **References** 359

1 INTRODUCTION

The balance of this chapter is based on an assessment of the relative significance of the earth sciences to the plant ecologist. For most ecological field studies an outline understanding of the geomorphological and geological

297

characteristics of a site is adequate. Soils demand more detailed consideration, since they are the immediate physical and nutritional support for the majority of plants, and are one of the direct controls on plant distribution and performance.

It is impossible to include a comprehensive coverage of soil science. The aspects considered are believed to be essential to most ecological investigations. The approach used is the 'pedological' concept of soil as a natural body which has evolved and is continuing to develop in response to physical and biotic influences of the environment. Within this idea of 'soil' as a three-dimensional dynamic continuum forming a thin surface skin over most of the earth, emphasis is placed on the morphological criteria applicable to field recognition of soil characteristics and to the classification of soil types. Attention is also given to general aspects of a range of analytical techniques which quantify soil characteristics.

No attempt has been made to discuss soil-dwelling organisms and their influence on soil-forming processes and on nutrient and energy cycling. Works which cover this field include Kühnelt (1961), Garret (1963), McLaren & Peterson (1967), Burges & Raw (1967), Wallwork (1970), Russell (1971), Phillipson (1971) and Parkinson *et al.* (1971). It is also impractical to give any discussion of the physico-chemical properties of soil as a complex mass or as individual particles although understanding of such properties is fundamental to intensive specialist laboratory or field studies of air, water and nutrient transfer between soil mineral and organic fabric, soil flora and fauna, and plant roots. Texts with particular emphasis on fertility questions are Russell (1974), Cooke (1967) and Buckman & Brady (1969). Soil formation and distribution are emphasized by Bunting (1967), Bridges (1970), Fitzpatrick (1971), Cruikshank (1972) and Buol *et al.* (1973).

2 SITE PHYSICAL CHARACTERISTICS

2.1 LOCATION

The location of a sampling or observation point should always be clearly identified at a scale of precision controlled by the nature of the particular study and dependent on whether subsequent observations may be needed. Where exact site identification or re-location is essential the only effective systems use fixed marker pegs or quantitative reference (bearings and distances) to adjacent permanent landscape features. The use of permanent quadrats in vegetation surveys is discussed in Chapter 3. Where a general location is acceptable, map grid references are suitable, or if adequate maps are unobtainable, sketch maps should be drawn as recommended in data tabulation for conservation sites in the International Biological Programme (Peterken,

1967). Sketch maps (e.g. Fig. 6.1a) emphasizing relevant natural and man-made features can be plotted from ground observation, or from air photographs which can supplement or substitute for maps. In hilly areas, height differences between adjacent features introduce horizontal scale distortions on air photographs which make their use in direct semi-quantitative mapping impossible (Firth, 1973) but in coastal or other flat areas, reasonably accurate mapping from air photos is possible without specialized photogrammetry (as in recent unpublished work by the Coastal Ecology Research Station of the Institute of Terrestrial Ecology, Natural Environment Research Council, UK).

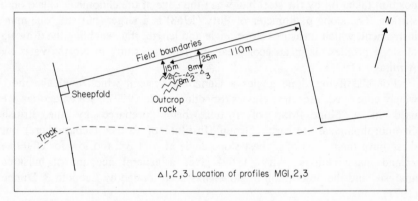

Figure 6.1a　An example of a site location sketch helpful in field data recording.

Appreciation of the virtues of precise site location must keep in mind the known scale of significant variability in soils. Major changes in soil type can be found within the order of a metre, and substantial chemical differences occur in samples of a morphologically 'uniform' soil taken at only the order of 10 cm apart. Problems caused by variability, sampling, and site identification are general for the ecologist but they are increased in soil work since any observation and/or sampling of a specific soil section modifies or destroys it, preventing further observations at the precise site.

2.2 PHYSIOGRAPHY

Clarke (e.g. 1971, 5th Edition) lists *altitude, slope, aspect, exposure, landform and microrelief* as key relief factors in soil description, and these should generally be recorded by the ecologist. Relief description for soil studies has also been discussed by Curtis *et al.* (1965).

2.2.1 Altitude, slope, aspect and exposure

The *altitude* of a site can be obtained with reasonable accuracy from maps where they are available, or measured directly with an altimeter, calibrated frequently against local bench-marks.

Slope angle should be given quantitatively rather than qualitatively since a slope which is 'moderate' to a worker in the plains might be described as 'gentle' or 'almost level' in mountain regions. One of the simplest instruments for measuring slope is the 'Dr. Dollar' clinometer (Cutrock Engineering Co., London) in which a steel ball-bearing runs in a semi-circular graduated groove between two thin perspex plates. Slope angle can be quickly read from the position taken up by the steel ball when the edge of the clinometer is laid on a surface. The slope pantometer of Pitty (1968) is a larger, but still 'one-man' instrument, which measures slope angle and length, thus enabling the drawing of slope profiles, as in an ecological and land-use study in North Wales by Armitage (1973).

For classifying slope angles a Canadian system, which could be more widely employed, uses nine classes (Sneddon *et al.*, 1972). Five categories are used in a Slope Map of Britain, being prepared by the British Geomorphological Research Group (Institute of British Geographers) but these have their class of highest slope angle at $>11 \cdot 3°$, too low for effective upland classifications. Pitty (1969) gives a general account of hillslope analysis, and the description of slopes has been treated by Leopold & Dunne (1971).

The *aspect* of a sloping site can be recorded as either a compass bearing or sector (e.g. N, N.E., E), by taking a reading parallel to the direction of, and facing outwards from, maximum slope.

Exposure, really a consequence of site relief factors such as altitude, slope and aspect, coupled with local climate, is more difficult to describe quantitatively. The degree of damage caused to standard 'tatter-flags' over measured time intervals can be used as a direct measure of exposure (Lines & Howell, 1963), or the relative exposure of sites may be estimated by considering climatic factors (Chapter 7), such as prevailing, dominant, and critical wind directions, mean and maximum wind speeds at these directions, and rainfall distribution and intensity, together with relief factors. A simple index of exposure based on relief factors alone has been applied in forestry (Pyatt *et al.*, 1969). A site 'topex' (topographic exposure) value is obtained as the sum of the angles of inclination of the skyline at the eight major compass points. The lower the topex value the higher the relative exposure of the site. This is a useful approach, particularly when comparing nearby sites within one locality, but would need the inclusion of climatic factors in the calculation to allow inter-regional quantitative comparisons.

Because of difficulties in assessing exposure directly Thomas (1973) used observations of damage to plants (tree growth forms) to grade sites into exposure classes.

2.2.2 Landform

Specialist geomorphological description and mapping of landforms is complex. In some studies the ecologist must take a similar specialist view. Suitable methods have been discussed by King (1966), while the classification of landforms in 'terrain units' has been treated by Beckett & Webster (1965) and Beckett *et al.* (1972). In most ecological work, *landform,* defined as the general character of the country around the sample site, can be recorded using descriptive phrases such as: mountain ridge crest; alluvial floor of river valley; or gently rolling low hills.

Clarke (1971), listed *micro-relief* (definable as the relief within 1 m of the site) as an important site factor, but what is more important is really *meso-relief,* or the ground shape in an area of about 10 m radius around the site. Terms such as: uniform slope; steep face of rocky outcrop; or small depression in hummocky terrain, can be used to supplement the landform description. Although such descriptive terminology may appear fundamentally unsatisfactory, it is generally adequate when coupled with quantitative values of altitude, slope and aspect. It can usefully be supported on data record cards by a sketch profile of the landform, of the type illustrated in Fig. 6.1b.

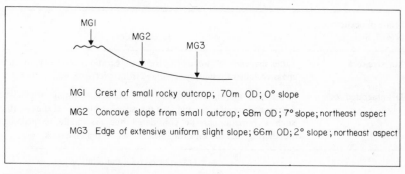

MGI   Crest of small rocky outcrop; 70m OD; 0° slope

MG2   Concave slope from small outcrop; 68m OD; 7° slope; northeast aspect

MG3   Edge of extensive uniform slight slope; 66m OD; 2° slope; northeast aspect

Figure 6.1b   An example of a landform (relief) sketch and qualitative description helpful in field data recording.

2.3 GEOLOGY

The ecologist should consider the geology of a site in three ways:

1. Stratigraphic age of the underlying rocks.
2. Their mode of occurrence.
3. Their lithology.

A basic understanding of geological principles can be obtained from Read & Watson (1962) and Holmes (1965).

2.3.1 Stratigraphy

Most geological maps use stratigraphic categories for mapping units, rather than rock properties more immediately relevant to soil scientists and ecologists, such as mineralogy, chemistry and particle size. Two sandstones of widely differing ages have more in common as soil parent materials than a sandstone and a limestone within the same stratigraphic unit. As age classes are the obvious feature of most geological maps, and there are occasions when age differences correlate with significant chemical or 'weatherability' properties in rocks of similar lithology, stratigraphic age should be included in any site description.

2.3.2 Mode of occurrence

Table 6.1 gives a simple classification of the form in which rock material exists at a study location. Further sub-division, such as within the drift group in

Table 6.1   A simple classification for rock substrate mode of occurrence.

Mode of occurrence of rock material	Rock *in situ*
Massive rock	Little evidence of weathering or disintegration; the rock forming a massive substrate sharply separated from overlying soil.
Disintegrated rock	Weathered and/or very fractured rock, but retaining original rock fabric, and usually passing into massive rock below. If thin (< 10 cm) such a horizon can be considered due essentially to pedological weathering, and the substrate then classed as massive rock.
	Transported rock material
Drift	Includes *glacial deposits* such as boulder-clay and glacial gravels, together with *periglacial deposits* such as head and scree. Where possible, these two categories and their sub-divisions should be distinguished.
Alluvium	Includes all water-deposited materials other than fluvio-glacial sands and gravels.
Aeolian deposits	Includes all wind-blown materials, such as dune sand or loess.

glaciated areas, may be appropriate, but the classes given are the essential distinctions to record. It is advisable, when transported material immediately underlies the soil at a site, to record, where possible, the characteristics of both this and the solid rock beneath it.

2.3.3 Lithology

Lithology is the description of a rock type from its fabric, mineralogy, chemistry and particle size, as determined from observation of hand specimens. Rock type at a point may be given on geological maps or in accompanying publications, but in many cases the ecologist will need to make a simple identification of rock types.

Table 6.2 provides a greatly simplified lithological classification. In the scheme for igneous and metamorphic rocks the sub-divisions are based on rock mineralogy and chemistry and on grain-size of constituent minerals. In part (ii) of the table, sedimentary rocks are sub-divided on grain-size and then on hardness, with separate groups for highly ferruginous or calcareous rocks. The simplification inevitably creates difficulties. Slate, included here with argillaceous sediments, is strictly a weakly metamorphosed shale, while other rocks overlap the boundaries of two classes, for example calcareous sandstones ('calcarenites'). The rock names listed are examples, not a comprehensive coverage of all rock types that can fit in the categories given.

Table 6.2 gives 14 categories into which rocks can be placed. Combining lithology, stratigraphy and mode of occurrence, typical descriptions would be: Hard argillaceous sediment (shale); Silurian; disintegrated rock *in situ*: Soft calcareous sediment (chalk); Cretaceous; massive rock *in situ*: or Acidic coarse-grained intrusive igneous rock (granite); transported drift (fluvio-glacial gravel).

2.4 CLIMATE

Chapter 7 deals with the measurement of climatic factors, many of which are important when considering soil development. As there are relatively few climatic recording stations, studies have often to be made in areas where local long-term measurement of a wide range of climatic variables has not been made in the past and will not be made in the future. Only generalizations, therefore, such as an extrapolated value for mean annual rainfall at the site, are usually recorded on a soil description sheet. This is inadequate when sites need to be contrasted in terms of their climatic variables. Annual and seasonal rainfall and temperature levels, wind-speeds and directions, radiation, and frost action are some key factors. Although short-term influences of climate can be of immediate significance to plants, and some factors such as incidence of frost or the distribution of rainfall throughout the year can be of direct importance to soils, the general nature of a soil profile may have been influenced by climate over hundreds or thousands of years. Even accurate data for present-day climates at a site are not necessarily adequate to explain soil differences, since the present climate may not be the same as that of the site over the period of soil development.

Table 6.2   An outline classification of rocks.

(i) *Igneous and metamorphic rocks*   Igneous rocks have formed by crystallization from a molten state. Metamorphic rocks have been substantially modified and recrystallized from pre-existing sedimentary or igneous rocks by subjection to high pressure and/or temperatures in the earth's crust.

Broad chemical/ mineralogical category	Acidic to intermediate rocks (much quartz, minor ferro-magnesian minerals)		Intermediate to basic rocks (much ferro-magnesian minerals, minor quartz)		Ultrabasic rocks (dominant ferro-magnesian minerals)	Silica rocks	Calcium/magnesium rocks
Grain size/ origin	Intrusive coarse-grained igneous and coarse-grained metamorphic rocks	Extrusive fine-grained igneous and fine-grained metamorphic rocks	Intrusive coarse-grained igneous and coarse-grained metamorphic rocks	Extrusive fine-grained igneous and fine-grained metamorphic rocks	Igneous and metamorphic rocks of high Fe/Mg content	Metamorphic rocks, composed of quartz	Metamorphic rocks, composed of carbonates
Examples of rock types within the rock category	Granite	Rhyolite, rhyolitic ash	Gabbro, dolerite	Basalt, pumice-tuff	Peridotite		
	Quartz-gneiss	Quartz-schist, mica-schist	Hornblende-gneiss	Chlorite-schist, calc-schist	Serpentine	Quartzite	Marble
	(i-i)	(i-ii)	(i-iii)	(i-iv)	(i-v)	(i-vi)	(i-vii)

(ii) *Sedimentary rocks* These have formed at surface temperatures and pressures by accumulation of water or wind-borne sediments derived from the weathering of pre-existing rocks.

Broad grain size chemical category	Arenaceous (sandy) sediments		Argillaceous (clayey) sediments		Ferruginous sediments	Calcareous sediments	
Fabric	Hard and/or consolidated	Soft and/or unconsolidated	Hard and/or consolidated	Soft and/or unconsolidated	—	Hard	Soft and/or unconsolidated
Examples of rock types within the rock category	Conglomerate, grit, hard sandstone	Soft sandstone, gravel, sand	Slate,* shale	Soft shale, clay, silt	Ironstone	Limestone	Chalk, marl
	(ii-i)	(ii-ii)	(ii-iii)	(ii-iv)	(ii-v)	(ii-vi)	(ii-vii)

* Slate is a weakly metamorphosed rock but lithologically is most conveniently grouped here in a simple classification.

3 FIELD DESCRIPTION OF SOILS

3.1 PREPARATION OF A SOIL FOR DESCRIPTION

Soil description and mapping procedures are given in USDA (1951), Clarke (editions 1936–1971) and Taylor & Pohlen (1962). The methods of the Soil Surveys in Great Britain are being covered in a Field Handbook, section one of which, description and sampling, has just appeared (Hodgson, 1974).

The basis for soil description and classification is the *soil profile* (3.2), a vertical section from ground surface to underlying pedologically unaltered parent material. An unequivocal definition of 'pedologically unaltered' is impossible to formulate without many qualifications but a commonsense judgement of where 'unaltered' parent material or underlying rock begins is usually possible. Sometimes the underlying rock is not the parent material, as where drift derived from one source overlies a different rock, or where the soil is in whole or part developed in a thin wind-blown stratum. Soil profiles can be studied in natural exposures but although these give useful background information, the unnatural condition of a long-established free face makes them liable to special moisture, chemical and biological régimes and therefore unsuitable for representative description and sampling. Recent excavations (field drains, pipe line courses, road works) can give a good picture of the three-dimensional variability of the soil profile but normally a soil profile pit (Fig. 6.2) must be dug. The possible size of such a pit depends on available resources and the nature of the study area. In some soft stoneless soils a manual coring auger can obtain an uncompressed profile without digging a pit. For a wider range of soils, an undisturbed core can be withdrawn using a power operated auger, for example mounted on a landrover (Wells, 1959). A mechanical excavator may occasionally be practical, but in most cases the soil pit will have to be dug with a spade.

The depth to which a profile pit is dug in Britain is normally about 1 m maximum, less if soil parent material or underlying hard rock are reached. In some situations, and in other regions, this depth will not be enough. To expose a face to 1 m, a pit of minimum surface area about 1 m × 60 cm is needed, but often a pit 60 cm square will do in Britain. In order to limit site damage, the surface soil and vegetation should so far as possible be removed in a way enabling later correct replacement, and excavated soil should be piled on a stout polythene sheet laid up to one face of the pit. Select the side of the pit receiving the maximum light as the face to be examined, and avoid trampling over this. When exposed fully, the study face should be carefully picked back a few centimetres further with a trowel or sheath knife to give a natural clean face on which observations and measurements can be made and from which samples can be taken. The cleaned face can be used immediately, or after allowing drying in order to make structural units more apparent. Photographs should include a suitable scale object against the photographed face.

3.2 THE SOIL PROFILE AND ITS RECORDING ON DATA SHEETS

Soils have been described by geological terms such as 'granite soil' or 'chalk soil', or by the nature of their surface humus, examples being 'mor' or 'mull' soils. They have been grouped from their location, examples being 'valley bottom soils' and 'steep slope soils'; or placed in simple morphological classes such as 'brown sandy soil' or 'heavy clay'; or in chemical categories such as 'low base status soil' or 'calcareous soil'. Descriptions such as these persist but are inadequate.

Section 4 discusses the difficulties preventing recommendation of a single universally applicable soil classification. The classifications used here rely on the concept of soil as a natural body which has developed into morphologically distinct types of *soil profiles* by the interaction of physical, chemical and biological processes on *parent materials,* usually mineral but occasionally organic. Soil profiles consist of sequences of one or more *horizons* (Fig. 6.2), which are more or less well-defined horizontally disposed layers distinguished from each other by properties such as colour, texture and structure.

Before discussing the information conventionally used to define soil horizons it must be considered how such data can best be recorded. A form of

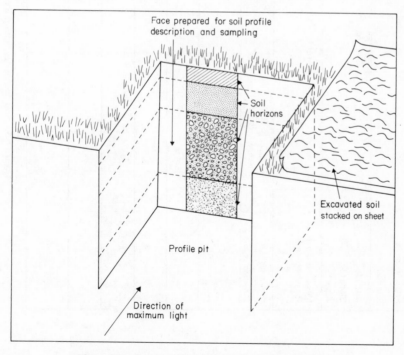

Figure 6.2 Cutaway schematic sketch of soil profile pit.

Profile number		Map symbol	Date of sampling						
Site location			Map reference						
Altitude Slope Aspect	Relief	Site drainage	Solid geology						Site rainfall
Major soil group (or sub-group)	Soil series Phase Type	Profile drainage	Soil parent material						Vegetation
Horizon depth, classi- fication,and sharpness of horizon boundary	Colour	Field texture	Stone quantity and type	Soil structure	Porosity	Handling consis- tency	Organic matter estimate	Root distri- bution	Field moisture	Earth- worms	Horizon number and notes

Figure 6.3 Soil profile description form suitable for recording site and horizon data with words and phrases.

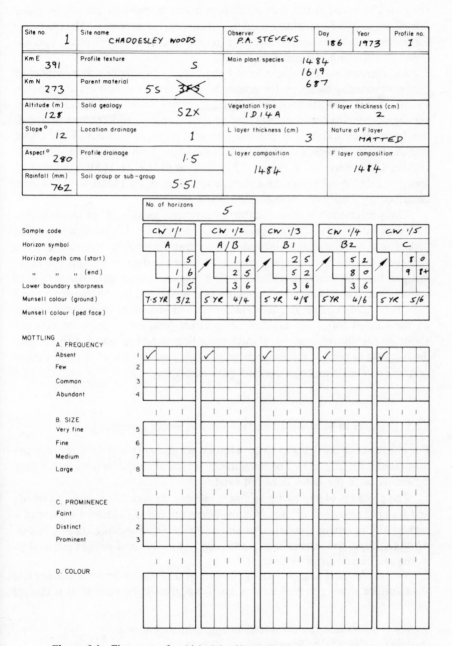

Figure 6.4 First part of a 'ticked box'/numerical entry type of soil profile description form.

the type illustrated in Fig. 6.3, employing entries in words, a slightly modified arrangement of a form widely used in soil survey, is convenient. The top of the form covers information on site and general profile characteristics, and the lower part contains horizon data.

An alternative method is a form completed by ticking one of a number of boxes, or entering numbers, for grades or measures of each observed property. Forms of this kind are under development by the British Soil Surveys in two versions depending on the detail required at the site, and another type (Fig. 6.4) has been tested by staff of the Institute of Terrestrial Ecology, U.K. Other approaches to numerical coding of profile data have been made (see e.g. Muir & Hardie, 1962; Barkham & Norris, 1970; John *et al.*, 1972). The use of punched feature-cards for soil profile data storage and retrieval on a local scale has been described by Rudeforth & Webster (1973).

The written form is most suitable for relatively small-scale or local studies, and directly gives a conventional written description for publications. The 'ticked box' method enables immediate data transference for computer storage and analysis, and is particularly useful for extensive data collection. The latter method has the merit of enforcing firm decisions regarding definitions of features because of the either/or nature of a box entry, but alternatively there are advantages in being able to show nuances of change as intergrade features on the written form. No amount of apparent increase in precision brought about by the chosen method of data recording should be allowed to obscure the fact that at present soil description still contains many features that cannot be easily defined in an objective and quantitative way.

3.2.1 Features to be observed and recorded in profile descriptions

For convenience the features to be recorded are given in the general order of the illustrated form (Fig. 6.3). The terminology is typical and acceptable rather than definitive, since the latter does not exist.

The profile is identified by an *index number.* its *location* is given in words and by *grid reference,* or sketch map. When the soil is described as part of a survey, the *mapping symbol* used should be given. *Sampling date* is noted. *Altitude, slope* and *aspect* are quantitatively entered, and *relief* described in words and sketch cross-section.

Site drainage is assessed as an estimate of the balance between surface and near-surface water received at, and lost from, the site by run-off. It is classed as:

normal—inflow and run-off waters in approximate balance
shedding—run-off significantly greater than inflow
receiving—inflow surface water likely to exceed run-off (a) run-off present but low, (b) no run-off.

Solid geology is entered from field observation or map knowledge, and *parent material* noted as identical to, or differing from solid geology.

It is occasionally implied in this chapter that some ecologists have limitations in their recording of soil data but the inadequate attention paid on the conventional soil description to *climate* and *vegetation* fully rebounds on the pedologist. Typically, only an estimate of annual rainfall and broad vegetation categories such 'chalk grassland', 'scrub birch woodland', or 'Sphagnum bog' are all the ecologist can expect to find.

Major Soil Group, Soil Series, Phase and *Type* may be entered in the field but are often done later in accord with the chosen classification system (4.2).

Profile drainage is a subjective classification based on observed water conditions and movement and soil colour, texture and structure. The recommended class divisions are:

Excessive—Soils with very low moisture retention occur in loose very permeable materials and in shallow soils overlying rocks. To an extent this category could cover shallow soils with low moisture capacity overlying indurated subsoils. These, however, are excessively drained in dry periods and waterlogged in wet periods because of low subsoil permeability so that they should be treated as a special class of complex drainage.

Free—In these soils there is moderate to good water retention but no waterlogging of soil pore space. There is thus no mottling of the matrix colour with grey or rusty colours within the depth to the parent material, to at least 75 cm. Of course, monochrome grey colours can occur in particular horizons of some freely-drained soils, and persistent greyish colours may occasionally be inherited from parent materials.

Imperfect—These soils have surface horizons dominated by the matrix colour of the freely drained soils on similar parent material, but with fine rusty mottles often along root channels, which pass below to stronger grey mottled horizons. Arbitrarily, 30 cm is taken as the minimum depth at which strong mottling resulting from prolonged waterlogging occurs in imperfectly drained soils.

Poor—Soils of poor drainage have dominantly grey colours, due to ferrous iron caused by reducing conditions as a result of waterlogging up to or near the surface, typically below a thin dark grey surface horizon with heavy rusty mottling. (In some particularly red-coloured parent materials the grey colours however develop only moderately, even in very poorly drained soils.)

Very poor—This drainage class is used for soils more or less permanently saturated with water to the surface (though not submerged). Such soils are often peaty throughout, or have surface peaty horizons overlying pale grey permanently waterlogged mineral soil.

The lower part of the form covers features to be noted in horizon descriptions and is based on Clarke (editions from 1936 to 1971) and USDA

(1951). *Guidelines for Soil Description* (FAO, 1968) recommends size classes for different horizon parameters, some of which are applied below. The FAO considered that soils can be best described using terms 'which have received wide acceptance among soil surveyors', and justifies basing its work on the USDA manual (1951) on the grounds that 'these ... terms enjoy wider use and acceptance than any other system'. The view that widely acceptable terminology is preferable at present to seeking greater precision through new but inadequately tested terms or measures is followed here.

Horizon depth, classification and boundaries. Horizon depths are preferably shown always simply from the surface downwards (e.g. 0–5, 5–18, 18–24 cm) rather than in the system of measuring both upwards and downwards from the lower boundary of organic surface horizons when these are present. In the field, or following laboratory analyses, classification of horizons is made by symbols (3.2.2). Sharpness of boundary is given as: sharp, change occurs over less than 2 cm; fairly sharp, change occurs between 2 and 5 cm; merging, clear change occurs over more than 5 cm. Whether the boundary is level or undulating is also noted.

Colour. Colour is identified either in the field or on field-moist samples at the laboratory, preferably by comparison with the standard colours of the Munsell system (Munsell Soil Color Charts, Munsell Color Co., Baltimore). Colour chips on these charts are graded by 'hue', 'value' and 'chroma', so that 2.5YR 5/4, reddish-brown, for example, is a complete 'Munsell colour'. Without a colour book, descriptive subjective names must be used but the use of the standard charts ensures greater uniformity. A Japanese version of the Munsell book is cheaper and has quite a good match of the different colour chips but is less weatherproof and does not give the internationally employed Munsell names. Mottling of the matrix colour with other colours is recorded as faint or strong; as fine (< 5 mm diameter), medium (5–15 mm) and large; and as occasional (< 5% of surface area occupied), frequent (5–25% of surface area) or abundant (e.g. faint, occasional, fine mottle 10YR 5/6, yellowish-brown).

Field texture is an assessment, by sight and feel, of the relative proportions in which primary particles of different size ranges occur in the soil. Laboratory analysis determines these proportions quantitatively (6.3.1) and field texture determinations are related to textural classes based on laboratory data. Standard textural analyses are carried out on the 'fine-earth' fraction of soils (i.e. on the fraction < 2 mm equivalent spherical diameter, after removal of 'stones' (> 2 mm)). Particle size ranges used are: *Sand* (2·0–0·05 mm), divisible into coarse sand (2·0–0·2 mm) with individual grains very obvious to the eye and giving a strongly gritty feel when rubbed between the fingers, and fine sand (0·2–0·05 mm), with grains still individually recognizable and giving a gritty feel; *silt* (0·05–0·002 mm), individual grains are not clearly recognizable but it gives a smooth and soapy feel with only slight stickiness;

clay (< 0·002 mm), feels sticky and is easily rolled between the fingers into coherent mouldable threads (if dried, considerable moistening and working may be needed to restore plasticity). The 0·05 mm limit is that of the widely used 'US' scale for the silt:fine sand boundary. The older 'International' limit, still often given, is at 0·02 mm, and recently Avery (1973) has proposed 0·06 mm as this limit, in accord with engineering practice.

Textural class boundaries are derived from a triangular diagram (e.g. USDA 1951; Fig. 6.8) of which the three corners represent 100% sand, silt and clay. Fields within this, chosen to relate to observed field properties, are given class names, and any individual soil is allotted to its class by plotting the laboratory particle size analyses on the diagram. Field texture estimates by feel, aided by previous practice with soils of known particle size distribution, can relate quite closely to the laboratory classes. In the field however, apparent texture is certainly influenced by stone content, especially of fine gravel, and by organic matter, so that qualifying terms such as gravelly or organic are used with texture class names. The texture classes of the triangular diagram are identified in the field by feel as indicated. *Sand*: loose when dry, and not at all sticky when wet. *Loamy Sand*: a small degree of cohesion and plasticity when wet. *Sandy Loam*: sand fraction obvious but moulds readily, when sufficiently moist, without stickiness. *Loam*: moulds readily when moist and sticks slightly to fingers, but sand fraction still obvious. *Silty Loam*: moderate plasticity but little stickiness; smooth soapy feel of silt is notable. *Sandy Clay Loam*: sufficient clay to be sticky when moist, but sand fraction is an obvious feature. *Clay Loam*: sticky when moist, but sand fraction still detectable. *Silty Clay Loam*: less sticky than silty clay or clay loam, with slight sand and a soapy feel due to the silt fraction. *Sandy Clay*: plastic and sticky when moistened, but sand fraction obvious, without modifying silt influence. *Clay*: very sticky when moist and very hard when dry. *Silty Clay*: very low sand fraction but the smooth silt character reduces the stickiness of the clay. *Silt*: dominated entirely by smooth soapy feel of silt fraction. *Organic*: high organic matter content, not fitting into any of the above classes.

The general terms *light, medium* and *heavy* are popularly used for sandy, loamy and clayey soils.

Stone quantity and type. The *quantity* of 'stones' (> 2·0 mm) can be indicated subjectively as none, occasional (< 5% by volume), frequent (5–25%) or abundant. It is quite straightforward to determine quantitatively the stone content by weight or volume after sieving in field or laboratory but this is not generally done. Stone size is given as small, medium or large (alternatively named gravel, stones, boulders), with size limits between these about 1 and 10 cm mean diameter. Stone shape can be rounded, subangular or angular. Lithology, e.g. slate, limestone etc., should be stated as for rocks (2.3.3).

Soil structure refers to the natural aggregates into which primary soil particles are combined. Many fine subdivisions have been proposed and quantitatively the shape, size and strength of aggregates can be determined in the laboratory (6.3.4) but the following terms are normally adequate: *Single grain, structureless*: no aggregation—e.g. in a sand. *Crumb*: roughly rounded, soft, porous agregates can be subdivided on size into small (< 3 mm), medium (3–6 mm) and large (> 6 mm). This is the dominant structure of medium textured freely-drained soils. *Granular*: units of crumb shape but relatively hard and poorly permeable. These are infrequent in most temperate zone soils but are typical of calcareous soils of Rendzina type. *Blocky* (or as a sometimes favoured synonym *Cloddy*); roughly equidimensional flat-faced, angular aggregates—(which may then themselves break down into crumb structure units)—graded as small (< 12 mm), medium (12–25 mm), large (> 25 mm). This structure occurs in medium and heavy textured soils of imperfect drainage, and is also often found in freely-drained soils of low organic matter compacted by stock or cultivation. *Prismatic*: angular flat-faced units with the vertical dimension much greater than the others. Occurs in heavy soils in seasonally waterlogged horizons, as a result of wetting and expansion alternating with drying and contraction. *Laminated*: a structure of horizontally elongated platy structural units. Occurs in some alluvial soils, particularly silts, and in soils which have inherited their structure from drift subjected to permafrost ice. *Massive, structureless*: no structural units obvious in a homogenous coherent horizon.

Structure strength may be given as: Weak; just visible in the profile face, and easily destroyed by handling. Moderate: not markedly distinct in the undisturbed soil but units remain intact when lightly handled. Strong: units obvious in undisturbed soil, and remain intact when the soil is handled, unless appreciable pressure is used.

Porosity is an assessment of size, quantity and distribution of cavities within the soil mass. It is a quality related to texture and structure and provides an estimate of ease of water movement. It really requires determining quantitatively but a subjective assessment can usefully supplement other recorded features. Classes used are: Poor; very dense fabric with few medium or larger interconnecting pores or fissures. Moderate; some interconnecting medium and large pores but a relatively dense fabric. Good: fabric of the soil has a generally strong interconnecting series of medium and large pores or occasionally an extensive permanent fissure system. Excessive: very open fabric with abundant large interconnecting pores.

Handling consistency is another supplementary quality, closely allied to texture as modified by organic matter content, stoniness and structure. Classes used are: *Loose*: singlegrain structure, no impression of any binding or sticky character. *Fluffy*: very friable, soft and approaching loose—but the material is aggregated, often being of small crumb structure and high organic matter. *Friable*: breaks easily into crumbs or other small aggregates but does not stick to the hand. *Sticky*: adheres to hands and implements. *Compact*: massive and firm without stickiness or ready fracture into small aggregates. *Indurated*: very hard to break with hands and to dig initially, although after digging the material may break down more rapidly. Applicable only to highly organic horizons—*Fibrous*: aggregates contain, and are held together largely by, root fibres or plant debris—or *Greasy*: soft slimy material with no obvious aggregates.

Organic matter estimate is a column intended for a broad indication of the quantity and nature of organic matter in each horizon. The quantity is estimated as very high, high, moderate or low, largely on a basis of soil colour, darker soils having a higher humus content (divisions at about 30, 15 and 5% organic matter by weight).

The nature of surface organic horizons can be described as: *Mor*: purely organic horizon with little or no mineral admixture. Sub-divisions frequently used (a modified form of these is given in Table 6.3) are L layer (freshly fallen litter), F layer (partially decomposed, but origin still recognizable), H layer (completely decomposed organic matter). *Moder*: an intergrade between mor and mull, dominated by relatively high quantities of decomposed organic matter but containing a substantial mineral admixture, either in very small crumb units which often show a gradation from darker to lighter zones passing from surface to core of the unit ('*mull-like moder*'), or as bleached sand grains. *Mull*: dominantly mineral material in which the organic matter is completely altered from its original fabric and is intimately and fully incorporated with the mineral soil, only a darker colour distinguishing it from horizons of lower organic matter. For peaty materials (0 horizons, see 3.2.2), a qualitative sub-division is: *Fibrous*: composed of recognizable plant residues, which retain their individual character when handled. *Pseudofibrous*: plant remains recognizable in initial appearance, but they are partially decomposed and do not retain their character when handled. *Amorphous*: recognizable plant remains are absent.

Root distribution. Detailed root counts by number or weight are not usually made by soil scientists. Usually only an estimate of quantity is given, without any quantitative limits in mind, as rare, few, frequent or abundant. Root type, whether fibrous, fleshy or woody, and any relationship of distribution to profile features, is also noted. Chapter 4 shows up the weakness of this limited approach and makes it clear that an improvement is needed in this aspect of soil data recording.

Field moisture. The moisture state, which may influence colour ·determination or structural assessment, is usually noted subjectively as dry, moist or wet.

Earthworms. The lack of ecological expertise of the average soil scientist concerned with field studies is again indicated by the limitation, in the conventional description, of soil faunal observations to the presence or absence of earthworms as a group. Usually again only an approximate quantitative

Table 6.3   A system of soil horizon symbols

Master horizon symbols

O   Horizon which, although it may contain some mineral admixture, is dominated by the organic fraction (loss-on-ignition values > 30%).

A   Horizon at or near the soil surface, consisting of an intimate mixture of organic and mineral material, that fails to fulfil the definition of the O master horizon (loss-on-ignition values < 30%).

E   Horizon from which sesquioxides (Fe and Al) and/or clay particles have been removed, occurring below O or A horizons.

B   Sub-surface horizon of mineral material, modified by physical, chemical and biological alteration so that it is differentiated by structure, colour or texture from horizons above or below.

C   Mineral material which has been little altered by pedological processes other than gleying (due to waterlogging) or the accumulation of secondary salts (typically of Ca or Na).

R   Unaltered rock, which, even when moist, is too hard to be dug with a spade.

Intergrades between master horizons may be indicated as A/B, B/C, etc. Only suffixes given below as applicable to all horizons would be applied to such intergrade horizons.

Numerical prefixes and suffixes

Arabic number prefixes (2-, 3-, etc.) are used to indicate buried soil profiles or horizons, that is to say the superimposition of horizons formed in two or more episodes of soil formation. By convention, 1 is omitted, so that, for example, a sand-dune section might have horizons A, C, 2A, 2C, 3A, 3C. This would indicate a superimposed sequence of three soils of AC type, developed at different periods of time.

Roman number prefixes (II-, III-, etc.) are used to indicate original geological (rather than pedological) discontinuities, within a soil profile which has developed during a single formative stage. For example, wind-blown sand overlying boulder-clay could give a soil with horizon sequence of A, E, B, IIC. These types of number symbol are clearly distinguished in written work, but must be made obvious in verbal presentation.

Arabic number suffixes (-1, -2, etc.) are used where a master horizon is subdivided on grounds other than those indicated by addition of letter suffixes. This is generally only where for analytical or other reasons a single horizon is divided when sampling. In such cases one might have a profile A1, A2, B1, B2, B3, C1. This convention can cause confusion with other horizon nomenclatures still used, in which number suffixes were employed in the way letters are here (e.g. an Ea horizon is equivalent to an A_2 horizon in other conventions). Care should therefore be taken when comparing horizons from the literature with those indexed as recommended here, and the usage being followed should always be indicated in published work where horizonation is important.

Subordinate horizon symbols

Suffixes applicable to any master horizon

g   Horizon showing evidence in structures and colour of the influence of moderate periods of waterlogging.

gg   Horizon dominated by structural and colour effects resulting from long-term waterlogging.

c   Horizon containing residual calcium carbonate.

k   Horizon containing deposits of secondary calcium carbonate.

n   Horizon containing excess of sodium in the exchangeable cations, or free sodium chloride.

x   Horizon having a massive consistency due to induration.

Suffixes applicable to O horizons

o   Horizon having a loss-on-ignition value > 60%.

l   Horizon of little altered plant remains ('fibrous').

f   Horizon of partially broken down and decomposed plant remains which are still recognisable to the naked eye ('pseudofibrous').

h   Horizon of decomposed humidified plant remains with no recognizable original structure ('amorphous').

p   Ploughed or otherwise cultivated horizon.

Suffixes applicable to A horizons

h   Horizon visibly darkened by having a high content of organic matter, but not fulfilling the requirement for the O master horizon.

he   As for h, but also including bleached sand grains or rock fragments.

p   Defined as for O horizons.

an   Surface horizon artificially deepened or modified by the addition of material by man.

Suffixes applicable to E horizons (Eluvial horizons)

a   Horizon from which sesquioxides have been translocated down the profile.

b   Horizon from which clay has been translocated down the profile.

Suffixes applicable to B horizons (Illuvial horizons)

h   Horizon with level of humic organic matter which is high compared to horizons above and below.

s   Horizon with levels of sesquioxides (Fe and/or Al) which are high compared to horizons above and below.

t   Horizon with level of clay which is high compared to horizons above and below.

fe   Iron pan.

Suffixes applicable to C horizons

r   Horizon predominantly composed of shattered or weathered material derived from underlying solid rock (R horizon).

---

measure of numbers of earthworms is given as few, frequent or abundant. In horizons where earthworm channels are conspicuous features, although the animals themselves are not seen, their presence is recorded.

Horizon number and notes. In this column soil horizons are numbered for subsequent reference, so that, for example, successive horizons of profile ME 78 would be numbered ME 78/1, /2, /3. The small space provided here also allows any other special characteristics not separately covered to be mentioned. A separate column for such features can be inserted on the description sheet, if likely to be frequently required (e.g. for 'secondary chemicals' in saline soils of arid areas).

3.2.2 Horizon symbols

Soils are described as profiles consisting of sequences of horizons. To assist soil comparison, correlation and classification, individual horizons can be indexed by symbols. Soils with similar horizon sequences can then be grouped together (4.2), regardless of differences in horizon thickness and morphological detail. A recommended set of horizon symbols is given in Table 6.3. No single internationally accepted system can be given, but that described, based on Kubiena (1953) and the International Society of Soil Science (1967), is of wide applicability.

An initial division is into six groups of *Master Horizons* indexed by capital letters, then these are subdivided by subscript small letters and numerals and by prefix numerals.

4 SOIL CLASSIFICATION

4.1 GENERAL CONSIDERATIONS OF SOIL CLASSIFICATION

The central problem in soil classification is imposing simple and rigid class boundaries upon a medium of three-dimensional variability, in which changes that are gradual rather than sharply delimited frequently do not proceed concurrently since they respond at different rates to different environmental controls. Theoretical and practical considerations of soil classification have been widely discussed (e.g. by Kubiena, 1953; Muir, 1962; USDA, 1938, 1960; Smith, 1965; Webster, 1968; Fitzpatrick, 1971; Buol *et al.*, 1973). No attempt is made here to cover the historical development of soil classifications or review comprehensively the range of available options. Many of the problems facing the soil scientist choosing a classification scheme are shared by ecologists concerned with vegetation classification (see Chapter 3).

It is assumed that a soil classification is of practical value to ecologists and that the most generally applicable schemes are based on profile morphology, which by definition, incorporates a wide range of chemical, physical and biological parameters. For some problems, prior knowledge may suggest that it is acceptable to classify soils simply in terms of a single parameter, such as surface horizon texture or organic matter content, but this is generally too limiting unless supplemental to a wider grouping.

Arguments are cogently put by Fitzpatrick (1971), against hierachical systems which accept explicit or implicit correlations between soil classes and pedogenetic trends related to environment. These arguments emphasize the inability of such systems to cover effectively a medium which is a continuum, rather than being evolutionary; the necessity they impose of forcing intergrade soils into a rigid class structure; and the inadequacy of a class name to communicate complete information about the soil. In spite of the validity of these objections in some degree there is sufficient justification for, and value, in local or regional 'natural' genetic systems, such as described for Europe by Kubiena (1953), to recommend their use.

Alternative approaches to classification are:

(i) The extended hierachical system of the USDA (1960, 1967) 7th Approximation, which introduced an original terminology with quantitative limits of field and laboratory parameters to determine some horizons and classes.

(ii) Fitzpatrick's scheme (1971), which defines each individual soil by a formula listing the nature of successive horizons, using letter symbols based on yet another new nomenclature for horizon definition, and their thickness, employing subscript numbers. Soil profiles are grouped into classes based on horizon sequence similarities, but without necessarily implying genetic or environmental relationships for classes which are selected for specific purposes.

(iii) Numerical methods of soil classification (Bidwell & Hole, 1964; Rayner, 1966, 1969; Campbell *et al.*, 1970; Cuanalo de la C. & Webster, 1970; Norris, 1971; Norris & Loveday, 1971; Norris, 1972; Moore *et al.*, 1972) use computer data handling to give classification clusters. Although many of the raw data employed remain subjective, these methods have a greater apparent objectivity in data handling but have not yet generally succeeded in achieving a clearer or more usable soil classification than given by a morphological and genetic approach.

4.2 SOIL CLASSIFICATION SYSTEMS

4.2.1 International soil classification

The USDA system (1960, 1967, and see Smith, 1965), is the most comprehensive attempt to classify world soils. It has not yet been clearly shown to be sufficiently superior to compel its universal adoption outside the USA and is complex for general use, but it should be considered when an ecologist wishes to define soils in detail by a method of international applicability. Practical applications in surveys in Tanzania and the Republic of the Sudan have been discussed by Mitchell (1973). The clearest discussion of the 10 Soil Orders of this system is that of Buol *et al.* (1973), and other summaries are given by Bridges (1970) and Cruikshank (1972). Application to British soils has recently been considered by Ragg & Clayden (1973).

The alternative internationally applicable system of Fitzpatrick (1971) has definite merits, but has not achieved wide acceptance.

Broad studies can use simple systems, such as that (after Aubert & Duchaufour, 1956; Aubert, 1965 and Duchaufour, 1956, 1970) described by Ball in Peterken (1967). This system has been applied by ecologists to classify soils in a world survey of areas of conservation importance. Comment by users has led to the version given in Table 6.4 being slightly amended from that previously published. Six soil 'categories' and a total of 14 'sub-categories' based on horizon sequences are included, and a key (Table 6.5) enables

Table 6.4 Outline world soil classification.

	SOIL CATEGORIES			
Soil category	Sub-category	Main profile type/s (horizon sequences)	Examples of soil sub-groups in each category	Sub-category symbol
Saline soils (S)	—	AngC AnBnC	Solonchak Solonetz	S_1 S_2
High sesquioxide (ferritic) soils (Fe)	—	ABC	Terra rossa, rotlehm, laterite, ferritic brown earth	Fe
Organic soils O	—	O	Peat, fen, bog	O
Well-drained non-saline non-ferritic soils with good profile development (F)	With calcareous surface or sub-surface horizon	AcC	Rendzina, chernozem, chestnut soil (kastanozem)	F_1
		ABCc	Brown calcareous soil, terra fusca	F_2
	Non-calcareous throughout profile	AC	Brown ranker, skeletal brown earth	F_3
		ABC	Brown earth (braunerde), sol brun acide, brown earth with gleying	F_4

		ABsC; OEaBh and/or BsC; Eb, Bt, C	Podzol, peaty podzol, brown podzolic soil (sol brun podzolique), sol lessivé	F_5
Poorly-drained non-saline non-ferritic soils with good profile development (P)	With calcareous surface or sub-surface horizon	AgBggcC	Calcareous gley, fen marl	P_1
	Non-calcareous throughout profile	AgBggC	Non-calcareous gley	P_2
Soils with weak profile development (I)	Lack of profile development controlled by climatic factors	AC	Serosem, burosem, desert soil	I_1
	Lack of profile development controlled by limitations of time available for profile development	AC	Unconsolidated parent materials; e.g. recent alluvium, raw warp soil, gray warp soil, regosol	I_2
		AC, OC	Massive rock parent materials; e.g. lithosols on rock surfaces and coarse boulder accumulation	I_3

Identification of the category to which a soil should be allocated may be assisted by the key below. (Where a soil cannot readily be placed in a single category, two or more must be given; e.g., a shallow AC profile on rock in a mountain area might be doubtfully placed between F_1 or F_3 and I_3.)

Table 6.5 Key to soil categories of Table 6.4.

1	Soil containing a high concentration of alkaline salts.	2
	Soil without high concentration of alkaline salts.	3
2	Saline gley soil with water-soluble salts in upper horizons.	S_1
	Saline soil with water-soluble salts in lower horizon and high exchangeable sodium in the surface.	S_2
3	Soil with high concentration of iron oxides.	Fe
	Soil with normal concentration of iron oxides.	4
4	Soil with dominantly organic surface horizon at least 50 cm deep. If total soil depth less than 50 cm, then surface organic horizon directly succeeded by unaltered rock.	O
	Soil without dominantly organic surface horizon or with organic surface horizon succeeded by mineral soil horizon at less than 50 cm depth.	5
5	Well drained (i.e. no evidence of strong impedance or waterlogging above 40 cm depth).	6
	Poorly drained (i.e. evidence in mottled colours or otherwise of strong impedance or waterlogging nearer surface than 40 cm).	11
6	Immature profile, that is with weakly developed and shallow soil formation, possibly with little biological activity.	7
	Well-developed horizon sequence with moderate to strong biological activity.	8
7	Immaturity resulting from climatic factors, e.g. very low rainfall and/or temperature.	I_1
	Immaturity resulting from lack of time for soil formation to proceed:	
	(a) on fine-textured material such as recent alluvium, dune-sands and eroded surfaces of fine-textured drift.	I_2
	(b) on very coarse-textured materials such as boulder accumulations, or on massive rock.	I_3
8	Calcareous in one or more soil horizons.	9
	Non-calcareous throughout profile.	10
9	Shallow or simple profiles of A horizons overlying parent material.	F_1
	A B C profiles.	F_2
10	Shallow or simple profiles of A horizons overlying parent material.	F_3
	A B C profiles.	F_4
	Profiles including BL, Bs or Bt horizons, showing accumulation of translocated humus, iron and aluminium sequioxides, or clay.	F_5
11	Calcareous in one or more soil horizons.	P_1
	Non-calcareous throughout profile.	P_2

selection of the appropriate sub-category from limited chemical and morphological data. For many purposes such systems are over-simplified.

4.2.2 National soil classification

For most ecological studies the classification of soils in a regional or national system is necessary. Kubiena (1953) gives a system applicable throughout Europe, while classifications of African soils are considered by D'Hoore in Moss (1968) and by Ahn (1970). A wide review of many national soil classification systems, as then used, is contained in the section on 'Soil Classification and Soil Fertility', pp. 278–551 in International Society of Soil Science (1962). In most countries, a national system or systems have been developed, e.g. Britain (Avery, 1973); Germany (Mückenhausen, 1962, 1965); France (Duchaufour, 1956, 1970); Australia (Northcote, 1965); Canada (Nat. Soil Surv. Comm., 1968); and for the U.S.A. prior to the major new U.S.D.A. (1960, 1967) system, Baldwin *et al.* (in U.S.D.A., 1938, pp. 979–1001) and Thorpe and Smith (1949).

Classification systems and terminology have been even more dynamic in their change of direction and emphasis than soil development is itself. They have thus added to the difficulties of dealing with soil complexity, for example in Britain soil class terminology has varied in detail from one regional soil memoir to the next.

As an example of a soil classification at a level of detail thought generally useful to ecologists, a scheme of British soils, proposed in Table 6.6, is based on Kubiena on past conventional British practice. Seven major soil groups are divided into 32 sub-groups named from what has become widely understood terminology. The definitions of the classification categories employed here and subsequently are: *Major Soil Group*—a class distinguished by the presence, and/or absence, of specified master horizons; *Soil Sub-group*—a class within a major Soil Group, defined by a particular horizon sequence; *Soil Series*—a class within a Sub-group, distinguished by formation from a specific soil parent material; *Soil Type*—a class within a Soil Series, defined by the textural class of the surface horizon, or of the profile; *Soil Phase*—a class within a Soil Series, defined by a correlation with a site or soil factor insufficient to cause a change at a higher classification level, but significant to soil performance; e.g. ground slope, or degree of stoniness. Figure 6.5 illustrates schematically the main horizon sequences of soils in these seven major soil groups.

The latest classification of British soils arising from soil survey experience (Avery, 1973) has 10 'Major Groups', 43 'Groups' and 117 'Sub-groups'. The definition of any sub-group must clearly be more precise when 117 classes are involved compared to 32, but more class limits create greater problems of decision. A balance must be assessed by the ecologist, in relation to each study, between the effort to be expended in soil data collection and classifica-

Table 6.6   A classification of British soils.

Major soil groups

(1)	*Raw mineral soil*	Soils lacking continuous lateral development of any master horizon other than C.
(2)	*Ranker*	Non-calcareous soils with O or A and C horizons but lacking B horizons. Some variants have incipient E horizons.
(3)	*Calcareous soil*	Soils with Ac and Cc horizons or with A, Bc and Cc horizons.
(4)	*Brown earth*	Mineral soils, generally of A, B (other than Bh or Bs) and C horizons, non-calcareous in the B horizon, but including variants with an Eb horizon.
(5)	*Gley*	Soils which have horizons of gg type at or near the surface.
(6)	*Podzolic soil*	Soils with Bs, and/or Bh horizons, which generally, but not always, underlie an E horizon.
(7)	*Organic soil*	Soils with no master horizons other than O present at a depth less than 50 cm.

Sub-groups

The horizon sequences shown are intended as examples, not to cover all variations possible within a sub-group.

Soil sub-groups		Horizons in profiles
1	RAW MINERAL SOIL	
1.1	Non-gleyed raw mineral soil	C
1.2	Gleyed raw mineral soil	Cgg
2	RANKER	
2.1	Peaty ranker	O, C or Cr
2.2	Humic ranker	Ah, C or Cr
2.3	Brown ranker	A, C or Cr
2.4	Podzol ranker	A, Ea, C or Cr; O, Ea, C or Cr
2.5	Gley ranker	A or Ag, Cg
3	CALCAREOUS SOIL	
3.1	Humic rendzina	Ahc, Cc, or Ccr
3.2	Rendzina	Ac, Cc or Ccr
3.3	Humic brown calcareous soil	Ah, Bc, Cc or Ccr
3.4	Brown calcareous soil	A, Bc, Cc or Ccr
3.5	Gleyed brown calcareous soil	A, Bcg, Ccg or Ccr
4	BROWN EARTH	
4.1	Eutrophic brown earth	A, B, C or Cr
4.2	Acid brown earth	A, B, C or Cr
4.3	Humic brown earth	Ah, B, C or Cr
4.4	Gleyed brown earth	A, Bg, Cg or Cr
4.5	Leached brown earth (sol lessivé)	A, Eb, Bt, C or Cr
4.6	Gleyed leached brown earth	A, Ebg, Btg, Cg or Cr

(The distinction between 'eutrophic' and 'acid' brown earth is an arbitrary one based on pH levels $>$ or $< 6\cdot0$ in the A horizon. It could also be applied to separate gleyed brown earths into two sub-classes.)

5 GLEY
5.1 Non-calcareous gley   Ag, Bgg, Cgg or Cr
5.2 Calcareous gley   As above but with Bggc and/or Cggc
5.3 Humic gley   Variants of the above sub-groups with Ah horizons
5.4 Peaty gley   Variants of the above sub-groups with O or Oo surface horizons
5.5 Podzolic gley   Ah, Eag, Bsg, Cgg

A distinction may be introduced between gleys resulting from a high water-table (ground-water gleys) and those due to impedance of surface water drainage through the profile (surface-water gleys.) Many gleys have characteristics partly attributable to both causes so that general use of the distinction is not recommended.

6 PODZOLIC SOIL
6.1 Brown podzolic soil   A or Ah, Bs, C or Cr
6.2 Iron podzol   O or A, Ea, Bs, C or Cr
6.3 Humus podzol   O or A, Ea, Bh, C
6.4 Humus-iron podzol   O or A, Ea, Bh, Bs, C
6.5 Peaty gley podzol   O, Eag, Bfe, Bs, C or Cg

7 ORGANIC SOIL
7.1 Eutrophic peaty soil   O
7.2 Acid peaty soil   O
7.3 Eutrophic peat   Oo
7.4 Acid peat   Oo

(Within the organic soil group, a IIC, IICgg or IIR horizon may be present between 50 and 100 cm. These should be noted and can give further divisions important to some forms of land-use or potential land-use. The distinction between 'eutrophic' and 'acid' soils is again an arbitrary one based, as above, on pH values of $>$ or $< 6\cdot0$.)

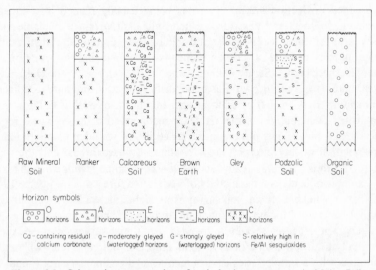

Figure 6.5 Schematic representation of main horizon sequences in Major Soil Groups in Britain.

tion and the precision of soil definition required for the communication of adequate information.

The highest categories of the system of Table 6.6 and that of Avery are equated as follows: Raw Mineral Soils = Terrestrial Raw Soils and Hydric Raw Soils; Ranker and Calcareous Soils (in part) = Lithomorphic Soils; Brown Earth and Calcareous Soils (in part) = Pelosols and Brown Soils; Podzolic Soils = Podzols; Gley Soils = Surface-Water Gley and Ground-Water Gley Soils; Organic Soils = Peat (Organic) Soils. Avery's system additionally has a group of Man-made Soils.

In summary, the ecologist should employ a national or regional system of reasonable but not excessive sub-division, using, if possible, generally understood terminology. Equivalence of class names in different languages (for example; Brown Earth, Sol Brun Acide and Braunerde; Brown Podzolic Soil and Sol Brun Podzolique) may be arguable in detail but such correlations generally carry sufficient uniformity of meaning for the national systems to be used with advantage. Examples of effective application of accepted national or regional terminologies for soil groups and sub-groups in ecosystem studies are the major study of the vegetation of the Scottish highlands by McVean & Ratcliffe (1962); a comparison of grassland types and soils in the Serengeti (Anderson & Talbot, 1965) and a study of spruce ecosystem-environmental relationships in British Columbia (Wali & Krajina, 1973).

4.2.3 Local soil classification

Soil surveys at scales of 1:63,360 or larger are mostly made at soil series level (as defined in Section 4.2.2), the series being usually named after locations where the soil was first described. When official surveys are available for the study area or for a climatically and geologically comparable district, these series names provide a compact means of conveying information. For example, the Denbigh Series, defined from North Wales, is, in the classification given here, an Acid Brown Earth on drift derived from hard argillaceous sedimentary rock (shale) of Silurian age. Initial use of both a full description and, when available, the series name, is recommended in ecological reports. The series name allows more concise repetition of soil identification and also extrapolation of any interpretation to other surveyed areas, and the former caters for those without ready access to the original definitive publication. An example of such usage is a study of a Malayan forest site by Poore (1968).

If no applicable official soil survey is available the ecologist can introduce symbols or names for his soil mapping and classification units at series level but geographic names should be avoided if national surveys can be expected eventually to produce an official regional map. Soil mapping units will, in some situations where simple series-environmental relationships are not found, need to be complexes of several series (see Section 5.1). More broad regional studies can use sub-groups or *association* mapping units. The latter in its most frequent application (see Soil Survey of Scotland publications) is a category comprising under one geographic name (e.g. the Countesswells Association of north-east Scotland) all sub-groups and series that have developed on a particular parent material. 'Association' has been used by others to cover soil classes associated with a particular landform or terrain unit, and these may include more than one soil parent material. For intensive studies, phases within series, based on characteristics such as soil depth, stoniness or site slope, or on soil chemical trends, may be the required mapping units.

5 SOIL MAPPING

5.1 GENERAL CONSIDERATIONS OF SURVEY PROCEDURE

For the ecologist who needs to map soils or to interpret existing soil maps, survey procedures are detailed in USDA (1951); Maignien (1970); and Clarke (1971).

The problems caused for classification by the three-dimensional variability of soils have been mentioned. Delimitation of mapping units and their boundaries is even more difficult. For vegetation (Chapter 3) the species composition is identifiable above the ground surface, whereas the most detailed soil survey can observe only a very small part of the total soil range. The same soil categories employed as classification units are also those employed as primary mapping units but these two usages differ in practice. The classification units are usually defined by ideal 'modal' profiles selected to show the distinguishing horizon criteria in their most clear-cut form. Buol *et al.* (1973) have pointed out that this creates a tendency among soil scientists to draw their modal profiles from examples of maximum horizon development for any particular class. This can involve selecting extremes from the range of characteristics found for a given soil class in the field, rather than mean profiles. Recently, more objective sampling studies of the range of properties within conventional series mapping units have begun to be carried out, such as by Courtney (1973), which test the suitability of the unit and/or the quality of mapping and can give a better definition of the true modal profile and the range of variation.

In practice the surveyor will frequently find soils that have profile characteristics intermediate between modal profiles of two or more classification units. Dependent on the time available for survey and the nature of the study, such soils must be grouped quite subjectively from experience or more objectively by quantitative classification based on observed and measured characteristics. Any mapping unit has to accept a range of variation, so far as possible explicitly defined, from the characteristics of the modal profile. This range must not be too great, otherwise the separate mapping units will differ as much internally as they do between units. In the most favourable circumstances, except for very large scale mapping, it is impractical to expect a situation in which many more than 75% of a sample of random points within a mapping unit have profiles which are closely correlatable with the soil or soils of the map legend.

Local circumstances control whether at a given scale of mapping it is possible to distinguish areas as occupied by a single series or whether the intricacy of soil pattern is such that the mapping units have to be complexes of more than one series. This is particularly likely to be so in areas of marked relief in the uplands. The nature of the problem of classifying and mapping soils in the field in such an area was shown in a study in the Rhinog mountains of North Wales (Mew & Ball, 1972) where profiles examined at points marked on a grid pattern (5.2.1) on air photographs were considered both independently and in relation to the complex mapping units of a 'free' survey (5.2.2) based on subjectively chosen transect lines traversing geological and relief boundaries. Of 80 profiles examined at grid pattern points, about 25% could not be clearly fitted into a simple sub-group classification similar to that given in Table. 6.6. Published surveys and reports generally emphasize the surveyor's final decisions rather than problems of mapping and classification, the described soils being chosen to fit as closely as possible to the classification finally adopted. Scientists familiar only with the literature are thus faced with a very different situation when involved in a field survey problem.

These comments could be taken to support the view that mapping of soil distribution is pointless and that the most that can be done, subject even here to accepting compression of the natural range of soil variation into an unnaturally limited number of classes, is to indicate the proportions in which these soil classes occur in an area, without claiming to give their distribution other than in generalities. This is an extreme view and, provided the problems and reservations are understood, a soil map does usually communicate usable and useful information by showing the distribution of units of reasonable internal consistency which broadly synthesize the interactions of primary

Table 6.7 Suggested sampling intensity for soil surveys of different scales.

Map scale	Sampling intensity (profiles per hectare)	Country	Source
1 : 100,000	1 per 500–1,200	France	Legros (1973)
1 : 63,360	1 per 5–10	UK (Berks.)	Burrough et al. (1971)
1 : 25,000	1 per 20–50	France	Legros (1973)
1 : 25,000	1 per 2	UK (Glos.)	Cope (1973)
1 : 25,000	1 per 1	UK (Berks.)	Burrough et al. (1971)
1 : 5,000	1 per 2–5	France	Legros (1973)

environmental factors. The precision of this information varies according to the nature of soil distribution locally and the mapping scale and effort possible.

The 'quality' of mapping, i.e. the efficiency with which soil maps delineate areas inside which the variation in soil properties is significantly less than it is between them, is a subject of much recent study (Burrough *et al.*, 1971; Bie & Beckett, 1971; Beckett & Burrough, 1971a, b; Legros, 1973). These studies have considered the required intensity of sampling for optimum map quality with acceptable effort, and figures are summarized in Table 6.7.

Clearly although these generalizations enable required effort in any survey to be approximately planned, there will be substantial differences between areas, dependent on their environmental complexity, and on how closely soil boundaries follow well-defined physiographic boundaries. The degree of variability in profile type that can occur over very short distances is illustrated by observations on Snowdon in North Wales by Ball *et al.* (1969). They recorded, in a number of complex mapping units, the soil sub-group at 20 points, spaced 3 m apart, on a cruciform transect pattern. In some units, >80% of the observations were of soils classifiable in one sub-group but in most mapping units soils attributable to four or more sub-groups were present as 10% or more of the sample. The only practical solution in such situations is to map soils in complex units, defined as closely as possible by association with parent material and/or landform.

The ecologist must exercise great caution against over-confident acceptance of published soil maps and in his claims for his own survey quality. Neither a brief review of the literature, nor a quick survey, will necessarily describe the soils adequately but it is even worse to leave the description of site factors and soil conditions for completion as a hurried afterthought before work is written up.

5.2 SURVEY TECHNIQUES

Pre-survey work and some practical points on survey methods will be briefly considered here.

Pre-survey data collection and interpretation are covered in outline by Clarke (1971) and Cruikshank (1972) and in more detail in USDA (1951). Climatic, physiographic and geological data are required, as described in Section 2.2.1. Air photographs can assist in preliminary delineation of landform units (Ball *et al.*, 1971) and in the initial planning of survey transects if the 'free survey' (5.2.1) approach is to be used (Mew & Ball, 1972). In these studies, and in the works of Goosen (1967); Vink (1968); Jarvis, R. (1962); Jarvis, M. (1969); Webster (1969); and Evans (1972), it has been emphasized that air photographs can seldom provide direct information on soils. They often show drainage patterns or the boundaries of peat and mineral soil areas very clearly, but their main advantages are to improve speed and efficiency at the planning stage, and to act as substitutes for maps in the field. The application of air photographs in soil survey has recently been reviewed by Carroll (1973).

The actual technique of soil mapping requires little comment. The excavation and recording of soils observed in profile pits has been considered and these act as key points from which the need for interpolated observations must be decided. During 'free' survey (5.2.1), a soil auger is the principal tool used to examine soils between profile pit localities.

Hand-operated coring augers can, in soils with few stones, provide a more-or-less undisturbed soil core, and variants of these are particularly used for specialized peat studies. Short lengths of metal tube with detachable rod handles can also be used as samplers of surface horizons (6.1.1) but the screw auger is the general tool used. Their threaded screw end is about 20 cm long and *c.* 2·5–3·0 cm diameter, and the overall auger length is about 90 cm, with a T-handle, used to screw the auger into, and pull it out from, the soil. Such augers are commercially available but are often locally constructed by welding a wood-auger bit to a steel rod of about 1·2 cm diameter, with a T-handle of metal or wood, to give a more robust and long-lasting tool than most commercial types. In use, the thread is screwed into the soil while moderate pressure is applied to the handle. The auger is withdrawn at about every 15 cm successive penetration by a steady upward pull so that retained soil may be examined. The thread can be rapidly cleaned between observations by spinning it between the fingers. The disadvantages of a screw auger are that it will not easily withdraw samples from very loose or dry soils; it obscures thin horizons; and it does not permit a full examination of undisturbed soil structure. However, it is quick and does negligible damage to a site. Auger T-handles are handy as seats of the shooting-stick type, but be wary of such use in deep peaty soils because there an auger will push in easily to its full depth with only light pressure!

5.2.1 'Free' survey

Free survey (e.g. as summarized recently by Burrough *et al.,* 1971) is the conventional soil mapping procedure in which observations are made at subjectively chosen locations as work proceeds, rather than being confined to pre-determined points. Prior to the main field survey, available topographic, geological and physiographic site data, and air photographs, with or without ground reconnaissance according to site accessibility, enable an assessment of land units or classes to be made. From this, with any knowledge from similar terrain, a provisional key can be drawn up of the soils which are expected to occur in the study area. Transects are planned to traverse the range of landform or land-use units and the different geological formations. A reasonable assessment of just what effort will be required for a particular survey can be made at this stage, and the survey itself carried out more economically than would be possible without the preliminary work.

The surveyor uses the chosen traverses as a foundation, but not a rigid control, for the selection of points where soil observations are to be made. The density of these observations and the proportion of auger to profile pit sites has to depend on the information added as survey proceeds. Modifications to the provisional map legend and the provisional site-soil correlations will also be needed as the survey proceeds.

5.2.2 'Grid' survey

In grid surveys soil observations are made only at predetermined points located from maps or air photographs, or from ground markers, at spacings chosen to relate to the required survey scale. Such surveys can, as with free surveys, be based on allocation of soils at the observation points into classes of a conventional classification system, and it is this usage which will be emphasized here. A grid survey can also provide data for numerically coded visual or measured parameters that can be used to classify the soils by quantitative techniques (4.1), or to plot the distribution of single properties. Trend-surface analysis of the distribution of single properties or combined indices, and other numerical interpretations of gridded data, are possible by computer (e.g. Norris, 1972). The advantages usually claimed for grid surveys are greater objectivity compared to free survey, and the lower degree of judgement and experience required by the surveyor. A regular grid pattern of sampling points is generally employed as being more convenient than a statistically random spread of points. Intersections of grid lines on maps are frequently used as the observation points, or a suitable network is drawn as an overlay on maps or air photographs.

In a regional survey of part of Pembrokeshire (South Wales) (Rudeforth, 1969; Rudeforth & Bradley, 1972) soil profiles were identified at intersections of a grid of 1 km spacing (1 observation per 100 hectares). Seven soil sub-groups were recognized but no attempt was made to insert boundaries between mapping units. Interpretation of the survey was based on the proportion in which different soil sub-groups were found in the region as a whole, and on the relationships between sub-group distribution and geologic substrate, landform unit, and land-use recorded at the same sampling points.

A test of grid survey as a basis for soil mapping in upland country has been carried out in the Rhinog mountains of North Wales (Mew & Ball, 1972). A grid of *c.* 300 m spacing (1 observation per *c.* 9 hectares), covering *c.* 12 km^2, had been marked on air photographs at 1 : 10,000 scale and the points then located on the ground for a pilot study of vegetation mapping

(Goodier & Grimes, 1970). Soil profiles were examined at these points and free survey was also carried out over the same area. When soil mapping units derived from free survey were compared with observations at the grid points, only half the soils assessed as 'important' in the former had been recorded at the latter, and 40% of soils at the latter were not considered to be of significant areal extent in the free survey. Alternative interpretations can be placed on these results but the tentative conclusion was that the 'subjective' free survey gave a better understanding of soils of the area and of their distribution than the grid survey.

Grid survey seems better adapted to broad regional studies which do not require maps as a product, or to close sampling of an experimental area, particularly when combined with ecological observations at the same points, than to intermediate scale detailed survey. If grid spacing is reduced to give more intensive coverage, the effort required is increased too much to be acceptable in all but small areas. A 'stratified grid', in which point spacing is adjusted according to local soil and terrain complexity, introduces similar subjective judgement to that of free survey and offers no improvement.

At the regional level of survey Beckett & Burrough (1971a, b) have made a comparative study of the quality of soil maps (at scales 1 : 20,000–1 : 100,000) produced by free and grid surveys (mainly at 100 m, some at 300 m spacings) in three areas in southern England. They found that the proportions of particular soil series in an area were better assessed by grid survey and that this gave greater 'purity' of soil mapping units on areas of complex soil-site relationships, although free survey was preferable where unit boundaries were related clearly to site features such as physiography. Scale of mapping also affected relative quality of maps by the two procedures. At larger scales, grid and free survey were nearly comparable, but free survey was superior at smaller scales (below 1 : 50,000). A very close grid of 50 m spacing (4 observations/ha) was used by Norris (1972) to study soil variability in a southern English woodland. Such intensity of survey is generally more relevant for analysis of variation of a single property than for mapping soils.

Each ecological study involving soil mapping requires individual consideration of the optimum scale and method of survey, and no single survey plan or standard recipe for sampling intensity is universally preferable.

6 SOIL ANALYSES

6.1 SAMPLING

6.1.1 Sampling techniques

The excavation of a soil profile pit was described in Section 3.1. Horizon sampling should be carried out by cutting, with a straight-sided 'planting' trowel, or sheath-knife, a slice of constant cross-sectional area from the full depth of the horizon, rather than by sampling centrally in each horizon as is sometimes recommended. Working upwards from the lowest horizon is generally quickest, as otherwise it is necessary to clean the face of the pit after each horizon is sampled. Sampling soils in their undisturbed field state as required for some analyses, is considered later.

Block-bottom bags made of bitumenized brown paper are most useful, and preferable to polythene, being easier to label and more resistant to cutting by stones. Care must however be taken with these (see Chapter 8, Section 2.1) if prolonged transport or storage is likely before the soils reach the laboratory. The usual range of routine chemical analyses can be carried out on quite small quantities of soil; on less than 50 g air-dry 'fine-earth' (<2 mm) material, or 150–200 g if particle-size analysis is also required. Unless transport is difficult, considerably larger samples are normally taken as standard, using bags of about $20 \times 15 \times 7 \cdot 5$ cm in size, giving an effective sample volume of *c.* 1,500 cc and a weight of *c.* 2,000 g, before drying and sieving treatment. The larger sample has some 'smoothing' effect in reducing variability, and gives a reserve of material for additional studies. Large and medium size stones can be hand-picked in the field to reduce weight. If quantitative data on stone content are required, then large samples are needed, and these determinations are preferably made in the field (6.3.1).

A straight-sided graduated planting trowel is also useful for obtaining near-surface samples, being well adapted to extracting samples of uniform cross-section and constant depth. Where soil horizons are quite uniform for considerable thicknesses, a screw auger can extract material suitable for analysis, but generally the auger mixes material too much to be suitable for other than observational sampling.

Sampling of surface zones can, if the soil contains few stones, be most easily carried out using tube corers. The simplest are made from short lengths (about 20 cm) of thick-walled (16 gauge, *c.* $1 \cdot 7$ mm) brass tubing (*c.* $2 \cdot 5$–5 cm diameter), with one end chamfered, and the other having holes drilled opposite each other to enable a short removeable rod to be used to give leverage to the corer as it is pressed into, and withdrawn from, the soil. The outer surfaces of the corer tubes are permanently marked at lengths required to control the depth or volume of sampling (e.g. 5, 10 and/or 15 cm or 100, 200 cc). The rod 'handle' can conveniently have a short plunger fixed at one end, of diameter slightly less than the tube bore, to enable it to be used to push the cores from the tube plunger into sample bags. Tube corers are particularly useful for collecting sub-samples of surface soil for compositing into large samples for analysis (6.1.2); for constant-volume sampling; and for obtaining undisturbed samples for certain physical determinations. *In situ* determinations can sometimes be made on samples retained in a simple tube corer but as soils cannot be extracted from these without compression they are not generally suitable for this latter purpose. Complex tube corers with an inner split container can be used to provide samples in field condition (e.g. Blake, in Black, 1965).

Minimally disturbed samples are required for determination of bulk density (6.3.2), or preparation of thin sections (6.4.1), and this sampling is best carried out from a profile pit, by inserting a sampler container into a vertical horizon face or horizontally cut-back horizon surface. Sampling tins which have removeable lids on two opposing sides can be made from galvanized steel or aluminium sheet. A standard size is about $8 \times 5 \times 5$ cm^3. With both lids removed, the four-sided frame of the tin is pushed (or gently hammered and cut) into the soil face, until the outer edges are flush with the soil surface. One lid is then placed over the exposed face and the tin carefully removed from the profile face by cutting around it with a knife. The orientation of the sample (top-bottom) is marked on the tin (frame rather than lid). After trimming the exposed soil flush with the edges the second lid can be pushed on. Such sampling may appear simple, but it is not always so straightforward in practice especially in stony soils, for which there is no easy solution. For sampling indurated soils, a piece removed with hammer or chisel may be packed in a tin. Larger tins of the same type can be used to collect profile sections for display, although practical sampling problems are greater. From such sections more permanent monoliths can be produced by impregnation of soil with a suitable resin (e.g. Wright, 1971).

An aspect of soil sampling related to analysis presentation and nutrient budget calculations (see Chapter 4, Section 4.1) is the collection of material so that analyses can be given on a volume basis, either directly as quantity/ml or for example as kg/ha (see e.g. Mehlich, 1972, 1973). Most soil analyses are made on a weight basis on air-dry soil which has passed a 2 mm sieve but direct sampling of measured soil volumes to give volume-weight conversion factors for each sample is possible. Weight basis results can be converted to a volume basis by bulk density measurements as discussed in Section 6.3.2.

6.1.2 Sampling programmes

Vegetation sampling for analysis has been treated in Chapters 4 and 8, and many general considerations mentioned there also apply to sampling soils. Although soil data should ideally be obtained by sampling sufficient profiles on a horizon basis to ensure comprehensive site coverage, this is not in general economically or practically feasible. Soil chemical variability has been found to be substantial even if samples are taken as close as 10 cm apart within one plant community on a soil which is uniform at series level (see e.g. Ball & Williams, 1968). This small-scale spatial variability includes the greater part of the total variability present over tens or hundreds of metres within a single soil-vegetation association. Significant variability has also been shown in physical properties such as bulk density and moisture parameters within one area of a single soil series (Nielsen *et al.,* 1972). Because of this, intensive sampling may be required to obtain results with desired precision.

Correlation of soil chemistry with plant species or community distribution can often be made from analyses of standard depth samples of the surface soil which can be collected more simply, quickly and less destructively than profile horizon samples. It is proposed therefore that, in many field plant-ecological studies, soil profile data should provide a foundation for an interpretation of surface soil chemistry determined from a more intensive sampling programme. The standard depths employed may be 0–5, 0–10, or 0–15 cm, the latter being generally most useful. From an agricultural viewpoint, nutrients in a 20 cm depth are often considered, as this approximates to modern ploughing depth, with a convenient volume of 2,000,000 dm^3 per hectare (Mehlich, 1972).

Standard depth sampling is particularly valid where the surface horizon has been found to be consistently thicker than the chosen sampling depth or where horizon change is completely merging over this depth range, and where the plant community has a high proportion of its active roots within the sampled zone. Standard depth sampling is least valid where narrow horizons vary rapidly in development and composition, or where active roots penetrate the soil much more deeply. Even in these cases however, there can be some merit in such sampling programmes, particularly where the question posed relates more to the effect of the vegetation on surface soil properties than to the effect of the total soil on the vegetation.

General considerations of soil chemical variability have been discussed by Peterson and Calvin (in Black, 1965); Ball *et al.* (1971) and Beckett & Webster (1971). The importance of spatial variability in determining and interpreting soil chemical data has been accepted in principle but the issue has been widely ignored in practice by soil scientists and ecologists alike.

Spatial variability is sometimes associated with specific vegetation patterns. Grubb *et al.* (1969) showed that a symmetrical pattern of soil pH change occurred in transects across individual plants of gorse *(Ulex europaeus),* heather *(Calluna vulgaris)* and heath *(Erica cineraea),* growing on a grass heath community on chalk-derived soils. A notable correlation of plant species with soil profile morphology as well as chemistry has been reported from New Zealand where 'egg-cup' podzols seem to have formed locally under individual Kauri *(Agathis australis)* trees from non-podzolized soils of the Yellow Earth group (Gibbs *et al.,* 1968).

The more general case is one of apparent random spatial variability in soil chemical parameters even in soils which are morphologically uniform and which support a single plant community. Ball & Williams (1968) determined the scale of this variability for 0–15 cm samples of an uncultivated Acid Brown Earth under upland grassland, and subsequently (Ball *et al.,* 1971) considered chemical variability in the same depth zone of Acid Peat soils. The

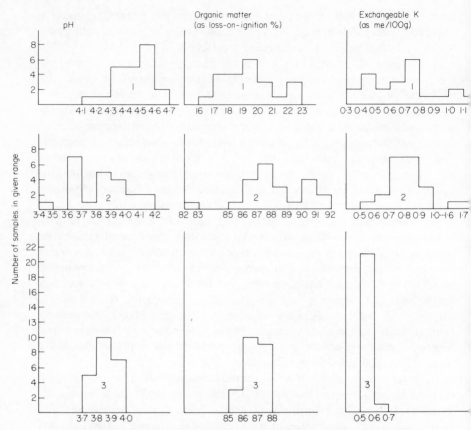

Figure 6.6 Range of some analysed quantities in 'uniform' uncultivated soils, sampled over 0–15 cm depth: 1—uncultivated upland acid brown earth on site of normal drainage, data from 22 samples taken on circumference of circle of 1 m radius; 2—uncultivated upland acid peat on site of shedding drainage, sampled as 1; 3—soil as 2, analyses of 22 aliquots taken from a single bulk sample.

results, summarized in Table 6.8, show chemical variability of the same general order in the two contrasting soil groups, and comparable to levels from other situations (e.g. Skene, 1960; Leo, 1963; Gallagher & Herlihy, 1963; Frankland *et al.*, 1963). Figure 6.6 shows as histograms the typical range in pH, organic matter and exchangeable potassium that is found within very short distances in a 'uniform' soil under 'uniform' vegetation. It will be seen that the range is comparable on two contrasting soils, in a mineral soil (Acid Brown Earth) on a site of normal drainage, and in an organic soil (Acid Peat) 'perched' on a site of shedding drainage. That this variability is not a result of sampling and analytical 'errors' can be seen from comparative range data for the same Acid Peat determined on analysis of aliquots from a single bulked sample of this soil.

Table 6.8 Chemical variability of morphologically uniform uncultivated soils.

Quantity		Coefficient of variation (%)	
		Acid brown earth—upland grassland[1]	Acid peat—upland bog[2]
pH (air-dry soil)		3	5
Loss-on-ignition		9	4
Exchangeable cations	K	35	42
	Na	6	11
	Ca	30	20
	Mg	41	11
	Mn	27	35

From the means of 22 separate analyses of 0–15 cm depth samples at each of 3 locations
[1] Within one area of an uncultivated soil on volcanic ash colluvium, North Wales
[2] From three different deep peat locations in North Wales.

From these data the number of samples needed to determine a mean value within a specified range from the estimated true mean is about 15 for pH to $\pm 2\%$ and loss-on-ignition to $\pm 5\%$; between 20 and 40 for exchangeable cations to $\pm 10\%$ (except for the generally much less variable Na); and generally higher than this (50–70 samples) for extractable $P_2O_5 \pm 10\%$. It is unwise therefore to assume that relatively small chemical differences are significant when they are based on small sample numbers, even when comparing sites which are quite homogenous from a soil type viewpoint.

Type profile analyses give the general chemical composition of a soil and allow internal comparisons to be made between horizons within the profile. Analysis of a composite sample increases the precision with which the true mean is estimated, but only if a number of samples are analysed independently can the mean, range and confidence limits be predicted. Standard statistical textbooks treat this aspect fully but it is often ignored by soil scientitsts, and sometimes by ecologists, in their use of soil data.

Clearly, an intensive ecological investigation of one site might justify a large-scale soil sampling programme to determine chemical factors precisely but, if many sites are involved, a compromise between effort and precision is essential.

Ball & Williams (1971) compared sampling programmes on an Acid Brown Earth against 'best' values determined from 275 samples covering a single block of contiguous plots, each sample being a composite of 10 sub-samples of the 0–15 soil depth. The least laborious programme involved determining mean and range from six plots only. The difference in effort involved was from 11 days field work + 55 days analysis-time to 0·2 days field work and 1·2 days analysis-time, giving a ratio of work of 50:1, but the 'quality' of information supplied at the two extremes was approximately 2:1, with the low effort sampling programme giving a result covering more than 50% of the total variation as assessed by the saturation sampling. Arising from this study, a sampling programme in which 10 sub-samples are combined from each of six plots, 6 m², marked about centres 20 m apart on a transect 100 m in length has subsequently been employed on sites including sand-dunes and chalk grassland. There is nothing sacrosanct about the dimensions or arrangement of this sampling design but something similar can be adapted to individual sites and problems to achieve a balance between sampling and analysis effort and data precision (see e.g. Falck, 1973).

In some situations chemical diversity is of a similar order of magnitude within a morphologically uniform soil class as within soil complexes. Profile type diversity, involving factors such as soil depth and drainage, is not necessarily reflected in greater diversity in the chemical factors that are usually measured in routine analysis. The soil properties not assessed by standard soil chemical determinations may be of crucial ecological importance. Plant community distribution can in some cases be better correlated with soils classified at the sub-group and series levels than with analysed chemical quantities. This emphasizes that the profile classification concept contains within it a wide range of variables and that in such cases one or more unmeasured or so far unmeasurable factor is significant rather than the standard chemical quantities.

6.2 SOIL CHEMISTRY

Chapter 8 considers methods of soil chemical analysis, so that only one or two general points are made here. Figure 6.7 illustrates the steps required to bring field samples of soil to the standardized condition at which the majority of conventional analyses are made.

The preceding section may have given an impression that chemical analyses of soils are not of particular value to the ecologist, but this would overstate the position. Technically, modern analyses, such as discussed in Chapter 8, give reproducible results when applied to homogenous samples, as illustrated by the relatively low variability of 'bulk' sample analyses given in Ball & Williams (1968) but ecologists and soil scientists must appreciate the limitations of extrapolation of such data, however reproducible, to field interpretations for large bodies of soil. These limitations include inherent chemical heterogeneity and the broader issue of how far the standard soil analyses give data which correlate with growth responses of the plants being studied. Oertli (1973), considering the use of chemical potentials as an assessment of nutrient availability to plants, has concluded that difficulties in choosing a relevant nutrient potential make 'it appear unlikely that a practical and scientifically-rigorous criterion for nutrient availability to crops will soon be introduced in soil management'.

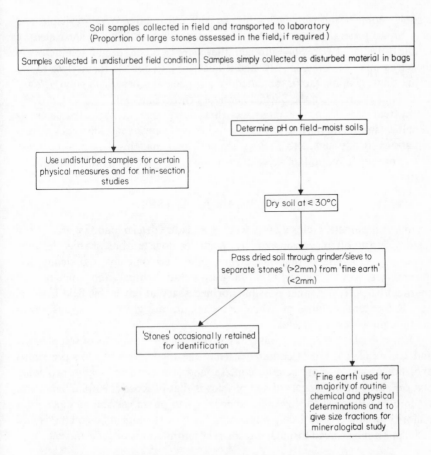

Figure 6.7 Outline flow chart for treatment of soil samples collected for analysis.

Together with the important two-volume compendium on soil analyses of all types published by the American Society of Agronomy (Black, 1965), and the FAO 'recipe-book' of Dewis & Freitas (1970), a useful recent text is that by Hesse (1971) who combines consideration of the theoretical background and a practical method for each chemical analysis with a discussion of why, in a soil context, it is useful. Hesse comments in his preface that 'the myth of soil analysis being the answer to the farmer's problem is not kept up' (in his book) and that 'the reader will repeatedly be reminded that field experiments are essential'. This caution is even more applicable for the ecologist than for the farmer. Prolonged and intensive studies of the relatively few important crop plants have only begun to give quantitative relationshps between soil chemical parameters and plant growth and productivity. For the majority of plant species which, in contrast to crop plants grown in monoculture, usually live in complex competition with each other, no such information is yet available, and only broad general principles of correlation are known. This is illustrated by Grime & Lloyd (1973) who, although they refer to other specific soil factors in their treatment of some species, find it necessary to confine themselves to pH as the single edaphic factor tabulated for all species in their *Ecological Atlas of Grassland Plants* from the Sheffield area of central England.

Epstein (1972) has emphasized that nutritional ecological studies or 'edaphic (soil-related) plant ecology', based on contrasting soil types, offers 'situations ideally suited to ecology studies'; and that this aspect of ecology, now 'entering a period of development', has not received 'the attention it merits'.

6.3 PHYSICAL ANALYSES

Chemical parameters, closely followed by moisture determinations, are the soil factors most often considered by plant ecologists, but it has become increasingly apparent that there are other ecologically significant soil parameters involving solid-liquid-gas relationships and mechanical characteristics. It is neither possible nor necessary to cover this field in detail here but general outlines of available techniques are given for aspects which are thought to be most useful.

Baver *et al.* (1972), after Baver (1956) emphasize principles of soil physics, and Childs (1969) treats comprehensively the theory of soil moisture status and movement. A most helpful, concise, and less mathematically forbidding text on agricultural aspects of soil physics is that of Rose (1966). Chapters in volume 1 of Black (1965) give broad coverage to principles and methods for a majority of soil physical determinations. Later developments of methods and concepts can be traced through the abstract journal *Soils and Fertilisers*.

6.3.1 Particle size distribution

Soil texture has previously been defined as a field assessment of the proportion in which particles of different size ranges are present in the soil. Particle size determination, less accurately but widely termed 'mechanical analysis' (see Day, in Black, 1965) gives a more reproducible basis for textural classification.

Normally, particle size analyses, as with other standard soil analyses, are made on the fine-earth fraction of the total soil, that is on material of less than 2 mm equivalent spherical diameter. Field soil properties are however affected by the size and quantity of larger particles (i.e. 'stones' by definition). When quantitative assessments of these are required, the larger stones, in samples that are relatively dry, stony and friable, can conveniently be weighed in the field. Soil can be passed through a 10 mm mesh nylon sieve, and weighing of separated fractions is simple using a portable folding alloy tube tripod from which a spring balance can be suspended to weigh material in polythene sacks.

Small 'stones' must normally be further separated from fine-earth in the laboratory as sieving of undried soils through smaller mesh sieves is not usually convenient in the field.

In the laboratory, soil samples are initially dried in air or in an oven at c. 30°C, then passed manually through a 2 mm sieve or through a mechanical grinder which combines a mild aggregate-crushing action with a sieve. This dried fine-earth fraction will generally include structural aggregates as well as primary particles and dispersal of these, normally in aqueous suspension, is essential before particle size analysis. Effective dispersion of aggregates in suspension may require chemical pre-treatment to remove organic matter and/or iron or other oxides and hydroxides acting as cementing media (Kunze, in Black, 1965). Where possible however, dispersion should be carried out without such pre-treatment, by taking 50 or 100 g of air-dry fine-earth in a bottle, and adding to it 400 ml of water and 25 ml of 5% 'Calgon' (sodium hexa-metaphosphate) at pH 9 as a dispersing agent. Mechanical shaking of the suspension for several hours is then required or alternatively an ultrasonic vibrator can be used (Watson, 1971; Pritchard, 1974). The objective is to ensure that all primary particles are dispersed in the suspension but without too vigorous handling which could break these down from their original size to smaller sizes during the treatment. Care must be taken to be aware of this possibility when dealing with soils containing fragile rock and mineral fragments.

For highly organic soils, particle size analysis of the mineral fraction requires initial destruction of the organic matter. In most soils this is done by carefully adding 30% hydrogen peroxide to the wetted soil, followed by a time interval for the reaction of peroxide with organic matter, carrying out this operation a sufficient number of times to ensure total destruction of organic matter, and completing the final digestion on a hotplate at 90°C. Hydrogen peroxide is effective only in an acid medium so that carbonates if present must also be removed by acid treatment prior to organic matter destruction, or an alternative to peroxide used. In general, when organic matter and/or carbonate levels are high, particle size analysis of the minor non-carbonate mineral fraction becomes an even more arbitrary procedure than usual, largely divorced from field properties of the soil, and of value mainly in studies relating to the origin of the mineral fraction rather than of ecological interest. It is therefore suggested that textural analysis which requires to be more quantitative than field assessment can be is most likely to be useful to ecologists in cases where minimal pre-treatments are needed.

After dispersal, the soil suspension is transferred to a litre cylinder, made up to the mark, and thoroughly mixed. The simplest method uses a 'Bouyoucos' hydrometer, inserted into the suspension at settling time intervals of 46 s for American silt + clay; 4 min, 48 s for International silt + clay and 5 h for clay, the values for material in suspension being read directly from the graduated stem in grams per litre. If 50 g of soil were taken, the read values are multiplied by 2 to give weight percentages. The settling times are calculated as for a pure water suspension at a given temperature and require small corrections for Calgon in the suspension and if the temperature of the suspension differs from 18°C. Hydrometer measurements are reasonably reproducible and of adequate precision for most purposes. It must be appreciated that the whole method allows various assumptions and uncertainties so that very careful efforts to be precise about the necessary corrections are unjustified. Nomographs from which settling times for different particle sizes and suspension temperatures can be obtained are given in several sources, including Tanner & Jackson (1948) and British Standards Institution (1967). Kaddah (1974) has compared the hydrometer and pipette methods.

Table 6.9 gives size classes used in soil particle size analyses. Conventionally, only three fractions: total sand, silt and clay, are distinguished. For more detailed work, a larger number of size classes is needed within the sand fraction, obtained by sieve separations. Cumulative percentage curves, in which particle sizes are plotted against the percentage of the total soil material having a smaller estimated diameter, enable visual comparison of particle size distribution. Actual size values are plotted on logarithmic paper or, more conveniently, a logarithmic size scale, such as the 'phi' scale given in Table 6.9, is used with ordinary paper.

Nomenclature of particle size ranges has inconsistencies such as those that plague much soil work. The table emphasizes the USDA (1951) limit for the silt-sand boundaries, but the older 'International' scale had an upper size limit for silt at 0·02 mm, and Avery (1973) has recommended a division at 0·06 mm. With routine methods, differences in value between measurements at nominal separations of 0·06 and 0·05 would not be significant in the great majority of soils. It must be borne in mind that, in particle size analysis, settling speeds in suspension are calculated on untrue assumptions, such as that soil particles are uniform spherical particles of uniform specific gravity. Similarly, in sieve analysis long lath-like fragments of larger volume can just pass the same sieve aperture as spherical particles of smaller volume. For these reasons, all soil particle size limits are 'equivalent spherical diameters'.

Sand, silt and clay proportions, as percentages of the fine earth, can be plotted on a triangular diagram and textural class names obtained. The U.S.D.A. system (1951) is illustrated (Fig. 6.8). The new variant in Hodgson (1974) differs in some class names and limits. Laboratory classes closely match field assessments by experienced observers but discrepancies can result from organic matter giving an impression of loamier texture, or abundant small stones one of lighter texture. So far as soil properties are concerned, field texture assessment is quick and readily applicable for description and classification of soils, but particle size analyses are more reproducible and less dependent on acquired expertise and confidence.

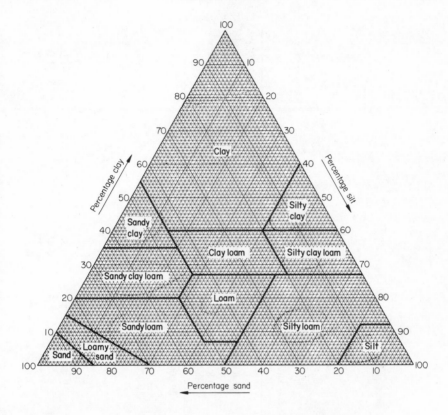

Figure 6.8 Soil texture class names (from U.S.D.A. 1951) based on proportions of sand, silt and clay fractions.

Particle size distribution gives a standardized means of comparing a basic soil characteristic related to key soil properties such as permeability. It has been found that reasonable estimates of available water capacity (the total quantity of water that a soil can store in a form accessible to plants, 6.3.5) can be made by calculation from analyses of particle size distribution and organic matter content (Salter *et al.*, 1966) and that this can also be done from field textural determinations (Salter & Williams, 1967).

6.3.2 Bulk density

Bulk density (e.g. Blake, in Black, 1965), or volume weight, is the ratio of the dry weight of a soil sample to the total volume it occupies in field condition (i.e. the moist volume of a soil). It can be used indirectly to assess differences in soil structure and porosity caused by natural processes or by management. Bulk density and porosity determinations can often be related to penetrability by

Table 6.9 Soil particle size classification.

Size fraction	British Standard sieve mesh numbers*	Mesh aperture of sieve (microns = 0·001 mm)	Assumed diameter of particles just passing through mesh (mm)	log₁₀ of particle diameter (microns)	'Phi' value** (−log₂ of particle diameter (mm)
Large stones > 100 mm	—	100,000 (10 cm)	100	5·0	−6·6
Medium stones 100–10 mm	—	10,000 (1 cm)	10	4·0	−3·3
Small stones 10–2 mm	8	2,057	2	3·3	−1·0
Coarse sand 2–0·2 mm	16	1,003	1	3·0	0
	30	500	0·5	2·7	1·0
	60	251	0·25	2·4	2·0
Alternatives	72	211	0·21	2·3	2·3
	85	178	0·18***	2·3	2·5

	B.S. mesh		Aperture (µm)	Aperture (mm)*		Phi**
Fine sand 0.2–0.05 mm	120		124	0.12	2.1	3.1
	150		106	0.1	2.0	3.3
	Fine sieves become increasingly delicate, expensive and unsuitable	240	63	0.07	1.8	3.8
		300	53	0.05	1.7	4.3
Silt (American) 0.05–0.002 mm		—	(50)	0.05	1.7	4.3
Silt (international) 0.02–0.002 mm		—	(20)	0.02	1.3	5.6
Clay < 0.002 mm		—	(2)	0.002	0.3	9.0

* The same mesh numbers in other systems (e.g. A.S.T.M. 1972) have different aperture sizes to those quoted for the British Standard sieves (as manufactured by Endecotts).

** 'Phi' values are conveniently used for plotting cumulative particle size curves on ordinary (non-logarithmic) paper (see Krumbein & Pettijohn, 1938).

*** Generally used as the preferred sieve for '0.2' mm.

roots (e.g. Mirreh & Ketcheson, 1972) or to the suitability for seed germination of a surface soil crust. Recently it has been shown that bulk density correlates well with available water, retained water and air capacity (Reeve *et al.*, 1973).

Bulk density is determined from the air-dry mass of a known field volume of soil. This volume can be obtained directly with the use of a volumetric corer (6.1.1) or an alternative method is to remove a quantity of soil as a disturbed sample, then determine the volume of the excavated hole by filling it with a measured volume of sand (British Standards Insitution, 1967) or by closely lining the cavity with an impermeable rubber or plastic film (Shipp & Matelski, 1964) and pouring water of measured volume in until the cavity is filled. These infill methods are particularly useful in gravelly soils for which core samples are impractical, but require transport or availability on the site of water, or sand, and volumetric measuring containers. Instrumental *in situ* field measurements of bulk density by radiation methods are possible (Blake, in Black, 1965; A.R.C., 1971—p. 7).

A direct laboratory method of determining bulk density applicable to small core samples, to samples of random shape cut from a transported block of soil, and to large soil-structural units, is to seal the outer surface of the sample with an impermeable resin, then immerse it consecutively into a series of liquids made up to known specific gravities until one is found in which it just floats (Campbell, 1973). The bulk density of small aggregates can be conveniently determined by measuring the volume change when a known weight of soil is mixed with a known volume of a fine powder (Bisal & Hinman, 1972).

Bulk density measurements are used to convert analyses made on a weight basis to presentation on a soil–volume basis. The volume occupied by a known weight of total soil and the proportions of total stones to fine earth in this soil are both needed to convert the standard presentation on a per hundred g fine earth basis to a field volume basis. Without measurements on each sample it is possible to give correction factors which are of the correct order of magnitude to be generally applicable. The close relationship between bulk density and organic matter content is relevant. Using weight/volume ratios of air-dry loosely-packed fine earth, rather than true field bulk densities of total soil, Ball *et al.* (1969) suggested conversion factors ranging from $\times 1 \cdot 0$ (0–5% organic matter), through $\times 0 \cdot 7$ (21–30% organic matter) to $\times 0 \cdot 3$ (91–100% organic matter) to give a relative volumetric basis from weight-based figures. Adams (1973) and Jeffrey (1970) give more precise considerations of the relationship between organic matter, true and bulk densities (see Chapter 4, Section 4.1).

6.3.3 Pore size distribution

Again the appropriate chapter in Black (1965), by Vomocil, gives a comprehensive review of this topic. Total porosity, which is the percentage of the bulk soil volume not occupied by solid material, can be calculated from the difference between volume of the undisturbed sample and the volume of soil solids or as:

$$100[(a-b)/a]$$

where $a=$ mean soil particle density and $b=$ soil bulk density. Mean soil particle density and soil solid volume can be determined from the volume of liquid displaced by a weighed, disaggregated dry soil sample.

Pore-size distribution is vital in consideration of water relationships in soils. A particular total porosity gives a very different pattern of water retention and release depending on whether it arises from few large pores or many small pores. The proportion of pores in different size grades is normally determined by measuring the quantities of water drained from a soil at a given suction pressure, assuming a capillary model to represent soil pore space. The volume of water extracted from saturated soil at a given suction is controlled by the proportion of total pore space occupied by pores of larger radius than that of the smallest capillary from which this suction is able to extract water.

The subject of moisture movement and pore-size relationships is complex and the reader is referred to Rose (1966), Childs (1969) and Black (1968) in addition to Black (1965).

Cary & Hayden (1973) have suggested an index for pore size distribution that is based on the change in soil water content as suction increases. The method requires laboratory measurement of soil moisture on samples subjected to three levels of suction. It appears to offer a useful basis for correlations between plant behaviour and soil porosity, by giving a single value to represent pore size distribution and therefore, indirectly, soil moisture characteristics.

6.3.4 Structural determinations

The primary particles of some soils either form a loose agglomeration of single grains or may occur as a compact mass. Both of these categories are termed structureless (3.2.1). The majority of soils have their primary particles arranged into aggregates, the size and shape of which provide the basis for subjective classification of soil structure (3.2.1). For greater precision, size distribution (Kemper & Chepil, in Black, 1965), and stability of aggregates (Kemper, in Black, 1965) require quantitative determination. De Boodt (1967) interprets structure more broadly and includes field description and sampling of soils with methods of determining particle size distribution, porosity and soil moisture, as well as a range of methods of structural determination *sensu stricto*.

For the determination of the size range of structural aggregates, a soil sample, preferably in a fairly dry state, is collected and transported carefully to the laboratories, dried thoroughly, and then passed through a nest of sieves on a shaker. Method V-41 of de Boodt (1967) uses sieves with apertures of 80, 40, 20, 10, 5 and 2 mm. Structural unit weights retained on the different sieves can then be plotted in cumulative curves as for primary particle size distributions (6.3.1), or calculated as single-value parameters (Kemper & Chepil, in Black, 1965). The size of aggregates is important to porosity and aeration of soils, and when reasonably large size is combined with stability, structural aggregates have a strong influence on resistance to erosion by wind or water.

Stability of aggregates is normally measured by determining the proportion of selected size fractions retained on a sieve after a standardized wetting procedure. Kemper has noted that limitation of stability tests to a single size group of aggregates (say 1–2 mm) is usually as effective as determination on several size classes. The method used to wet the aggregates is adjusted according to the purpose for which stability is required to be measured. Immersion wetting simulates surface irrigation, and the stability of aggregates wetted in this way relates to their performance as a surface soil. Sub-surface wetting is more closely imitated when aggregates adsorb water gently under tension or vacuum. Methods of this type are discussed by Williams *et al.* (1966) and ones in which the stability of individual crumbs is assessed by a simulated rainfall 'falling drop' technique are given by Low (1954, repeated in de Boodt, 1967). The limitations of conventional techniques are discussed by Emerson (1954) who suggests a method in which, after aggregates are saturated by a strong NaCl solution, their stability is assessed by leaching them with a gradually reducing strength of NaCl until they collapse. In clay-rich soils a method of determining aggregate stability has been suggested that is dependent on an index obtained from proportions of clay obtained on dispersal with and without chemical pre-treatment and dispersing media (Harris, 1971). Resistance of dry aggregates to wind erosion can be assessed (Chepil, 1958, 1962) by determining the rate at which a chosen aggregate size class breaks down with time on prolonged dry rotary sieving. Bryan (1971) has made a comprehensive study of the accuracy with which various indices of aggregate stability reflect differences between soils.

In general the standardization of results in the whole field of structural unit measurement between different laboratories, even when the same technique is employed, has proved difficult. One of a wide number of possible variations on the general principles can be adopted and used as an internally consistent system, using a fair level of replication, and only assuming that major differences in size ranges and stabilities are significant.

6.3.5 Soil moisture

The determination of soil water status is one of the most important fields of soil analysis to the ecologist. General discussions of the subject are to be found in the chapters on water in Black (1968), Marshall (1959) and Rose (1966), while more detailed discussions and descriptions are given in the relevant chapters in Black (1965). Four main aspects of water relationships in soils are involved; the rate at which water moves into and infiltrates through soils; the quantity of water in a soil; the force with which the water is held; and, for the botanist, related to all these aspects, the availability of soil moisture to plants. Chapter 5 discusses (Section 4) the water relationships of plants from the physiological viewpoint.

(a) *Infiltration and movement of water through the soil*

Bertrand (in Black, 1965) emphasizes that many of the methods proposed for the study of infiltration have been developed for special purposes and may not be widely applicable. Childs (1969) points out that, although laboratory methods are capable of refinement and control, and are suitable for the study of principles, field determinations are to be preferred when the results need to be applied to field situations.

A distinction must be made between determinations in soils that are already saturated with water and those in which this condition must be achieved artificially. Methods for both situations are described by Boersma (in Black, 1965). Infiltration rates in a saturated soil can be measured, either by pumping down the water level and observing the consequent effect upon lowering of the water table in relation to a known pumping rate, or by recording the rate of recovery of the water table when pumping has been stopped. Where the soil is not fully saturated the simplest method involves measurement of the infiltration rate from an artificially flooded surface. A metal cylinder of quite large diameter is driven partially into the soil and allowed in part to project above the soil surface. After initial saturation the rate of infiltration into the soil is observed, as a determination of the time for a known volume of water added to the tube to pass through a known surface area (e.g. see Hills, 1970). This procedure can be repeated with the cylinder inserted to different depths or into different horizons of the soil profile. Knapp (1973) has described the complexities involved in a fully instrumented installation for determining water flow in the soil on a horizon basis.

Soil leachate studies using lysimeters to measure rates of flow of water through soils in the field have been described by several authors (e.g. Bourgeois & Lavkulich, 1972a, 1972b, and see also Chapter 4, Section 4.3.1). Reynolds (1966) has described the use of fluorescent dyes to demonstrate and measure patterns of percolation of rainwater through the soil.

Where the depth, or fluctuations, of the water table are of interest sets of access tubes can be inserted to different depths in the soil. Such installations and examples of their use in the study of the water table are described by Fourt (1961), Thomasson & Robson (1967) and Thomasson (1971). Rutter (1955), working upon the relationship between wet heath vegetation and the depth of the water table, employed batteries of five tubes driven into the ground to different depths at intervals along a transect. The height of the water in each of the tubes was measured with an electrical probe and expressed as height above some common datum. The differences in height represented differences in hydrostatic potential and lines of equipotential were drawn on a section along the transect (Fig. 6.9). Potential water movement can be indicated by drawing

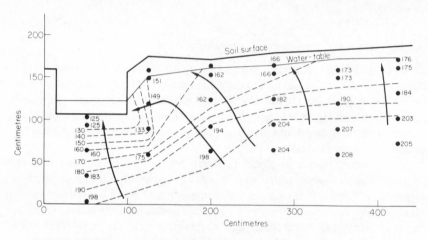

Figure 6.9 Measured hydrostatic potentials, lines of equipotential and ground-water flow, in vicinity of ditch (from Rutter, 1955).

arrows at right angles to the equipotential lines in the direction of falling potential. If satisfactory measurements of permeability can be obtained then estimates of flow rates can be calculated. Where measurement of the depth of the water table is required it is important that the access tube or soil pit does not go too deep and tap a source of higher hydrostatic potential, when the water will rise higher in the pit than in the surrounding soil. Ideally the water table should be measured by means of tubes inserted to different depths and the water table taken as the level obtained in the shallowest tube containing free water. This will not always be possible when a number of sites are to be monitored, and the possible errors in determining water tables in some soils by access tubes should be recognized (e.g. Visvalingam, 1964).

(b) *Determination of the water content and status of soil*

Actual moisture content would seem at first to be a much simpler determination than considerations of water flow through soil but this in fact is far from straightforward. Comprehensive accounts of the determination of water content; the physical condition of soil water and the energy with which it is held; and the concepts of soil water capacity and availability; are given by Gardner, Richards, and Klute and Peters respectively (all in Black, 1965). Clearly only an outline of some aspects can be included here.

Soil moisture content is generally defined as the water content of the soil that can be driven off by heating a sample at 105°C. It can be expressed in terms of volume or dry weight of soil. Although less careful sampling is required to determine soil moisture on a dry weight basis, it is often considered more meaningful to determine moisture content on a soil volume basis. In order to express soil moisture upon a volume basis (*cf.* soil nutrients) it is necessary to make the determinations on samples of known volume or to know the bulk density of the soil (see Section 6.3.2 and Chapter 4, Section 4.1).

Gravimetric methods

Gravimetric sampling and analysis is the simplest method of measuring soil moisture and provides the standard reference technique for calibrating other methods. A soil sample, if possible of known field volume, is taken, weighed at the earliest opportunity, and then reweighed after it has been dried in an oven at 105°C. The advantages of the method are that it is cheap, simple and requires no calibration; the disadvantages are that each sample differs physically because of soil heterogeneity, and, where regular measurements are required, sampling can produce damage and physical changes within the experimental site. Reynolds (1970) has discussed methodological aspects of the gravimetric determination of soil water.

Neutron scatter

The neutron scatter technique of measuring soil moisture (reviewed by Visvalingam & Tandy, 1972; Bell, 1973) depends upon the fact that when fast high energy neutrons are emitted into the soil they collide elastically with the various atomic nuclei with a loss of kinetic energy. This loss is greatest when a neutron collides with a particle of similar mass. The hydrogen nuclei are the particles nearest in mass to the neutron, and the rate of conversion of fast neutrons to slow (thermal) neutrons is proportional to the hydrogen content of the soil. Although hydrogen occurs in organic matter and in the structural water of clay and hydroxide minerals, for other than peat soils most of the soil hydrogen is contained in water so that the count rate of slow neutrons obtained is proportional to the soil moisture content. Calibration of the method must be carried out, for each soil and soil horizon studied, against gravimetric determinations of soil moisture. Once calibrated, a fast neutron source and slow neutron detector can be lowered into permanently installed access tubes to provide rapid and repeatable measures of moisture content. The method will give only approximate values if the sphere of influence of the probe, approximately 120 mm in wet soil and 200 mm in very dry soil, breaks the soil surface. The apparatus, although portable, is expensive, and, because of its nature, must be used with care and according to the relevant safety regulations.

Other methods for measuring soil moisture

A wide range of methods other than those described have been used to measure soil moisture (see e.g. the review by Shaw & Arble, 1959). These include the electrical capacitance of the soil (de Plater, 1955; Thomas, 1966), its thermal conductivity (e.g. Cornish *et al.,* 1973), the resistance of a soil to penetration, and a variety of chemical reactions.

(c) *Availability of soil water*

Water availability to plants depends not only on the quantity of water in a soil but also on on the energy with which this water is held by the soil matrix. Water is held in the soil pores under a soil moisture tension, only pressures (or suctions) in excess of this tension being able to extract the water. The energy required to remove water from a soil pore is greater the smaller the pore, so that two soils with different pore-size distributions have different water availabilities at any given moisture content, and, as a single soil dries, successive equal increments of water require more energy for their withdrawal because the remaining water is held in smaller pores and as films on particles. Thus considerations of water relationships require not only measures of the absolute water content, but of the force with which this water is held. Schofield (1935) introduced, by analogy with pH, the term pF as $\log_{10}$ (suction in cm of water), where 'suction' is the force needed to remove water from a soil. A pF curve is a plot of water content against suction which gives a picture of the moisture release characteristics of a soil over its range of 'available water'. Suction levels range from $\leqslant 0$ when the soil is saturated (pF 0), to *c.* 10^7 cm of water suction (pF 7·0) when the soil is oven-dry. Individual soils or soil horizons have their individual pF curves.

The range of soil water available to plants extends between two arbitrarily defined states. Considering surface increments of water to a soil, these percolate through the soil and progressively fill pore spaces. When further supply of water to the soil surface ceases, water drains away rapidly from the larger pores to a state in which gravity drainage rate falls off sharply and tends to zero; at this stage a soil is said to be at 'field capacity'. This corresponds to *c.* one-third atmosphere pressure (pF *c.* 2·0). Plants can abstract water over a range of water suctions until 'wilting point' is reached at a tension of about 15 atmospheres (pF *c.* 4·0). The quantity of water between these somewhat indeterminate 'points' is termed the 'available water capacity'. The wilting point varies with the plant species and the unsaturated conductivity of the soil. There is no fixed relationship between soil moisture tension and the soil moisture content expressed in gravimetric terms. The relation depends upon the soil and whether the soil is wetting or drying. Calibration curves used to convert tensions to water content must be constructed for both situations.

The measurement of pF (or tension) curves in the laboratory requires a series of methods to cover the full range of soil moisture tensions. The water content at tensions up to almost an atmosphere (pF 3) can be measured on a suction plate apparatus in which the soil sample is placed in close contact with a fine sintered disc, and moisture equilibrium is then established with water at a known tension. After equilibrium at each stage, the moisture content of the soil is determined gravimetrically. Pressure membrane apparatus can be used to measure water contents at much higher tensions. The actual range of such apparatus depends upon the particular model in use but the principle is most useful for tensions in the range of the wilting point (15 atmospheres). The soil sample rests on a cellulose membrane supported by a sintered bronze disc in a stainless steel pressure chamber. Pressure from a gas cylinder is applied to the chamber to produce a pressure difference across the cellulose membrane. The soil sample is then allowed to equilibrate with water at atmospheric pressure below the membrane. The measurement of water contents at tensions in excess of the wilting point can be carried out in vacuum desiccators (Croney *et al.*, 1952) by allowing the soil samples to equilibrate in atmospheres of known relative humidity.

The measurement of soil moisture tension in the field depends upon the range of tensions that will be encountered. The basic type of tensiometer consists of a porous pot buried in the soil, filled with air-free water and connected to a manometer or vacuum gauge. Direct continuity exists between the water column in the tensiometer and the soil, with soil moisture tension being measured directly by the apparatus. Tensiometers will yield accurate results over a range from zero to about 0·8 atmosphere (pF 0 to about 3) but at higher tensions the system breaks down when air enters the apparatus through the porous pot. A number of commercial tensiometers are available but most are designed essentially for agricultural use. Practical details concerning the use of tensiometers can be found in a number of papers (e.g. Richards, 1965; Croney *et al.*, 1952), and a tensiometer installation that can be assembled from readily available components is described by Webster (1966). A small tensiometer suitable for assessment of short-term changes over small distances has been described by Rogers (1974). Manually read tensiometers are cheap and simple to operate, but can be automated by means of a suitable pressure transducer. Such an installation may be only of limited value as tensiometers require regular inspection and servicing to remove any air bubbles if reliable results are to be obtained.

Soil water tension can be determined over a wide range by measurement of vapour pressure *in situ,* or in the laboratory, using thermocouple psychrometers (Richards in Black, 1965; Globus, 1972). Spanner (1951) used this technique to measure water potentials in plants. Modifications, and applications to comparative plant-soil studies, have been described by other workers (e.g. Rawlins, 1966; Rawlins *et al.,* 1968; Lang, 1967; Brown & Haveren, 1972; Easter & Sosebee, 1974).

A cheap semi-quantitative assessment of moisture status, related to soil moisture tension, can be obtained from resistance blocks buried in the soil (Bouyoucos, 1953, 1954). These blocks are made of porous material and contain a pair of electrodes. The moisture content of the buried blocks depends upon the passage of water from the soil into the porous medium of the resistance block, and this depends both on moisture content and soil water tension. The water content of the block is determined by measuring the electrical resistance between the electrodes and having a calibration of resistance against moisture. Gypsum ($CaSO_4 . 2H_2O$) blocks are most useful in dryer soils (tensions greater than 1 atmosphere), but blocks made of nylon or fibreglass (Bouyoucos & Mick, 1948; Bouyoucos, 1972) have somewhat better characteristics at lower tensions. The accurate calibration of resistance blocks presents a number of problems and they are probably best used as general indicators of soil moisture status rather than as indicators of absolute soil moisture. Errors in the method are caused by lack of contact between block and soil, by the slow response time, hysteresis effects and by variations in soil water salinity in the case of nylon blocks. The construction of a simple AC bridge suitable for measuring resistance blocks in the field is described by Hinson & Kitching (1964).

Rutter & Sands (1958) used a modification of a method described by Davis & Slater (1942) to measure soil moisture in their study of the water balance of *Pinus sylvestris.* Wedge shaped blocks of gypsum were inserted into cavities lined with filter paper, at the bottom of tubes sunk into the soil. The moisture contents of the wedges were in equilibrium with the soil and determined by weighing as and when required. A variant using filter paper in contact with the soil has been examined (Al-Khafaf & Hanks, 1974).

Indirect determination of availability of soil water to plants

Available water capacity, as defined above, as the quantity of water that can be held in a soil between field capacity and wilting point, can be estimated indirectly from data for particle size distribution and organic matter content of a soil (Salter & Williams, 1965a, b; Salter *et al.,* 1966; Salter & Williams, 1967).

6.3.6 Soil accretion and erosion

Although the instability of soil surfaces due to accumulation of soil-forming sediments, or to erosion of existing soils, may interest ecologists, techniques for study are geomorphological (see e.g. King, 1966) and cannot be considered in detail here. Carson & Kirkby (1972) included detailed consideration of the instability of soil masses and of surface-water erosion on slopes, emphasizing the importance of the work of soil mechanics engineers, both in theory and methodology. Many techniques used in soil erosion control (see e.g. Hudson, 1971) that are of vital land-use importance in many regions can be significant for ecologists where erosion is a natural hazard.

Substantial rates of accretion giving a more-or-less continuous build-up of soil can occur through sedimentation, as in salt marshes, or by wind-blow, as in the formation of dunes. Accretion rates can be studied relatively and indirectly by stratigraphic examination of buried horizons and textural changes, and directly by sediment trapping, by the movement of tracers, and by periodic re-examination of markers fixed on or below a known surface level. Laboratory methods for studying relative erodibility can compare soils under different vegetation and land-use (e.g. Selby, 1970).

Erosion of material from one facet of a landform unit, sometimes involving localized deposition elsewhere, can result from mass-movements such as landslips, creep of the soil mantle on a slope (Young, 1960) and mass flow during seasonal thaw of frozen ground; or from particulate movement such as slope-wash of water-eroded material (Bryan, 1968, 1968–69, 1970); wind-blow; and sorting of size fractions by differential vertical and lateral heave due to frost. Active wind and frost-induced erosion, by particulate and mass-movements, can separately or together create regular and irregular patterning of vegetated and bare ground. Washburn (1973) gives a world text on frost-patterned ground; Ball & Goodier (1970, 1974) discuss patterned ground features in Snowdonia and Shetland. This patterning is an important local ecological factor, and the inter-action of climate, soil and micro-relief can be associated with regular distribution of plant species within the vegetated zones. This is true of frost and wind-affected ground described in studies such as the Cairngorm work of Watt & Jones (1948), Metcalfe (1950), and Burges (1951); and also of areas where mass-movement and biotic influences give micro-relief patterning on slopes, such as the terracettes discussed by Carson & Kirkby (1972).

Soil patterning due to sorting and movement under former different climatic conditions can also have a profound effect upon the present vegetation, as in the Breckland grass heathlands of eastern England where species such as *Calluna* are present over the areas of deeper sandy drift that occur in solifluction patterns, while species characteristic of calcareous soils overlie the chalky drift (Watt *et al.*, 1966).

6.3.7 Mechanical strength

Measurements of soil mechanical strength are primarily the concern of civil engineers (e.g. British Standards Institution, 1967; A.S.T.M., 1970) in relation to the capacity of soil material to sustain engineering loads, but can have ecological application.

Compactibility (Felt, in Black, 1965) is generally considered by engineers with regard to controlling the porosity and cohesion of a soil to achieve maximum mechanical strength and stability. Compaction can be produced by animal and human activity on areas of ecological interest, and methods for compactibility measurement may be applicable in the study of such areas.

Stress and shear strength measurements (Barber & Sallberg, in Black, 1965) can be of interest where soil movement is of ecological importance, as on unstable slopes under different types of vegetation.

The most useful measurement of soil strength for the pedologist and ecologist is probably the resistance of the soil to penetration. A rod or cane can be pushed into a soil as a rough penetrometer to measure the depth of the characteristic very non-cohesive open structure of frost-heaved soils, or of peat over mineral soil. Quantitatively, penetrometer tests of various levels of sophistication have been developed to measure the force needed to achieve a standard degree of penetration with a probe (Davidson, in Black, 1965; Reynolds, 1971; Sanglerat, 1972; Soane, 1973). Although such measures are clearly related to other parameters such as porosity and bulk-density, they provide a quick method of examining the effects of trampling and machinery.

6.4 MINERALOGICAL AND FABRIC ANALYSES

Laboratory investigations of the mineralogy of soils, and of the inter-relationships of component particles, aggregates and voids investigated on a micro-scale by thin-section techniques, are of specialist pedological interest. Thin-section studies can provide information directly applicable by soil ecologists but in general the plant ecologist need only be aware of the range of techniques that are available.

6.4.1 Thin-section techniques

Mineralogical and fabric studies of rocks using the polarizing microscope are the classic laboratory technique of the petrologist. Mineral identification of individual grains, or of constituent minerals in rock thin sections, depends on a knowledge of the optical properties of rock-forming minerals (e.g. Kerr, 1959). Thin sections can be prepared directly from massive rocks but friable rocks require an initial impregnation with a resin compound to give adequate mechanical strength during grinding and polishing of a section.

Impregnation is even more essential before thin sections of soils can be made. Microscopic study of soils was first widely discussed by Kubiena (1938) who later (1953, 1970) used the results of these methods to interpret soil-forming processes and assist soil classification. The most comprehensive modern view of soil fabric studies is given in the work of Brewer (1964). The development of a soil sequence in relation to weathering, leaching, and vegetation type has been studied by thin section methods by Barratt (1965). Although thin-sections provide one of the few direct means of detailed study of the inter-relationships of the different components of the soil body in their undisturbed state, there is not yet a sufficient body of data to allow their use in pedological studies as comprehensively as is possible for rock sections in petrology.

Direct observation of the fine detail of the relationship between biological and physical soil components is possible through thin-sections. An early example of this is the illustration by Hepple & Burges (1956) of a fungal hypha bridging a soil pore as it crosses between humus coatings on two quartz sand-grains. Barratt (1964, 1968–69) deals with the classification of soil humus forms from their micro-fabric, and also has related humus forms, classified from a micro-morphological approach, to land-use differences on a number of New Zealand soils (1968). Anderson & Bouma (1973) have correlated hydraulic conductivity of soil horizons with measurements made on thin-sections. The combination of thin-sections and electron probe micro-analysis allows the possibility of studying chemical differentiation on a micro-scale (Cescas *et al.*, 1968; Jenkins, 1970). A further discussion of the use of soil thin sections is given in Chapter 4, Section 3.6.1.

6.4.2 Sand-grain mineralogy

The main purposes of qualitative and quantitative identification of the mineral species of the sand fraction of soils are to determine the nature of the soil parent material, to follow the effects of weathering within a soil profile, and to compare profiles derived from the same parent material (Cady, in Black, 1965). Dispersion of soil aggregates into their primary particles, as required for determination of particle size distribution (6.3.1), is an essential preliminary to mineralogical investigations. Size fractions are separated by pipetting or decanting and are then centrifuged. Coarser fractions can be obtained by washing the material on sieves. It may be necessary to clean this material by removing organic matter, carbonates or iron hydroxide cements and coatings (e.g. Kunze, in Black, 1965).

A separated sand fraction is sub-divided into 'light' and 'heavy' minerals by flotation in a heavy liquid, usually bromoform. The light fraction, dominated by quartz, is larger in quantity, but less diagnostic of parent material origin, though occasionally the shape or surface character of grains can be helpful. The subordinate heavy fraction generally contains a wider range of minerals among which are those types which can be used to identify the parent material. Study of sand grain material is covered comprehensively by Milner (1962), while Marshall (1964) includes a chapter reviewing the mineralogy and chemistry of the major minerals of the sand and silt fractions of soils.

6.4.3 Silt and clay mineralogy

The microscope techniques applicable to mineralogy of sand grains can be used, though less easily because the particles are smaller, for examination of the silt fraction. This can also be investigated by the non-optical methods needed for study of the very small particles which compose the soil clay fraction. Clay minerals are not simply smaller particles of those minerals that occur in the sand fraction, but are a suite of generally sheet-structure alumino-silicates (see e.g. Deer *et al.,* Vol. III, 1962). Some crystalline oxides and hydroxides and some amorphous minerals contribute to the soil clay assemblage and are locally important, but the clay fraction is dominated by small flakes of sheet-structure minerals. The silt size fraction is intermediate, in both composition and size, containing small particles of the minerals found in the sand fraction together often with 'large' crystals of the clay minerals.

Clay minerals and their properties and effects on soil character are discussed in detail by Grim (1962, 1968), their chemistry is reviewed by Weaver & Pollard (1973), and soil clays are considered in a seminar edited by Rich & Kunze (1964). Clay minerals have a more positive role in soils than the sand material since, with organic matter, they possess the property of cation exchange and contribute to the retention of nutrient cations in a form relatively resistant to leaching, but accessible to the solid-solute-root interchanges of the soil-plant nutrient cycle. Identification of the main groups of soil clay minerals is mainly by X-ray diffraction (Brown, 1961) supported by differential thermal analysis (Mackenzie, 1957). Quantitative analysis of the proportion of different clays in mixed assemblages (the general case for soil clays) depends on arbitrary methods and has relative rather than absolute meaning. Clay mineral identification and the study of clay properties can assist understanding of nutrient régimes in soils, and of some other important soil properties, such as the swelling and contraction which occurs with 'expanding' clays on wetting and drying. Knowledge of the clay mineral species and interstratified types present in a particular soil does not however usually have any direct application by the plant ecologist.

7 CONCLUSION

The soil with its biotic and abiotic characteristics is certainly a world, as Russell (1971) described it. This world is not of course a closed unit because its properties depend on, and interact with, the physical aspects of its external environment, and the flora and fauna which live on, as well as in, it. To the plant ecologist the soil is primarily a physical and chemical support system for plants. It is not a simple system, but is a complex, shadowy and quantitatively ill-defined medium that can be an embarassment to the research worker who wants quantitative answers to all his questions.

There are two main approaches: either an attempt can be made to simplify the soil world by studying individual fragments in isolation, hoping for a subsequent synthesis of the jigsaw; or one accepts a present ignorance of detail and aims for a better general but almost intuitive grasp of broad principles and relationships, until new concepts and techniques enable a better understanding of the whole. For the plant ecologist, the former approach includes a study of the plant under simpler and experimental culture conditions which may help to explain its behaviour in a real soil. The latter approach accepts that we can at present, in many cases, more convincingly correlate plant distribution and performance with soils classified at sub-group or series level, than we can with specific soil quantitative data. This is because the concept of the soil class contains within it factors which we can and do measure, as well as factors that we either only appreciate but cannot measure, and inter-relationships that have not yet been grasped. Both lines of investigation are necessary for the plant ecologist and the soil scientist who have an extensive, complex, and only marginally explored field in the study of soils and their relationships with plant distribution and growth.

8 REFERENCES

ADAMS W.A. (1973) The effect of organic matter on the bulk and true densities of some uncultivated podzolic soils. *J. Soil Sci.* **24,** 10–17.

AHN P.M. (1970) West African Soils (*West African Agriculture Vol. 1*) Oxford Univ. Press.

AL-KHAFAF S. & HANKS R.J. (1974) Evaluation of the filter paper method for estimating soil water potential. *Soil Sci.* **117,** 194–199.

ANDERSON G.D. & TALBOT L.M. (1965) Soil factors affecting the distribution of the grassland types and their utilization by wild animals on the Serengeti Plains, Tanganyika. *J. Ecol.* **53,** 33–56.

ANDERSON J.L. & BOUMA J. (1973) Relationships between saturated hydraulic conductivity and morphometric data of an argillic horizon. *Soil Sci. Soc. Amer. Proc.* **37,** 408–413.

A.R.C. (1971) *Annual report of the Agricultural Research Council,* 1970–71. London, H.M.S.O.

ARMITAGE P.L. (1973) *Aber Mountain: A land and land-use study.* M.Sc. thesis, University College, London.

A.S.T.M. (1970) *Special Procedures for Testing Soil and Rock for Engineering Purposes.* American Society for Testing and Materials. Philadelphia.

A.S.T.M. (1972) *Manual on Test Sieving Methods.* American Society for Testing and Materials. Philadelphia.

AUBERT G. & DUCHAUFOUR PH. (1956) Projet de classification des sols. *6th International Congress of Soil Science,* Paris. Vol. E, pp. 597–604.

AUBERT G. (1965) Classification des sols. *Pédologie–Cahiers ORSTOM* III, **3**, 269–288.

AVERY B.W. (1973) Soil classification in the Soil Survey of England and Wales. *J. Soil Sci.* **24**, 324–338.

BALL D.F. & GOODIER R. (1970) Morphology and distribution of features resulting from frost-action in Snowdonia. *Fld. Stud.* **3**, 193–218.

BALL D.F. & GOODIER R. (1974) Ronas Hill: a preliminary account of its ground pattern features resulting from the action of frost and wind. In R. Goodier (ed.) *The Natural Environment of Shetland* pp. 89–106. Edinburgh, Nature Conservancy Council.

BALL D.F. & WILLIAMS W.M. (1968) Variability of soil chemical properties in two uncultivated Brown Earths. *J. Soil Sci.* **19**, 379–391.

BALL D.F. & WILLIAMS W.M. (1971) Further studies on variability of soil chemical properties: Efficiency of sampling programmes on an uncultivated Brown Earth. *J. Soil Sci.* **22**, 60–68.

BALL D.F., HORNUNG M. & MEW G. (1971) The use of aerial photography in the study of geomorphology and soils of upland areas. In R. Goodier (ed.) *The Application of Aerial Photography to the Work of the Nature Conservancy* pp. 66–77. Edinburgh, Nature Conservancy Council.

BALL D.F., MEW G. & MACPHEE W.S.G. (1969) Soils of Snowdon. *Fld. Stud.* **3**, 69–107.

BALL D.F., WILLIAMS W.M. & HORNUNG M. (1971) Variability of chemical properties in 'uniform' soils. *Welsh Soils Discussion Group, Report No.* **11**, 31–40.

BARKHAM J.P. & NORRIS J.M. (1970) Multivariate procedures in an investigation of vegetation and soil relations of two beech woodlands, Cotswold Hills, England. *Ecology* **51**, 630–639.

BARRATT B.C. (1964) A classification of humus forms and microfabrics of temperate grasslands. *J. Soil Sci.* **15**, 342–356.

BARRATT B.C. (1965) Micro-morphology of some yellow-brown earths and podzols of New Zealand. *N.Z. J. agric. Res.* **8**, 997–1042.

BARRATT B.C. (1968) Micromorphological observations on the effects of land-use differences on some New Zealand soils. *N.Z. J. agric. Res.* **11**, 101–130.

BARRATT B.C. (1968/69) A revised classification and nomenclature of microscopic soil materials with particular reference to organic components. *Geoderma* **2**, 257–271.

BAVER L.D. (1956) *Soil Physics* (3rd edition). Wiley.

BAVER L.D., GARDNER W.H. & GARDNER W.L. (1972) *Soil Physics* (4th edition). Wiley.

BECKETT P.H.T. & BURROUGH P.A. (1971a) The relation between cost and utility in soil survey. IV. Comparisons of the utilities of soil maps produced by different soil survey procedures and to different scales. *J. Soil Sci.* **22**, 466–480.

BECKETT P.H.T. & BURROUGH P.A. (1971b) The relation between cost and utility in soil survey. V. The cost-effectiveness of different soil survey procedures. *J. Soil Sci.* **22**, 481–489.

BECKETT P.H.T. & WEBSTER R. (1965) *A classification system for terrain. MEXE Report No.* **872**, Christchurch, Hampshire, M.I. Eng. Exp. Est.

BECKETT P.H.T. & WEBSTER R. (1971) Soil variability: A review. *Soils Fertil., Harpenden.* **34**, 1–15.

BECKETT P.H.T., MCNEIL G.M. & MITCHELL C.W. (1972) Terrain evaluation by means of a data bank. *Geogrl. J.* **138**, 430–456.

BELL J.P. (1969) A new design principle for neutron soil moisture gauges: The 'Wallingford' neutron probe. *Soil Sci.* **108**, 160–164.

BELL J.P. (1973) Neutron Probe Practice, *Inst. Hydrol., Rep. No. 19*. Wallingford, England.

BIDWELL O.W. & HOLE F.D. (1964) An experiment in the numerical classification of some Kansas Soils. *Soil Sci. Soc. Amer. Proc.* **28**, 263–268.

BIE S.W. & BECKETT P.H.T. (1971) Quality control in soil survey. I. The choice of mapping unit. *J. Soil Sci.* **22**, 32–49.

BISAL F. & HINMAN W.C. (1972) A method for estimating the apparent density of soil aggregates. *Can. J. Soil Sci.* **52**, 513–514.

BLACK C.A. (editor-in-chief) (1965) *Methods of Soil Analysis.* Amer. Soc. Agron.

BLACK C.A. (1968) *Soil-Plant Relationships.* Wiley.

BOURGEOIS W.W. & LAVKULICH L.M. (1972a) A study of forest soils and leachates on sloping topography using a tension lysimeter. *Can. J. Soil Sci.* **52,** 375–391.

BOURGEOIS W.W. & LAVKULICH L.M. (1972b) Application of acrylic plastic tension lysimeters to sloping land. *Can. J. Soil Sci.* **52,** 288–290.

BOUYOUCOS G.J. (1953) More durable plaster of paris moisture blocks. *J. Soil Sci.* **76,** 447–451.

BOUYOUCOS G.J. (1954) New type electrode for plaster of paris moisture blocks. *J. Soil Sci.* **78,** 339–342.

BOUYOUCOS G. (1972) A new electrical soil-moisture measuring unit. *J. Soil Sci.* **114,** 493.

BOUYOUCOS G.J. & MICK A.H. (1948) Fabric absorption unit for continuous measurement of soil moisture in the field. *J. Soil Sci.* **66,** 217–232.

BREWER R. (1964) *Fabric and Mineral Analysis of Soils.* Wiley.

BRIDGES E.M. (1970) *World Soils.* Cambridge University Press.

BRITISH STANDARDS INSTITUTION (1967) *Methods of Testing Soils for Engineering Purposes.* British Standard 1377, London.

BROWN G. (1961) *The X-ray Identification and Crystal Structure of Clay Minerals.* London, Miner. Soc.

BROWN R.W. & VAN HAVEREN B.P. (editors) (1972) *Psychrometry in Water Relations Research.* Utah State University.

BRYAN R.B. (1968) Development of laboratory instrumentation for the study of soil erodibility. *Earth Sci. J.* **2,** 38–50.

BRYAN R.B. (1968/69) The development, use and efficiency of indices of soil erodibility. *Geoderma* **2,** 5–26.

BRYAN R.B. (1970) An improved rainfall simulator for use in erosion research. *Can. J. Earth Sci.* **7,** 1552–1561.

BRYAN R. B. (1971) The efficiency of aggregation indices in the comparison of some English and Canadian soils. *J. Soil Sci.* **22,** 166–178.

BUCKMAN H.O. & BRADY N.C. (1969) *The Nature and Properties of Soils.* Macmillan.

BUNTING B.T. (1967) *The Geography of Soil.* London, Hutchinson.

BURGES A. (1951) The ecology of the Cairngorms. III. The *Empetrum-Vaccinium* zone. *J. Ecol.* **39,** 271–284.

BURGES A. & RAW F. (1967) *Soil Biology.* Academic Press.

BUOL S.W., HOLE F.D. & MCCRACKEN R.J. (1973) *Soil Genesis and Classification.* Iowa State University Press.

BURROUGH P.A., BECKETT P.H.T. & JARVIS M.G. (1971) The relation between cost and utility in soil survey (I–III). *J. Soil Sci.* **22,** 359–394.

CAMPBELL D.J. (1973) A flotation method for the rapid measurement of the wet bulk density of soil clods. *J. Soil Sci.* **24,** 239–243.

CAMPBELL N.A., MULCAHY M.J. & MCARTHUR W.M. (1970) Numerical classification of soil profiles on the basis of field morphological properties. *Aust. J. Agric. Res.* **8,** 43–58.

CARROLL D.M. (1973) Remote sensing techniques and their application to soil science. Pt. I. The photographic sensors. *Soils Fertil., Harpenden* **36,** 259–266.

CARSON M.A. & KIRKBY M.J. (1972) *Hillslope Form and Process.* Oxford University Press.

CARY J.W. & HAYDEN C.W. (1973) An index for soil pore size distribution. *Geoderma* **9,** 249–251.

CESCAS M.P., TYNER E.H. & GRAY L.J. (1968) The electron microprobe X-ray analyzer and its use in soil investigations. *Advan. Agron.* **20,** 153–198.

CHEPIL W.S. (1958) *Soil Conditions that Influence Wind Erosion. Tech. Bull. 1185,* Washington, United States Dept. of Agric.

CHEPIL W.S. (1962) A compact rotary sieve and the importance of dry sieving in physical soil analysis. *Soil Sci. Soc. Amer. Proc.* **26,** 4–6.

CHILDS E.C. (1969) *An Introduction to the Physical Basis of Soil Water Phenomena.* Wiley.

CLARKE G.R. (1971) *The Study of Soil in the Field* (5th edition, assisted by P. Beckett). Oxford, Clarendon Press.

COOKE G.W. (1967) *The Control of Soil Fertility.* Crosby Lockwood.

COPE D.W. (1973) *Soils in Gloucestershire* I. *Soil Survey Record No. 13.* Harpenden, Soil Survey of England and Wales.

CORNISH P.M., LARYEA K.B. & BRIDGE B.J. (1973) A non-destructive method of following moisture content and temperature changes in soils using thermistors. *Soil Sci.* **115**, 309–314.

COURTNEY F.M. (1973) A taxonometric study of the Sherborne soil mapping unit. *Trans. Inst. Brit. Geog.* **58**, 113–124.

CRONEY D., COLEMAN J.D. & BRIDGE P.M. (1952) The suction of moisture held in soil and other porous materials. *D.S.I.R. Road Res. Tech. paper 24.* H.M.S.O.

CRUIKSHANK J.G. (1972) *Soil Geography.* David & Charles.

CUANALO DE LA C, H.E. & WEBSTER R. (1970) A comparative study of numerical classification and ordination of soil profiles in a locality near Oxford. Pt. 1. Analysis of 85 sites. *J. Soil Sci.* **21**, 340–352.

CURTIS L.F., DOORNKAMP J.C. & GREGORY K.J. (1965) The description of relief in field studies of soils. *J. Soil Sci.* **16**, 16–30.

DAVIS W.E. & SLATER C.S. (1942) A direct weighing method for subsequent measurements of soil moisture under field conditions. *J. Amer. Soc. Agron.* **34**, 285.

DE BOODT M. (Secretary-general, editing committee) (1967) *West-European Methods for Soil Structure Determination.* Gent, Belgium.

DEER W.A., HOWIE R.A. & ZUSSMAN J. (1962) *Rock-forming Minerals* (5 vols.) Longmans.

DEWIS J. & FREITAS F. (1970) *Physical and Chemical Methods of Soil and Water Analysis.* Rome, F.A.O.

DUCHAUFOUR PH. (1956) *Pédologie: Applications Forestières et Agricoles.* Nancy, Ecole Nat. des Eaux et Forêts.

DUCHAUFOUR PH. (1970) *Précis de Pédologie* (3rd edition). Nancy, Ecole Nat. des Eaux et Forêts.

EASTER S.J. & SOSEBEE R.E. (1974) Use of thermocouple psychrometry in field studies of soil-plant-water relationships. *Plant and Soil* **40**, 707–712.

EMERSON W.W. (1954) The determination of the stability of soil crumbs. *J. Soil Sci.* **5**, 233–250.

EPSTEIN E. (1972) *Mineral Nutrition of Plants: Principles and Perspectives.* Wiley.

EVANS R. (1972) Air photographs for soil survey in lowland England: Soil patterns. *Photogrammetric Rec.* **7**, 302–322.

FALCK J. (1973) A sampling method for quantitative determination of plant nutrient content of the forest floor. [Swedish with English summary.] *Research Notes No. 1.* Stockholm, Royal College of Forestry.

F.A.O. (1968) *Guidelines for Soil Description.* Rome.

FIRTH J.G. (1973) The aerial photograph—not a map yet more than a map. *Scott. For.* **27**, 336–344.

FITZPATRICK E.A. (1971) *Pedology. A Systematic Approach to Soil Science.* Edinburgh, Oliver & Boyd.

FOURT D.F. (1961) The drainage of a heavy clay site. *Forestry Commission. Report on Forest Research for the year ended March, 1960,* pp. 137–150. London, H.M.S.O.

FRANKLAND J.C., OVINGTON J.D. & MACRAE C. (1963) Spatial and seasonal variations in soil, litter and ground vegetation in some Lake District woodlands. *J. Ecol.* **51**, 97–112.

GALLAGHER P.A. & HERLIHY M. (1963) An evaluation of errors associated with soil testing. *Irish J. agric. Res.* **2**, 149–167.

GARRET S.D. (1963) *Soil Fungi and Soil Fertility.* Pergamon.

GIBBS H.S., COWIE J.D. & PULLAR W.A. (1968) *Soils of New Zealand.* Pt. 1. Soil Bureau Bulletin 26, New Zealand Dept. of Sci. and Indust. Res., Auckland.

GLOBUS A.M. (1972) Design, operation and temperature sensitivity of a thermocouple psychrometric moisture potentiometer based on the Peltier effect. *Soviet Soil Sci.* **4,** 745–752.

GOODIER R. & GRIMES B.H. (1970) The interpretation and mapping of vegetation and other ground surface features from air photographs of mountainous areas in North Wales. *Photogramm. Rec.* **6,** 553–566.

GOOSEN D. (1967) Aerial photo interpretation in soil survey. *Soils Bull. No. 6.* Rome, F.A.O.

GRAY T.R.G. & WILLIAMS S.T. (1971) *Soil Micro-organisms.* Oliver & Boyd.

GRIM R.E. (1962) *Applied Clay Mineralogy.* McGraw-Hill.

GRIM R.E. (1968) *Clay Mineralogy (2nd Edition).* McGraw-Hill.

GRIME J.P. & LLOYD P.S. (1973) *An Ecological Atlas of Grassland Plants.* Arnold.

GRUBB P.J., GREEN H.E. & MERRIFIELD R.C.J. (1969) The ecology of chalk heath: its relevance to the calcicole-calcifuge and soil acidification problems. *J. Ecol.* **57,** 175–212.

HARRIS S. (1971) Index of structure: evaluation of a modified method of determining aggregate stability. *Geoderma* **6,** 155–162.

HEPPLE S. & BURGES A. (1956) Sectioning of soil. *Nature, Lond.* **177,** 1186.

HESSE P.R. (1971) *A Textbook of Soil Chemical Analysis.* John Murray.

HILLS R.C. (1970) The determination of the infiltration capacity of field soils using the cylinder infiltrometer. *Technical Bulletin 3.* British Geomorphological Research Group, Univ. East Anglia, Norwich.

HINSON W.H. & KITCHING R.A. (1964) A readily constructed transistorised instrument for electrical resistance measurement in biological research. *J. appl. ecol.* **1,** 301–305.

HODGSON J.M. (editor) (1974) *Soil Survey Field Handbook. Technical Monograph 5.* Harpenden, Soil Survey of England and Wales.

HOLMES A. (1965) *Principles of Physical Geology.* Nelson.

HUDSON N. (1971) *Soil Conservation.* Batsford.

INTERNATIONAL SOCIETY OF SOIL SCIENCE (1962) *Transactions of Commissions IV and V joint meeting,* New Zealand.

INTERNATIONAL SOCIETY OF SOIL SCIENCE (1967) Proposal for a uniform system of soil horizon designations. *Bull. Int. Soc. Soil Sci. No. 31,* 4–7.

JARVIS M.G. (1969) Terrain and soil in North Berkshire. *Geogrl. J.* **135,** 398–403.

JARVIS R.A. (1962) The use of photo-interpretation for detailed soil mapping. *Symp. Photo Interpretation, Delft, Working Group 31,* 177–182.

JEFFREY D.W. (1970) A note on the use of ignition loss as a means for the approximate estimation of bulk density. *J. Ecol.* **58,** 297–299.

JENKINS D.A. (1970) Micromorphological heterogeneity of soil. *Welsh Soils Discussion Group, Report 11,* 1–11.

JOHN M.K., LAVKULICH L.M. & ZOOST M.A. (1972) Representation of soil data for the computerized filing system used in British Columbia. *Can. J. Soil Sci.* **52,** 293–300.

KADDAH M.T. (1974) The hydrometer method for detailed particle size analysis. I Graphical interpretation of hydrometer readings and test of method. *Soil Sci.* **118,** 102–108.

KERR P.F. (1959) *Optical Mineralogy.* McGraw-Hill.

KING C.A.M. (1966) *Techniques in Geomorphology.* Arnold.

KNAPP B.J. (1973) A system for the field measurement of soil water movement. *Technical Bulletin 9.* British Geomorphological Research Group, Univ. East Anglia, Norwich.

KRUMBEIN W.C. & PETTIJOHN F.J. (1938) *Manual of Sedimentary Petrography.* New York, Appleton-Century Crofts.

KUBIENA W.L. (1938) *Micropedology.* Ames, Iowa, Collegiate Press.

KUBIENA W.L. (1953) *Soils of Europe.* Murby.

KUBIENA W.L. (1970) *Micromorphological Features of Soil Geography.* New Jersey, Rutgers Univ. Press.

KÜHNELT W. (trans. N. Walker) (1961) *Soil Biology.* Faber & Faber.

LANG A.R.G. (1967) Psychrometric measurement of soil water potential *in situ* under cotton plants. *Soil Sci.* **106**, 460–464.

LEGROS J.-P. (1973) Précision des cartes pédologiques: La notion de finesse de caractérisation. *Sci. Sol.* **2**, 115–128.

LEO M.W.M. (1963) Heterogeneity of soil of agricultural land in relation to soil sampling. *J. agric. Fd. Chem.* **II**, 432–435.

LEOPOLD L.B. & DUNNE T. (1971) Field method for hillslope description. *Technical Bulletin 7*, British Geomorphological Research Group, Univ. East Anglia, Norwich.

LINES R. & HOWELL R.S. (1963) The use of flags to estimate the relative exposure of trial plantations. *Forestry Commission Forest Record 51*. London, H.M.S.O.

LOW A.J. (1954) The study of soil structure in the field and in the laboratory. *J. Soil Sci.* **5**, 57–74.

MACKENZIE R.C. (1957) *The Differential Thermal Investigation of Clays*. London, Miner. Soc.

MAIGNIEN R. (1970) *Manuel de Prospection Pédologique*. Paris, O.R.S.T.O.M.

MARSHALL C.E. (1964) *The Physical Chemistry and Mineralogy of Soils. Vol. I. Soil Materials*. Wiley.

MARSHALL T.J. (1959) Relations between water and soil. *Commonwealth Agric. Bur. Tech. Comm. 50*, Harpenden.

MCLAREN A.D. & PETERSON G.H. (editors) (1967) *Soil Biochemistry*. Vol. 1. Arnold.

MCVEAN D.N. & RATCLIFFE D.A. (1962) *Plant communities of the Scottish Highlands. Monograph No. 1, The Nature Conservancy*. London, H.M.S.O.

MEHLICH A. (1972) Uniformity of expressing soil test results. A case for calculating results on a volume basis. *Commun. Soil Sci. Plant Anal.* **3**, 417–424.

MEHLICH A. (1973) Uniformity of soil test results as influenced by volume weight. *Commun. Soil Sci. Plant Anal.* **4**, 475–486.

METCALFE G. (1950) The ecology of the Cairngorms. II. The Mountain *Callunetum*. *J.Ecol.* **38**, 46–74.

MEW G. & BALL D.F. (1972) Grid sampling and air photography in upland soil mapping. An investigation in the Rhinog mountains of North Wales. *Geogrl. J.* **138**, 8–14.

MILNER H.B. (1962) *Sedimentary Petrography*. Allen & Unwin.

MIRREH H.F. & KETCHESON J.W. (1972) Influence of soil bulk density and matric pressure on soil resistance to penetration. *Can. J. Soil Sci.* **52**, 477–483.

MITCHELL C.W. (1973) Soil classification with particular reference to the Seventh Approximation. *J. Soil Sci.* **24**, 411–420.

MOORE A.W., RUSSELL J.S. & WARD W.T. (1972) Numerical analysis of soils: a comparison of three soil profile models with field classifications. *J. Soil Sci.* **23**, 193–209.

MOSS R.P. (editor) (1968) *The Soil Resources of Tropical Africa*. Cambridge University Press.

MÜCKENHAUSEN E. (1962) *Entstehung, Eigenschaften und Systematik der Böden der Bundesrepublik Deutschland*. Frankfurt, D.L.G.

MÜCKENHAUSEN E. (1965) The soil classification system of the Federal Republic of Germany. *Pédologie, Gent*. Special No. 3, 57–89.

MUIR J.W. (1962) The general principles of classification with reference to soils. *J. Soil Sci.* **13**, 22–30.

MUIR J.W. & HARDIE H.G.M. (1962) A punched-card system for soil profiles. *J. Soil Sci.* **13**, 249–253.

NATIONAL SOIL SURVEY COMMITTEE OF CANADA (1968) *Proceedings of the 7th Meeting of N.S.S.C. Canada*. University of Alberta.

NIELSEN D.R., BIGGAR J.W. & COREY J.C. (1972) Application of flow theory to field situations. *Soil Sci.* **113**, 254–264.

NORRIS J.M. (1971) The application of multivariate analysis to soil studies. I. Grouping of soils using different properties. *J. Soil Sci.* **22**, 69–80.

NORRIS J.M. (1972) The application of multivariate analysis to soil studies. III. Soil variation. *J. Soil Sci.* **23**, 62–75.

NORRIS J.M. & LOVEDAY J. (1971) The application of multivariate analysis to soil studies. II. The allocation of soil profiles to established groups: A comparison of soil survey and computer methods. *J. Soil Sci.* **22**, 395–400.

NORTHCOTE K.H. (1965) *A Factual Key for the Recognition of Australian Soils.* Divisional Report 2/65. Adelaide, C.S.I.R.O.

OERTLI J.J. (1973) The use of chemical potentials to express nutrient availabilities. *Geoderma* **9**, 81–95.

PARKINSON D., GRAY T.R.G. & WILLIAMS S.T. *Methods for Studying the Ecology of Soil Micro-organisms.* I.B.P. Handbook No. 19. Blackwell Scientific Publications.

PETERKEN G.F. (1967) *Guide to the Check Sheet for I.B.P. Areas.* I.B.P. Handbook No. 4. Blackwell Scientific Publications.

PHILLIPSON J. (editor) (1971) *Methods of Study in Quantitative Soil Ecology.* I.B.P. Handbook No. 18. Blackwell Scientific Publications.

PITTY A.F. (1968) A simple device for the field measurement of hillslopes. *J. Geol.* **76**, 717–720.

PITTY A.F. (1969) A scheme for hillslope analysis. *Occ. Papers in Geog. 9, University of Hull.*

DE PLATER C.V. (1955) A portable capacitance-type soil moisture meter. *Soil Sci.* **80**, 391–395.

POORE M.E.D. (1968) Studies in Malaysian rain forest. I. The forest on Triassic sediments in Jengka Forest Reserve. *J. Ecol.* **56**, 143–196.

PRITCHARD D.T. (1974) A method for soil particle-size analysis using ultrasonic disaggregation. *J. Soil Sci.* **25**, 34–40.

PYATT D.G., HARRISON D. & FORD A.S. (1969) *Guide to site types in forests of North and Mid-Wales. Forest Record No. 69.* Forestry Commission. London, H.M.S.O.

RAGG J.M. & CLAYDEN B. (1973) *The classification of some British soils according to the comprehensive system of the United States. Technical Monograph 3.* Harpenden, Soil Survey of England and Wales.

RAWLINS S.L. (1966) Theory for thermocouple psychrometers used to measure water potential in soil and plant samples. *Agric. Met.* **3**, 293–310.

RAWLINS S.L., GARDNER W.R. & DALTON F.N. (1968) *In situ* measurement of soil and plant leaf water potential. *Soil Sci. Soc. Amer. Proc.* **32**, 468–470.

RAYNER J.H. (1966) Classification of soils by numerical methods. *J. Soil Sci.* **17**, 79–92.

RAYNER J.H. (1969) The numerical approach to soil systematics. In J.G. Sheals (ed.) *The Soil Ecosystem.* London, The Systematics Association.

READ H.H. & WATSON J. (1962) *Introduction to Geology. I. Principles.* Macmillan.

REEVE M.J., SMITH P.D. & THOMASSON A.J. (1973) The effect of density on water retention properties of field soils. *J. Soil Sci.* **24**, 355–367.

REYNOLDS E.R.C. (1966) The percolation of rainwater through soil demonstrated by fluorescent dyes. *J. Soil Sci.* **17**, 127–132.

REYNOLDS S.G. (1970) The gravimetric method of soil moisture determination. Pt. I. A study of equipment and methodological problems. *J. Hydrol.* **11**, 258–273.

REYNOLDS S.G. (1971) The Alafua penetrometer for measuring soil crust strength. *Trop. Agric., Trin.* **48**, 365–366.

RICH C.I. & KUNZE G.W. (editors) (1964) *Soil Clay Mineralogy. A Symposium.* Univ. Nth. Carolina Press.

RICHARDS L.A. (1949) Methods of measuring soil moisture tension. *Soil Sci.* **68**, 95–112.

ROGERS J.S. (1974) Small laboratory tensiometers for field and laboratory studies. *Soil Sci. Soc. Amer. Proc.* **38**, 690–691.

ROSE C.W. (1966) *Agricultural Physics.* Pergamon.

RUDEFORTH C.C. (1969) Quantitative soil surveying. *Welsh Soils Discussion Group. Report 10,* 42–48.

RUDEFORTH C.C. & BRADLEY R.I. (1972) *Soils, classification and land use of West and Central Pembrokeshire. Special Survey No. 6.* Harpenden, Soil Survey of England and Wales.

RUDEFORTH C.C. & WEBSTER R. (1973) Indexing and display of soil survey data by means of feature cards and Boolean maps. *Geoderma* **9**, 229–248.

RUSSELL SIR E.J. (1971) *The World of the Soil.* Collins.

RUSSELL E.W. (1974) *Soil Conditions and Plant Growth (10th edition).* Longmans.

RUTTER A.J. (1955) The composition of wet heath vegetation in relation to the water table. *J. Ecol.* **43,** 507–543.

RUTTER A.J. & SANDS K. (1958) The relation of leaf water deficit to soil moisture tension in *Pinus sylvestris* L. I. The effect of soil moisture on diurnal changes in water balance. *New Phytol.* **57,** 50–65.

SALTER P.J., BERRY G. & WILLIAMS J.B. (1966) The influence of texture on the moisture characteristics of soils. III. Quantitative relationships between particle size composition and available-water capacity. *J. Soil Sci.* **17,** 93–98.

SALTER P.J. & WILLIAMS J.B. (1965a) The influence of texture on the moisture characteristics of soils. I. A critical comparison of techniques for determining the available-water capacity and moisture characteristic curve of a soil. *J. Soil Sci.* **16,** 1–15.

SALTER P.J. & WILLIAMS J.B. (1965b) The influence of texture on the moisture characteristics of soils. II. Available-water capacity and moisture release characteristics. *J. Soil Sci.* **16,** 310–317.

SALTER P.J. & WILLIAMS J.B. (1967) The influence of texture on the moisture characteristics of soils. *J. Soil Sci.* **18,** 174–181.

SANGLERAT G. (1972) *The Penetrometer and Soil Exploration.* Elsevier.

SCHOFIELD R.K. (1935) The pF of water in soil. *Trans. 3rd Int. Cong. Soil Sci.* **2,** 37–48.

SELBY M.J. (1970) A flume for studying the relative erodibility of soils and sediments. *Earth Sci. J.* **4,** 32–35.

SHAW M.D. & ARBLE W.C. (1959) Bibliography on methods for determining soil moisture. *Engng. Res. Bull. 13–78,* Pennsylvania State Univ.

SHIPP R.F. & MATELSKI R.P. (1964) Bulk density and coarse fragment determinations on some Pennsylvania soils. *Soil Sci.* **99,** 392–397.

SKENE J.K.M. (1960) Sampling errors in the evaluation of soil potassium and pH. *J. Aust. Inst. Agric. Sci.* **26,** 353–354.

SMITH G.D. (1965) Lectures on soil classification. *Pédologie, Gent,* special number 4.

SNEDDON J.I., CUKOR N. & FARSHAD L. (1972) A technique for rapidly determining topographic class from topograpic maps. *Can. J. Soil Sci.* **52,** 518–519.

SOANE B.D. (1973) Techniques for measuring changes in the packing state and core resistance of soil after the passage of wheels and tracks. *J. Soil Sci.* **24,** 311–323.

SPANNER D.C. (1951) The Peltier effect and its use in measurement of suction pressure. *J. exp. Bot.* **2,** 145–168.

TANNER C.B. & JACKSON M.L. (1948) Nomographs of sedimentation times for soil particles under gravity or centrifugal acceleration. *Soil Sci. Soc. Am. Proc.* **12,** 60–65.

TAYLOR N.H. & POHLEN I.J. (1962) *Soil Survey method. Soil Bur. Bull. 25.* Dept. of Sci. Ind. Res., New Zealand.

THOMAS A.M. (1966) *In situ* measurement of moisture in soil and similar substances by 'fringe' capacitance. *J. scient. Instrum.* **43,** 21–27.

THOMAS T.M. (1973) Tree deformation by wind in Wales. *Weather* **28,** 46–58.

THOMASSON A.J. (1971) Soil water regimes. *Welsh Soils Discussion Group, Report 12,* 96–105.

THOMASSON A.J. & ROBSON J.D. (1967) The moisture regimes of soils developed on Keuper Marl. *J. Soil Sci.* **18,** 329–340.

THORPE J. & SMITH G.D. (1949) Higher categories of soil classification: Order, Sub-order, and Great Soil Groups. *Soil Sci.* **67,** 117–126.

U.S.D.A. (1938) *Soils and men. Yearbook of Agriculture.* Washington, United States Dept. of Agric.

U.S.D.A. (1951) *Soil survey manual. Agric. Handbook No. 18.* Washington, United States Dept. of Agric.

U.S.D.A. (1960) *Soil Classification: A Comprehensive System. 7th Approximation.* Washington, United States Dept. of Agric.

U.S.D.A. (1967) *Supplement to 7th Approximation.* Washington, United States Dept. of Agric. Soil Conservation Service.

VINK A.P.A. (1968) Aerial photographs and the soil sciences. In *Aerial Surveys and Integrated Studies.* Paris, U.N.E.S.C.O.

VISVALINGAM M. & TANDY J.D. (1972) The neutron method for measuring soil moisture content—a review. *J. Soil Sci.* **23,** 499–511.

VISVALINGHAM M. (1974) Well-point techniques and the shallow water-table in boulder clay. *J. Soil Sci.* **25,** 505–516.

WALI M.K. & KRAJINA V.J. (1973) Vegetation–environment relationships of some sub-boreal spruce zone ecosystems in British Columbia. *Vegetatio* **26,** 237–381.

WALLWORK J.A. (1970) *The Ecology of Soil Animals.* McGraw-Hill.

WASHBURN A.L. (1973) *Periglacial Processes and Environments.* Edward Arnold.

WATSON J.R. (1971) Ultrasonic vibration as a method of soil dispersion. *Soils Fertil., Harpenden* **34,** 127–134.

WATT A.S. & JONES E.W. (1948) The ecology of the Cairngorms. I. The environment and the altitudinal zonation of the vegetation. *J. Ecol.* **36,** 283–304.

WATT A.S., PERRIN R.M.S. & WEST R.G. (1966) Patterned ground in Breckland: structure and composition. *J. Ecol.* **54,** 239–258.

WEAVER C.E. & POLLARD L.D. (1973) *The Chemistry of Clay Minerals.* Elsevier.

WEBSTER R. (1966) The measurement of soil water tension in the field. *New Phytol.* **65,** 249–258.

WEBSTER R. (1968) Fundamental objections to the 7th Approximation. *J. Soil Sci.* **19,** 354–366.

WEBSTER R. (1969) Aerial photography in soil and land survey. *Welsh Soils Discussion Group, Report 10,* 49–55.

WELLS C.B. (1959) Core samplers for soil profiles. *J. agric. Engng Res.* **4,** 260–266.

WILLIAMS B.G., GREENLAND D.J., LINDSTROM G.R. & QUIRK J.P. (1966) Techniques for the determination of soil aggregates. *Soil Sci.* **101,** 157–163.

WRIGHT M.J. (1971) The preparation of soil monoliths for the 9th International Congress of Soil Science, Adelaide, Australia. *Geoderma* **5,** 151–159.

YOUNG A. (1960) Soil movement by denudational processes on slopes. *Nature, Lond.* **188,** 120–122.

CHAPTER 7

CLIMATOLOGY AND ENVIRONMENTAL MEASUREMENT

R.B. PAINTER

1   **Introduction**   369

2   **Solar energy**   371
2.1   Solar radiation   372
2.2   Ultraviolet and infra-red   374
2.3   Visible radiation   376
2.4   Net radiation   376

3   **Temperature**   377
3.1   Liquid in glass thermometer   378
3.2   Deformation thermometer   379
3.3   Resistance wire   380
3.4   Thermistors   380
3.5   Thermocouples   381

4   **Precipitation**   381
4.1   Raingauges   382
    4.1.1   Errors in rainfall measurement   382
    4.1.2   Mean areal rainfall   385
4.2   Snow   385
4.3   Dew   386

5   **Interception loss**   387
5.1   Gross precipitation   388
5.2   Net precipitation   388
5.3   Throughfall   388

5.4   Stemflow   389
5.5   Duration of wetness   389

6   **Humidity**   391
6.1   Psychrometry   392
6.2   Hygroscopic elements   394
6.3   Condensation method   395

7   **Airflow**   395
7.1   Wind speed   396
7.2   Wind direction   397

8   **Evaporation**   399
8.1   Water balance   399
8.2   Energy balance   400
8.3   Vapour transfer   402
8.4   Predictive equations   403

9   **Soil moisture**   404

10   **Runoff**   404
10.1   Water balance   405
10.2   Lateral subsurface flow   405
10.3   Stream gauging   406
    10.3.1   V notch weir   407
    10.3.2   Velocity–area method   408
    10.3.3   Gulp dilution gauging   408

11   **References**   409

1 INTRODUCTION

Climate is a fundamental factor in determining the occurrence and growth of plants. This was tacitly assumed for many years on the basis of qualitative observations, and subsequently proved by measurement. Plants in turn will significantly affect the microclimate.

The relationship of macroclimate with plant distribution is demonstrated by Perring & Walters (1962); close correlations exist between the plant distribution and the extremes of temperature, humidity and precipitation excess or deficit. Many workers, including Hawksworth & Rose (1971), Gilbert (1970), have shown that the distribution of lichens is closely related to atmospheric pollution which may be considered as a man-made modification to climate. Many field studies in plant ecology such as nutrient and energy balances, require measurements of environmental variables, and others such as shelterbelt studies examine the interaction of climate and plant. This increasing need of plant ecologists to quantify environmental measurements has resulted in a number of relevant publications, notably Wadsworth (1968) and Monteith (1972).

It is important at the outset to recognize the validity of the comment by Gates (1962), 'that there must be a reason for measuring a particular climatic factor, not just because the apparatus is available, or one cannot think of anything else to do'. When this criterion is met, it is equally important to remember that any field instrument whether a simple 'home-made' type or one bought commercially must be capable of operating under field conditions. This seemingly naïve statement is important as many instruments operate satisfactorily in a warm, dry, clean laboratory but fail when used in a cold, wet, dirty environment. The automatic weather station (Fig. 7.1) developed by

Figure 7.1 Automatic weather station developed by Institute of Hydrology.

the Institute of Hydrology is an example of good design, providing measurements of seven meteorological variables on magnetic tape at five minute intervals. Loggers are serviced in the laboratory and are never opened in the field, thus excluding dirt and damp. Although these standards may not be possible with all instrumentation, the principles involved should be followed. In addition when planning any instrumentation every effort should be made to discuss possible problems with experienced field workers; if this advice is followed time and money will be saved.

SOLAR ENERGY

The sun emits radiation as a black body at around 6,000°K, and as a result the solar spectrum extends from 0·3–3·0 µm, with a sharp peak at 0·55 µm. In contrast the earth radiates in the long wave region with the characteristics of a black body at 250°K, and its spectrum extends from about 3 to 100 µm. Thus solar and terrestrial radiation can be separated according to their waveband widths, as shown in Fig. 2.

Figure 7.2   Distribution of solar and terrestrial radiation.

At the earth's surface, solar radiation includes the direct component from the sun, and a diffuse component scattered downward by air and water droplets. Outside the atmosphere the intensity of the sun's energy (called the solar constant) is 1,400 joules m^{-2} s^{-1}, and in the tropics this figure is reduced by some 25% at the ground surface. In the United Kingdom the sun's intensity in midsummer is about 875 joules m^{-2} s^{-1}, though this is increased by up to 20% over short periods, due to reflection from white clouds. Daily total radiation is a function of intensity and daylength, and in the United Kingdom is of the order 25×10^6 joules m^{-2} in summer and $4·2 \times 10^6$ joules m^{-2} in winter. Some of this incoming radiation will be reflected and the ratio of reflected to incoming, termed the albedo, defines the solar energy not available for absorption by plants.

The exchange of long wave radiation at the earth's surface has two components; a downward component emitted by the atmosphere, its value dependent on air temperature and vapour pressure, and an upward component emitted by the ground, its value dependent on surface temperature. Although daily totals of both components are large, some $(21–25) \times 10^6$ joules m^{-2}, their difference is only of the order 4×10^6 joules m^{-2}.

The difference between long and short wave radiation gained by the surface, and that lost by reflection and long wave emission is called the net radiation, which is important as it represents the energy available for evaporation and for heating both soil and air.

Radiation measurements may be divided into 4 groups:

(a) solar or short wave,
(b) ultraviolet and infra-red,
(c) visible,
(d) net.

In ecology, these measurements are made for two main reasons: studies of climatic energy budgets, and studies of photosynthesis and of biological energy budgets (see Chapters 4 and 5). All measurements may need to be made above and below a canopy, and the instruments used will differ in size and geometry. Above the canopy where radiation is fairly uniform, a small sensing element will suffice and has the advantage of lightness and portability. Within the canopy the radiation can be very uneven due to foliage movement and changes in the sun's position. In this situation a greater area must be sampled in order to obtain spatial integration and larger sensing elements have been developed for this purpose.

2.1 SOLAR RADIATION

The most commonly used sensors for measuring solar radiation are based either on thermoelectric principles, or use the distortions of bimetallic strip.

Thermoelectric instruments are based on thermopiles which are a group of thermocouples arranged in series so that the two sets of opposing junctions are at different temperatures. The hot junction is blackened and exposed to the radiation, and the opposing set is either arranged in good thermal contact with the body of the instrument as in the Kipp instrument (Fig. 7.3) or grouped under a white reflecting area as in the American Eppley instrument. In a well designed instrument the resulting temperature difference, and hence its emf output, is a linear function of the radiation's intensity. The instruments are restricted to the $0\cdot3-3$ μm band by surrounding the sensor with a glass dome, which also protects the sensing surface from wind and rain.

Both instruments have good directional response provided the glass dome is of uniform thickness, and they can be turned to face the ground in order to measure reflected short wave radiation. The Kipp instrument is stable and reliable but its weight often precludes its use above or within high canopies; the Eppley though a little lighter tends to be less robust. Where an integrated value of uneven radiation is required, a tube solarimeter should be used, and Szeicz

Figure 7.3   Kipp solarimeter.

et al. (1964) describe an instrument comprising a thermopile 25 mm wide and 900 mm long contained within a glass tube, which provides radiation measurements averaged over an area of 225 cm². Later, Szeicz (1965) described a miniature version of this instrument for use in grass and herbage.

The most commonly used instrument based on the distortion of bimetallic strip is the actinograph shown in Fig. 7.4. This works mechanically by recording the difference in expansion of three metallic strips, one of which is blackened, exposed to radiation, and fixed to the frame to compensate for changes in air temperature; the other stwo strips are shielded. Bending of the exposed strip is recorded on a chart, but there is a significant lag in the instruments's response to changes in radiation. Provided it is frequently recalibrated the bimetallic actinograph is satisfactory for calculating weekly and daily totals; it is however of limited value in detailed studies.

Other methods for measuring solar radiation include photoelectric devices which are described in the section on visible radiation. In the tropics, distillation techniques are often employed, in which the radiant energy is measured by the amount of liquid distilled from a blackened copper sphere into a graduated glass tube. Photochemical methods are more applicable to the measurement of the ultraviolet part of the spectrum and are discussed in Section 2.2.

Figure 7.4   Bimetallic actinograph.

2.2 ULTRAVIOLET AND INFRA-RED

A few studies necessitate measurements of energy in the ultraviolet range
(0·3–0·4 μm) and in the near infra-red range (0·7–3·0 μm). Ultraviolet
radiation has been detected and measured in three main ways: biologically,
physically and chemically, as outlined by Platt & Griffiths (1965). Biological
methods give non-linear results and are also difficult to reproduce, while
physical methods which are either photoelectric, fluorescent or radiometric,
are unsuitable for absolute measurement or precise work. The chemical
method which compares the effect of ultraviolet on chemicals such as acetone,
with standards, produces fairly precise and reproducible results, though the
method is suitable only for long-term estimates. For greater accuracy a
solarimeter modified for sensitivity to only the ultraviolet range, can be used.
Radiation in the near infra-red range can be detected and measured directly
using a clear filter and a Schott RG8 filter successively over a solarimeter.
Szeicz (1966) describes the Rothamsted bandpass solarimeter which could be
so modified to monitor the near infra-red. This instrument enables account to
be taken of varying positions of the sun during the day; larger versions of this
instrument can be built for measurements within a canopy.

Figure 7.5 Campbell-Stokes sunshine recorder.

2.3 VISIBLE RADIATION

Radiation in the visible part of the spectrum is generally measured by translating radiant energy into a heating or photoelectric effect. Instruments within these groups are normally classified as sunshine recorders and photocells respectively. Sunshine recorders are used 'to measure the duration of bright sunshine', and the Campbell–Stokes instrument (Fig. 7.5) is most commonly used. This defines periods of bright sunshine, from a trace burnt by the sun on a card. It consists of a section of a spherical bowl, within which a glass sphere is positioned concentrically. The sun shines through this sphere onto a card, which is changed from straight to curved during the year, according to the sun's position. The instrument must be installed on a very rigid base with the correct orientation, and there must be no obstructions to its exposure.

In the USA the Marvin 'sunshine' recorder is normally used, which is a differential air thermometer using one clear and one blackened bulb. Their differential heating causes a mercury column in the black bulb to expand and make an electrical contact which is recorded on a chart. The Marvin recorder also measures diffuse radiation and could therefore be better described as a 'daylight recorder'.

Most photoelectrically based instruments utilize photovoltaic cells, though photoemissive and photoresistive instruments have been used in radiation studies. Selenium cells, Powell & Heath (1964), are popular due to their low costs, but their limitations of low sensitivity and tendency to fatigue, have led to increased use of silicon solar cells, Biggs *et al.* (1971). Selenium cells comprise a layer of the oxide from which electrons are liberated by the action of light. These electrons flow over and accumulate in a transparent metal film on the face of the cell. When a low resistance galvanometer is connected between the film and the metal base, a current, proportional to the light intensity is induced and can be recorded. The cell is sensitive to the waveband $0.3-0.7\,\mu m$ and has the advantages of robustness, instantaneous response and no external power requirement. Unfortunately like all photocells, it has a non-uniform spectral response, and its use is therefore limited to situations where the spectral balance of the measured radiation remains constant.

2.4 NET RADIATION

Because net radiation is the difference between two large, and sometimes similar, opposing fluxes in all wavelengths, its precise measurement is more difficult than for a single radiation flux. Normally it is measured by arranging thermopiles to determine the temperature difference between the top and bottom surfaces of blackened discs which are exposed to radiation. The influence of air movement must be excluded from the instrument, as wind considerably alters the output of the thermopile per unit net radiation due to convective currents. This problem is normally overcome by covering both the top and bottom surfaces with thin polythene hemispheres. These are prevented from collapse by passing dry air or nitrogen through them under slight pressure; this also minimizes condensation. Glass hemispheres cannot be used as this would restrict radiation measurements to the 0·3—3 μm band.

Funk (1959) describes the most useful field instrument (Fig. 7.6), which

Figure 7.6　Funk net radiation sensor.

can be mounted from the top of trees for measurements over the canopy. Monteith & Szeicz (1962) describe an unshielded and unventilated instrument, requiring no power or nitrogen supply, which gives an output of sufficient accuracy to provide weekly totals of net radiation. Within the canopy, tubes or a number of point measurements are necessary to obtain spatial representation.

3 TEMPERATURE

Measurements of temperature are fundamental to many ecological studies including those dealing with the energy exchange pattern between vegetation and the atmosphere, and in the determination of the response of vegetal growth to its environment. Section 8 will demonstrate its use in the determination of evaporation rates.

Temperature measurements for ecological studies have been made with six basic types of thermometer:

(a) liquid in glass,
(b) deformation,
(c) resistance wire,
(d) thermistor,
(e) thermocouples,
(f) sucrose inversion (see Chapter 9).

The liquid in glass thermometer gives an instantaneous measure and is non-recording, while the others will provide a continuous record of temperature. Whichever type of thermometer is used, the temperature it indicates is a summation of convective heat exchange with the air flow past it, and radiative exchange with other surfaces such as the earth, sky and sun. The true air temperature is measured only when thermometer and air are in thermal equilibrium. Temperature errors arising from direct radiation can be diminished by screening the sensor and by coating both sensor and screen with a matt white paint. Radiative exchange can also be decreased and convective exchange increased by using small sensors, though integration may be necessary if their response is rapid.

3.1 LIQUID IN GLASS THERMOMETER

Liquid in glass thermometers are used at all standard meteorological stations, for measuring maximum and minimum daily temperatures and in psychrometry for humidity determination (see Section 6.1). The most common maximum thermometer is mercury-in-glass, with a plug or constriction in the bore below the lowest graduation; when the temperature falls the mercury column is not drawn back past this constriction, provided the thermometer lies within 10° of the horizontal. The instrument is reset by shaking, to move the mercury back past the constriction.

Minimum temperature is normally measured by a glass thermometer filled with a colourless organic liquid containing a small index just below the liquid surface. As the temperature falls this index is dragged back by the liquid's surface tension effect; as the temperature rises the liquid flows past the index, leaving it stationary. The upper end of the index denotes the minimum temperature since the instrument was last reset by tilting it until the index comes in contact with the liquid surface.

Generally both maximum and minimum thermometers are mounted in a screen to prevent direct radiation and to ensure that they are adequately ventilated; obviously this makes them unsuited to microclimatic measurement.

3.2 DEFORMATION THERMOMETERS

Two types of deformation thermometer, bimetallic strip and the Bourdon tube, are used in environmental studies.

Bimetallic strip thermometers are based on the fact that differential expansion of two narrow sheets of different metals joined along their flat faces, produces bending of this element. By making the strip into a spiral, the bending is magnified, and small temperature changes can be recorded by a pointer attached to the end of the spiral. Normally the strip is made from invar and brass sheets, with a coefficient of expansion ratio of $1:20$. The bimetallic strip is the sensing element used in many thermographs (Fig. 7.7), but these are unsuitable for microclimatic measurements, due to radiation exchange between the large case and its surroundings; in addition they have a slow response time, and in very humid conditions may stick.

Figure 7.7 Thermograph.

The Bourdon tube is a closed curved tube of elliptical cross section completely filled with an organic liquid. Changes in temperature cause the liquid to expand or contract, which alters the radius of curvature of the tube; this movement is magnified and recorded by a lever and pen arrangement. The Bourdon tube enables the temperature to be recorded away from the instrument.

3.3 RESISTANCE WIRE

Wire resistance thermometers are based on the increase of resistance of most metals caused by an increase in temperature. Nickel or platinum are most commonly used, as within the temperature range $-25\,°C$ to $50\,°C$, they both show less than 0.2% departure from linearity, to give negligible errors for most environmental studies. Their high stability makes wire sensors ideal for absolute measurement, and platinum sensors, a few millimetres long and stable to $\pm0.01\,°C$, are ideal for microclimatic studies; they can also be used to measure temperature differences provided care is taken with lead compensation. Because they are linear, several wire sensors may be connected in series to provide a spatial average of temperature.

3.4 THERMISTORS

Thermistors are solid state semi-conductors which exhibit large non-linear changes of resistance with temperature. They are composed of sintered mixtures of very pure metallic oxides such as nickel and copper, in the form of beads, rods or discs. Depending on which oxide is used their resistance will vary with temperature by 3 to 6% per $°C$, and most commercial versions have a negative temperature coefficient with a near logarithmic response in the range $-25\,°C$ to $+50\,°C$.

Beads with diameters of less than 1 mm, respond rapidly to changes of temperature, though if required their thermal capacity can be increased by sealing them inside a suitable container. The high temperature coefficient and low current requirement of the thermistor make them suitable for use with portable battery-operated Wheatstone Bridge circuits (Fig. 7.8). However, most resistance recorders are designed for use with positive temperature coefficient

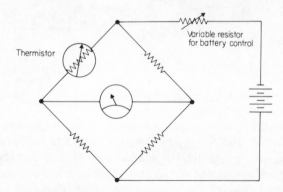

Figure 7.8   Simple Wheatstone Bridge circuit for use with thermistors.

sensors and give a left to right deflection with increasing temperature. To obtain this deflection thermistors must be placed in the opposite arm of the bridge circuit to normal, and the recorder zero may require adjustment.

Although previously unsuited for continuous recording, due to long-term instability, thermistors which operate satisfactorily in the field for many months are now commercially available. A further advantage of thermistors is their small size, which enables point measurements to be made; they can be attached to leaf surfaces to give measurements of plant temperature (Idle, 1968).

3.5 THERMOCOUPLES

To measure the ambient temperature using a thermocouple, one junction must be held at a constant known reference temperature, and the ambient temperature will be the sum of this reference temperature and the temperature difference between the two junctions. Slowly melting ice in distilled water contained in a thermos flask gives a stable reference for fieldwork, though only for a short time. The response of a thermocouple is almost linear but as the output is small, an amplifier is required to actuate a recorder.

Because of the need to maintain a constant reference temperature and to screen long leads, thermocouples are not the best choice for measuring the ambient temperature. They are however very satisfactory for measuring temperature differences, and are easy and inexpensive to make. When made of fine wire, their size and small thermal capacity, make them very suitable for measuring differences in temperature between leaves and the air (Warren Wilson, 1957). Several sets of thermocouples wired in series give a spatial average of the leaf–air temperature difference.

4 PRECIPITATION

Precipitation of water from the atmosphere occurs as rain, snow, hail, sleet, dew, rime or mist interception. Since all life is dependent on water availability the measurement of precipitation is fundamental to many ecological studies. Precipitation is a major component of the water balance calculation, which is often used in studies of the nutrient cycle, and in the determination of the water consumption by different vegetation. For measurement purposes precipitation may be divided into three groups; rain, hail and sleet, snow, and dew. When measuring rain, hail and sleet it is normally sufficient to make a limited number of point measurements and from these determine the areal average; snow, however, often exhibits large and frequent spatial variations and a similar number of point measurements is rarely adequate (Blyth & Painter, 1974).

Dew and mist interception account for a small proportion of total precipitation but in certain dry areas they may have a critical effect on the vegetation type.

4.1 RAINGAUGES

A raingauge is a canister containing a funnel which directs the rain caught either into a storage container (non-recording) or into a measuring mechanism (recording). The gauge rim defines the catching area of the gauge and should be sharp edged and made of durable material that will not distort. Precipitation depth is equal to the volume caught divided by the catching area. In non-recording gauges the funnel leads to a narrow necked bottle and the catch is measured by emptying it into a tapered measuring cylinder. Gauges are set at standard heights above the ground, though this varies with the country: in the United Kingdom the height is 305 mm.

Recording raingauges are normally based on either a tilting syphon mechanism or on tipping buckets. Figure 7.9 shows the Dines tilting syphon gauge, which is in common use in the United Kingdom. Rain funnelled into the container causes the float to rise, and this moves a pen up a paper chart on a clock driven drum. At a preset level (usually equivalent to 5 mm of precipitation) the float actuates a trip mechanism which allows the container to tip forward and syphon its contents. When empty it tips back into position and the cycle restarts. Rain falling during the syphoning period will not be recorded and in cold weather it is advisable to lag the container and provide a battery-powered heat source, to prevent freezing.

The Rimco gauge (Fig. 7.10) exhibits many of the best features of tipping bucket raingauges. Rain is funnelled into a gold plated bucket and when full its weight alters the balance between the buckets, and they tip. This trips a reed switch which records a pulse on either an electromechanical counter or a logger. Standard bucket sizes equivalent to 0·25 and 0·5 mm of precipitation are available. Like the tilting siphon gauge, most tipping bucket gauges suffer errors due to rainfall lost during the tipping movement. Excessive loss is however prevented in the Rimco gauge by limiting the maximum input rate to the buckets by a siphon; this control ensures that all tips are recorded even in very intense rain. All recording gauges should be accompanied by a storage gauge in order to provide a check on the total quantity of rainfall and to ensure continuity of record during recording gauge failures.

4.1.1 Errors in rainfall measurement

Point measurements using any raingauge are beset with error sources, and the assumption commonly made that the gauge catching the most rain is the best is not necessarily valid, because insplashing can occur. Kurtyka (1953) listed

Figure 7.9   Dines recording raingauge with open scale chart.

six error sources and their approximate sizes, and concluded that exposure is the most serious. Any gauge is an obstacle to airflow and the caused local acceleration results in the transport of raindrops across the orifice and not into it. To measure the resulting loss, gauge catch at standard height has been compared with the catch from a gauge installed with its rim at ground level, and surrounded by a non-splash surface (Fig. 7.11). This latter gauge is largely

Figure 7.10   Mechanism of Rimco tipping bucket raingauge.

free from the effects of wind and is generally considered to give the most accurate measurements of point rainfall, particularly in over-exposed sites such as moorland.

Most sources of error are negative, causing the gauge to under-register. Splash out of a gauge can be a considerable problem where shallow funnelled gauges are used and rainfall intensities are high. Gauge funnels should be smooth to minimize adherence of raindrops, and evaporation from storage gauges, particularly in hot climates, should be minimized.

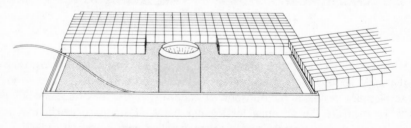

Figure 7.11   Ground level raingauge.

4.1.2 Mean areal rainfall

The amount of rainfall at any point is determined by the geographical location and by a number of topographic factors such as altitude, aspect and exposure. Because of the resultant spatial variations of rainfall a raingauge network is necessary to determine the mean areal total. The gauges in the network can either be sited randomly and their totals later 'weighted' in some way to account for their uneven distribution, or these effects can be accounted for in the network design by the use of domain theory in which the effects of altitude, aspect and slope are accounted for in the siting of the gauges.

If a random distribution is used, the areal rainfall may be found either by the arithmetic mean, by the use of Thiessen polygons, or by plotting isohyets which are lines of equal rainfall. Derivation of areal rainfall from the arithmetic mean is only likely to be satisfactory where gauge distribution is even, the terrain is flat and rainfall distribution is fairly even. The Thiessen polygon method makes allowance for an uneven gauge distribution, and also enables data from adjacent areas to be utilized. In this method perpendicular bisectors are drawn through the straight lines between adjacent gauges, leaving each gauge in the centre of a polygon. Each gauge's rainfall total is multiplied by the percentage of the total area represented by its surrounding polygon, and the sum of these products gives the average rainfall. In areas of high relief the Thiessen polygon method is unlikely to be accurate; in this situation the plotting of isohyets offers a better solution. Its success, however, depends on the ability of the operator to take account of relief, aspect, direction of storm movement, etc., when plotting the isohyets from the gauge totals.

The use of domain theory in setting up gauge networks will remove this subjectivity; the mean areal rainfall is simply the sum of the gauge totals multiplied by the percentage of the total area represented by their own domain.

4.2 SNOW

Although separation of precipitation into rain and snow is seldom necessary in ecological studies, the accurate measurement of the water equivalent of snow is necessary, particularly where it constitutes an appreciable proportion of total precipitation. Traditionally the water equivalent is obtained by catching the snow as it falls, in a gauge, or by manual measurements of depth and density after it has reached the ground and been redistributed. Both methods suffer large errors; in the first case a gauge will under register due to flakes being blown past the gauge, while a manual survey would have to be very intensive to successfully sample the large and frequent variations of snow depth. Gauge installations often use a Nipher shield, shown in Fig. 7.12, and invariably require a heating element. Manual surveys are carried out along a number of

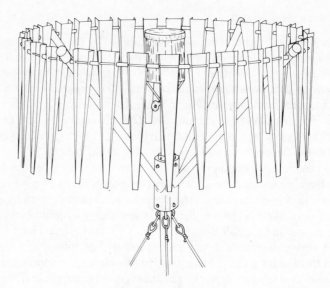

Figure 7.12   Nipher shield.

straight lines called snow courses, taking depth and density readings at 8–10 points per course, with a corer.

Recent developments in snow measurement include the snow pillow, measurement of natural soil radiation, and stereo photography. The snow pillow, described by Tollan (1970), is a large bag of water containing antifreeze; the water is displaced by the weight of snow lying on it, and the displacement operates a float and counterweight which in turn produces an ink on paper trace. Errors are caused when partial melting and refreezing of the snow pack, causes the latter to bridge over the pillow. Attenuation of the earth's natural radioactivity by the snowpack has been monitored (Bissell & Peck, 1973), but at present this attractive solution to the sampling problem can only be used in deep packs. The use of aircraft to provide stereo photographs of the ground before and after snowfall is described by Smith *et al.* (1967), but in the United Kingdom where daily readings are necessary the use of terrestrial photogrammetry would be preferable.

4.3 DEW

Dew occurs on surfaces which have a lower temperature than the dew point of the adjacent air. In arid regions, dew may constitute an important factor in plant growth, because although its quantity may be small it occurs at a time of minimal evaporation, thus increasing its relative effectiveness. The persistence of dew is a major contributing factor in the occurrence of plant diseases such as apple scab and potato blight (World Meteorological Organisation, 1963).

The Hiltner dew balance is typical of the many instruments developed to measure dew deposition. The sensing surface is a circular nylon filter which is fixed to the beam of the balance; all changes of weight are mechanically transmitted to a recorder drum. In order to reduce the oscillations of the sensing surface caused by wind, an oil damping system is built into the balance, and a wind screen can be placed around the surface.

5 INTERCEPTION LOSS

When precipitation occurs, some part of it will be stored on the vegetation, and subsequently evaporated to the atmosphere. This quantity is termed the 'interception loss' and is the difference between precipitation above the vegetation (gross precipitation), and that reaching the ground as throughfall and stemflow; the sum of these two quantities is called the net precipitation. Hamilton & Rowe (1949) defined throughfall as that part of precipitation to reach the ground directly through gaps in the canopy or as drip from leaves and stems, and stemflow as that reaching the ground by running down the stems.

As a result of interception the exposed surface of vegetation becomes covered with a film or drops of water, and under this condition transpiration does not occur. Thus in high rainfall areas, evaporation of intercepted water will represent a considerable proportion of the total evaporation component. In arid areas the redistribution of rainfall by stemflow around the base of a tree may be vital to its survival, and Johnston (1964) points out that many plants can absorb moisture directly into their leaves, a particularly useful characteristic during periods of soil moisture deficit.

Very many studies, summarized by Rutter (1968), have been made of the amount of precipitation intercepted by a particular species in a particular climate. The majority of these studies have been carried out for tree and scrub, but increasingly the interception characteristics of grass and low vegetation have been examined (Leyton *et al.* 1968). If the results of any study are to be transferable from one climate to another then the process of interception must be understood and a physically based model built.

Normally, measurements are made of the precipitation above the canopy, throughfall, stemflow and duration of foliage wetness; limited direct measurements have also been made, e.g. by weighing foliage before and after precipitation (Rutter, 1963). In studies of grass and herb communities it is impracticable to measure throughfall and stemflow separately, and hence the net precipitation is measured.

5.1 GROSS PRECIPITATION

The measurement of gross precipitation over low vegetation should follow the criteria laid down in Section 4.1. In woodland, gross precipitation may be measured either in clearings or by gauges mounted above the canopy. In clearings it is standard practice to site gauges a minimum of twice the tree height from the trees to minimize interference.

Above canopy gauges must be designed to reduce the effect of increased windspeed and turbulence on gauge catch, and generally this is achieved, either by shielding them or by streamlining their form. A comparison of the performance of unshielded, shielded and steamlined gauges, is given by Reynolds (1964).

5.2 NET PRECIPITATION

The measurement of net precipitation rather than its components, throughfall and stemflow, is necessary in areas of grass and herb cover. Usually the surface of an area of ground is sealed with polythene sheet, and the net precipitation measured as surface runoff, using a tipping bucket device or a float and counterweight system driving a chart recorder.

Some studies have used a polythene sheet stretched below an entire tree crown, though separate measurements of stemflow and throughfall are likely to be more useful, particularly in studies of nutrient cycling (Chapter 4, 4.2.1).

5.3 THROUGHFALL

Under most plant covers, there are three kinds of spatial variation in throughfall:

(a) a systematic variation below individual plants,
(b) variations within the general pattern of (a), caused by differences in crown size, height, shelter, etc.,
(c) gaps in the canopy through which precipitation penetrates directly to the ground.

All three sources of variation, together with any topographic variations in a site, must be accounted for when sampling. The accuracy of areal throughfall estimates can be improved to a limited extent, by increasing the number of gauges used, though the estimate from a given number of gauges is improved if half the gauges are moved after a preset quantity of rain. Alternatively the use of larger gauges will increase accuracy, but if standard gauges only are available their catch can be related to their position under the crown of the trees by regression analysis and the resulting equation applied to the whole area.

Increasing the size of gauge is probably the most practicable solution. This can be achieved by using large annular gauges round the trunks of trees, or long narrow troughs. Troughs sited randomly are probably the most effective means of measuring throughfall, though they require frequent attention to prevent blockage by falling vegetation. In addition, substantial 'wetting up' may be required before their contents are discharged to a measuring device, which must be capable of coping with considerable volumes of water.

5.4 STEMFLOW

Water moving down the stems of trees, is usually measured by intercepting it near the ground with a polythene or metal channel wound round the tree, sealed with bitumastic paste (Fig. 7.13). The collection of stemflow samples is important in studies of the movement of nutrients through an ecosystem, and is considered in Chapter 4 (4.2.1). If trees vary considerably in size then it may be necessary to stratify them before sampling on a random basis. When random sampling only is used the standard error of the mean stemflow for the area of interest is likely to be high, but this is usually unimportant as stemflows are normally small in relation to throughfall. Few measurements of stemflow in low vegetation have been made, though Beard (1962) showed that some 50% of the net precipitation under grass resulted from stemflow.

5.5 DURATION OF WETNESS

To relate the interception process to climate, measurements must be made of the duration of wetness of the vegetation, as this determines the available interception storage at the beginning of each period of precipitation. When the canopy is dry available storage is equal to the total storage. A commonly used instrument described by Hirst (1957), assumes that the wetting and drying behaviour of a slightly curved polystyrene surface is similar to that of natural plant surfaces. Any wetness on the polystyrene surface causes a weight change which is transmitted to a recording pen. Though an oil-damping system is provided to reduce oscillations due to wind, it is insufficient in exposed locations. This oscillation and the bulk of the instrument restrict its use to below the canopy, and excludes the upper part of the canopy where the most frequent changes between dry and wet conditions occur.

Changes in the resistance from infinite to finite of the elements of a piece of 'veroboard' mounted in a canopy, reflect the wetness of the vegetation. When dry, no circuit is made and the resistance is infinite; giving zero current; when

Figure 7.13 Stemflow gauge and tipping bucket counter.

wet, a circuit is formed between the lengths of copper strip and a finite current is recorded. This method is satisfactory for instantaneous measurements, but when continuous measurements are required polarity effects must be eliminated.

Various methods have been devised for measuring changes in wetness of the actual plant surface, based on changes in resistance between electrodes applied to the leaf surface. Damage to the leaf tissue requires that the instrument must be moved frequently and, to overcome this problem Leyton *et al*. (1968) describe a method of painting silver electrodes on a thin plastic film, which is fastened against the leaf (Fig. 7.14).

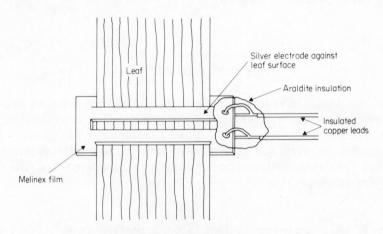

Figure 7.14 Leaf wetness sensor.

Although not ideal for all leaf types, its use avoids damage and minimizes interference with the microclimate around the leaf.

6 HUMIDITY

All surface air contains moisture, the amount normally increasing with temperature. The general term 'humidity' refers to this moisture content, which may be defined in a number of ways. The most common is relative humidity which is the ratio (expressed as a percentage) of the actual vapour pressure (e) of the air, to the saturation vapour pressure (e_s) at the same temperature. Another useful definition is dew point which is the temperature at which further cooling would cause dew formation if a surface or nucleus was available.

The level of humidity in the air effects the transpiration rates of plants (see Section 8), and may also partly control their growth rates. In addition many crop destructive tropical insects breed most rapidly under certain humidity conditions (Flitters, 1965).

Three of the most common methods for measuring humidity are:

(a) thermodynamic—the psychrometer, where a thermometer bulb is kept moist so that evaporation into the free air causes cooling;

(b) hygroscopic—using an element capable of taking in moisture from the air and causing either a change in length, a chemical change, or an electrical change;

(c) condensation—a surface is cooled until condensation takes place, when the dew point temperature is measured.

6.1 PSYCHROMETRY

Before describing the various psychrometers available, it is necessary to give an outline of the theory of the wet bulb thermometer. If air at temperature T, pressure p and vapour pressure e flows past a thermometer bulb covered with wet muslin, the water will be evaporated into the air until equilibrium is reached. For this equilibrium state, consideration of heat interchange gives the following equations:

$$e = e' - a.p. \, (T' - T)$$

where T' and e' are temperature and vapour pressure of air leaving the bulb, and a is the psychrometric constant. The value of a varies with wind speeds of up to $2 \cdot 5$ m s^{-1}, where its value is $0 \cdot 364$; hence for accurate work the wet bulb must be adequately ventilated. Tables and graphs relating T and T', the dry and wet bulb temperatures, are available, though pressure corrections are necessary at altitudes other than sea level.

Manually read psychrometers normally use mercury in glass thermometers, either in a screen, in the whirling psychrometer, or in the Assman model. When recording is required thermocouples (Monteith, 1954) or resistance psychrometers (Penman & Long, 1949) are used.

The screen model is simply two ordinary mercury in glass thermometers hung vertically in a screen, the wet bulb covered by a close fitting muslin bag, from which a wick leads to a water source a few inches away. For single point measurements in the field the hand aspirated model (Fig. 7.15) is commonly used. In this instrument the thermometers are mounted in ducts along which air is drawn by a hand wound fan. Because it draws air from a distance of $0 \cdot 25$ m, it cannot be used to determine temperatures and hence humidity gradients near the ground.

All mercury in glass psychrometers suffer from a number of error sources

Figure 7.15   Hand aspirated psychrometer.

which all cause too high a reading for humidity. Heat conduction down the stem is difficult to eliminate, but can be reduced by covering an appreciable length of stem with the wet muslin. Fine muslin must be used and must be changed regularly to prevent it becoming dirty; similarly the bulb should not become encrusted with dirt. Finally it must be stressed that a true wet bulb depression cannot be obtained in calm conditions, and that adequate ventilation must be provided.

The use of thermocouples in psychrometry was described by Monteith (1954). In addition to the problems given above, errors will arise if the moisture-cooled air from the wet bulb affects the dry couple reading, In addition to ensure full wet bulb depression the moist covering must extend at least 10 mm on either side of the junction.

Jehn (1948) discussed the use of thermistors in psychrometry, and showed that satisfactory results can be obtained by using a thermistor dipped in distilled water prior to reading; subsequent developments in thermistors (see Section 3.4) make them particularly attractive for use in psychrometry. Resistance thermometers should be robust for use as psychrometer elements, and industrial resistance thermometers sheathed in stainless steel have proved successful on the Wallingford automatic weather station (Strangeways, 1972).

6.2 HYGROSCOPIC ELEMENTS

The most widely used hygroscopic element is the human hair, which on exposure to water vapour, changes length due to the absorption of moisture. Other elements which change weight or colour, or whose electrical properties vary on contact with water vapour are also available.

Human hair is the most satisfactory element and produces a 2·5% change in length over a change in humidity from zero to 100%. This change is non-linear which causes cramping at the top end of the scale and to counteract this a cam is used to move the pen arm on standard hygrographs. Because the temperature coefficient is small, errors due to temperature changes can be neglected in most environmental studies. The standard hygrograph (Fig. 7.16)

Figure 7.16   Standard hygrograph.

has a bundle of hairs arranged horizontally, though some manufacturers supply models with the hairs supported vertically and connected directly to the pen arm. Although this minimizes friction, the resulting non-linearity of scale is a disadvantage. Whichever hair hygrograph is used the hair bundle should be washed weekly in distilled water, and the instrument should be recalibrated at frequent intervals.

A commercial hygrometer, the Shaw meter, gives an instantaneous humidity reading at a precision of 3% full scale deflection, over a relative humidity range 0–99%. The sensor comprises a 10 micron thick dialectric layer, the capacitance of which is changed by the entry of water. Protection against the entry of contaminants is good, and the instrument is very robust;

the sensor must not however become saturated, as its calibration would then change.

A very simple field hygrometer was described by Solomon (1945), based on the change of colour that results when cobaltous salts are exposed to moist air. Cobaltous chloride paper is blue at low humidities and pink at high humidities, with a series of lilac colours between. A set of standards is prepared by exposing cobaltous chloride papers over constant humidity solutions, then sealing them in liquid paraffin between opal and clear glass. Exposure in the field should be at least 30 minutes, and at high humidities an exposure of 2 hours may be required. With care, relative humidities in the limited range 40–70% can be obtained to within 2%, and to 5% outside this range.

6.3 CONDENSATION METHOD

In the condensation method the air is cooled and the temperature at which water vapour starts to condense is measured. Because the dew point is obtained, the method is known as dew point hygrometry (or frost point when low enough temperatures are considered).

Although simple in theory, there are field problems in obtaining a representative sample of the air. Condensation is normally induced on a silvered-surface or a metal mirror cooled by the evaporation of ether; the temperature of the surface is measured by a thermocouple attached to the underside. Condensation is normally detected by a photoelectric cell which gives a reproducible and objective answer. Simple field measurements using the human eye can be made, but the dew point temperature should be taken as the average of the temperatures at which dew forms and vanishes. Whichever form of dew point hygrometer is used, it is important to ensure that no temperature gradient exists across the condensation surface.

7 AIRFLOW

Air in motion possesses density, viscosity and specific heat, and is the vehicle for the turbulent transfer of momentum, heat, water vapour, gases, pollutants, pollen, spores and seeds. Many processes in ecology are dependent in part on these factors; run of wind is necessary to the computation of evaporation using the Penman equation (see Section 8.4), while shelter belt studies (Caborn, 1965) require measurements of both wind speed and direction.

The type, precision and duration of airflow measurements varies according to use. In the case of windspeed, this ranges from daily run of wind, to continuous recordings necessary to the study of diurnal variations. Similarly when examining wind direction a cumulative measure of its distribution by compass points is obtained with relative ease, whereas a continuous record can only be obtained with fairly elaborate instrumentation.

7.1 WIND SPEED

Wind speed is measured by anemometers, which are of three main types:

(a) mechanical, which depend on either the rotation of a cup assembly or propeller, or wind pressure on a suitably mounted plate;
(b) pressure tube, in which air flows through the instrument; the wind speed is found from the relationship between the dynamic and static pressure of air;
(c) thermal and thermoelectric devices based on the cooling power of the air.

Cup anemometers are normally used for macro-climatic measurements, while micro studies are usually made with thermal or thermoelectric devices.

The cup anemometer (Fig. 7.17) consists of three conical or hemispherical cups with beaded edges, mounted on a rotor. Cup size and the ratio of cup diameter to the diameter of the circle described by the cups vary considerably in commercial models. The response speeds of most instruments lie in the range $0 \cdot 1 - 1 \cdot 3$ m s^{-1}, and at low speeds most cup anemometers tend to un-

Figure 7.17   Cup anemometer.

derestimate and even stall, while in fluctuating winds over-running may occur. Excessive friction in the rotor spindle causes insensitivity, but as in most environmental instrumentation a compromise between sensitivity and ruggedness must be reached. There is a near linear relation between the rate of cup rotation and wind speed, and revolutions can either be counted mechanically or made to produce either a pulse after a pre-set increment of wind run, or an A.C. current which is rectified to give a galvanometer reading.

The other mechanical systems sometimes used in the field are the vane anemometer and the pressure plate. In fluctuating winds the vane anemometer gives errors of up to 15%, though in the more steady conditions found under canopies it is more satisfactory. Although the pressure plate is more precise, its bulk excludes it from most field uses.

Thermal and thermoelectric devices whose rate of cooling is related to wind speed, are the other group of sensors to find widespread application in ecological studies. Under canopies where windspeeds are low, simple instruments such as the Kata thermometer (Hill *et al.*, 1916) have been used in situations where the air temperature is also measured. Hot wire anemometers have been increasingly used in environmental studies and these are particularly suited to low velocity work in crops. Their theory is given by Tanner (1963) and may be summarized as follows. When a pure metal wire (commonly, platinum, nickel, tungsten or platinum-iridium) is heated by an electric circuit and then exposed to the air, the wire can either be maintained at constant temperature and hence constant resistance by varying the current, or allowed to cool naturally. In the first case the resulting current is a measure of wind speed, and in the second case the resultant variations in the wire's resistance are related to wind speed.

Because commercial instruments tend to be expensive, many models have been built up for individual projects. Caborn (1957) described an instrument for use in a confined space, and Long (1957) constructed a 'hot bulb' thermometer which was robust, had a fairly stable calibration and used a smaller heating current than hot wire systems. Thermal systems cannot be used in rain because of the cooling effect, and it is important that their temperature is sufficiently high to make variations in the ambient temperature negligible.

7.2 WIND DIRECTION

Measurements of wind direction are generally made with the wind vane which can give either spot readings or be linked into a mechnical or electrical recording system.

In the mechanical system it is difficult to site the pen near to the vane, and to provide the time record. An instrument described by Platt & Griffiths

(1965) overcomes both these problems; here the chart is wound round a drum fixed to the revolving spindle of the vane, and the pen is connected to a collar moving vertically on a threaded rod linked to the clock drive. Care must be taken to ensure that the drum and clock container do not interfere with the flow of air. By mounting the pen on the clock drive, discontinuities at either end of the scale, due to the pen reaching the end of the paper, are avoided.

Electrical recorders are commonly based on varying resistance, or sometimes on self-synchronizing transmitting motors. In the first system an indicator is moved across a resistance wire in an electrical circuit according to the direction of the wind. If a constant voltage is maintained the current recorded will be a function of resistance and hence of wind direction. Where only an approximate measurement of wind direction is required, continuous recording through the entire 360° is unnecessary. Instead the vane can be made to carry a brush within a circle of contacts representing compass directions; the closure of a particular contact can then be recorded on paper chart or a logger. Alternatively wind direction can be sensed by means of a circular printed circuit card containing a number of radially mounted reed switches. These are activated by a magnet connected to the vane's shaft, and connected into a network of fixed resistors across which a constant voltage is fed; the current measured indicates wind direction. This system is used by the Institute of Hydrology, in automatic weather stations, and is shown in Fig. 7.18.

Figure 7.18 Wind direction sensor.

8 EVAPORATION

Evaporation may be defined as the movement of water in the form of vapour from natural surfaces, regardless of whether the water source is in soil or vegetation. Soil and plants may be considered as alternative paths through which water flows on its way from the soil water reservoir to the evaporating surface, from which it diffuses into the air. In the case of vegetation this basic pattern is modified by the heterogenuity of the effective surface and by the biologically controlled variability of the resistance to internal flow. The loss of water from plants is discussed in greater detail in Chapter 5 (4.3).

Three main factors determine evaporation rates: availability of energy, the vapour pressure gradient between water at the evaporating surface and the air, and resistance to the movement of the water vapour from the vegetal surface. Sufficient energy must be present at the evaporating surface to meet the latent heat demand. This energy is provided by radiation from the sun, sky and clouds and by sensible heat transfer from the adjacent air and soil. Under steady conditions the three factors adjust to produce a particular rate of evaporation. A change in any single factor does not necessarily produce a proportional change in evaporation, but will be associated with a change in the other factors, finally establishing a new evaporation rate.

There are three basic methods for measuring evaporation from natural surfaces:

(a) by difference from the water balance,
(b) from the determination of the latent heat term in the energy balance,
(c) by measuring the net upward flow of water vapour in the air layers near the ground.

In addition methods have been developed using combinations of (b) and (c) and a number of empirical equations have been derived, relating evaporation to readily measured meteorological variables.

8.1 WATER BALANCE APPROACH

The complex water balance may be simplified to an equation:

$$P - R - E - \Delta M = 0 \qquad \text{(equation 8.1)}$$

where P is precipitation,
  R is runoff,
  E is evaporation,
  ΔM is change in soil moisture over the time period considered.

This equation is generally applicable from entire catchments to small plots; normally all components except E are measured, to give the evaporation by difference. However, changes in groundwater storage below the zone of root influence may cause significant errors.

For small plots the water balance is generally determined by measurements in a lysimeter (Pelton, 1961; Harrold, 1966). This is a large container filled with soil and growing vegetation, which fits into, but does not touch, a lined pit in the ground. The top surface of the container is flush with the surrounding ground. A weighing mechanism is fixed in the bottom of the pit, and the container rests on this mechanism, thus enabling changes of weight to be recorded. After a settling down period, the precipitation input and drainage from the base, together with changes of lysimeter weight give the evaporation by difference.

Lysimeters vary in size from small laboratory units to large containers several metres in diameter. To operate a lysimeter successfully it is necessary to minimize disturbances due to the lysimeter itself and to ensure that its local situation is typical of the general situation being sampled. They are therefore suited to areas of well watered homogenous vegetation, and less applicable to areas where the species are mixed or have an uneven distribution. When conditions are good lysimetry can detect evaporation differences of 0·25 mm, which is sufficient to reveal diurnal variations.

Where vegetation is homogenous, it is possible during periods of dry weather to obtain estimates of evaporation from soil moisture measurements alone. Under these conditions the terms P & R in equation 8.1 disappear, though an unknown quantity of vertical drainage could still occur. Evaporation is then equal to the decrease in soil water storage in the root zone during the time interval chosen, adjusted if possible for any net gain from or loss to the underlying soil. Accuracies better than 2·5 mm cannot be expected from this method, and hence diurnal changes cannot be determined.

The water balance equation can also be applied on a catchment basis, where R is the streamflow (see Section 10). Unless the moisture content of the soil is monitored frequently on a network basis see section (9), equation 8.1 can only be used on an annual basis when the effect of neglecting changes in soil moisture is minimized, or over periods between similar soil moisture conditions (Rutter, 1964).

8.2 ENERGY BALANCE

Net radiation absorbed by plant and soil surfaces by energy exchange is both dissipated as sensible and latent heat, and used in metabolic processes. These processes are represented in equation 8.2.1.

$$R_n - H - l.E - G - a.A = 0 \qquad \text{(equation 8.2.1)}$$

where R_n is net radiation,

H is sensible heat exchange with atmosphere,

l is latent heat of vapourization of water,

E is evaporation,

G is sensible heat exchange with vegetation and soil,

a is chemical energy storage coefficient,

A is the net rate of photosynthesis.

Over periods of days, R_n, H and $l.E$ are the significant components; net changes in G from day to day are small, and $a.A$ seldom amounts to more than 2% of R_n. Where periods of hours are considered, G must also be measured, as heat exchange with the bare soil is a considerable proportion of net radiation during periods of maximum vertical heat flow. As R_n, G, and where necessary A (Chapter 5), can be measured, then the determination of the term $H/l.E$ enables $l.E$ to be found.

The vertical transfer of both sensible and latent heat occurs mainly by turbulent eddy movement in the lower atmosphere. Each net flux results from the vertical temperature gradient, specific humidity gradient and coefficients relating to the eddy movement. It can be shown that:

$$\frac{H}{l.E} = \frac{C_p}{l} \frac{(\Delta T)}{(\Delta q)}$$

(equation 8.2.2)

where C_p is the specific heat of air at constant pressure,

$\dfrac{(\Delta T)}{(\Delta q)}$ is the ratio of the differences of temperature and specific humidity over a given height interval.

Strictly temperature should be corrected for change of pressure with height, but in studies made over a few metres this correction is negligible. The ratio of sensible to latent heat (equation 8.2.2) is called the Bowen ratio. This equation shows that the evaporation rate may be found from measurements of net radiation R_n soil heat flux G, and the temperature and humidity differences over the same height interval above the vegetation.

Net radiation and humidity measurements were described in Sections 2 and 6 respectively. Soil heat flux can be measured directly using small plates (Philip, 1961) similar in construction to net radiometer elements. These are buried flat in the ground near the surface and the temperature difference across their faces gives a measure of the heat flow perpendicular to them. Alternatively soil heat flux can be determined from changes in the temperature profile of the soil, provided its thermal capacity is known.

The energy balance approach gives best results when used over structurally uniform vegetation, and in these conditions can give evaporation estimates to within 10%. This enables the evaporation in periods as short as one hour to be determined, but requires sophisticated logging equipment to provide on-line computation. Figure 7.19 shows the scale of instrumentation at the Institute of Hydrology's site in Thetford Forest, required for a study of the evaporation process over tall vegetation.

Figure 7.19   Thetford Forest experimental site.

8.3 VAPOUR TRANSFER METHOD

A direct estimate of evaporation is obtained by measuring the vertical water vapour flux over the area of study. Under natural conditions this is achieved by the integration of the product of the instantaneous departure of vertical air flow and water vapour content from their mean values at a point. The rapid and sizeable variations of both variables necessitates very delicate instrumentation, and although its theory was described by Swinbank (1951) measurements have usually been limited to periods of the order of one hour.

Indirect estimates of the vertical vapour flux can be made from measurements of the vertical humidity and using a transfer coefficient obtained either empirically or derived from turbulence theory.

8.4 PREDICTIVE EQUATIONS

A number of fairly empirical procedures have been derived to provide estimates of the evaporation from open water surfaces, using standard meteorological measurements such as air temperature, humidity and windspeed. Many of these procedures have subsequently been modified to deal with evaporation estimates from vegetation and bare soil; the most soundly based equation was proposed by Penman (1948).

The Penman equation combines the energy balance method with aerodynamic theory, and though more complicated than most other formulae of its type, it is solved using standard meteorological data. Initially derived for free water evaporation, it is:

$$E = \frac{(\Delta R_n + \gamma Ea)}{\Delta + \gamma} \qquad \text{(equation 8.4.1)}$$

where Δ is the slope, in mm mercury per °C, of the saturation vapour pressure curve at the bulk air temperature T_a,
R_n is net radiation,
γ is the psychometric constant,

$$E_a = f(u) \cdot (e^\circ(T_a) - e_a)$$

$e^0 (T_a)$ is saturation vapour pressure and e_a is actual vapour pressure, at air temperature T_a.

The original form of $f(u)$ was

$$f(u) = 0 \cdot 35 \quad 1 + \frac{u}{100}$$

but was later modified to

$$f(u) = 0 \cdot 35 \quad 0 \cdot 5 + \frac{u}{100}$$

where u is the run of wind in miles per day.

Penman related free water evaporation (E_w) to the potential evaporation (E_f) from an extensive area of dense well watered physiologically active short green crop, as:

$$E_f = f . E_w$$
(equation 8.4.2)

where f ranges from 0·6 to 0·8 depending on length of day and the season.

More realistic functions than equation 8.4.2 have subsequently been developed, based on the resistance characteristics of the crop. Although the introduction of crop surface and soil water parameters provides a more valid means for determining actual evaporation, both characteristics must be derived empirically from existing data or determined for each crop–soil combination. The most-used modification to the Penman equation was proposed by Montieth (1965) and requires measurements of the available energy, air temperature and vapour pressure at a known height, preferably at hourly intervals, together with an estimate of the aerodynamic resistance, and a measurement or estimate of the stomatal resistance. These resistances are often found empirically; the stomatal resistance can also be measured by using a porometer on individual leaves, but it is difficult to average leaf values into crop values. The aerodynamic resistance can also be found from boundary layer theory.

9 SOIL MOISTURE

The determination of soil moisture is important in many branches of plant ecology and is especially important when considering the water relations of a particular site or area. The methods available for the measurement, and a discussion of other aspects of soil moisture are given in Chapter 6 (Section 6.3.5).

10 RUNOFF

Runoff may be simply defined as the part of precipitation which sooner or later enters a stream channel; generally the routes taken by runoff fall into three groups: surface runoff, subsurface flow and groundwater flow. Surface runoff refers to flow over the ground surface, while subsurface flow results from precipitation infiltrating vertically through the soil–air interface, and then running laterally on meeting an 'impermeable' layer. Groundwater flow is derived from the water table which may have its upper surface in either the drift or solid geology. Clear boundaries cannot normally be drawn between these routes as runoff may start by one route and enter the stream by another. Similarly the characteristics of the runoff within each group will vary considerably according to the properties of the material over or through which it moves.

Where the boundaries of an area of ecological study coincide with defined hydrological boundaries, and the area contains a surface channel, runoff measurement is relatively straightforward using a gauging structure, velocity-area methods, or chemical dilution techniques. However, many ecological projects are based on experimental sites lacking surface channels and easily defined hydrological boundaries, where virtually all runoff is generated below ground. In these circumstances runoff is often estimated by difference from the water balance equation.

Lateral subsurface flow may be measured by intercepting it in a vertical cut at the lower end of a sloping site.

Measurements of runoff are fundamental to studies of nutrient circulation (see Chapter 4, Section 4.3.1), where it constitutes a major vehicle for the transfer of nutrients out of a site.

10.1 WATER BALANCE

The use of the water balance equation as a means of determining evaporation, was described in Section 8.1; conversely to estimate runoff from ungauged catchments, precipitation and changes in soil moisture storage are measured and evaporation is estimated from an equation of the Penman type. Over short periods, errors in the evaporation estimate may be of the order 20% which could lead to similar errors in the runoff estimate. For nutrient studies where weekly runoff totals are required these errors should be reduced to around 10%. Errors of this order are generally acceptable, but considerable instrumentation including perhaps the expensive neutron probe, is required.

10.2 LATERAL SUBSURFACE FLOW

Where the site of study slopes and the majority of runoff is generated laterally above an impeding layer in the soil, direct measurements of this flow can give reasonable estimates of the total runoff. Whipkey (1965) describes an installation for this purpose, as outlined in Fig. 7.20. Pits must be drained continually, or pumped out intermittently when water reaches a preset depth. Collecting troughs are formed from plastic guttering and good contact between soil and trough must be ensured, as flow is often by water films or via capillary pores. The flow from the trough may be taken into an oil drum, where a float and counterweight system is used with a chart recorder. Alternatively the trough can lead to a large tipping bucket device such as shown in Fig. 7.13, and the consequent pulses put on to an electromechanical counter or magnetic tape.

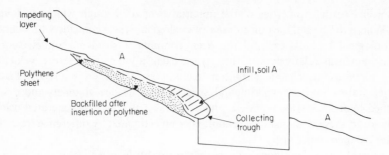

Figure 7.20   Measurement of lateral subsurface flow.

10.3 STREAM GAUGING

To measure the flow in a stream two separate operations are necessary; the first is to measure the level of the water surface relative to a fixed datum, and the second is to relate this level to streamflow.

Measurements of stream level may be made periodically or be continuous, the choice depending on the use to which the data will be put and the nature of the catchment. Where the annual average runoff is to be found, daily

Figure 7.21   Stream level recorder.

measurements may be adequate on large permeable catchments in which changes in streamflow are slow; on small catchments with impermeable soils, hourly or three hourly measurements may be required for the same purpose.

Periodic level measurements are made visually using gauging staffs, which are graduated scales set in the water. Continuous records are normally obtained using a float and counterweight system mounted over a stilling well which is either set in the stream or connected to it by an inlet pipe. The float and counterweight system is normally linked to a recorder (Fig. 7.21) or a punched paper tape output. As its name suggests the stilling well minimizes variations in the water surface; when used, the inlet pipe must be of sufficient diameter to enable well level to follow stream level rapidly, but small enough to damp out surface oscillations in the stream.

Translation of stream level into flow is achieved in three ways: gauging structures, velocity area methods and chemical dilution gauging, all of which are discussed at length in Institution of Water Engineers (1969). The present account will therefore concentrate on a simple and cheap approach to each of these methods—the V notch weir, the use of floats for velocity measurements and gulp dilution gauging.

10.3.1 V-notch weir

A V notch weir comprises a sharp-edged triangular notch, usually 90°, cut into a vertical metal plate, installed in a watertight structure across the stream channel (Figure 7.22). Marine ply containing a notch edged by a sharpened

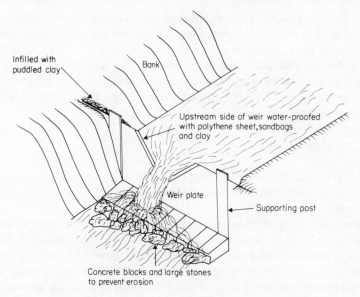

Figure 7.22 V-notch weir. (After Gregory & Walling, 1971.)

metal strip is often used, and this can be simply installed in a stream using sandbags and polythene sheet to provide a solid watertight base. The discharge (Q) is related to the depth of water (h) above the apex of the notch by the equation:

$$Q = 0.015h^{2.48} \text{ litres per second where } h \text{ is in cm}$$

Sediment gathering upstream of the notch must be cleared regularly and although it may not be possible to measure very high discharges in this way because of drowning by downstream levels, the V-notch offers a useful solution where resources are limited.

10.3.2 Velocity–area method

The velocity–area method requires measurements of the area and mean velocity of flow at a chosen point along a stream. Current meters are used to measure velocities in hydrological studies where flow conditions are not excessively turbulent, but with care the use of floats can give the mean velocity to within 10%. Coloured wooden blocks are thrown into the stream and their time of travel over a given reach is measured. This gives the surface velocity which is divided by the factor 1·2 to give the average velocity at that point; multiplying the result by the mean cross-sectional area of the reach gives the discharge.

10.3.3 Gulp dilution gauging

In gulp dilution gauging a known mass of common salt or other cheap, non-toxic chemical is dissolved in water and then thrown into the stream. Samples of the streamwater are taken every 30 seconds for some 15 to 30 minutes, from a downstream point where mixing of chemical and streamwater is adjudged to have taken place. The samples are analysed in the laboratory for one of the ions contained in the injected chemical, and the streamflow is found from the equation:

$$Q = \cdot \frac{M}{\int c \cdot dt}$$

where M is mass of injected tracer (e.g. Na^+ or Cl^-),

C is concentration above natural level at time t at downstream sampling point.

Thorough mixing of chemical and streamwater is vital and for this reason the technique is most applicable to turbulent flow. The length of stream required to achieve mixing is commonly found by throwing a small amount of fluorescene into the stream, choosing the downstream sampling point where the water is evenly coloured over its width. If mixing is adequate accuracies of 2–3% can be achieved by this method.

11 REFERENCES

BEARD J.S. (1962) Rainfall interception by grass. *J. S. Afr. For. Ass.* **42,** 12–15.

BIGGS W. *et. al.* (1971) Photosynthesis light sensor and meter. *Ecol.* **52,** 125–131.

BISSELL V.C. & PECK E.L. (1973) An aerial measurement of snow water equivalent by terrestrial gamma radiation survey. *Bull. Int. Ass. Hydrol. Sci.* **XVIII,** 47–62.

BLYTH K. & PAINTER R.B. (1974) Analysis of snow distribution using terrestrial photogrammetry. *Proc. U.S./IHD Symp,* on Advanced Concepts & Techniques in the Study of Snow and Ice Resources, Monterey, California.

CABORN J.M. (1957) Shelterbelts and microclimate. *Bull. For. Comm., London, No. 29.*

CABORN J.M. (1965) *Shelterbelts and Windbreaks,* 288. London, Faber & Faber.

FLITTERS N.E. (1965) Programming relative humidity in combination with fluctuating temperatures: the influence of relative humidity on development of tropical fruit flies and other insects. Section 1, No. 8, 65–72. In E.J. Amdur (ed.) *Humidity and Moisture,* Vol. 2. Applications. New York, Reinhold Publ. Corpn.

FUNK J.R. (1959) Improved polythene shielded net radiometer. *J. Sci. Instrum.* **36,** 267–270.

GATES D.M. (1962) *Energy Exchange in the Biosphere.* New York, Harper & Row.

GILBERT O.L. (1970) Further studies in the effect of SO_2 on lichens and bryophytes. *New Phytol.* **69,** 605–727.

GREGORY K.S. & WALLING D.E. (1971) Field measurements in the drainage basin. *Geography,* lvi **4,** 277–292.

HAMILTON E.L. & ROWE P.B. (1949) Rainfall interception by Chaparral in California. *Calif. For. Range Exp. Sta.* 1–46.

HARROLD L.L. (1966) Measuring evapotranspiration by lysimetry. In *Evaporation and its Role in Water Resources Management.* Michigan, Amer. Soc. Agr. Engrs.

HAWKSWORTH D.L. & ROSE F. (1970) Quantitative scale for estimating sulphur dioxide air pollution in England and Wales using Epiphytic Lichens. *Nature* **227,** 145.

HILL L., GRIFFITHS O.W. & FLACK M. (1916) Kata thermometers. *Phil. Trans. Roy. Soc. B* **207,** 186.

HIRST J.M. (1957) A simplified surface wetness recorder. *Pl. Path.* **6,** 57–61.

IDLE D.B. (1968) The measurement of apparent surface temperature. In R.M. Wadsworth (ed.) *The Measurement of Environmental Factors in Terrestrial Ecology,* pp. 47–58. Oxford, Blackwell Scientific Publications.

INSTITUTION OF WATER ENGINEERS (1969) *Proc. Symp. River Flow Measurement,* 187. Loughborough.

JEHN K.H. (1948) *Texas Univ. Elec. Res. Lab. No. 082, 055, Rpt. No. 20.*

JOHNSTON R.D. (1964) Water relations of Pinus Radiata under plantation conditions. *Aust. J. Bot.* **12,** 111–124.

KURTYKA J.C. (1953) *Precipitation Measurements Study, Rpt. of Invest.* 20, State Water Survey Div., Urbana, 111.

LEYTON L., REYNOLDS E.R.C. & THOMPSON F.B. (1968) Interception of rainfall by trees and moorland vegetation. In R.M. Wadsworth (ed.) *The Measurement of Environmental Factors in Terrestrial Ecology,* pp. 97–108. Oxford, Blackwell Scientific Publications.

LONG I.F. (1957) Instruments for micrometeorology. *Quart. J. Roy. Met. Soc.* **83,** 202–214.

MONTEITH J.L. (1954) Error and accuracy in thermocouple psychrometry. *Proc. Phys. Soc., London* **67,** 217.

MONTEITH J.L. (1965) Evaporation and environment. *Symp. Soc. Exptal. Biol.* **29,** 205–234.

MONTEITH J.L. & SZEICZ G. (1962) Simple devices for radiation measurements and integration. *Arch. Met. Geophs. Bioklim, B* **11,** 491–500.

MONTEITH J.L. (1972) *Surveys of Instruments for Micrometeorology.* I.B.P. Handbook No. 22, pp. 263. Oxford, Blackwell Scientific Publications.

PELTON W.L. (1961) The use of lysimetric methods to measure evapotranspiration. *Proc. Hydrol. Symp.* **2**, 106–134. Ottawa.

PENMAN H.L. (1948) Natural evaporation from open water, bare soil, and grass. *Proc. Roy. Soc., London A* **193**, 120.

PENMAN H.L. & LONG I.F. (1949) A portable thermistor bridge for micrometeorology among growing crops. *J. Sci. Instrum.* **26**, 77–80.

PERRING F.H. & WALTERS S.M. (1962) Interpretation of maps. *Atlas of British Flora* XX–XXIII.

PHILIP J.R. (1961) The theory of heat flux meters. *J. Geophys. Res.* **66**, 571–579.

PLATT R.B. & GRIFFITHS J.G. (1965) *Environmental Measurement and Interpretation*, p. 235. New York, Reinhold Publ. Corpn.

POWELL M.C. & HEATH O.V.S. (1964) A simple and inexpensive integrating photometer. *J. exp. Bot.* **15**, 187–191.

REYNOLDS E.R.C. (1964) The accuracy of raingauges. *Met. Mag., London* **93**, 65–70.

RUTTER A.J. (1963) Studies in the water relations of *Pinus sylvestris* in plantation conditions. I. Measurements of rainfall and interception. *J. Ecol.* **51**, 191–203.

RUTTER A.J. (1964) Studies in the water relation of *Pinus sylvestris* in canopy conditions. II. The annual cycle of soil moisture change and derived estimates of evaporation. *J. Appl. Ecol.* **1**, 29–44.

RUTTER A.J. (1968) The water consumption of forests. In T.T. Kozlowski (ed.) *Water Deficits and Plant Growth*. Academic Press.

SMITH F.M., COOPER C.F. & CHAPMAN E.G. (1967) Measuring snow depths by aerial photogrammetry. *Proc. 35th Western Snow Conf., Boise*, 66–72.

SOLOMON·M.E. (1945) The use of cobalt salts as indicators of humidity and moisture. *Ann. Appl. Biol.* **32**, 75.

STRANGEWAYS I.C. (1972) Automatic weather stations for network use. *Weather, Oct.* 403–408.

SWINBANK W.C. (1951) The measurement of vertical transfer of heat and water vapour by eddies in the lower atmosphere. *J. Meteorol.* **8**, 135–145.

SZEICZ G. (1965) A miniature tube solarimeter. *J. Appl. Ecol.* **2**, 145–147.

SZEICZ G. (1966) Field measurements of energy in the 0·4–0·7 micron range. In R. Bainbridge, G.C. Evans & O. Rackham (eds.) *Light as an Ecological Factor*, pp. 41–51. Oxford. Blackwell Scientific Publications.

SZEICZ G., MONTEITH J.L. & DOS SANTOS J.M. (1964) Tube solarimeter to measure radiation among plants. *J. Appl. Ecol.* **1**, 169–174.

TANNER C.B. (1963) Basic instrumentation and measurements for plant environment and micrometeorology. *Soils Bull., No. 6*, Coll. Agric. Univ., Winconsin, U.S.A.

TOLLAN A. (1970) Experience with snow pillows in Norway. *Bull. Int. Ass. Sci. Hydrol.* XV, **2**, 66–72.

WADSWORTH R.M. (ed.) (1968) *The Measurement of Environmental Factors in Terrestrial Ecology*, British Ecological Society Symposium 8. Oxford, Blackwell Scientific Publications.

WARREN WILSON J. (1957) Observations on the temperatures of Arctic plants and their environment. *J. Ecol.* **45**, 499–531.

WHIPKEY R.Z. (1965) Subsurface stormflow from forested slopes. *Bull. Int. Ass. Sci. Hydrol.* X, **2**, 74–85.

WORLD METEOROLOGICAL ORGANISATION (1963) The influence of weather conditions on the occurrence of apple scab. *Tech. Note* **65**, Geneva.

CHAPTER 8

CHEMICAL ANALYSIS

S.E. ALLEN, H.M. GRIMSHAW, J.A. PARKINSON, C. QUARMBY and J.D. ROBERTS

1   Introduction   412
1.1   Expression of results   413
1.2   Accuracy and precision of analyses   413

2   Soils   414
2.1   Collection   414
2.2   Sampling equipment   414
2.3   Transport and storage   415
2.4   Drying   415
2.5   Sieving and grinding   415
2.6   Preparation of solution for analysis   416
   2.6.1   Choice of extractant (for mineral cations)   417
   2.6.2   Conditions of extraction   418
   2.6.3   Extraction of common nutrient cations   418
   2.6.4   Cation Exchange Capacity (CEC)   419
   2.6.5   Notes on extraction procedures for other common elements   420
   a. Aluminium
   b. Iron
   c. Nitrogen
   d. Phosphorus

3   Plant materials   421
3.1   Sample collection   421
3.2   Transport and storage   422
3.3   Drying   422
3.4   Grinding and sieving   423
3.5   Preparation of solution for analysis   423
   3.5.1   Dry ashing   423
   3.5.2   Acid digestion   424

4   Animal tissues   426
4.1   Collection and storage of tissues   426
4.2   Drying and grinding   427
4.3   Preparation of sample solution   427

5   Waters   427
5.1   Collection and storage   428
5.2   Preliminary tests   429
   5.2.1   Total suspended solids   429
   5.2.2   Total dissolved solids   429
   5.2.3   Total organic matter   429
   5.2.4   pH   429
   5.2.5   Conductivity   429
   5.2.6   Acidity and alkalinity   430
   5.2.7   Dissolved oxygen   431
   5.2.8   Dissolved carbon dioxide   431
5.3   Preparation of solution and further analysis   431
   5.3.1   Dissolved constituents   431
   5.3.2   Total constituents   432

6   Non-specific tests   432
6.1   Moisture   432
6.2   Loss on ignition and crude ash   433
6.3   Silica-free ash (plant material)   434
6.4   pH (soils, waters)   434
6.5   Redox potential (soils)   434

7   Determination of individual elements   435
7.1   General notes   435
   7.1.1   Instrumental techniques   435
   7.1.2   Sample solutions and standards   436
   7.1.3   Calibration curves   436
   7.1.4   Calculations   436
   a. Flame emission and atomic absorption   437
   b. Colorimetry   438
   c. Titrations   438
7.2   Typical nutrient levels in soils and plant materials   438
7.3   Calcium   438
   7.3.1   Atomic absorption   438
   7.3.2   EDTA titration   440

7.4   Carbon   441
      7.4.1   Organic carbon   441
      a. Wet oxidation   442
      b. Rapid titration   444
      7.4.2   Carbonate-carbon   445
7.5   Iron   445
      7.5.1   Atomic absorption   445
      7.5.2   Colorimetry   446
7.6   Magnesium   447
      7.6.1   Atomic absorption   448
      7.6.2   EDTA titration   448
7.7   Manganese   449
      7.7.1   Atomic absorption   449
      7.7.2   Colorimetry   450
7.8   Nitrogen   451
      7.8.1   Total organic nitrogen   451
      7.8.2   Organic nitrogen in
              waters   453

7.8.3   Ammonium nitrogen   454
a. Distillation   454
b. Colorimetry   456
7.8.4   Nitrate nitrogen   456
a. Distillation   456
b. Colorimetry   457
c. Selective ion-electrode   458
7.9   Phosphorus   458
      7.9.1   Colorimetry   458
7.10 Potassium   460
      7.10.1 Flame emission   460
7.11 Silicon   461
      7.11.1 Colorimetry   461
7.12 Sodium   462
      7.12.1 Flame emission   462

8   References   463

1 GENERAL INTRODUCTION

This chapter mainly deals with the chemical analysis of soils, plant materials and waters which are of most interest to the ecologist. Many of the methods are also applicable to animal materials after slight modifications. The information given should be sufficient for the non-chemist to attempt the analysis himself, but references to additional sources of information are also given where appropriate.

The procedures mainly concern the principal inorganic nutrient elements, organic constituents being omitted as being beyond the scope of this text. Some physical analyses are dealt with elsewhere in the handbook.

Before attempting any chemical analyses the following general points should be noted.

1. In most methods at least two blank determinations should be carried through in the procedure to allow for any background contamination introduced from chemicals, filter papers, etc. The mean of these values should be subtracted from the sample result.

2. It is good practice to include in any run of samples a reference material whose chemical composition is known. This will monitor any gross errors, particularly bias, which may occur.

3. Analytical grade reagents and chemicals should be used where possible. Many salts which are used to prepare standard solutions require drying at 105°C for 3 hours before use.

4. Glass-distilled or deionized water should always be used. Deionized water should have a conductivity value $< 0 \cdot 2 \, \mu S \, cm^{-1}$.

5. Glass and plastic ware used in the procedures should be scrupulously clean to avoid contamination. Washing with detergent (preferably phosphorus-free) followed by several rinses with distilled or deionized water will suffice in most instances. More powerful cleaning agents include chromic acid and alcoholic caustic potash solution. Borosilicate glass should be used where possible.

6. Sources of contamination must be identified and eliminated. Some of the more common examples include:

   (a) Metallic dust from grinding and sieving.
   (b) Dust from general cleaning operations.
   (c) Powder and rust from decorations and fittings.
   (d) Soap, washing materials and cosmetics.
   (e) Laboratory reagents from other tests.

1.1 EXPRESSION OF RESULTS

1. SI units should be used throughout. A description of this system is given by the Royal Society (1971). In this text litre is used in place of dm^3. The other main change is that the unit of conductance is now the siemen (S) which is identical with the former mho.

2. Total constitutents should be expressed as % or $\mu g\ g^{-1}$ in soils and plant materials and $mg\ l^{-1}$ in waters.

3. Parts per million (ppm) should be avoided in quoting results and if employed at all should be confined to internal use in the laboratory.

4. Soil extractable data is normally quoted as $mg\ 100\ g^{-1}$ but can be expressed in terms of equivalents, i.e. $ml\ 100g^{-1}$ which is obtained by dividing $mg\ 100\ g^{-1}$ by the equivalent weight of the element in question.

1.2 ACCURACY AND PRECISION OF ANALYSES

Accuracy is defined as the closeness of a result to the 'true' or 'absolute' value. Any bias in a method will reduce its accuracy. Precision is the spread of results within a group of replicate determinations carried out together. Reproducibility is the spread of results within a group of replicate determinations carried out over a period of days or weeks.

Both precision and reproducibility are governed by random errors and may be calculated statistically. Further information on this and other statistical aspects of experimental work are given by Grimshaw & Lindley (1974). See also 2.1 below. Some values for the precision of individual methods are given in the text on the assumption that optimum operating conditions prevail.

Some of the factors causing inaccuracy or lack of precision are mentioned where necessary for the different methods. Many are associated with inter-element interferences and separate tests may sometimes be needed to check on this effect. The method of adding known amounts of different elements to a sample solution is an effective way of checking for recovery under operating conditions.

2 SOILS

Soil is an extremely complex medium and its variability causes many problems in analysis and in interpretation of analytical results. Ball discusses some of these problems in Chapter 6. Reference books dealing with the analysis of soils include Piper (1950), Metson (1956), Jackson (1958), Black (1965), Hesse (1971) and Allen *et al.* (1974).

There is no generalized technique for the handling of soils in chemical analysis but certain basic principles underlay the variety of procedures in use. These will be discussed under each sub-heading as appropriate.

2.1 COLLECTION

In general soil sampling methods depend on the type of study being carried out. In experimental studies, sampling should be randomized to minimize bias and to facilitate judgement of the significance of the results. Two or three replicate samples will suffice in most cases but in some experimental designs five or even ten may be required. This may result in large numbers of samples for analysis. The number may be reduced by bulking but this will lead to loss of information on variability. In survey work systematic rather than random sampling may be employed by utilizing a grid or series of transects over the chosen area, but it is important to ensure adequate replication at each sampling point.

The purpose of the investigation also determines to some extent the depth of sampling. Often rooting depth (10 to 15 cm) will be sufficient but occasionally profile sampling to greater depths is required. This involves sampling at specific depths or according to the soil horizon.

2.2 SAMPLING EQUIPMENT

For results which are to be expressed on a weight basis a trowel or similar tool is sufficient. For analysis on a volume basis soil corers or augers are preferable but not always suitable. Methods for measuring bulk density of soils where augers cannot be used are described in Chapter 6. Care must be taken to guard against inter-horizon contamination particularly in sandy soils. This problem can become accentuated when using augers or corers.

Polythene bags are frequently used for the collection of soil samples. Although they are suitable for short-term use there is a danger of incubation effects particularly when left sealed in a warm atmosphere. Paper bags with waterproof liners were widely used before polythene became available. Plastic and aluminium containers are sometimes chosen but the possibility of contamination particularly from metal must not be overlooked.

2.3 TRANSPORT AND STORAGE

Transport times should be kept short and the samples cool during transit, to minimize possible deterioration. Frozen 'cold packs' with the samples packed in insulated containers are effective for maintaining low temperatures. In the laboratory prolonged storage of fresh soil samples should be avoided if possible. If this is unavoidable then the samples should be kept at a temperature just above freezing because the exchange properties of a soil can be affected by freeze storage (Allen & Grimshaw, 1962). Labile constituents such as inorganic nitrogen fractions and values such as pH should be determined immediately on receipt but many other estimations can be made using the air-dried material.

2.4 DRYING

In practice there are no ideal drying conditions. The most satisfactory procedure for air-drying is to leave the thinly spread sample at 40°C in an air-circulated oven overnight. True air-drying on a laboratory bench is laborious and time consuming, and the extra time needed to reach equilibrium may result in marked changes due to bacterial and enzymic activity. When analysis for extractable constituents is required fresh or air-dried material must be used and high temperatures (105°C oven-dried) avoided as these cause irreversible changes in the lattice structure of the soil minerals. Some recommendations for drying are given in Table 8.1.

2.5 SIEVING AND GRINDING

Fresh material is handled as received, although it is usual to pick out large stones and roots before proceeding. Air dry material should be passed through a 2 mm mesh sieve, aggregates being broken gently because some minerals are very soft. Machines are available to standardize this operation. A suitable model is marketed by D. Mackay of Brittania Works, Cambridge in which stainless steel bars inside a cylindrical cage drilled with 2 mm holes crush the

Table 8.1 Procedures for the initial treatment of soils.

Analyses required	Recommended treatment
Redox potential, pH	Field examination
NO_3^-, NH_4^+, pH Peat extractions	Minimum storage at +1°C, then examine fresh
Extraction of mineral elements and P Cation exchange capacity	Air-dry at less than 40°C, then lightly crush through 2 mm sieve
Loss on ignition Total N, P, C	Dry at 105°C then grind finely

soil as the cylinder rotates. Stainless steel sieves are recommended to reduce contamination. Brass should be avoided, particularly if copper and zinc are to be determined.

The soil will need to be well mixed after sieving and in addition some bulk reduction may be necessary. Successive mixing, heaping and quartering with retention of opposite quarters is the standard procedure for this purpose. There are also a number of mechanical aids which are suitable for sample division.

Soil passed through 2 mm should be used for the determination of extractable nutrients, but a further grinding stage will be needed for other procedures particularly if only a fraction of a gram of sample is taken. In this case a representative fraction should be ground to pass 100 or even 250 mesh (Kleeman, 1967) using a mortar and pestle, ball or swing mill or an abrasive grinder. The grinding equipment should be made of agate, hardened steel or tungsten carbide.

2.6 PREPARATION OF SOLUTION FOR ANALYSIS

The total concentrations of mineral elements in soils are generally of greater interest to pedologists than to ecologists and methods are not therefore dealt with here. This does not apply to carbon, nitrogen and phosphorus because large proportions of each are organically bound resulting from biological processes and so a total (organic) content is often required in ecological studies.

A small proportion of each mineral nutrient cation is present in true solution or held in the ionic form by negative organic and inorganic adsorption sites in the colloidal complex. This is the fraction most readily available for plant uptake and an estimate of it is much more informative for nutritional studies than the total content. Adsorbed cations are readily exchanged for another cation supplied in excess by a chemical extractant. The displaced cations are then measured in the filtered extract.

This is the basis of soil extraction methods which provide a value often recorded as 'exchangeable cation'. However, in practice the action of an extractant is somewhat arbitrary and inevitably releases some of the less available element from inorganic salts and/or organic matter depending on the extraction conditions notably the pH of the extractant (see 2.6.1). For this reason it seems best to use the term 'extractable cation' and to always specify the extractant employed. However, in determining the total displaced cations the term 'cation exchange capacity' is retained and is discussed further in 2.6.4.

Extractants are widely used in agricultural work, particularly for predicting fertilizer requirements and their use is justified because values from specific extractants have been correlated with crop uptake on various soil types. Ecologists do not normally have such information for native species and extractable data should be regarded as no more than a guide to the nutrient status of natural sites.

The only commonly extracted anions are phosphate and nitrate. The available fraction is largely in solution and amounts are very low in many natural soils. Inorganic fixation is an important complication for phosphate-phosphorus and in general specific extractants are used for this nutrient (see 2.6.5).

2.6.1 Choice of extractant (for mineral cations)

Two widely used extractants are M ammonium acetate at pH 7 and (in Britain) Morgan's reagent which is 10% sodium acetate in 3% acetic acid. These rely upon ammonium (NH_4^+) and sodium (Na^+) ions respectively to displace nutrient cations from the soil exchange complex. Another frequently used extractant is 2·5% acetic acid.

Although ionic displacement is the basis for their action, all these extractants, but notably 2·5% acetic acid have a solubilizing effect. It is this property which more particularly determines the choice.

A good extractant should dissolve a minimum of both mineral salts and organic matter. The former is best achieved by extractants with high pH values and the latter with low pH extractants. Hence neutral extractants such as ammonium acetate at pH 7 are generally the best overall. Calcareous soils, however, always require an extractant with a high pH value to minimize the dissolution of calcium carbonate. Ammonium acetate at pH 9 can be used. Organic soils such as peat are best extracted with ammonium acetate (pH 7), taking care that the extraction conditions are optimal (see below).

In practice the two ammonium acetate extractants can be employed for a range of non-saline soil types but 2·5% acetic acid should be confined to non-calcareous, non-saline mineral soils even though it is more convenient to prepare and, unlike ammonium acetate, is suitable for phosphorus (see 2.6.5).

All three are described below but Morgan's reagent is omitted. This extractant is a useful alternative to ammonium acetate at pH 7 and can also be used for phosphorus but its very high salt content is a drawback for some flame equipment.

2.6.2 Conditions of extraction

It is preferable but not essential to extract fresh material. Air-dry sieved material, however, is more convenient to handle and more homogeneous and is therefore often used. Peat, however, should always be extracted fresh. Oven-dried material (105°C) is not suitable.

The ratio of extractant to soil is important. If the proportion of soil is too high then re-absorption of the released nutrients becomes significant. This is particularly likely in the case of clay soils. If the proportion of extractant is too high then the quantities extracted are too low for satisfactory measurement. A ratio of about 25:1 is suggested for air-dried soil, and 10:1 for fresh soil. Ideally the extractant is allowed to pass through soil columns but in practice a shaker is recommended. Peat is more conveniently extracted using a leaching procedure carried out on a funnel lined with filter paper as shaking results in a dense colloidal suspension which is difficult to filter.

Sufficient time should be allowed for the extraction to reach equilibrium but slow dissolution of rock minerals may continue after this stage has been reached. Periods of half to one hour are usual.

It is most important to keep all the conditions constant for samples whose results are to be subsequently compared.

2.6.3 Extraction of common nutrient cations

PROCEDURE

1. Weigh 10 g sieved air-dried soil into a 500 ml polythene bottle.
2. Add 250 ml extractant and shake for one hour on a rotary end-over-end shaker.
3. Filter through No. 44 filter paper rejecting first few ml of filtrate.
4. Prepare blank solutions following the same procedure.
5. Use the filtrate to determine cations by the methods given in 7.

REAGENTS

1. Ammonium acetate, M. (pH 7)—Add 575 ml of glacial acetic acid and 600 ml ammonia solution (0·880) to about 2 litres water. Mix and dilute to 10 litres. Check that the pH is $7·0 \pm 0·1$. If necessary adjust with acetic acid or ammonia solution.

2. Acetic acid 2·5% v/v—Dilute 250 ml glacial acetic acid to 10 litres with water.

3. Ammonium acetate, M. (pH 9)—Prepare as for the pH 7 extractant above but include 740 ml ammonia solution (0·880).

Note

The 2·5% acetic acid extract may also be used for the determination of extractable phosphorus (see 2.6.5).

2.6.4 Cation exchange capacity

The total content of cations which can be displaced from the soil complex is termed the Cation Exchange Capacity (CEC). M ammonium acetate at pH 7 is generally employed for the initial displacement and the excess ammonium ions are removed by washing. The adsorbed ammonium is then itself displaced by the cation of another leaching solution in which the ammonium-nitrogen is later determined. Although this procedure has limitations similar to those of other extraction methods the CEC value is often a useful indicator of soil nutrient status.

PROCEDURE

1. Extract the soil with M ammonium acetate (pH 7) as given above and filter.

2. Wash the residue on the filter paper repeatedly with 50 ml portions of 60% industrial spirit until excess ammonium acetate has been removed. Add 10 ml 10% w/v NH_4 Cl solution to the first portion of industrial spirit. Test filtrate for chloride (with $AgNO_3$ solution). When clear of chloride excess ammonium acetate is assumed to have been removed.

3. Leach the residue with about 30 ml of 1% KCl solution followed by several further portions of a 5% KCl solution, collecting the leachate in a vessel marked at 250 ml (500 ml for fresh samples). Allow each portion to pass through before adding the next. Collect the leachate up to the mark.

4. Prepare blank leachates following the same procedure.

5. Determine NH_4^+-N in an aliquot of the leachate by distillation and titration with M/140 HCl as given in 7.8.3.

REAGENTS

1. Ammonium acetate, M.(pH 7)—prepare as given above.

2. Industrial spirit, 60% (v/v).

3. Potassium chloride, 5% (w/v).

Other reagents are as listed in 7.8.3.

CALCULATION

If

$$1 \text{ ml M}/140 \text{ HCl} = 0{\cdot}1 \text{ mg NH}_4^+\text{-N}$$

then CEC

$$\text{CEC (me 100 g}^{-1}) = \frac{\text{titre HCl (ml)} \times \text{leachate volume (ml)}}{1{\cdot}4 \times \text{aliquot (ml)} \times \text{sample wt (g)}}$$

Correct to dry weight where necessary.

Total exchangeable bases and exchangeable hydrogen.

It is sometimes of value to estimate the total exchangeable bases (TEB) where 'base' is understood to exclude hydrogen. The TEB can be obtained by summation of the individual metal cations or directly by ignition of the ammonium acetate extract, dissolution in excess standard acid and back titration with standard alkali (Bray & Willhite, 1929). If the TEB is expressed as a percentage of the CEC it is known as % base saturation.

Exchangeable hydrogen is sometimes measured although it is difficult to define in terms of soil processes. Its role in soil acidity is partly bound up with that of aluminium. It is probably best determined simply as CEC-TEB. Direct methods seem to be of limited value and are normally based on a titration of the soil filtrate back to the initial pH (7·0) or use of the pH of the filtrate (Brown, 1943).

2.6.5 Notes on extraction procedures for other common elements

a. Aluminium

The method of extraction used depends on the purpose of which the results are required: for profile studies, 3% oxalic acid is recommended as described by Ball & Beaumont (1972); for nutrient studies, M ammonium acetate at pH 4·8 is more suitable.

b. Iron

3% oxalic acid is used as extractant for investigations into the development of soil profiles. Soil nutrient studies involving extractions for iron are rarely carried out.

c. Nitrogen

Water is normally used for the extraction of nitrate-N and 6% sodium chloride solution for ammonium-N. If both ions are to be determined in the same sample it is convenient to use 6% sodium chloride solution for both. Nitrite-N is extracted similarly to nitrate-N. In all cases extract fresh soil. Weigh 25 g, add 180 ml extractant and stir for 10 minutes. Centrifuge until supernatant liquid is clear.

d. Phosphorus

There are several extractants suitable for phosphorus some of which are claimed to extract 'available phosphorus', but in practice all release some fixed phosphorus. Three extractants are in common use for phosphate-phosphorus:

a. Truog's reagent (Truog, 1930).

0.001 M H_2SO_4 buffered at pH 3 with $(NH_4)_2SO_4$. Suitable for neutral and acid soils. Use air-dried material and an extractant: soil ratio of $200:1$. A 30-minute shaking time is required. Not suitable for calcareous soils.

b. Acetic acid, 2.5% v/v (as in 2.6.3). Not suitable for calcareous soils.

c. Olsen's reagent (Olsen *et al.*, 1954).

0.5M $NaHCO_3$ adjusted to pH 8.5 with NaOH. This is used for calcareous soils only. A ratio of $20:1$ extractant to soil is required and the extraction is carried out on the air-dry material. A 30-minute shaking time is recommended. Some organic phosphorus may be extracted by this high pH extractant but the colorimetric procedure given in 7.9 estimates only phosphate-phosphorus.

Methods suitable for the determination of soil organic phosphorus are referred to in 7.9.1.

3 PLANT MATERIALS

This section deals with the initial treatment of vegetation and litter and preparation of the solution for subsequent determination of the individual elements. Certain organic soils such as peats may be treated in the same way. General texts dealing with the analysis of plant materials include Piper (1950), Chapman & Pratt (1961), AOAC (1970) and Allen *et al.* (1974). Although the procedures are basically suitable for animal tissues there are some exceptions, which are discussed in 4.

3.1 SAMPLE COLLECTION

Some aspects of the sampling of plant communities have already been discussed in Chapter 4, Section 3.3. Several points must be taken into account:

a. Choice of component for analysis—in general the green photosynthetic parts will give the most relevant information on the nutrient status of the plant.

b. Age of plant or component—for most purposes the current year's growth is adequate.

c. Sampling time—the composition changes with the season but many elements reach a relatively constant level over the late July to early August period.

d. Site aspect and degree of shading which have minor effects.

e. Statistical aspects including randomization and replication.

f. Field contamination due to soil or to mammal and bird droppings.

Methods of sampling are similar to those used in forestry and agriculture. High level pruners are suitable for tree leaves. Shears or secateurs are suitable for grassland and similar herbage, although 'hand snatch' samples may be considered adequate for some purposes. In the latter case polythene gloves should be worn to avoid contamination from perspiration.

3.2 TRANSPORT AND STORAGE

Transport and storage problems are generally similar to those encountered when dealing with soils. Estimates of total nutrients do not normally require analysis of fresh plant material, so it is advisable to dry the samples as soon as possible. However, if long-term storage of fresh material is required then freeze storage at below—$10°C$ is recommended.

Contamination of vegetation by soil splash, airborne dust and animal materials can be troublesome. Removal by gentle agitation in water using ionic vibrators is sometimes recommended but such a treatment can result in some nutrient loss. It is generally preferable to confine cleansing to the wiping of the sample material with a very weak solution of detergent in water. It is necessary to check that the detergent itself will not introduce contamination. Steyn (1959) reports on the extent of elemental losses during cleaning.

3.3 DRYING

Unless labile constituents are to be examined oven-drying is usually acceptable for vegetation. The choice lies between high temperatures for a short period or lower temperatures for a longer time. With the former there is a risk of volatilization losses, and with the latter the possibility of increased microbial activity in the initial stages. For plant material drying at $40°C$ for several hours until the sample is friable enough for grinding is recommended when total nutrient elements are to be determined. Just prior to analysis a small sub-sample of the ground material should be dried at $105°C$ for three hours or an air-dry moisture determined so that analytical values may be corrected to a dry weight basis.

3.4 GRINDING AND SIEVING

Samples are ground to produce a sufficiently homogeneous material for a representative sample to be taken for analysis. Unlike soils there is no standard mesh size, but between 0·4 and 0·7 mm is convenient for most purposes.

The types of grinding machines available include:

a. Beater Cross Mill—best for woody and bulky samples but can cause differential loss of dust.
b. Disc Grinder Mill—very effective for leaves but tends to overheat.
c. Cutting Mill—used only for small leaf samples but gives a high recovery.
d. Ball Mill—gives a high recovery but a longer grinding time is necessary.

All grinders can contaminate samples depending on the materials used for the grinding enclosure. Fresh material is best treated in a macerator, e.g. Waring blender or a simple Petter-Elverheim homogenizer.

3.5 PREPARATION OF SOLUTION FOR ANALYSIS

Analysis for total nutrient elements in plant materials requires the complete breakdown or oxidation of all organic matter. The two basic methods are dry ashing and acid digestion.

3.5.1 Dry ashing

Whilst not being wholly convenient the method has two advantages:

a. Relatively large sample weights can be handled.
b. The residue is taken up in hydrochloric acid which is an excellent solvent and very suitable for subsequent analysis.

The main disadvantage is that it is very difficult to remove all the carbon without some loss of volatile mineral nutrients, notably phosphorus in non-calcareous samples. This may be prevented by adding an alkaline salt, such as calcium or magnesium acetate, but these elements themselves may have to be estimated. There is also a slight loss of heavy metals or trace elements by retention on silica (Likens & Bormann, 1970).

The usual temperature for ashing is 450°C to 500°C for times varying from an hour at 500°C to several hours at 450°C. The muffle furnace should not be preheated, but should be started from cold with the samples inside. This prevents sudden deflagration. Nitric acid added at the dissolution stage facilitates the oxidation of ferrous salts.

PROCEDURE

Suitable for determinations of sodium, potassium, calcium, magnesium, iron, and manganese, but not for phosphorus unless the material is calcareous.

1. Weigh 0·5 g dry ground sample into a dry acid-washed porcelain crucible.
2. Ignite at 500°C for 1 hour. Make sure air has access to the muffle furnace and that the chimney vent is open.
3. Allow to cool, add 5 ml HCl (1 + 1) and warm to dissolve the residue.
4. Add 0·5 ml of conc HNO_3 and evaporate to dryness on a boiling water bath. (If silica is required continue heating for one hour to dehydrate the residue.)
5. Add 2 ml HCl (1 + 1), swirl, warm slightly to dissolve the residue and dilute.
6. Filter through a suitable paper into a 100 ml volumetric flask, wash the residue and dilute the filtrate to volume. This solution contains 1% HCl (v/v).
7. Prepare blank solutions following the same procedure.

3.5.2 Acid digestion

Acid digestion of an organic material is an oxidizing system and has the advantage that phosphorus and in some cases, even nitrogen can be determined on the final solution along with other nutrients. In general the procedure is preferable to dry ashing. Some points to note are outlined below.

a. In some circumstances the use of perchloric acid for oxidizing organic materials can be hazardous because of the explosive nature of some perchlorates. Recommendations for the use of perchloric acid are given by the Society for Analytical Chemistry (1959, 1960) and by Gawen (1965). In the procedure given in the text sulphuric acid is added to the digestion mixture to reduce the possibility of drying out. Any interference in the subsequent analysis can be compensated for by inclusion of sulphuric acid in the standard solutions.

b. As a general rule powerful oxidizing agents such as nitric and perchloric acids will complete the oxidation quickly but will drive off nitrogen. Less powerful oxidants such as hydrogen peroxide in sulphuric acid are slower and require a catalyst, but nitrogen is retained during the reaction.

c. Large sample weights are not easily handled so the method is less suitable than dry ashing for trace elements.

d. Fatty materials, such as certain seeds, may be difficult to oxidize. Prior digestion with nitric acid is then recommended.

If the procedures given below are followed then the loss of nutrients by the formation of insoluble salts such as perchlorates and calcium sulphate should not occur, even with nutrient rich samples. The loss of iron as ferric sulphate is possible but diluting the cold digest slightly and bringing the solution to the boil will re-dissolve the precipitate. Loss of manganese in the oxidized form will only take place in manganese-rich samples and this can be reduced by dilution and boiling with a few drops of sulphurous acid.

Two acid digestion systems which have been proved suitable for use with a wide variety of plant and litter samples are given below.

Nitric acid — perchloric acid — sulphuric acid
(Suitable for determination of sodium, potassium, calcium, magnesium, iron, manganese, and phosphorus but not for nitrogen.)

PROCEDURE

1. Weigh 0·2 g of dry ground sample into a 50 ml Kjeldahl flask.
2. Add 1 ml 60% $HClO_4$
       5 ml conc HNO_3
       1 ml conc H_2SO_4
3. Digest at moderate heat until white fumes are evolved.
4. Heat strongly for a few minutes to drive off most of the perchloric acid and allow to cool. The acid digest should now be colourless or occasionally pink (due to manganese). If yellow or brown reheat with a few drops of $HClO_4$.
5. Dilute the solution and boil for a few minutes if iron is required. Filter through a No. 44 Whatman filter paper and dilute to volume. (Dilution to 100 ml will yield a 1% H_2SO_4 solution which is convenient for use in subsequent analysis.)
6. Prepare blank solutions following the same procedure.

Hydrogen peroxide — sulphuric acid
(Suitable for nitrogen in addition to all the metals listed for the previous system. Selenium is included as a catalyst and a lithium salt to raise the boiling point.)

PROCEDURE

1. Prepare digestion mixture. Add 0·42 g Se and 14 g of $Li_2SO_4.H_2O$ to 350 ml of 100 vol H_2O_2. Mix well and add *with care* 420 ml of conc H_2SO_4. Cool the mixture during the addition of the acid.
2. Weigh 0·4 g of dry ground sample into a suitable Kjeldahl flask.
3. Add 4·4 ml of digestion mixture.
4. Digest at moderate heat until the initial reaction subsides.
5. Continue the digestion until a clear and almost colourless solution is obtained.

6. Dilute the solution and boil if iron is to be determined.

7. Filter through a suitable paper and dilute to volume. (Dilution to 50 ml will yield a 5% H_2SO_4 solution but as 1% is desirable for most analyses a further five fold dilution is necessary.)

8. Prepare blank solutions following the same procedure.

In both systems recovery of the silica from the residue on the filter paper (by ignition) allows an approximate determination of silica.

If system (a) is employed then nitrogen must be determined following a separate digestion. The Kjeldahl method given in 7.8 is suitable for soils and plant materials.

If system (b) is used nitrogen is estimated along with the other elements. However, this method should not be used for nitrogen in soils because selenium is rather less efficient than mercuric oxide as a catalyst for the breakdown of soil organic matter.

4 ANIMAL TISSUES

Many of the procedures which are used for the analysis of vegetation are suitable for animal tissue and faecal materials. However, the initial treatment of this type of sample creates problems. For example the maceration of a mixture of fatty tissue, bone and fur to produce a homogenous sample for analysis is particularly difficult and in these circumstances the examination of individual organs or components is often advisable. Samples of small invertebrates are generally easier to process providing sufficient material is available for analysis.

In a book of this nature it is only possible to highlight some of the principal difficulties and suggest general treatments which may be tried. For further information about the examination of animal materials contact should be made with research organisations specializing in animal health and nutrition.

4.1 COLLECTION AND STORAGE

Meaningful sampling of an animal population requires specific techniques which are outside the scope of this chapter. Statistical considerations apply and it is relatively difficult to take a representative sample of a population. Among factors to be considered are:

    a. Age of individual and population.
    b. Health of individuals and population.
    c. Size of population.
    d. Distribution of population.
    e. Seasonal effects.

It is essential to dry and analyse samples as soon as possible after collection. If delays are unavoidable deep freezing is recommended for storage.

4.2 DRYING AND GRINDING

It is more satisfactory to freeze-dry than to oven-dry animal tissues because freeze-drying minimizes the loss of volatile components. Freeze-dried material can be chopped finely and macerated. The high fat content of animal tissue however, makes it difficult to grind in the type of mill used for plant material. Where facilities are available animal tissue can be handled relatively easily if the sample is first cooled to a very low temperature using liquid nitrogen. Small amounts of fresh material can be homogenized in a blender or mincing machine. The necessity to obtain a homogeneous material means that only small mammals can be handled whole. In the case of larger animals it is usual to select particular organs or tissues for analysis.

4.3 PREPARATION OF SAMPLE SOLUTION

Dry ashing and acid digestion are both applicable but certain limitations should be noted. There is a risk of loss of phosphorus when dry ashing and with both acid digestion systems calcium in many tissues will precipitate as calcium sulphate unless small weights are taken. It is also difficult to digest fatty materials, although pretreatment with nitric acid is helpful. The Kjeldahl technique should be used for nitrogen digestion. Certain animal products such as droppings or gut contents can generally be treated in a similar way to vegetation. However, these materials contain breakdown products in a very labile state, and should be subjected to the minimum of treatment particularly in terms of oven drying (Greenhill, 1960).

5 WATERS

Waters relevant to terrestrial ecology include rainwater, run-off waters, canopy leachates and stem flow waters. The analytical techniques described here are also applicable to open freshwater systems for estimations of total or dissolved salts, but the study of chemical equilibria in lakes and pools forms a limnological rather than a terrestrial ecological problem. Marine ecosystems are outside the scope of this book although many of the analytical techniques can be applied after certain modifications.

Further information on water analysis can be obtained from Standard Methods for the Examination of Waters (1971) and from the Department of the Environment (1972). Books specializing in the requirements of limnologists include Mackereth (1963) and Golterman (1969). The collection of water samples was discussed by Ciaccio (1971). The examination of waters is also considered by Allen *et al.* (1974).

5.1 COLLECTION AND STORAGE

It is not always easy to collect water samples such that the subsequent analyses will give nutrient information. The types of water listed at the beginning of 5 all require specialized techniques apart from stream water which may be collected by immersing a polythene bottle to about one third depth.

The sampling of rainwater causes particular difficulty and the following points should be considered:

a. Choice of site. This can be critical especially in hilly, wooded or built-up areas. Meteorological Office criteria should be observed.

b. A random distribution of gauges on the site is necessary to get unbiased results.

c. Replication must be adequate to get an estimate of variability and to overcome the effects of any gross contamination by bird droppings.

d. The gauge lip should be at least one metre from the ground to avoid soil splash.

e. The funnel and collection vessel must be of glass or polythene. (The standard Meteorological Office rain gauge should not be used.)

f. Collections should be made daily.

Even with those precautions it is still possible to get misleading analytical values and care is needed in interpreting results (Gore, 1968).

Some preservation treatment is needed for water samples unless analysis can be carried out immediately or within hours of the sample being taken. Storage at a temperature just above freezing point is a safe treatment for short periods. Freeze storage is suitable for most mineral elements and phosphorus but is less desirable when iron, aluminium and silica are to be determined. Chemical preservatives, such as mercuric chloride will inhibit microbial changes but their use may result in subsequent analytical interferences. Many of these preservatives are also highly toxic to humans and should be handled most carefully. Heron (1962) suggests the impregnation of polythene storage bottles with iodine to prevent the bacterial growth and uptake of phosphorus. Addition of a few drops of hydrochloric or preferably sulphuric acid to the samples lowers the pH to about 1 and largely prevents bacterial growth. This treatment is also valuable in stabilizing ammonium-nitrogen and in preventing precipitation of iron and other heavy metals. On the other hand dissolved and colloidal organic matter may be precipitated at a low pH and absorb nutrient ions that would otherwise remain in solution. The use of chloroform or toluene as a bacteriostat is widespread but is not always fully effective.

5.2 PRELIMINARY TESTS

There are a number of useful preliminary tests, specific to waters, which should be carried out as soon as possible after collection, and performed on the unfiltered sample prior to any treatment.

5.2.1 Total suspended solids

Dry and weigh a glass fibre filter paper. Pass a known volume of the sample through the filter paper on a funnel, dry and reweigh the paper. Calculate percentage total suspended solids.

5.2.2 Total dissolved solids

Weigh a dry evaporating dish. Evaporate to dryness an aliquot of the filtrate following determination of total suspended solids. Reweigh the evaporating dish and calculate the percentage total dissolved solids.

5.2.3 Total organic matter

Weigh a previously ignited porcelain evaporating dish. Add a known volume of the unfiltered water and evaporate to dryness. Reweigh the dish. Ignite the residue at 500°C for 1 hour. Again reweigh. The difference between dish and dry residue and dish and ignited residue is a measure of the organic matter.

5.2.4 pH

It is advisable to use a pH meter (see 6.4). Colorimetric methods are less suitable particularly for coloured samples and test papers are of very little value.

5.2.5 Conductivity

The specific conductance, or conductivity of a water sample is a measure of the total ionic concentration. Specific conductivity, the reciprocal of specific resistance, is measured in a conductivity cell by a low voltage A.C. wheatstone bridge. A number of commercial instruments are available. Samples are not filtered or treated before measurement.

The conductivities of most fresh waters are low and are expressed in μS cm^{-1} rather than the older mho/cm. A conductivity cell is calibrated against a suitable potassium chloride solution (0·01 M KCl = 1412 μS cm^{-1} or 0·001 M KCl = 147 μS cm^{-1} both at 25°C). The cell constant (C) can then be calculated from the measured resistance of the standard solution in ohms (R_{KCl}) and the specific conductance of the same solution (K_{KCl}) at 25°C.

$$C = R_{KCl} \times K_{KCl}$$

Conductivity varies with temperature so that results must be expressed for a standard temperature (usually 20°C or 25°C). This can be done by bringing the sample to the standard temperature, by a compensating circuit in the conductivity meter, or by calculation.

$$K_{20} = \frac{C}{R_t} \cdot f$$

where K_{20} = specific conductivity at 20°C,
R_t = measured resistance of the sample at t°C,
$f = 1·02^{20-t}$

In acid waters the hydrogen ions can contribute a major proportion of the total conductivity, and it is often ecologically more meaningful to subtract the conductivity due to hydrogen ions leaving a corrected conductivity (K_{CORR}) that is more representative of the plant nutrients in solution.

$$K_{CORR} = K_{20} - (K_{H^+})_{20}$$

This approach has been described by Sjörs (1950) and used by other workers as a convenient method of investigating the base status and overall nutritional relationships of peat ecosystems under field conditions.

5.2.6 Alkalinity and acidity

Alkalinity is largely a measure of carbonate and bicarbonate ions and also hydroxide in the most alkaline waters. Phosphate and silicate also contribute but are not often present in significant amounts. Both alkalinity and acidity can be determined by titrating 100 ml water sample against standard acid or alkali as follows:

pH of water	Titrant	Indicator	Result
< 4·5	0·01 M NaOH	Methyl orange	'acidity'
4·5–8·3	0·01 M HCl	Methyl orange	'total alkalinity'
> 8·3		Phenolphthalein	'OH⁻ alkalinity'

Calculate and express the results in terms of mg l^{-1} $CaCO_3$. It is usual for both the phenolphthalein and methyl orange titrations to be carried out successively on the same sample so that the approximate composition of the total alkalinity may be deduced. For further details see Standard Methods for the Examination of Waters (1971) and Department of the Environment (1972).

5.2.7 Dissolved oxygen

Because of the difficulties of excluding air during sample collection and subsequent handling, direct measurement in the field is preferable. Several instruments can be used for this purpose but the electrode developed by Mackereth (1964) is especially convenient because of its portability and simplicity in use. It is available from several manufacturers. The classical Winkler titration method (fully described in Department of the Environment, 1972), is still in frequent use and does not require any special instrumentation.

5.2.8 Dissolved carbon dioxide

The most suitable procedure is the nomograph method (Standard Methods for the Examination of Water, 1971), whereby the carbon dioxide figure is derived from the total solids content, the bicarbonate alkalinity, pH and temperature measurements.

5.3 PREPARATION OF SOLUTION FOR FURTHER ANALYSIS

5.3.1 Dissolved constituents

Ideally anions and cations are in true solution and each element can be determined directly after removal of suspended matter by filtration. Nevertheless coloured or turbid filtrates cause problems if colorimetric procedures are to be used. Samples may be clarified by flocculation but this technique may lead to loss of ions by absorption. Destruction of colour by evaporation and oxidation will lead to total rather than dissolved constitutents being determined.

The concentrations of some metal ions in solution may be too low to be conveniently measured. In these cases extraction into a small volume of organic solvent following complexing with APDC (ammonium pyrrolidine dithiocarbamate), is a suitable concentration step. (Brooks *et al.,* 1967, Christian & Feldman, 1970).

Concentration by evaporation of the sample is not advisable as this may increase the viscosity and colour. Collective values for anions are of some value and can be estimated as alkalinity (for weak acid salts) or by the ion exchange method of Mackereth (1963) for strong acid salts.

5.3.2 Total constituents

Determinations of total concentrations are carried out after evaporation and oxidation of the organic matter. The method is usually applied to metals and to phosphorus. It is not applicable to nitrogen in water and it is usually of greater value to estimate the organic and inorganic fractions of this element separately.

Procedure for destruction of organic matter
1. Measure a suitable aliquot of unfiltered water into a conical flask with small base (Taylor's pattern) or round bottomed flask.
2. Add 0·5 ml conc H_2SO_4.
3. Boil until white fumes appear but do not allow to dry.
4. Add 1 ml 60% $HClO_4$ and 2 ml conc HNO_3.
5. Digest to white fumes, and continue for a few minutes only. Take care not to allow the digest to boil dry.
6. Dilute, filter through a No. 44 Whatman filter and dilute to volume. (Dilution to 50 ml will yield a 1% H_2SO_4 solution which is convenient for use in subsequent analyses).
7. Prepare blank solutions following the same procedure (see note).

Note   Blank determinations should not be based on the evaporation of deionized or distilled water in place of the water sample since this introduces an additional factor. It is preferable to allow the blanks to commence at the digestion stage.

6 NON-SPECIFIC TESTS

The tests discussed and described below are mainly applicable to soil and plant materials. Tests which are applicable to water only are described in 5.

6.1 MOISTURE

As it is not possible to remove all the water without loss of volatile components or some breakdown of organic matter, it is difficult to define the dry state. A loss of volatiles can occur even at 40°C whilst at higher drying temperatures some of the structural water associated with organic compounds in plants and soils and in the mineral lattice framework in soils will be retained (Mitchell, 1951). For practical purposes these errors are of little significance in determining fresh moisture except when the moisture content is very low. Net dry weights of experimental material may however be slightly affected and prolonged drying at 105°C should be avoided.

PROCEDURE

1. Dry a suitable container with lid at 105°C for a few minutes. Cool in a desiccator.
2. Weigh the container empty and then with a convenient amount of fresh sample. Take enough to be representative of the sample but do not pack the container too tightly.
3. Remove the lid and dry at 105°C for 3 hours in an air-circulated oven. Cool in desiccator and reweigh.
4. Repeat drying for further one hour periods until constant weight is reached.

Calculate as percentage moisture if the figure is required as an analytical result, or as percentage dry matter if it is to be used to convert other results to a dry weight basis. The procedure is the same for the moisture content of air-dry material, but this figure is chiefly used only for correction purposes.

6.2 LOSS ON IGNITION AND CRUDE ASH

It is difficult to remove the last traces of carbon without loss of volatile inorganic components so there remains a basic problem similar to that described for moisture (6.1), namely that of defining the residual mineral state. Generally the errors involved are relatively slight except for very low values of loss on ignition (soils) or crude ash (plant materials). The problem should not however, be overlooked if the loss on ignition figure is regarded as an approximate measure of organic matter (7.4).

PROCEDURE

1. Weigh a porcelain crucible, add about 1 g of sieved dry soil or ground dry vegetation sample and weigh again.
2. Ignite in a muffle furnace at 500°C. Leave soils for about 4 hours but vegetation, litter and peat need only about 1 hour. Allow the muffle furnace with the samples inside to come to temperature from the cold, to avoid deflagration. Ensure adequate access of air.
3. Remove the crucible, cool in a desiccator and weigh.
4. Obtain net loss in weight for soils and net ash weight for vegetation samples. Calculate both as a percentage of the dry weight.

In most cases loss on ignition or crude ash will be required along with a moisture determination and the sequence of events will be, weigh fresh or air-dry sample, dry in oven, reweigh, ignite and reweigh. This avoids the necessity of weighing out oven-dry material which is susceptible to weighing errors.

6.3 SILICA-FREE (plant material)

PROCEDURE

1. Ignite a known weight of dry sample as given above (6.2).
2. Weigh the cooled residue and obtain net weight of ash.
3. Add 3 ml HCl (1 + 1) to the residue and warm to dissolve. Then add 0·5 ml HNO_3 if filtrate is to be analysed. Evaporate dry and heat on a steam bath for one hour to dehydrate the silica.
4. Add 2 ml HCl (1 + 1), warm, dilute, filter the solution and collect the silica on the filter paper.
5. Ignite the paper in a previously weighed crucible and then reweigh the crucible and residue.
6. Obtain the net weight of the silica and calculate as approximate percent silica in the original material.

$$\% \text{ Silica free ash} = \% \text{ crude ash} - \% \text{ silica.}$$

6.4 pH (soils, waters)

The pH values of waters and water extracts of soils are easily determined (particularly with a pH meter) but are less easy to interpret. In particular the influence of aluminium on the pH values of soils must not be overlooked (Coulter, 1969).

PROCEDURE

1. Operate the pH meter as given in the maker's instructions, using glass and calomel electrodes and temperature compensation if provided. A dual glass-calomal electrode is more robust and is recommended for field use.
2. Standardize the meter with two buffers.
3. Use a water to soil ratio of about 2 : 1.
4. Mix the soil and water by stirring and allow to settle for about 20 minutes. Measure the pH of the supernatant liquid.
5. Ensure the temperature of the buffers and the sample solutions are the same and set the temperature compensator accordingly.

Further details are not given here, the reader is referred to Bohn (1971) and Hesse (1971).

6.5 REDOX POTENTIAL (soils)

There are some important oxidation-reduction systems in soils which govern, for example, the balance between ammonium nitrogen and nitrate-nitrogen or the mobilities of iron and manganese. The redox potential, usually determined in soils as a check for possible reducing conditions, is easily measured using a pH meter with a millivolt scale and with a platinum electrode in place of the glass electrode. Alternatively, special meters and electrode units are available. Further details are not given here the reader is referred to Bohn (1971) and Hesse (1971).

7 DETERMINATION OF INDIVIDUAL ELEMENTS

7.1 GENERAL NOTES

7.1.1 Instrumental techniques

The two analytical techniques used for the determination of most of the elements in this section are atomic absorption (with flame emission) and colorimetry. They are recommended as being rapid and technically easy to apply, whilst at the same time being sufficiently accurate for the purpose and in most cases relatively free from interferences. Some gravimetric and volumetric methods are also given. Colorimetric methods for a wide range of elements are described by Snell & Snell (1959), whilst procedures for metals are dealt with by Sandell (1959). Some practical hints on absorption spectrometry are given by Edisbury (1966).

The principles and applications of atomic absorption spectroscopy have been fully discussed by Christian & Feldman (1970), Ramirez Munoz (1968), Elwell & Gidley (1966) and Price (1972). Specialist applications and later developments are dealt with by L'vov (1970) and Dagnall & Kirkbright (1970). Further details are given for specific elements later but it is convenient to mention certain features here.

It is important to operate all instruments according to the manufacturer's instructions. This is particularly so for flame instruments where such details as gas flows, flame lighting procedure, etc., are specific for particular models. The correct burner must be fitted according to the gases used. Some models also have a special burner for aspirating soil extracts such as ammonium acetate which rapidly clog normal burners with carbon. Various causes of error in flame work include:

i. Blockages in the nebulizer system.
ii. Changes in air and fuel rates.
iii. Drift in the monochromator.
iv. Low pressures in the fuel supply.
v. Hollow cathode lamp dift.
vi. Flame and chemical interferences.

These sources of error are discussed in the references provided, but details for controlling the more important interferences particularly for calcium and magnesium are given at the appropriate places in the sections that follow.

7.1.2 Sample solutions and standards

Solutions should be prepared by the methods given in 2.6, 3.5, 4.3 or 5.3. Digest solutions are diluted to volume to give acid concentrations of 1% or 5% (v/v) but only 1% solutions should be analysed by the procedures given. High levels of acid soon corrode neubulizers and burners of flame equipment and may give unnecessary problems of pH control in some colorimetric methods.

Standards are prepared as given in the procedures for the individual elements. In both flame and colorimetric methods the standards must be run with the samples and should always contain equivalent levels of the background components (acid, soil extractant, etc.). In flame work composite standards containing two or more elements are permissible but not wholly recommended because it is difficult to avoid mutual contamination during their preparation.

In flame methods any sample solutions which have to be diluted below 1% (v/v) for specific purposes should include sufficient background components to restore the original levels. This avoids the need to prepare extra sets of standards. Special dilutions must always be prepared for determining calcium and magnesium by atomic absorption (see 7.3.1 and 7.6.1). These solutions and their standards must include lanthanum to control interferences as well as the necessary background components. Details of these dilutions are summarized in Table 8.2.

7.1.3 Calibration curves

In the colorimetric methods described in the calibration curves are linear over the ranges quoted, but in the flame systems the curves are only linear in the lower ranges becoming non-linear at higher concentrations.

7.1.4 Calculations

To avoid repetition the calculation stages are not given for individual methods. The majority of the procedures involve flame emission, atomic absorption or colorimetry and typical calculations can be given in standard formats as a guide. The following points must be taken into account for each class of calculation.

i. Blank values must be read from the graph and subtracted before the final calculation.

ii. Factors for dilution or concentration must be included where necessary. There will always be a dilution factor in determining calcium and magnesium by atomic absorption because of the necessity to introduce lanthanum.

iii. The final value should be corrected to a dry weight basis where necessary.

iv. Results should be expressed as indicated in Section 1.1.

Table 8.2 Dilution of sample solution prepared by ashing, digestion or extraction.

Solution preparation	Analysis for:	Reagents required	Composition of solution for analysis	Text references
Dry ashing	Cations (except Ca, Mg)	10% HCl	1% HCl	
	Ca + Mg by atomic abs	10% HCl 5,000 mg l^{-1} La	1%HCl + 400 mg l^{-1} La	Prepn of La soln given in 7.3
$HNO_3/HClO_4/H_2SO_4$	Cations + P (except Ca, Mg)	10% H_2SO_4	1% H_2SO_4	
	Ca + Mg by atomic abs	10% H_2SO_4 5,000 mg l^{-1} La	1% H_2SO_4 + 400 mg l^{-1} La (vegn) or 800 mg l^{-1} La (soil)	Prepn of La soln given in 7.3
H_2SO_4/H_2O_2 digestion	Cations P + N (except Ca, Mg)	5% Digest blank (see Note 1)	1% H_2SO_4 + Se + Li_2SO_4	
	Ca + Mg by atomic abs	5% Digest blank (see Note 1) 5,000 mg l^{-1} La	1% H_2SO_4 + Se + Li_2SO_4 + 400 mg l^{-1} La	Prepn of La soln given in 7.3
2·5% HOAc soil extracts	Cations + P (except Ca, Mg)	10% HOAc	2·5% HOAc	
	Ca + Mg by atomic abs	10% HOAc 10% H_2SO_4 5,000 mg l^{-1} La	2·5% HOAc 1% H_2SO_4 800 mg l^{-1} La	Prepn of La soln given in 7.3
M NH_4OAc	Cations (except Ca, Mg)	4M NH_4OAc	M NH_4OAc	Prepare 4M soln at 4 fold concn given in 2.6.3
	Ca + Mg by atomic abs	4M NH_4OAc 10% H_2SO_4 5,000 mg l^{-1} La	M NH_4OAc + 1% H_2SO_4 + 800 mg l^{-1} La	Prepn of La soln given in 7.3
Kjeldahl digestion	N only	—	5% H_2SO_4	See Note 2

Notes
1. Use digestion blank diluted to one fifth the final volume. Digests should be refluxed for at least 30 minutes to prevent precipitation of selenium on dilution.
2. Kjeldahl digests are not normally diluted due to the difficulty of preparing large quantities of digestion blank. If samples are outside the normal standard range, it is preferable to redigest using reduced sample weights.
3. All acids prepared on volume to volume basis.

The calculations are as follows:

(a) Flame emission and atomic absorption

If $C = $ mg l^{-1} element is obtained from the calibration curve then for

Plant materials:

$$\text{Total element (\%)} = \frac{C \,(\text{mg } l^{-1}) \times \text{solution volume (ml)}}{10^4 \times \text{sample wt (g)}}$$

Soils:

$$\text{Extractable element (mg 100 g}^{-1}) = \text{as above} \times 10^3$$

Waters:

$$\text{Dissolved element (mg l}^{-1}) = C(\text{mg l}^{-1})$$

(b) Colorimetry

If C = mg element per aliquot is obtained from the calibration curve then for

Plant materials:

$$\text{Total element (\%)} = \frac{C\,(\text{mg}) \times \text{solution volume (ml)}}{10 \times \text{aliquot (ml)} \times \text{sample wt (g)}}$$

Soils:

$$\text{Extractable element (mg 100 g}^{-1}) = \text{as above} \times 10^3$$

Waters:

$$\text{Dissolved element (mg l}^{-1}) = \frac{C\,(\text{mg}) \times 10^3}{\text{aliquot (ml)}}$$

(c) Titrations

Calculations for most of the titrations methods are as given for colorimetry except that C (mg) is replaced by titre (ml) $\times A$ where A = mg element per ml titrant. The CEC calculation (2.6.4) is expressed as me 100 g^{-1} and is given in full. Sections 5.2 and 7.4 contain calculations based on molarities and these are also given in full.

7.2 TYPICAL NUTRIENT LEVELS IN SOILS AND PLANT MATERIALS

Some typical ranges of nutrient levels found in ecological materials are shown in Table 8.3 but results outside the limits of these ranges may well occur. Further information on the chemical composition of ecological materials is given by Allen *et al.* (1974).

7.3 CALCIUM

7.3.1 Atomic absorption method

Atomic absorption is a rapid and sensitive method for the determination of calcium. Interference from aluminium, phosphorus and iron is controlled by

Table 8.3 Typical concentrations of the main nutrient elements in soil and plant materials (dry basis).

Element	Plant material[1] % element	Soil (extractable element in mg 100 g[-1]) Low	High
K	0·5 to 4	10	30
Ca	0·2 to 5	10	200[2]
Mg	0·1 to 0·5	5	30
Mn	0·01 to 0·2	0·1	2
NH$_4$-N	—	0·5	2
NO$_3$-N	—	0·2	1
PO$_4$-P	—	0·2	2
		% element	
Total N	0·8 to 3	0·1 to 0·5[3]	
Total P	0·1 to 0·3	0·02 to 0·2	

Notes
1. Photosynthetic tissues only. Fertilized materials and agricultural samples excluded.
2. May be much higher in calcareous soils.
3. Higher in many organic soils.

the addition of lanthanum as a releasing agent. Flame emission is less satisfactory as it lacks sensitivity unless a very hot flame is used.

PROCEDURE

1. Prepare the sample solution as described in 2.6, 3.5, 4.3 or 5.3 and dilute to volume as appropriate (see note).
2. Dilute and include sufficient lanthanum chloride in the dilution to give 400 mg l^{-1} La in plant solutions and waters and 800 mg l^{-1} La in soil extracts. Where necessary (and including all soil extracts) add sufficient 10% sulphuric acid to ensure the final acid concentration is 1% (v/v).
3. Switch on the instrument and allow the hollow cathode lamp adequate time to stabilise.
4. Select the absorption line at 422·7 nm and adjust instrument settings, gas and air flows, etc., as recommended in the manufacturer's instructions. An air-acetylene flame is suitable for calcium.
5. Adjust scale reading to a suitable value with the top standard after setting the baseline with the zero standard.
6. Check zero and top standard readings for stability. Aspirate a range of standards to prepare a calibration curve.
7. Aspirate samples. Check stability of calibration from time to time using top and zero standards.

8. Flush frequently with water and for at least one minute after each batch of samples.

9. Construct a calibration curve, read off sample concentrations, correct for blank readings and calculate concentrations in original samples (see 7.1.4a).

REAGENTS AND STANDARDS

1. Primary standard solution (1,000 mg l^{-1} Ca).

Dissolve 2·4973 g dry $CaCO_3$ in 200 ml water containing 5 ml conc HCl, boil to drive off CO_2, cool, and dilute to 1 litre with water.

2. Secondary standard solutions.

Prepare a range of standards by suitable dilution of the primary standard. The top standard should not be greater than 100 mg l^{-1}. Include sufficient $LaCl_3$. $7H_2O$ solution in each standard to give a final concentration of 400 mg l^{-1} La for plant solutions and 800 mg l^{-1} for soil extracts. Add 10% H_2SO_4 to each standard to give a final concentration of 1%. Also include soil extractant where necessary (7.1.2).

3. Lanthanum chloride solution (5,000 mg l^{-1} La).

Dissolve 6·6837 g of $LaCl_3$. $7H_2O$ in water. Add 1 ml 2 M HCl and dilute to 500 ml.

4. Sulphuric acid, 10% v/v.

Note Calcium and magnesium must be run on solutions specially diluted to contain lanthanum to control interferences. These diluted solutions are unsuitable for the determination of other elements.

7.3.2 EDTA titration method

The method involves the use of EDTA (the di-sodium salt of ethylene diamine tetra-acetic acid) which forms stable complexes with many elements at specific pH values. Various indicators are in use for calcium determinations but in many cases the end-point is not sharp and they are best used in conjunction with a photoelectric titrator.

Interference from heavy metals is not serious with most organic materials, but with soils it can be significant and can cause difficulty with the determination of the end point. The high dilution recommended in the method minimizes end-point interferences but limits the value of the procedure if the calcium level is low.

About 10 μg calcium can be determined, but reproducibility at normal levels is unlikely to be better than ± 5% for ecological materials.

PROCEDURE

1. Prepare the sample solution as described in 2.6, 3.5, 4.3 or 5.3 and dilute to volume where appropriate.

2. Mix 5 ml M sodium hydroxide with indicator (0.1 g murexide, 5 drops calcon or 5 drops glyoxal) and dilute to about 100 ml with water. This gives a bluish reference end-point.

3. Standardize the EDTA solution as follows: Pipette 10 ml calcium standard into a titration flask and add water, sodium hydroxide and indicator as above. Titrate with EDTA solution until the colour matches that of the reference end-point.

4. Pipette an aliquot (up to 5 ml) of the sample solution into a titration flask, add water, sodium hydroxide and the indicator. Titrate with the EDTA solution as above.

5. Carry out blank determinations following the same procedure and subtract from the sample values. Calculate the calcium content of the original material (see 7.1.4c).

REAGENTS AND STANDARDS

1. Calcium standard (100 mg l^{-1} Ca).
Dissolve 0·2497 g dry $CaCO_3$ in water containing approximately 1 ml conc HCl.
Boil to drive off the CO_2, cool and dilute to 1 litre.
2. EDTA solution (1 ml = 0·1 mg Ca).
Dissolve 0·931 g of di-sodium ethylene diamine tetra-acetate in 1 litre of water and standardize the solution by titrating against the Ca standard (see above).
3. Sodium hydroxide, M. (40·01 g l^{-1}).
4. Indicators.
   (a) Murexide: Grind together 0·1 g murexide and 50 g NaCl in a mortar and store in a dark bottle.
   (b) Calcon: Dissolve 20 mg calcon in 50 ml methanol. Prepare fresh weekly.
   (c) Glyoxal: Dissolve 0·20 g glyoxal-bis-(2-hydroxyanil) in 50 ml methanol.

7.4 CARBON

7.4.1 Organic carbon

The methods available for the determination of organic carbon in soils and plant materials fall into two main categories, absolute methods and 'rapid' methods. The two absolute procedures are dry combustion and wet oxidation.

Dry combustion involves burning the sample under carefully controlled conditions and measuring the carbon dioxide evolved. Although combustion trains can readily be set up in the laboratory it is normal to use commercial equipment some of which is partly or fully automatic. The principles are fully discussed by Ingram (1948, 1956).

Wet oxidation methods use strong acids and oxidizing agents to break down organic matter and liberate carbon dioxide. Rapid methods are also based on an acid oxidation system but only partly oxidize the organic matter so some form of calibration should be employed.

If desired an approximate estimate of organic carbon can be obtained from the percentage loss-on-ignition which is a crude measure of organic matter. One frequently used factor is based on the assumption that soil organic matter contains 58% carbon but Ball (1964) and Howard (1966) query the validity of this assumption.

(a) Wet oxidation method

The method of acid oxidation described by Clark & Ogg (1942) has been coupled with a gravimetric estimation of carbon dioxide (Shaw, 1959 and Allison, 1960).

The method is applicable to soils and plant materials and at normal levels of carbon it is possible to achieve a precision of $\pm 1\%$.

PROCEDURE

A diagram of the apparatus is given in Fig. 8.1.

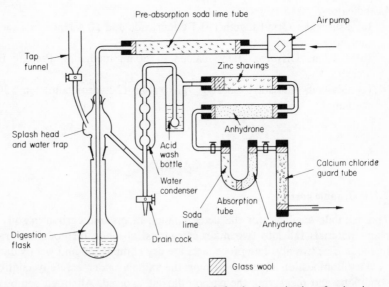

Figure 8.1 Acid oxidation apparatus suitable for the determination of carbon in soils and plant materials.

1. Fill the pre-absorption tube with 4 to 10 mesh soda lime.

2. Pass air through the apparatus for about 30 minutes prior to starting the assay.

3. During this 30-minute period fill the absorption U-tube with the 10–16 mesh soda-lime and magnesium perchlorate and close the taps. (The use of self-indicating soda-lime will indicate when the U-tubes have to be refilled.)

4. Weigh the U-tube.

5. Weigh the sample into the digestion flask (0·10 g into a 50 ml flask for plant materials, but 0·5 g to 1·0 g into a 100 ml flask for most soils, oven-dry material being used in all cases).

6. Add 3 g $K_2Cr_2O_7$ and 3 ml water and mix.

7. Attach the flask and the U-tubes to the apparatus and allow air to pass through at about 30 to 35 ml per minute. (About 2 bubbles per second using the sulphuric acid bottle as an indicator of flow rate.)

8. Measure 25 ml of the digestion mixture into the separating funnel and run it slowly in to the digestion flask, closing the taps just before the last drops run in to exclude air.

9. Heat the flask continuously for about 15 minutes using a small burner.

10. Maintain the air flow and after 15 minutes heating continue passing air through the apparatus for a further 20 minutes.

11. Discontinue the flow, close all taps, disconnect the U-tubes and reweigh them.

CALCULATION

$$C (\%) = \frac{\text{increase in wt of U-tubes (g)} \times 27 \cdot 29}{\text{sample wt (g)}}$$

Subtract blanks and correct to dry weight as necessary.

REAGENTS

1. Soda-lime, 4 to 10 mesh self-indicating granules.

2. Soda-lime, 10 to 16 mesh self-indicating granules (for use in absorption tubes).

3. Magnesium perchlorate, anhydrous, ('anhydrone').

4. Potassium dichromate, powdered.

5. Zinc wire.

6. Digestion mixture.

Mix 3 parts conc H_2SO_4 with 2 parts conc H_3PO_4 (S.G. 1.75).

Note Wet oxidation methods are subject to errors arising from the presence of carbonates in soils. Carbonates should be removed by a dilute hydrochloric acid pre-treatment or determined as carbonate-carbon (see 7.4.2 below) and subtracted from the wet oxidation value.

(b) Rapid titration method

The method of Tinsley (1950) was shown by Bremner & Jenkinson (1960) to give quantitative results for plant materials but for other natural organic materials and for soils the recoveries were, with some exceptions, slightly low. The recoveries from 15 soils ranged from 87 to 106% averaging 96·3% but it was felt the discrepancies were not sufficiently consistent to justify a correction factor. Hence the titration values for each soil type should be compared where possible with those of an absolute method and a correction factor computed. The reproducibility at the normal levels will probably not be better than ± 3%.

PROCEDURE

1. Weigh 50 mg of sample into a 250 ml conical flask fitted with a cold finger condenser.
2. Add 25 ml of the chromic acid mixture and boil gently for one hour.
3. Allow to cool and add 100 ml water.
4. Add 5 ml of the indicator solution and titrate the unused dichromate with ferrous ammonium sulphate solution. When the first colour change occurs, add a further 2·5 ml chromic acid mixture and complete the titration dropwise with care.

CALCULATION

If T ml of 0·4 M ferrous ammonium sulphate are used in the titration then:

$$\text{Carbon (\%)} = \frac{(27\cdot5 - T)\ \text{ml} \times 0\cdot12}{\text{sample wt (g)}}$$

Subtract blanks and correct to dry weight as necessary.

REAGENTS

1. Chromic acid mixture 0·0667 M.
Add 19·86 g $K_2Cr_2O_7$ and 200 ml H_3PO_4 (S.G. 1.75) to 400 ml conc H_2SO_4 and dilute to 1 litre.
2. Ferrous ammonium sulphate, 0·4 M.
Dissolve 156·86 g ferrous ammonium sulphate $((NH_4)_2SO_4 . FeSO_4 . 6H_2O)$ in water, add 20 ml conc H_2SO_4 and dilute to 1 litre. Make up fresh daily.
3. Indicator solution.
Dissolve 5 g $BaCl_2 . 2H_2O$ and 0·3 g barium diphenylamine sulphonate in 100 ml water.

7.4.2 Carbonate-carbon

Inorganic carbon is usually present as the carbonates of calcium or magnesium. A number of methods of analysis are available including the gravimetric method of Allison (1960) and that described by Bascomb (1961), involving the reaction of dilute hydrochloric acid with the carbonate in the soil and the measurement of the evolved carbon dioxide using a calcimeter. For soils with a low carbonate content the simple apparatus described by Pitwell (1968) is quite effective. In the method given below the soil is treated with standard dilute hydrochloric acid and the excess acid estimated by titration with sodium hydroxide solution. Results obtained may be on the high side as the hydrochloric acid reacts with other soil constituents. Reproducibility at the normal levels will probably not be better than $\pm 5\%$.

PROCEDURE

1. Weigh approximately 5 g finely ground sample into 200 ml beaker flask.
2. Add 10 ml of 0·5 M hydrochloric acid $(=A)$.
3. Cover the flask with a watch glass and simmer on a steam bath for approximately 10 minutes. Cool.
4. Filter the solution through a suitable paper and wash the residue.
5. Titrate the excess acid in the filtrate against 0·1 M sodium hydroxide $(=B)$ using phenolphthalein solution as indicator. (If the excess hydrochloric acid is small, take a larger initial volume of acid.)

CALCULATION

$$CO_3^{2-}-C\ (\%) = \frac{(5A-B)(\text{ml}) \times 0\cdot06}{\text{sample wt (g)}}$$

Subtract blanks and correct to dry weight as necessary.

REAGENTS

1. Hydrochloric acid, 0·5 M.
2. Sodium hydroxide, 0·1 M.
3. Phenolphthalein, 1% w/v.
Dissolve 0·5 g phenolphthalein in 50 ml industrial spirit.

7.5 IRON

7.5.1 Atomic absorption method

Allan (1959, 1961) and David (1962) found atomic absorption to be satisfactory for the determination of iron in biological materials. The sulphuric acid present in the digests will depress the absorption reading (Curtis, 1969) but this is compensated for by the addition of sulphuric acid to the standards. For soil extracts checks should be made on each soil type for interference by sulphate, and if necessary standards should be compensated accordingly. The method is sensitive to approximately 0.01 μg ml^{-1} iron and reproducibility at the normal levels should be of the order of $\pm 3\%$.

PROCEDURE

1. Prepare the sample solutions as described in 2.6, 3.5, 4.3 or 5.3 and dilute to volume as appropriate (see note).
2. From this point follow the procedure outlined in 7.3.1 (beginning at step 3).
3. The wavelength setting for iron is 248.3 nm (note there are two less sensitive lines close to this one).

Reagents and standards
1. Primary standard solution (100 mg l^{-1} Fe).
Dissolve 0.1 g clean untarnished Fe wire in about 10 ml of warm 10% H_2SO_4. When cool dilute to 1 litre.
2. Secondary standard solutions.
Prepare a range from 0 to 5 mg l^{-1} Fe by suitable dilution of the primary standard. Include acid or soil extractant as described in 7.1.2 and 2.6.5 before diluting to volume.

Note Digests which contain sulphuric acid as the final acid require the addition of 10 to 15 ml water and bringing to the boil before filtering and diluting to volume. This is due to the fact that ferric sulphate is insoluble in anhydrous sulphuric acid and precipitates from solution during the final stages of the digestion.

7.5.2 Colorimetric method

The method described below uses a sulphonated form of bathophenanthroline and was developed by Quarmby & Grimshaw (1967) from the procedure of Riley & Williams (1959).
   The method is applicable to most samples without modification and will detect as little as 2 μg iron. At normal levels reproducibility is $\pm 2\%$.

PROCEDURE

1. Prepare the sample solutions as described in 2.6, 3.5, 4.3 or 5.3 and dilute to volume where appropriate (see note 2).
2. Measure suitable sample aliquots not exceeding 20 ml, into 50 ml volumetric flasks.

3. Pipette aliquots of 0 to 30 ml of the secondary standard solution into 50 ml volumetric flasks to give a standard range from 0 to 0·03 mg iron.

4. Add acid or soil extractant to the standards to match the sample aliquots. From this point treat standards and samples in the same way.

5. Add 16 ml of the combined reagent and dilute to volume (see note 1).

6. Measure the optical density at 536 nm or with a yellow filter using water as a reference.

7. Construct a calibration curve from the optical densities of the standard solutions and use it to obtain mg Fe in the sample aliquots. Subtract blank values. Calculate concentrations of iron in the original material (see 7.1.4b).

REAGENTS AND STANDARDS

1. Primary standard solution (100 mg l^{-1} Fe).
Prepare as described in 7.5.1.

2. Secondary standard solution (1 mg l^{-1}).
Dilute the primary standard solution 100 times with water.
Prepare fresh at regular intervals.

3. Sulphonated bathophenanthroline reagent.
Add 4·0 ml fuming H_2SO_4 (containing 20% SO_3) to 0·4 g bathophenanthroline (4:7-diphenyl-1:10 phenanthroline). Stir until dissolved and allow to stand for 30 minutes. Pour into 400 ml water.
Neutralize with NH_4OH to between pH 4 and 5 and finally dilute to 1 litre.

4. Sodium acetate hydrate, 33% w/v.
Dissolve 330 g $CH_3COONa.3H_2O$ in water and dilute to 1 litre.

5. Hydroxylamine hydrochloride, 2·5% w/v.

6. Combined reagent.
Mix 33% sodium acetate solution, sulphonated bathophenanthroline reagent and 2·5% hydroxylamine hydrochloride in the ratio 4:3:1.

Notes   1. The sodium acetate buffer is sufficient to control acid levels up to the equivalent of 20 ml of 1% (v/v) H_2SO_4.
2. Refer to the note under 7.5.1 concerning the treatment of sample digests.

7.6 MAGNESIUM

The atomic absorption method is preferable to any other procedure for this element. An EDTA method, in which magnesium is determined by difference is less sensitive and subject to interferences. Colorimetric methods are less satisfactory and the standard gravimetric method which involves precipitation as magnesium ammonium phosphate is tedious for routine work.

7.6.1 Atomic absorption method

With a suitable instrument 0.005 µg ml^{-1} magnesium can be detected and at normal levels reproducibility of the order of $\pm 1\%$ should be possible. Lanthanum chloride is included in all solutions as a releasing agent to overcome interferences particularly from phosphate.

PROCEDURE

1. Prepare the sample solution as described in 2.6, 3.5, 4.3 or 5.3, and make up to volume where appropriate (see note).
2. From this point follow the procedure outlined in 7.3.1 (starting at step 2).
3. The wavelength setting for magnesium is $285 \cdot 2$ nm.

REAGENTS AND STANDARDS

1. Primary standard solution (100 mg l^{-1} Mg).
Dissolve $1 \cdot 0136$ g $MgSO_4 . 7H_2O$ in water containing about 1 ml H_2SO_4. Dilute to 1 litre.
2. Secondary standard solutions.
Dilute the primary standard solution to obtain a range of standards between 0 and 3 mg l^{-1} Mg. Include sufficient stock $LaCl_3 . 7H_2O$ solution to give a final concentration of 400 mg l^{-1} La for plant digests and 800 mg l^{-1} for soil extracts. Add 10% H_2SO_4 to each standard to give a final concentration of 1% (v/v). Also include soil extractant where necessary (7.1.2).
3. Lanthanum chloride solution ($5,000$ mg l^{-1} La).
Dissolve $6 \cdot 6837$ g $LaCl_3 . 7H_2O$ in water containing 1 ml 2 M HCl and dilute to 500 ml.
4. Sulphuric acid, 10% v/v.

Note Calcium and magnesium must be run on solutions specially diluted to contain lanthanum to control interferences. The solutions are unsuitable for the determination of other elements.

7.6.2 EDTA titration method

This method of magnesium determination involves determining both calcium and magnesium together and calcium separately (7.3.2). Magnesium is then obtained by difference. Direct titrations for magnesium are not recommended for many soil or plant materials owing to interference problems.

Levels of magnesium as low as 30 µg can be determined but if difficulty is experienced with the end point a precision better than about $\pm 5\%$ should not be expected even at normal levels.

PROCEDURE

1. Prepare the sample solution as given in 2.6, 3.5, 4.3 or 5.3 and dilute to volume where appropriate.
2. Standardize the EDTA solution as follows. Pipette 10 ml magnesium standard into a titration flask and dilute to about 100 ml with water. Add 15 ml buffer solution, 2 ml triethanolamine and 10 drops of indicator solution. Titrate with EDTA solution from red to clear blue.
3. Pipette an aliquot (up to 5 ml) of sample solution into a titration flask. Dilute and add buffer solution, triethanolamine and indicator solution and titrate with EDTA solution as above.
4. Carry out blank determinations in the same way and subtract from the sample readings. Calculate the content of calcium + magnesium in the original material (see 7.1.4c).
5. Determine calcium as recommended in 7.3.2.
6. Subtract the calcium result from that of calcium + magnesium to give the value for magnesium.

REAGENTS AND STANDARDS

1. Primary standard solution (100 mg l^{-1} Mg)
2. EDTA solution (1 ml = 0·1 mg Mg).
Prepare the solution as given in 7.3.2 for calcium. Standardize against the magnesium standard as described above. For the dual titration of calcium and magnesium the EDTA solution should be standardized against a standard solution containing both elements.
3. Indicator solution.
Dissolve 0·25 g eriochrome black T in 50 ml industrial spirit—prepare fresh weekly.
4. Triethanolamine.
5. Buffer solution.
Dissolve 67·5 g NH_4Cl in water, add 570 ml of 0·88 NH_3 solution and dilute to 1 litre.

7.7 MANGANESE

7.7.1 Atomic absorption method

The method is sensitive and convenient and is relatively free from interferences but it is essential that the standards and samples contain manganese in the same valency state. The valency state will be most uniform in acid digest solutions following oxidation but it may be variable within a soil extract.

The application of atomic absorption to the determination of manganese is discussed by Christian & Feldman (1970). Most instruments will determine $0.05\ \mu g\ ml^{-1}$ or less of manganese and at normal concentrations the reproducibility is about $\pm 2\%$.

1. Prepare the sample solutions as in 2.6, 3.5, 4.3 or 5.3 and dilute to volume where appropriate.
2. From this point follow the procedure outlined in 7.3.1 (beginning at step 3).
3. The wavelength setting for manganese is 279.5 nm.

REAGENTS AND STANDARDS

1. Manganese standards.
Primary standard solution ($100\ mg\ l^{-1}$ Mn).
Dissolve 0.4060 g $MnSO_4.4H_2O$ in water containing 1 ml conc H_2SO_4 and dilute to 1 litre.
2. Secondary standard solutions.
Prepare a range from 0 to $2\ mg\ l^{-1}$ Mn by suitable dilution of the primary standard. Include acid or soil extractant as given in 7.1.2 before making up to volume.

7.7.2 Colorimetric method

Manganese reacts with formaldoxime in alkaline solution to form a brown complex. Iron and copper interfere by complexing with the formaldoxime, but the complexes are broken down by warming the solution. Perchloric acid is added to prevent early fading of the colour. N-hydroxyethyl-ethylene-diamine-triacetic acid (HEEDTA) is added to prevent formation of a precipitate in the presence of excess phosphate (Bradfield, 1957).

The method will detect as little as $5\ \mu g$ manganese in solution. It is applicable to most sample materials but waters may have to be concentrated. At normal levels the reproducibility is about $\pm 2\%$.

PROCEDURE

1. Prepare the sample solution as described in 2.6, 3.5, 4.3 or 5.3 and dilute to volume where appropriate.
2. Pipette a suitable aliquot of the sample solution (usually between 10 and 25 ml) into a 50 ml volumetric flask.
3. Pipette 0 to 6 ml of the secondary standard solution into 50 ml volumetric flasks, to give a standard range of 0 to 0.03 mg manganese.

4. Add acid or extractant to the standard solutions according to the aliquot or sample solution taken.

From this point treat standards and samples in the same way.

5. Add 5 ml perchloric acid and 5 ml HEEDTA and mix well.

6. Adjust the pH to between 8 and 10 using 10% sodium hydroxide.

7. Add 1·5 ml formaldoxime followed immediately by a further 2 ml 10% sodium hydroxide solution.

8. Dilute to the base of the neck (but not to volume). Mix and allow to stand in a water bath at 65°C for 2 hours.

9. Cool, dilute to volume, mix well, and measure the optical density at 450 nm using water as the reference.

10. Plot a calibration curve from the optical density values of the standards and use it to obtain mg manganese in sample solutions. Subtract blank values. Calculate manganese concentrations in the original material (see 7.1.4b).

REAGENTS AND STANDARDS

1. Primary standard solution (100 mg l^{-1} Mn).
Prepare as described in 7.7.1.

2. Secondary standard solution (5 mg l^{-1} Mn).
Dilute the primary standard 20 times with water.

3. Formaldoxime reagent.
Dissolve 20 g paraformaldehyde and 55 g hydroxylamine sulphate $(NH_2OH)_2H_2SO_4$ in boiling water and dilute to 100 ml when cold.
Dilute 10 times with water immediately before use.

4. Sodium hydroxide, 10% w/v.

5. Perchloric acid, 10% v/v.
Prepare immediately before use.

6. Sodium salt of N-hydroxyethyl-ethylene-diamine-triacetic acid (HEEDTA), 10% w/v.

7.8 NITROGEN

7.8.1 Total organic nitrogen

Organic nitrogen in biological materials can be converted to ammonium-nitrogen by the Kjeldahl digestion procedure. In this digestion the sample is heated with sulphuric acid in the presence of a catalyst until the organic matter is destroyed. Sodium or potassium sulphate included in the digestion mixture to elevate the temperature to the optimum required for the reaction. This digestion procedure is often coupled with a subsequent distillation stage for recovery of the ammonium-nitrogen. A semi-micro version of this technique is given below.

An alternative digestion for nitrogen which is only suitable for plant materials is the sulphuric acid-hydrogen peroxide system described in 3.5.2. It has the advantage that other elements can be determined in the same solution.

Nitrogen levels as low as 10 μg can be measured and with care, reproducibility should be about ± 1%.

PROCEDURE

I Digestion stage

1. Weigh a suitable quantity of dried ground sample into a 50 ml round bottomed Kjeldahl flask.

Suitable quantities are:

0·100 g—plant materials, peat and litter;

0·250 g—organic rich soils;

0·500 g—soils low in organic matter.

2. Add 2 g of potassium sulphate—mercuric oxide mixture followed by 3 ml concentrated sulphuric acid. Run the acid slowly down the neck of the flask whilst rotating the flask.

3. Heat the bulb of the flask gently on a digestion rack until the frothing ceases. Gas or electric heating can be used.

4. When the frothing has stopped increase the heat; the sulphuric acid should now reflux down the neck of the flask.

5. Continue heating until the solution becomes colourless or pale yellow/green and for a further period ranging from 15 minutes for plants through 30 minutes for peat, litter, and animal tissue to 1 hour for soils.

6. Remove from heat, allow the flask to cool and dilute with water. Dilute to 50 ml unless the entire solution is to be taken for distillation.

7. Prepare blank solutions following the same procedure.

II Distillation stage

8. Set up a steam distillation apparatus as shown in Fig. 8.2b. Use a steam generator containing ammonia-free water (see note).

9. Pass steam through the apparatus for 30 minutes. Check the steam blank by collecting 20 to 30 ml distillate and titrating with M/140 HCl as given below. The steam blank should not require more than 0·1 ml acid.

10. Transfer the entire sample solution or an aliquot to the reaction chamber and add 12 ml of alkali mixture.

11. Commence distillation immediately and collect 25 ml distillate in a suitable receiver containing 5 ml of boric acid—indicator solution.

12. Titrate the distillate with M/140 hydrochloric acid to a pale grey (neutral) end point using a clean microburette. The colour passes through a transient blue stage just before the end point.

13. Occasionally check that recovery is satisfactory by taking an aliquot of the standard ammonium chloride solution in place of the sample.

14. Subtract blank values from the sample titrations where necessary. Calculate percentage nitrogen in the original material (see 7.1.4c).

REAGENTS AND STANDARDS

1. Standard ammonium chloride (100 mg l^{-1} NH$_4^+$-N).
Dissolve 0·1910 g dry NH$_4$Cl in water and dilute to 500 ml.
Add 1 to 2 ml chloroform as a preservative.
2. Hydrochloric acid, M/140. (1 ml = 0·1 mg NH$_4^+$-N).
Prepare 0·1 M HCl and standardize it against 0·05 M Na$_2$CO$_3$ solution using 0·1% bromo-phenol blue as indicator. When standardized dilute to exactly M/140.
3. Potassium sulphate—mercuric oxide mixture.
Mix K$_2$SO$_4$ and HgO (NH$_3$ free grades) in the ratio of 20:1. (Tablets of this preparation are available and in the method given above 2 g tablets are used).
4. Sulphuric acid, conc (NH$_3$ free grade).
5. Sodium hydroxide—sodium thiosulphate mixture.
Dissolve 500 g NaOH and 25 g sodium thiosulphate in water with care, cool and dilute to 1 litre.
6. Boric acid indicator solution.
Dissolve 0·350 g bromo-cresol-green in 10 ml industrial spirit.
Add 1 ml of 0·5 M NaOH and dilute to approximately 200 ml with water. Add 750 mg 4-nitrophenol (dissolved in a few ml of industrial spirit) and 22 ml of 1% DYRETO*. Make up to 250 ml and mix well. This stock indicator solution is stable.
(Other indicator solutions based on methyl red can be used for the titration but the above indicator mixture has been found to give an end point which is easy to detect.)
Dissolve 20 g boric acid in water, add 15 ml of indicator stock solution and dilute to 1 litre.

Note Deionized water is preferable to distilled water since the latter often gives a high blank value. Ammonium ions can be removed from distilled water by shaking with a strong cation exchange resin (Zerolit 225 or Amberlite IR 120). Another possibility is to use mains tap water since in some areas it has a very low ammonium-nitrogen content.

7.8.2 Organic nitrogen in waters

The method involves use of Devarda's alloy and magnesium oxide to reduce nitrite and nitrate to ammonia which is then eliminated before digestion of the

---

* Dye available from The Vanguard Manufacturing Company, Maidenhead, Berkshire.

remaining organic nitrogen. It is applicable to most water samples and levels down to 10 μg can be measured. Reproducibility at normal levels is likely to be in the region of ± 2%.

PROCEDURE

1. Measure a suitable volume of the water sample into a 500 ml conical flask (Taylor's pattern) or round bottomed flask. Take up to 250 ml depending on the ammonia concentration.
2. Add 0·05 g magnesium oxide and 0·1 g Devarda's alloy.
3. Boil down to a few ml but not to dryness.
4. Add 3 ml of concentrated sulphuric acid and 0·04 g mercuric oxide.
5. Digest until white fumes are evolved but again do not allow to boil dry.
6. Prepare blank solutions following the same procedure (see note in 5.3.2).
7. Continue with step 8 in the procedure of 7.8.1 adding 12 ml alkali mixture for the distillation.

REAGENTS AND STANDARDS

1. Standard ammonium chloride solution (100 mg l^{-1} NH_4^+-N). ⎫
2. Standard hydrochloric acid solution, M/140.   ⎪ Prepare
3. Boric acid—indicator solution.   ⎬ as in
4. Sodium hydroxide—sodium thiosulphate mixture.   ⎪ 7.8.1
5. Devarda's alloy powder.   ⎭

The commercially available Devarda's alloy should be finely ground before use. It contains 50% Cu, 45% Al and 5% Zn.
6. Magnesium oxide, powder.
7. Mercuric oxide.
8. Sulphuric acid, conc.

7.8.3 Ammonium nitrogen

Two methods are described: a semi-micro distillation method and a colorimetric method involving the use of the Nessler reagent.

(a) Distillation method

PROCEDURE

1. Set up a distillation apparatus as shown in Fig. 8.2a and run steam blanks as given in 7.8.1.
2. Transfer an appropriate amount of the soil extract or water sample into the distillation flask and add 0·2 g of magnesium oxide.
3. Without delay attach the flask to the apparatus and commence distillation.

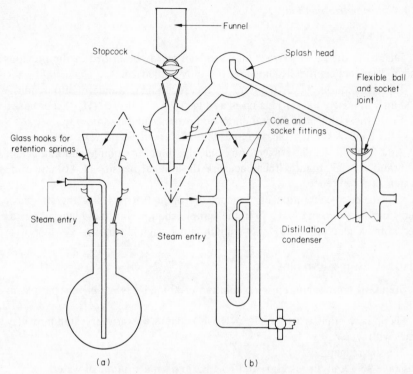

Figure 8.2 Apparatus suitable for (a) macro, and (b) semi-micro distillation of solutions containing ammonium-nitrogen.

4. Collect 50 ml distillate in a suitable receiver containing 5 ml of the boric acid indicator solution.

5. Titrate the distillate with M/140 HCl to a pale neutral (grey) end point using a microburette. The colour passes through a transient blue stage just before the end point.

6. Occasionally check that recovery is complete by taking an aliquot of the standard ammonium chloride solution in place of the sample.

7. Carry out blank determinations using the same procedure and subtract where necessary (see note in 5.3.2).

Calculate the ammonium-nitrogen content of the original sample (see 7.1.4c).

REAGENTS AND STANDARDS

1. Standard ammonium chloride (100 mg l^{-1} NH$_4^+$-N).⎫ Prepare
2. Standard hydrochloric acid, M/140.⎬ as in
3. Boric acid—indicator solution.⎭ 7.8.1.
4. Magnesium oxide powder.
5. Colorimetric method.

(b) Colorimetric method

PROCEDURE

1. Distil the digest or extract or water sample as described in the previous method and collect the distillate in a 100 ml volumetric flask.
2. Pipette aliquots of 0 to 50 ml of the secondary standard into separate 100 ml volumetric flasks. This gives a range of 0 to $0 \cdot 1$ mg NH_4^+-N per flask.
3. From this point treat the standards and samples in the same way.
4. Add 2 ml of sodium tartrate and dilute to about 90 ml with water.
5. Add 2 ml of Nessler reagent, mix quickly, dilute to volume and mix again.
6. Stand for 25 minutes then measure the optical density at 410 nm using water as a reference.
7. Construct a calibration curve from the optical densities of the standards and from it obtain mg NH_4^+-N in the sample aliquot. Subtract blank values as necessary. Calculate the NH_4^+-N content in the original sample.

REAGENTS AND STANDARDS

1. Standard ammonium chloride solution (100 mg l^{-1} NH_4^+-N): prepare as in 7.8.1.
2. Secondary standard (2 mg l^{-1} NH_4^+-N): dilute the primary standard fifty times with water.
3. Nessler reagent.
Dissolve 98 g anhydrous HgI_2 and 70 g KI in a small volume of water.
Dissolve 160 g NaOH in about 200 ml water.
Mix the two solutions, with care, and dilute to 1 litre.
Allow to stand for at least one and preferably several days.
Decant off the clear solution for use.
4. Sodium tartrate, 10% w/v.

7.8.4 Nitrate-nitrogen

Methods described here include a distillation procedure similar to that for ammonium nitrogen, and a colorimetric procedure involving phenoldisulphonic acid.

(a) Distillation method

The reproducibility and limits of detection are similar to the other distillation methods for nitrogen compounds.

PROCEDURE

1. Set up the distillation apparatus as shown in Fig. 8.2a and run steam blanks as given in 7.8.1.

2. Transfer a suitable volume of soil extract or water sample to the distillation flask and add 0·2 g MgO.

Suitable volumes are:

20 to 50 ml soil extracts;

50 to 100 ml stream and lake waters;

200 ml rain water.

3. Connect the flask to the apparatus and commence distillation.

4. Collect about 100 ml distillate and discard (see note below).

5. Remove distillation flask, add 0·4 g Devarda's alloy and reconnect immediately.

6. Commence steaming again and collect the distillate in a suitable receiver containing 5 ml boric acid-indicator solution.

7. Titrate with M/140 HCl as given in 7.8.2.

8. Carry out blank determinations and subtract where necessary. Calculate the concentration of NO_3^--N in the original sample (see 7.1.4c).

REAGENTS AND STANDARDS

1. Standard ammonium chloride solution (100 mg l^{-1} NH_4^+-N). ⎫ Prepare
2. Standard hydrochloric acid, M/140.   ⎬ as in
3. Boric acid—indicator solution.   ⎭ 7.8.2.
4. Magnesium oxide powder.
5. Devarda's alloy, finely powdered.

Note Total inorganic nitrogen can be obtained by adding the magnesium oxide and the Devarda's alloy at the same time and collecting the total distillate. Ammonium-nitrogen may separately be determined if the distillate at Step 4 is collected in 5 ml boric acid—indicator solution. This is titrated as described in 7.8.2.

(b) Colorimetric method

The method will measure to about 2 μg of nitrate-nitrogen provided there is no interference. Reproducibility is probably not better than ± 5%.

PROCEDURE

1. Pipette an aliquot of the soil extract or water sample into a suitable evaporating basin.

2. Transfer 0 to 5 ml aliquots of the secondary standard into separate borosilicate evaporating basins, giving a range of 0 to 0·05 mg NO_3^--N. (If the soils were extracted with 6% NaCl add the appropriate amount to the standard aliquots.)

3. From this point treat the standards and samples in the same way.

4. Evaporate to dryness on a steam bath, but do not bake.

5. When cool add 2 ml phenoldisulphonic acid and swirl rapidly.

6. Allow to stand for 10 minutes and then add 20 ml water.

7. Add ammonium hydroxide with care, until the pH is between 10 and 11, and then filter through a 9 cm Whatman No. 541 filter paper into a 50 ml volumetric flask.

8. Dilute the solutions to volume and mix well.

9. Measure the optical density at 410 nm using water as a reference.

10. Construct a calibration curve for the standards and from it obtain the mg NO_3^--N in the sample aliquots. Subtract blank values. Calculate the NO_3^--N content in the original sample (see 7.1.4b).

REAGENTS AND STANDARDS

1. Primary standard solution (100 mg l^{-1} NO_3-N).

Dissolve 0·72 g dry KNO_3 in water and dilute to 1 litre.

2. Secondary standard solution (10 mg l^{-1} NO_3^--N)—dilute the primary standard ten times with water.

3. Phenoldisulphonic acid.

Dissolve 25 g phenol in 150 ml conc H_2SO_4.

Add 75 ml fuming H_2SO_4 containing 15% free SO_3 and stir thoroughly.

Heat for 2 hours in a boiling water bath with a glass bubble in the neck of the flask.

Store in a dark container.

4. Ammonium hydroxide (1 + 1).

(c) Selective ion electrode method

Nitrate-nitrogen can sometimes be determined in waters and soil extracts using a selective ion electrode. However, its use is generally limited to concentrations of nitrate-nitrogen in excess of about 0·5 mg l^{-1}. If the ionic concentration is sufficiently high the technique is particularly appropriate for use in the field.

Use of the equipment is described in the manufacturers' instructions and other information can be obtained from Andelman (1968), Carlson & Keeney (1971), Summerfeldt *et al.* (1971) and Moody & Thomas (1971).

7.9 PHOSPHORUS

7.9.1 Colorimetric method

The most common method for the determination of phosphorus is that based on the development of molybdenum blue which is described here. The reaction estimates orthophosphate and after the mixed acid or hydrogen-peroxide digestions all the phosphorus will be in this form, but a fairly large proportion of soil total phosphorus is present in an organic form. A number of methods are available for the determination of organic phosphorus in soils and they have been reviewed by Black & Goring (1953) and Williams *et al.* (1970).

The molybdenum blue method will measure amounts down to about 1 μg phosphate-phosphorus and with care the reproducibility at normal levels can be ± 1%.

PROCEDURE

1. Prepare the sample solution as described in 2.6.5, 3.5, 4.3 or 5.3 and dilute to volume where appropriate.
2. Pipette a suitable aliquot of sample solution into a 50 ml volumetric flask (usually up to 10 ml of vegetation solutions) (see notes 1 and 2).
3. Pipette aliquots of 0 to 15 ml of secondary standard solution into separate 50 ml volumetric flasks. This gives a standard range from 0 to 0·03 mg phosphorus. Include in each flask either digest acid or soil extractant to match the sample aliquots.
From this point treat standards and samples in the same way.
4. Add water to each flask until it is about two thirds full.
5. Add 2 ml ammonium molybdate reagent and mix.
6. Add 2 ml stannous chloride reagent, mix and dilute to volume.
7. Time from this stage and after 30 minutes measure the optical density at 700 nm using water as a reference.
8. Construct a calibration curve for the standards and use it to determine the mg phosphorus in the sample aliquots. Subtract blank values as necessary. Calculate the phosphorus content in the orginal samples (see 7.1.4b).

REAGENTS AND STANDARDS

1. Primary standard solution (100 mg l^{-1} P). Dissolve 0·4393 g dry KH_2PO_4 in water and dilute to 1 litre.
2. Secondary standard solution (2 mg l^{-1} P).
Dilute the primary standard solution 50 times with water. Make up fresh at intervals.
3. Ammonium molybdate—sulphuric acid reagent.
Dissolve 25 g ammonium molybdate $(NH_4)_6Mo_7O_{24}.4H_2O$ in about 200 ml water in a beaker. It may be necessary to warm slightly to dissolve.
Carefully add 280 ml conc H_2SO_4 to about 400 ml water with mixing and cooling.
Filter the molybdate solution into the acid solution, mix and cool.
Finally dilute to 1 litre and store the solution in the dark.

4. Stannous chloride reagent.

Dissolve $0 \cdot 5$ g $SnCl_2 . 2H_2O$ in 250 ml of 2% v/v HCl. Prepare immediately before use.

Notes   1. If Olsen's extractant has been used (see 2.6.5) then the sample aliquots should be neutralized with dilute H_2SO_4 until just yellow to $0 \cdot 1$% p-nitrophenol indicator. Brown-coloured aliquots are best neutralized by adding a pre-determined amount of dilute H_2SO_4 to each flask.

2. For samples in which the phosphorus level is very low (e.g. natural waters) the above procedure can be modified by including an extraction procedure in which the phospho-molybdate complex is extracted into n-butanol and the colour developed in the organic phase (see Allen *et al.*, 1974).

7.10 POTASSIUM

7.10.1 Flame emission method

PROCEDURE

1. Prepare the sample solutions as described in 2.6, 3.5, 4.3 or 5.3 and dilute to volume where appropriate.

2. Select the emission line at 766 nm or use a potassium filter and adjust the instrument settings and gas pressures according to the manufacturers instructions.

3. Adjust scale reading to a suitable value with the top standard after setting the baseline with the zero standard.

4. Check zero and top standard readings for stability. Aspirate a range of standards to prepare a calibration curve.

5. Aspirate samples. Check stability of the calibration from time to time by reading top and zero standards.

6. Flush frequently with water and for at least one minute after every batch of samples.

7. Construct a calibration curve, read off sample concentrations, correct for blank readings, and calculate concentrations of potassium in original samples (see 7.1.4a).

REAGENTS AND STANDARDS

1. Primary standard solutions (1,000 mg l^{-1} K).

Dissolve $1 \cdot 9068$ g KCl in deionized water and make up to 1 litre.

Check the Na content (see the notes below).

2. Secondary standard solutions.

Prepare a range of standards by suitable dilution of the primary standard. Include acid or soil extractant as given in 7.1.2 before diluting to volume.

Notes 1. Some commercially available flame photometers are dual channel instruments which allows potassium and sodium to be measured simultaneously. Standards must therefore contain both elements and the sodium level in the potassium primary standard should not exceed 1 mg l^{-1} or significant errors in the sodium analysis can occur.

2. For most sample solutions prepared from ecological materials the three ranges 0 to 10 mg l^{-1}, 0 to 25 mg l^{-1} or 0 to 100 mg l^{-1} will be adequate. Reproducibility for routine work should be about $\pm 1\%$.

7.11 SILICON

7.11.1 Colorimetric method

Methods available include a simple gravimetric procedure for approximate silica and a colorimetric procedure based on the formation of heteropoly-blue colour. An approximate method was mentioned earlier in 6.3. The colorimetric procedure which is described here is taken from Morrison & Wilson (1963). It is suitable for use with plant materials, soils and waters and can measure silicon levels of about 0·05 μg and the reproducibility at normal concentrations will be about $\pm 2\%$.

PROCEDURE

I Preparation of the sample solution

1. Weigh 0·6 g sodium carbonate and 0·05 g sodium peroxide into a platinum crucible.

2. Add 0·10 g ground sample, and cover with 0·3 g sodium carbonate.

3. Fuse at 950°C for 15 minutes in a muffle furnace. Whilst still hot, withdraw the crucible and swirl it round to wash down the sides.

4. Cool and immerse the crucible in water in a polypropylene beaker.

5. Add 3 ml hydrochloric acid and heat for half-an-hour on a steam bath.

6. Cool, transfer the contents to a volumetric flask, wash the crucible and dilute to volume. Mix and immediately transfer to a polythene bottle.

7. Carry out blank fusions along with the samples using the same procedure.

II Colorimetric procedure

8. Pipette a suitable aliquot of sample solution into a 50 ml flask.

9. Neutralize the solutions by the addition of a few drops of M sodium hydroxide using phenolphthalein as an indicator.

10. Pipette 0 to 6 ml of secondary standard solution into volumetric flasks to give a range of standards of 0 to 0·3 mg silicon.

11. Neutralize the standard aliquots with 1% of hydrochloric acid solution.

12. From this point treat standard and sample aliquots in the same way.

13. Add 1·25 ml acid molybdate reagent. Mix and leave for 10 minutes.

14. Add 1·25 ml tartaric acid. Mix and leave for 5 minutes.

15. Add 1·0 ml reducing solution. Mix and dilute to volume with distilled water. (This is preferred to deionized water which often contains silicon.)

16. Allow the flasks to stand for 15 minutes to ensure maximum colour development.

17. Measure the optical density of each solution at 810 nm using water as a reference.

18. Construct a calibration curve for the standard range and from it obtain the mg silicon in the sample aliquots. Correct for blank values.

Calculate the silicon content of the original material (see 7.1.4b).

REAGENTS AND STANDARDS

1. Primary standard (100 mg l^{-1} Si).

Fuse 0·2139 g dry SiO_2 powder with 1 g anhydrous Na_2CO_3 in a platinum crucible at 950°C until a clear melt is obtained. Cool, and immerse the whole in water contained in a polypropylene beaker. Warm to dissolve, cool and dilute to 1 litre, washing the crucible carefully. Store in a polythene bottle.

2. Secondary standard solution (5 mg l^{-1} Si).

Dilute the primary standard solution 20 times. Store in a polythene bottle.

3. Ammonium molybdate—sulphuric acid reagent.

Dissolve 89 g ammonium molybdate $(NH_4)_6Mo_7O_{24}.4H_2O$ in about 800 ml water. Dilute 62 ml conc H_2SO_4 to about 150 ml by adding it carefully to water. Allow to cool and add the acid to the molybdate solution. Dilute to 1 litre.

4. Tartaric acid, 28% w/v.

5. Reducing solution.

Dissolve 2·4 g $Na_2SO_3.7H_2O$ and 0·2 g 1-amino-2-naphthol-4-sulphonic acid in about 70 ml water. Add 14 g $K_2S_2O_5$, shake to dissolve and dilute to 100 ml. Prepare fresh weekly.

6. Hydrochloric acid, 1% v/v.

7.12 SODIUM

7.12.1 Flame emission method

PROCEDURE

The steps set out for potassium (7.10.1) should be followed with the exception that the 599 nm emission line or a separate filter are used for sodium.

REAGENTS AND STANDARDS

1. Primary standard solution (1,000 mg l^{-1} Na).
Dissolve 2·5420 g of dry NaCl in deionized water and make up to 1 litre.
Secondary standard solutions.
Prepare a range of standards by suitable dilution of the primary standard.
Include acid or soil extractant as given in 7.1.2 before diluting to volume. Store the solutions in borosilicate or plastic containers.

Notes   1. Some commercially available flame photometers are dual channel instruments which allows potassium and sodium to be measured simultaneously. Standards must therefore contain both elements and the sodium level in the potassium primary standard should not exceed 1 mg l^{-1} or significant errors in the sodium analysis can occur.

2. For much of the type of work described in this chapter, three ranges of standards will be found convenient, namely 0 to 5 mg l^{-1}, 0 to 25 mg and 0 to 100 mg l^{-1}. At these levels reproducibility for routine work using a standard commercial instrument of average performance should be about ± 1%.

3. Due to the very widespread use of sodium both as a laboratory reagent and in many products outside the laboratory, blank values will generally be higher than for other elements and particular care should be taken to ensure that they are representative and reproducible.

8 REFERENCES

ALLAN J.E. (1959) Determination of iron and manganese by atomic absorption. *Spectrochim. Acta.* **10**, 800–806.

ALLEN J.E. (1961) The determination of copper by atomic absorption spectrophotometry. *Spectrochim. Acta.* **17**, 459–466.

ALLEN S.E. (ed.) (1974) *Chemical Analysis of Ecological Materials.* Oxford, Blackwell.

ALLEN S.E. & GRIMSHAW H.M. (1962) Effect of low-temperature storage on the extractable nutrient ions in soils. *J. Sci. Fd Agric.* **13**, 525–529.

ALLISON L.E. (1960) Wet-combustion apparatus and procedure for organic and inorganic carbon in soil. *Proc. Soil Sci. Soc. Am.* **24**, 36–40.

ANDELMAN J.E. (1968) Ion-selective electrodes: theory and applications in water analysis. *J. Wat. Pollut. Control Fed.* **40**, 1844–1860.

ASSOC. OF OFFIC. ANALYTICAL CHEMISTS (1970) *Official Methods of Analysis.* Washington.

BALL D.F. (1964) Loss-on-ignition as an estimate of organic matter and organic carbon in non-calcareous soils. *J. Soil Sci.* **15**, 84–92.

BALL D.F. & BEAUMONT P. (1972) Vertical distribution of extractable iron and aluminium in soil profiles from a brown earth-peaty podzol association. *J. Soil Sci.* **23**, 298–308.

BASCOMB C.L. (1961) A calcimeter for routine use on soil samples. *Chemy. Ind.* **45**, 1826–1827.

BATES R.G. (1964) *Determination of pH: Theory and Practice.* New York, John Wiley & Sons.

BLACK C.A. (ed.) (1965) *Methods of Soil Analysis.* Vol. 2, Amer. Soc. of Agronomy.

BLACK C.A. & GORING C.A.I. (1953) Organic phosphorus in soils. In W.H. Pierre & A.G. Norman (eds.) *Soil and Fertilizer Phosphate in Crop Nutrition.* New York & London, Academic Press.

BOHN H.L. (1971) Redox potentials. *Soil Sci.* **112,** 39–45.

BRADFIELD E.G. (1957) An improved formaldoxime method for the determination of manganese in plant material. *Analyst. Lond.* **82,** 254–257.

BRAY R.H. & WILLHITE F.M. (1929) Determination of total replaceable bases in soils. *Ind. Engng Chem. Analyt. Edn.* **1,** 144.

BREMNER J.M. & JENKINSON D.S. (1960) Determination of organic carbon in soil. II. Effect of carbonized materials. *J. Soil Sci.* **11,** 403–408.

BROOKS R.R., PRESLEY B.J. & KAPLAN I.R. (1967) Ammonium tetramethylenedithio carbamate-isobutyl methyl ketone extraction system for the determination of trace elements (cobalt, copper, iron, nickel, lead and zinc) in saline waters by atomic-absorption spectrophotometry. *Talanta* **14,** 809–816.

BROWN I.C. (1943) A rapid method of determining exchangeable hydrogen and total exchangeable bases of soils. *Soil Sci.* **56,** 353–57.

CARLSON R.M. & KEENEY D.R. (1971) In L.M. Walsh (ed.) *Instrumental Methods for Analysis of Soils and Plant Tissue.* Madison, Wisconsin, Soil Sci. Soc. Amer. Inc.

CHAPMAN H.D. & PRATT P.F. (1961) *Methods of Analysis for Soils, Plants and Waters.* University of California.

CHRISTIAN G.D. & FELDMAN F.J. (1970) *Atomic Absorption Spectroscopy: Applications in Agriculture, Biology and Medicine.* Interscience.

CIACCIO L.L. (ed.) (1971) *Water and Water Pollution Handbook.* Dekker.

CLARK N.A. & OGG C.L. (1942) A wet combustion method for determining total carbon in soils. *Soil Sci.* **53,** 27–35.

COULTER B.S. (1969) The chemistry of hydrogen and aluminium ions in soils, clay minerals and resins. *Soils Fertil.* **32,** 215–223.

CURTIS K.E. (1969) Interferences in the determination of iron by atomic-absorption spectrophotometry in an air-acetylene flame. *Analyst. Lond.* **94,** 1068–1071.

DAGNALL R.M. & KIRKBRIGHT G.F. (eds.) (1970) *Atomic Absorption Spectroscopy.* Plenary lectures presented at International Atomic Absorption Spectroscopy Conference, Sheffield, UK, July 1969.

DAVID D.J. (1962) *Atomic Absorption Newsletter.* December, 1962.

DEPARTMENT OF THE ENVIRONMENT (1972) *Analysis of Raw, Potable and Waste Waters.* H.M.S.O.

EDISBURY J.R. (1966) *Practical Hints on Absorption Spectrometry.* Hilger & Watts.

ELWELL W.T. & GIDLEY J.A.F. (1966) *Atomic Absorption Spectrophotometry.* Pergamon Press.

GAWEN D. (1965) Handling of perchloric acid in laboratories and fume cupboards. *Lab. Pract.* **14,** 1397–1398, 1409.

GOLTERMAN H. (1969) *Methods for the Chemical Analysis of Freshwaters.* I.B.P. Handbook No. 8. Oxford, Blackwell Scientific Publications.

GORE A.J.P. (1968) The supply of six elements by rain to an upland peat area. *J.Ecol.* **56,** 483–495.

GREENHILL W.L. (1960) Determination of the dry weight of herbage by drying methods. *J. Br. Grassland Soc.* **15,** 48–54.

GRIMSHAW H.M. & LINDLEY D.K. (1974) S.E. Allen (ed.) *Chemical Analysis of Ecological Materials.* Oxford, Blackwell.

HERON J. (1962) Determination of phosphate in water after storage in polyethylene. *Limnol. and Oceanogr.* **7,** 316–321.

HESSE P.R. (1971) *A Textbook of Soil Chemical Analysis.* Murray.

HOWARD P.J.A. (1966) The carbon-organic matter factor in various soil types. *Oikos.* **15,** 229–236.

INGRAM G. (1948) A critical examination of the empty tube combustion method. *Analyst.* **73,** 548–551.

INGRAM G. (1956) The rapid micro-combustion procedure. *Chemy Ind.* 103–107.

JACKSON M.L. (1958) *Soil Chemical Analysis.* Prentice-Hall.

KLEEMAN A.W. (1967) Sampling error in the chemical analysis of rocks. *J. Geol. Soc. Aust.* **14,** 43–48.

LICKENS G.E. & BORMANN F.H. (1970) Chemical analyses of plant tissues from the Hubbard Brook ecosystem in New Hampshire. *Bull. Yale Univer. Sch. For. No. 79.* 25 pp.

L'VOV B.V. (1970) *Atomic Absorption Spectrochemical Analysis.* Hilger.

MACKERETH F.J.H. (1963) Water analysis for limnologists. *Freshwater Biological Ass. Scientific Publication No. 21.*

MACKERETH F.J.H. (1964) Improved galvanic cell for determination of oxygen concentration in fluids. *J. Scient. Instrum.* **41,** 38–41.

METSON A.J. (1956) Methods of chemical analysis for soil survey samples. *NZ, Dept. Sci. and Ind. Res. Soil Biol. Bull. 12.*

MITCHELL J. (JNR.) (1951) Karl Fischer reagent titration. *Analyt. Chem.* **23,** 1069–1075.

MOODY G.J. & THOMAS J.R. (1971) *Selective Ion Sensitive Electrodes.* Merrow Publishing Co.

MORRISON I.R. & WILSON A.L. (1963) The absorptiometric determination of silicon in water. Part II. Method for determining 'reactive' silicon in power-station waters. *Analyst, Lond.* **88,** 100–104.

OLSEN S.R., COLE C.V., WATANABE F.S. & DEAN L.A. (1954) Estimation of available phosphorus in soils by extraction with sodium bicarbonate. *U.S. Dept. of Agric.* Circular 939.

PIPER C.S. (1950) *Soil and Plant Analysis.* University of Adelaide.

PITTWELL L.R. (1968) Apparatus for the analysis of small quantities of carbonates. *Mikrochimica Acta.* 903–904.

PRICE W.J. (1972) *Analytical Atomic Absorption Spectrometry.* Heydon & Son.

QUARMBY C. & GRIMSHAW H.M. (1967) A rapid method for the determination of iron in plant material with application of automatic analysis to the colorimetric procedure. *Analyst. Lond.* **92,** 305–310.

RAMIREZ-MUNOZ J. (1968) *Atomic Absorption Spectroscopy.* Elsevier.

RILEY J.P. & WILLIAMS H.P. (1959) Micro-analysis of silicate and carbonate minerals. III. Determination of silica, phosphoric acid and metal oxides. *Mikrochim. Acta.* 804–824.

ROYAL SOCIETY SYMBOLS COMMITTEE (1971) *Quantities, Units, and Symbols.* The Royal Society.

SANDELL E.D. (1959) *Colorimetric Determinations of Traces of Metals.* Interscience.

SHAW K. (1959) Determination of organic carbon in soil and plant material. *J. Soil Sci.* **10,** 316–326.

SJORS H. (1950) On the relation between vegetation and electrolytes in north Swedish mire waters. *Oikos* **2,** 241–258.

SMALL J. (1954) *Modern Aspect of pH.* Bailliere, Tindall & Cos.

SNELL F.D. & SNELL C.T. (1959) *Colorimetric Methods of Analysis.* Vol. IIA. Van Nostrand.

SOCIETY FOR ANALYTICAL CHEMISTRY (1959) *Notes on Perchloric Acid and its Handling in Analytical Work.* Analytical Methods Committee. *Analyst. Lond.* **84,** 214–216.

SOCIETY FOR ANALYTICAL CHEMISTRY (1960) Methods for the destruction of organic matter. Analytical Methods Committee. *Analyst. Lond.* **85,** 643–656.

STANDARD METHODS FOR THE EXAMINATION OF WATER AND WASTEWATER (1971) American Public Health Association.

STEYN W.J.A. (1959) A statistical study of the errors involved in the sampling and chemical analysis of soils and plants with particular reference to citrus and pineapples. Ph.D. Thesis, Rhodes University, South Africa.

SUMMERFELDT T.G., MILNE R.A. & KOZUB G.C. (1971) Use of the nitrate-specific ion electrode for the determination of nitrate-nitrogen in surface and ground water. *Comm. Soil Sci. Plant Anal.* **2,** 414–420.

TINSLEY J. (1950) The determination of organic carbon in soils by dichromate mixtures. *Trans. 4th Int. Cong. Soil Sci.* **1,** 161–164.

TRUOG E. (1930) The determinations of the readily available phosphorus of soils. *J. Am. Soc. Agron.* **22,** 874.

WILLIAMS J.D.H., SYERS J.K., WALKER T.W. & REX R.W. (1970) A comparison of methods for the determination of soil organic phosphorus. *Soil Sci.* **110,** 13–18.

CHAPTER 9

DATA COLLECTION SYSTEMS

C.R. RAFAREL and G.P. BRUNSDON

1 **Introduction** 467

2 **Recording methods** 468
2.1 Integration 469
   2.1.1 Electrical integrators 469
   2.1.2 Chemical integrators 470
2.2 Chart recorders and event counters 471
   2.2.1 Chart recorders 471
   2.2.2 Event counters 471
2.3 Magnetic tape data loggers 472

3 **General design and construction** 473
3.1 Power supplies 474
   3.1.1 Power sources 474
   3.1.2 Stabilization 474
3.2 Timing and control 476
   3.2.1 Clocks 476
   3.2.2 Control and scanning 476
3.3 Interfacing 480
   3.3.1 Bridge circuits 480
   3.3.2 Operational amplifier circuits 482

3.3.3 Analogue–digital converters 488
3.4 Transducers 491

4 **Practical magnetic tape systems** 492
4.1 Basic data logging system 492
4.2 Systems in use 494
   4.2.1 Field system 494
   4.2.2 Laboratory system 496
   4.2.3 Commercial equipment 497
4.3 Practical applications 498
   4.3.1 Climatological stations 498
   4.3.2 Evaporation experiments 499
   4.3.3 Bore hole logging 500
   4.3.4 Meteorological site data collection 501

5 **Translation and data handling** 502
5.1 Translation of data from charts 502
5.2 Translation of data from magnetic tape loggers 504

6 **References** 506

1 INTRODUCTION

Experimental, field or laboratory projects often require monitoring of a large number of scientific variables for use in later evaluation or in automatic control of a particular experimental process. Because of the number of data sources involved and the period over which measurements must be taken, a data logging system is often essential. Logging systems are expensive and the economics must first be assessed before the human operator is finally superseded by the latest sophisticated equipment. The amount of data from even a small logging system can be considerable and can involve needless expenditure of valuable scientific time in laborious analysis of results, unless there has been careful planning for the translation and handling of the data.

An automatic data logging system is most usefully employed either where large amounts of data have to be collected in a short space of time or where the data has to be gathered over a long period, possibly from some remote areas.

The automatic data logging requirements of the ecologist can be divided into two types, firstly there is the requirement for a long-term reliable system capable of multichannel use and with the ability to accept and record signals from a number of different types of transducer and secondly, there is the need for a less sophisticated lower cost recording system. The use of low cost equipment is often important where a high degree of replication is required but it must be remembered that a greater amount of human supervision is required and that performance is often strictly limited. In many cases low cost equipment can be produced in the laboratory workshop and the ability to replicate data from a number of sites or over a wide area can compensate for the reduced accuracy and reliability that may be a feature of the system. In such systems, because the cost per data channel is low, it is often feasible to use a channel to monitor a standard input signal during the normal recording sequence. Any deviation from true which the recorded signal produces can then be used as a basis for applying error corrections to the data.

2 RECORDING METHODS

There are three methods by which the output from a sensor, or transducer, can be recorded automatically with respect to the time. The signal can be recorded continuously, as on a chart recorder, and is perhaps of greatest use where it is necessary to determine maximum or minimum values or the rate of change of a variable. The method is not recommended where mean values are required because of the difficulty and labour involved in extracting and translating data from graphical records. The use of an integrator sums the output from a sensor over some particular period of time at the end of which the output can be recorded. This approach is particularly useful where mean values are required. Integration has the advantage that subsequent data processing may be reduced but has the disadvantage that it is difficult to accurately integrate an analogue signal electronically over periods greater than an hour. This difficulty does not apply to the summation of events where digital methods can be used to store data prior to recording. The third method which can be used to record data is to interrogate the sensor at intervals and to record the output. This method is perhaps the one most commonly used and is employed in commercial data logging systems. It is important that the rate of interrogation be related to the rate of change of the variable or it may be difficult to interpret the results that are recorded.

2.1 INTEGRATION

2.1.1 Electrical integrators

A number of miniature electrochemical elapsed time indicators have become commercially available in the past few years (Mercron and Curtis meters, etc.) and have provided a cheap and reliable method of integrating small electrical currents with respect to time. While these devices are primarily intended for use with constant current sources to measure elapsed time they can also be used to integrate currents that vary over a wide range provided certain circuit design precautions are observed. Examples of their use can be found in an integrating pyranometer as described by Federer & Tanner (1965) and a device for measuring the average temperature of water, soil or air, Brown (1973) which is further discussed in Section 3.3.1.

These devices are based on the principle of the mercury coulometer, two threads of mercury in a precision glass capillary separated by a gap of liquid electrolyte. As a current passes through the system, mercury is transferred across the gap from the anode to the cathode resulting in the movement of the gap towards the positive electrode. The rate of transfer of mercury, and hence rate of movement of the gap is proportional to current flowing through the cell, the total movement of the gap over a period of time is proportional to the time current integral. In practice this is not always so, especially when operated over a wide current range. The internal resistance of the device varies with current flow and ambient temperature; individual integrators that have been measured vary in internal resistance from about 700 ohms at 150 µA to about 2,000 ohms at 15 µA, and have a temperature coefficient of about 40 ohms per degree Centrigrade at low currents. This variation in resistance can be largely overcome by feeding the device from a high resistance source. A resistance of about 10,000 ohms in series with the integrator will buffer variations in internal resistance but a higher driving voltage will be required to maintain the current flow.

These devices should be handled with care as the electrolyte gap can become separated by mechanical shock causing marked changes in their electrical characteristics. The electrolyte must not be driven right to the end of the capillary or the unit will be ruined. The full scale time current integral for the Mercron (type X) is 6 mAh and for Curtis meters 5 mAh. The current passed should not exceed the manufacturer's recommendations and for the Mercron this is 150 µA. The distance moved by the electrolyte gap can be measured with the stage vernier on a low power microscope as demonstrated by Tanner *et al.* (1963), a fuse holder cemented to a microscope slide makes a suitable holder for mounting Mercron type devices for measurement. The bubble can be reset by reversing the device on the recording apparatus or by means of a separate electrical resetting system.

2.1.2 Chemical integrators

The effect of light intensity or temperature upon the rate of a chemical reaction can be used to provide an integrated measure of these two variables. Photgraphy is perhaps the best known example of chemical integration by means of light. Pearsall & Hewitt (1933) used the effect of light upon potassium iodide to measure the light penetration in Lake Windermere, Atkins & Poole (1930) used uranyl oxalate and Dore (1958) anthracene to obtain integrated measurements of light by photochemical means.

The rate at which a sucrose solution is hydrolysed (inverted) to glucose and fructose is dependent upon temperature and the pH of the solution. Pallman *et al.* (1940) introduced this principle in their method (Sucrose inversion) for obtaining measurements of mean temperature over known periods of time. The method has since been described in detail by Berthet (1960) and by Lee (1969). A great advantage of the method lies in the ease with which the degree of inversion can be measured by polarimetric analysis.

About 10 to 15 ml samples of an acidified-buffered sucrose solution are sealed in tubes and exposed in the measurement site after which the degree of sucrose inversion is measured in terms of change of optical rotation by means of a laboratory polarimeter. The sensitivity of the method can be controlled for particular applications by adjusting the pH of the solution. Lee (1969) relates effective mean temperature to pH and optical rotation in the following equation.

$$T_e = \frac{a}{pH + b - \log t + \log[\log(R_o - R_\infty) - \log(R_t - R_\infty)]}$$

where T_e = effective mean temperature (°K) over period t
             pH = pH of sucrose solution
               t = time of exposure (days)
            R_o = optical rotation of solution at start
           R_∞ = optical rotation of solution at infinity
                  (i.e. when no further measurable change occurs)
            R_t = optical rotation of solution at time t
          a and b are constants

Lee gives values for a and b but it has been found advisable to calibrate individual batches of solution at a series of constant temperatures and to calculate the constants by solving a series of simultaneous equations (Chapman *pers. comm.*).

The response of chemical reactions to temperature are not linear but exponential so the temperature measured by this method is a logarithmic and not an arithmetic mean. The difference between the two will be small unless fluctuations in temperature are large. If required Lee (*loc. cit.*) provides a correction that can be applied to make the result approximate to an arithmetic mean. Lee quotes the precision of the method as $\pm 0.02\,°C$ over two weekly periods and accuracies of between 0.4 and $1.0\,°C$ depending upon non-linearity effects and ambient temperatures.

2.2 CHART RECORDERS AND EVENT COUNTERS

2.2.1 Chart recorders

There are two main types of chart recorder, the potentiometric recorder and the moving coil chopper bar recorder. Potentiometric recorders operate by comparing a signal produced by the position of the recording pen with the input signal. A servo system then moves the pen until these two signals are equal. This type of recorder tends to be expensive and requires a generous power supply. The chopper bar recorder is normally preferred for field use because of its robustness and economic power requirements. It consists of a moving coil movement to which the input signal is fed and a chart drive motor available in a range of d.c. voltages. The needle, or pointer, carries a stylus and is positioned between pressure sensitive paper and a chopper bar. This bar 'chops' the stylus against the paper leaving a mark recording the position of the pointer. The chopping speed is linked to the chart speed, a continuous trace being produced provided the chopping rate is fast compared with the chart speed and that the signal does not vary too rapidly.

A characteristic of the chopper bar recorder is the non-linearity of the scale across the chart due to the chopper bar striking a pointer pivoting from a central fixing point. The charts supplied for use with these recorders correct for this so that manual translation of data causes no problem, but if a linear chart reader is used then a suitable correction must be applied. (See Section 5.1.) Two commercially available examples of chopper bar recorder are the Smiths Industries Miniscript and the Rustrak recorders. They are available in a range of sensitivities and chart speeds to suit particular applications. The accuracy to be expected from this type of recorder is about 2% of the full scale deflection.

2.2.2 Event counters

The Post Office type counters and other impulse counters consist of an electromagnet and an armature activating a digital mechanical counter each time the magnet is energized. The use of this type of counter is virtually restricted to counting events over a single time period as it is impossible to record the output automatically other than perhaps by photographic means. A development of the simple Post Office type counter is the Unidec counter (English Numbering Machines Limited) where single decade counters of modular design can be arranged into groups to count any number of digits.

The Unidec gives a visual indication of the count and also has an internal system of contacts that can provide an electrical output corresponding to the count to feed a recording device. Unidec counters are also available fitted with print wheels and complete printing modules can be obtained to give a print out of the count.

Stepping switches and the Post Office type uniselectors can be used for counting events. The signal from the event to be recorded is used to energize a solenoid in the switch and advance it one position for each event. These switches do not have a visual readout but they can easily be wired up to provide an electrical readout. Each contact is wired to the next through a resistor and a constant voltage is applied across the resistance chain so formed. The wiper contact then taps off a voltage proportional to its position. Automatic rain gauges using tipping bucket principles can be used with this system, the output from the stepping switch being used to feed a chart recorder , which interrogates the switch, say once daily. The disadvantage with stepping switches, especially in field use, is that dust and humidity adversely affect their performance and great care must be taken to keep the units well sealed and regularly cleaned.

Event counters provide a cheap means of recording the number of times that some event takes place and an example of their use can be seen in a device used to count people using a footpath that has been described by Bayfield & Pickrell (1971).

2.3 MAGNETIC TAPE DATA LOGGERS

Magnetic tape data logging systems in general fall into two groups; large mains-operated sophisticated laboratory logger and the small battery-operated field data logger. In the laboratory high resolution and precision and a considerable degree of control is essential and such loggers are generally capable of handling a large number of input channels. The form of incoming information may be analogue or digital and there may be a facility for two output devices to operate simultaneously. Power requirements and cost are relatively high and these systems need good environmental operating conditions. Should it be necessary to establish a large experimental system in the field using such equipment, it would require to be suitably housed in a warm dry caravan and if mains is not available must be provided by a diesel generating set which has good load regulation and frequency stability. Throughout the running period of such logging system an operator must always be in attendance.

More recently there has been a considerable demand for simple field data loggers which can be left unattended for long periods collecting data at prescribed intervals. Since these loggers must be powered from their own batteries, the power consumption must be low. To meet this requirement small, compact, incremental magnetic tape recorders with only a small number of input channels have been developed. Unfortunately earlier developments of this type were not notable for their reliability which is dependent on simplicity of design and careful choice of components.

The improved reliability of data loggers currently available is largely dependent on the integrated circuit elements which provide a basic logic family for designing a data logging system. Specification and tolerances of these logging packages are closely controlled during manufacture and their use substantially reduces the complexity of the system while the reduction in size improves component density and facilitates packaging. The use of digital techniques contributes substantially to noise immunity while stepping motors have notable advantages in reliability not only for field recording on magnetic tape but also for punched paper tape machines. The compact data cassette is cheap, small and simple in operation for field loggers. Since modern digital voltmeters now have rapid digitizing times it is not necessary to hold a transducer signal during a measurement period. The 'end of scan' signal is generated from the last channel in the scan drive circuit and is rooted to the system control so that the logger may be closed down until the next set of readings is required.

3 GENERAL DESIGN AND CONSTRUCTION

A basic data recording system consists of a transducer, an interface circuit and a recorder. Associated with these are the timing, scanning, control and power supply circuitry. The transducer or sensor converts a physical or environmental variable into an electrical signal which can be fed to the interface circuitry. The interface processes the signal to give the required ranging and linearization and give an output compatible with the recording technique used.

Many of the variables that are of interest to the ecologist change relatively slowly and continuous recordings are not required, consequently recordings can be made at selected intervals initiated by a clock in the timing circuit. Hourly or daily recordings are sometimes sufficient resulting in a saving of battery power. When multichannel recording is needed and the recorder is basically a single channel instrument then a multiplexing or switching device is placed between the transducers and the recorder. This scanning circuit scans the output of each transducer in turn and presents its output for recording. The control unit controls the overall operation of the system switching the power supplies on and off and generally controlling the sequence of events during a recording period. Finally a power supply, adequately stabilized, is required to run the system and the equipment must be housed to withstand the climatic conditions that will be experienced in the field.

3.1 POWER SUPPLIES

3.1.1 Power sources

Most data recording systems require a power source; in rare cases a mains supply may be available and problems arising from battery operation do not occur. These problems are due to the finite capacity of batteries and the fact that they are temperate sensitive. The capacity of the battery determine the power that can be used before replacement or recharging is necessary. Manufacturers publish discharge curves for their batteries showing the variation of output voltage with time for different discharge rates. The drop in voltage is due to changes in the internal resistance of the battery limiting the current that can be drawn without causing a serious voltage drop. The operating temperature affects the capacity and output voltage of battery systems and must be considered in the design of equipment that can be used in the field.

Several types of battery suitable for operating field equipment are available Smith (1973), the particular type chosen and the need for additional stabilization will depend upon the particular equipment and application. Characteristics and details of various types of battery are shown in Table 9.1.

3.1.2 Stabilization

Power supplies for field equipment often need stabilization so that a constant output voltage is provided regardless of battery condition and ambient temperature. Mercury cells may sometimes be used without stabilization in equipment requiring only low current but it is good policy to buffer them from extremes of temperature. In the case of mercury cells the simplest method is to house them in a waterproof case and bury under 15 to 20 cm of soil.

Electronic voltage stabilization can range from a simple zener diode circuit to a complex series voltage regulator. The design of a particular circuit should be able to compensate for the worst conditions, usually when batteries are nearly discharged with equipment drawing maximum current at low ambient temperatures. For the sake of economy of operation the current used to operate the stabilizer circuit should be as low as possible. It is often possible to operate some components of the recording system from an unregulated supply; relays, stepping switches and electric motors all offer large and fluctuating loads to a power supply and if they can be operated from an unregulated supply the problem of voltage stabilization for the remainder of the equipment is made easier. A number of integrated circuit regulators are available which when combined with a few extra components make excellent stabilizers for field equipment at relatively low cost, Hnatek (1973), Mammano (1971) and Ross (1973), give details of some typical circuits.

Table 9.1 Battery characteristics.

	Voltage per cell	Capacity range available	Discharge characteristics	Operating temperature range	Charge retention characteristics at 20°C	Rechargeable	Remarks
Mercury	1·35	75–2,400 mAh	Almost constant voltage until nearly discharged	Up to 70°C low temperature cells available	Up to 2½ years	No	High energy to weight ratio. Good temperature stability.
Alkaline–Manganese	1·5 (Nominal)	125 mAh–10 Ah	Not as good as mercury cells but better than ordinary dry cells	–20–+70°C	Up to 2½ years	No	Available in standard dry cell replacement sizes. Can supply large currents.
Nickel–Cadmium (sealed type)	1·25 (Nominal)	0·45 Ah–500 Ah	Voltage constant over 80% of discharge curve	–40–+60°C	30% reduction after 1 month	Yes	Sealed unit requiring no maintenance. Can be used in any position. Suitable for large continuous current drains. Standard dry cell replacement sizes available. Low energy to weight ratio.
Lead–Acid (car type)	2·0 (Nominal)	Mainly in 1–60 Ah range	Voltage not constant over discharge curve	Capacity can be reduced by 70% at –10°C	50% reduction after 3 months	Yes	Liable to spillage. Requires regular maintenance.
Lead–Acid (sealed type)*	2·0 (Nominal)	0·9–36 Ah	Voltage constant over majority of discharge curve	–30–+50°C	50% reduction after 16 months	Yes	Sealed unit requiring no maintenance can be used in any position.

* Sonnenschein Dry Fit.

3.2 TIMING AND CONTROL

3.2.1 Clocks

This unit takes the form of either a mechanical clock electrically wound, or in more sophisticated systems, a crystal controlled digital clock. The former is normally used in small field data loggers where power requirements are not at a premium and since this is a simple device offers the best advantages. The latter clock, however, has a considerably better accuracy and where power is available it is a natural choice. The recent advances in integrated circuit technology using the latest 'COSMOS' series of circuits has made it possible to use a crystal with suitable frequency division which will provide accurate time with extremely low power consumption. This type will almost certainly in the future replace mechanical escapements used in field loggers at present. The basic requirement of this type of unit is to command the logger to operate at predetermined intervals but can, in larger logging systems, also provide the necessary real time statement in the output data. Pre-set time intervals are determined by either switches or a patch pin system. A numerical display of time is sometimes provided for convenience in all systems where an operator is present and power requirements are not an embarrassment.

The Smiths Industries 'Sectronic' clock movement is an example of an inexpensive battery operated clock with mechanical escapement. The clock, powered by a C11 type dry cell battery, will run for a year without battery change. An electrical signal can be obtained at the required time by actuating a reed switch with a small magnet mounted on a thin plastic arm and rotated at its centre by the clock hand drive spindle. The other end of the arm should have a weight attached to counter-balance the magnet. By using different numbers of reed switches and magnets and by choice of the hour hand or minute hand drive spindle a variety of periods can be chosen ranging from quarter-hourly to once every 12 hours.

3.2.2 Control and scanning

The purpose of the scanning unit is to switch each transducer in turn to the measuring unit. There are many design requirements which must be considered with scanning devices, for example, speed of operation, reliability, operating life, cross talk, electrical noise and environmental working conditions. Firstly, there is the electrical mechanical type generally known as a stepping switch. This unit consists of a rotary switch with a possible 100 wafer contacts arranged around the stator periphery. The central switching wiper rotates between each wafer and is driven by a d.c. stepping motor. Heavy gold plating is used for these contacts in order to improve switching at low levels and reduce noise. However, the life of these units is rather limited and speed of operation is comparatively slow. With careful design rotary switches of reasonable speeds are available but because of the meticulous construction, tend to be rather expensive.

The dry reed, mercury wetted and diaphragm operated relays are suitable for most scanning designs and have proved to be popular in many systems. They are used in relatively high speed systems with rates of up to 100 steps per second. The reed switch which has thin reed-like contact plates is hermetically sealed within a glass envelope. Terminals are provided at the two extremities of the glass envelope and, therefore, the whole switch works well in any hostile environment. Actuation of the switch is accomplished by the use of a magnetic coil wound as a solenoid with the reed switch as the centre core. When the coil is energized, the magnetic field closes the contact. Five milli-second closure times are possible but bounce and contact noise is apparent. This disadvantage is much reduced with the diaphragm operated reed and completely eliminated with the mercury wetted relays. Although long life can be expected with all these types of relays, the wetted relay is outstanding in this application.

Due to the use of iron in the construction of these switches, when connecting copper wire to these devices in any scanning system, thermoelectric e.m.f.s are generated and in some cases can be quite high ($35 \mu V/°C$) making it essential to avoid thermal gradients near these junctions. Compensation, however, may be achieved by using two reeds in both feed and return path of the transducer cable. This is rather important since it is possible to measure to within one μV resolution with modern Digital Voltmeters (D.V.M.s).

Finally, the semiconductor type scanning switch should hold many possibilities such as high speed operation and reliability—to mention but a few. However, techniques in this field are only now beginning to evolve as suitable simple devices. These refer to metal-oxide-semiconductor-transistor devices (MOST) which offer high and low 'turn-on' 'turn-off' characteristics respectively. Such devices should become popular within the next few years, and their use and characteristics are described by Litus *et al.* (1972) and Thomas (1973).

A method of pulse generation is needed in logging equipment to provide timing and drive sources for the control and scanning circuitry. In the more sophisticated loggers pulses of millisecond duration may be needed whilst in scanning circuits used with chart recorder systems, several seconds may be needed. To give short duration repetitive square wave pulses an astable transistor multivibrator can be used of the type shown in Fig. 9.1. The pulse rate being governed by the values of C and R. $f = 1·4/RC$ for a symmetrical square wave. Malmstadt & Enke (1969) show how integrated circuit logic gates can also be wired to provide square wave pulses, see Fig. 9.2.

The drive unit for scanning in data logging systems usually consists of a Johnson type counter which is effectively a shift register type which first fills up with 'ones' and then with 'noughts' which by simple decoding, can then drive the respective output devices. The advantage of this type of counter is

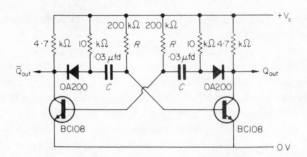

Figure 9.1 4 ms pulse length astable multivibrator circuit.

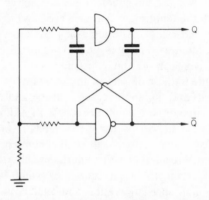

Figure 9.2 Astable multivibrator constructed with logic gates.

that it only requires half the number of counter stages per scanning devices which for design makes an economical choice. The counter is always under the control of the control module and certain facilities such as skip channels and scanning between pre-set high and low channels are usually accommodated. The counter position is often made available to the output device for channel identification purposes.

Pamplin (1967) has described how a unijunction transistor (UJT) connected as a relaxation oscillator can provide short pulses with repetition rates from milliseconds to many seconds. The longer delays being useful for driving scanning circuits in association with chart recording systems. Figure 9.3 shows a simple relaxation oscillator and stepping switch drive circuit using a Thyristor. The period of oscillation is given by approximately $0.8\, C_1 R_3$. For accurate temperature stabilization R_2 should be calculated for the particular UJT used from the equation.

$$R_2 \simeq \frac{0.4 \cdot Rbbo}{n \cdot V_1} + \frac{(1-n) \cdot R_1}{n}$$

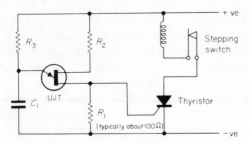

Figure 9.3 UJT oscillator and stepping switch drive circuit.

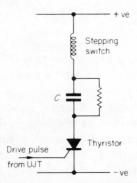

Figure 9.4 Capacitor coupled stepping switch drive circuit.

where V_1 is the supply voltage

$n =$ UJT intrinsic standoff ratio

$Rbbo =$ UJT interbase resistance

The thyristor's anode voltage has to be briefly reduced to zero after the drive pulse to switch the thyristor off. This is best done, where possible, by using internal contacts in the stepping switch which open when the switch is energized, the contacts being wired in series with the anode load. Where contacts are not available a capacitor can be used Fig. 9.4. The resistance across the capacitor must be a high enough value to ensure the switch is not held on between drive pulses and small enough to discharge C between drive pulses.

Integrated circuit programmable timer/counters are available capable of producing timing signals over periods ranging from micro-seconds to days. They consist of a time base oscillator, a programmable counter and control circuitry. The time delay is set by an external CR network and by programming the counter to give an output after the programmed number of counts.

The control unit controls the complete working system of the logger and ensures that one particular part has completed its operation before commencing the next. Here faster devices must wait for the slower ones. The common pulse to start operations is received from the clock unit whilst those for stop are generated by an end of scan pulse in the scanning unit. The control unit ensures that the required output peripheral is switched in if the logger is not under continuous operation. Facilities such as a step command is, therefore, incorporated so that the logger can be advanced channel by channel under manual control. This is used for the selection of individual sensors when setting up requirements are needed.

3.3 INTERFACING

Interfaces are units that are used as connectors between two devices; for example, between sensor and D.V.M. or D.V.M. and output periphery or translator and computer. They provide the necessary amplification, switching, storage decoding, matching linearization and bridge circuitry which will allow the required operation between two units. Input interfaces are connected between sensor and D.V.M. and vary in design, depending on the type of transducer used. For example, if we take a simple platinum resistance thermometer to be connected to a D.V.M., this will require a Wheatstone bridge circuit. A stable excitation voltage will be necessary to feed this bridge and the resulting low level output signal will possibly need amplification before the signal can be presented to this unit. All these requirements are provided for from within the interface itself. Similarly the interface requirements between translator and computer need circuits to accept serial data from the translator which is converted into parallel form so that it can be accepted directly to the input highway of the computer. Timing is important in this instance, together with suitable matching circuits, bi-stable latches and storage devices, all of which forms part of an interface unit. Thermistors are now commonly used for temperature measurement which have the advantage of greater sensitivity and having short-time responses. These elements, however, are now normally operated in a bridge network and because of their logarithmic characteristics, linearization circuits must also be incorporated. This is accommodated for and provides part of the interface unit.

3.3.1 Bridge circuits

Bridge circuits, especially of the Wheatstone bridge type are often used in interface circuits. Their applications include backing-off circuits to provide artificial zero points, linearization of thermistor characteristics and differential measurements in temperature and strain gauge work.

There are two basic ways a bridge can be operated. It can be balanced to give a zero or null output by adjusting the bridge circuit values either manually or by means of a servo system, in this mode the reading obtained from the bridge is independent of the bridge supply voltage. Examples of this method of operation are to be found in many measuring instruments, see soil moisture resistance blocks Chapter 6, Section 6.3.5, and conductivity measurements Chapter 8, Section 5.2.5. The other method of operation is to produce an output signal from the bridge which is dependent on the amount of out of balance occurring. The bridge is permanently balanced for one particular value of resistance and any departure from this value results in an output. This mode is the one normally used in interface circuits but it suffers from the disadvantage of its output being dependent on the bridge supply voltage, especially away from the balance point. To overcome this difficulty a stabilized bridge supply voltage should be used. Bridge circuits and the linearity of their response when feeding into an operational amplifier for instrumentation applications are described by Clayton (1971).

Thermistors (Hyde, 1971) are often used in ecology for temperature measurements their main disadvantage being their non-linear resistance v. temperature characteristics which follows the law

$$R_T = Ae^{B/T}$$

where T = temperature of thermistor °K
  B = characteristic temperature °K
  A = constant.

A simple method of obtaining a very nearly linear response over a limited temperature range can be achieved by shunting the thermistor with a fixed resistance equal to the thermistor's resistance at the mid point of the working temperature range. An approximately linear resistance/temperature response of the combination can be achieved over a 30°C range, for details see STC Information Sheet (1960).

In applications requiring greater accuracy and the ability to provide a zero output at some predetermined temperature a bridge circuit should be used. Linear bridge circuits and values have been described by Beakley (1951), Bowman (1970), Hole (1971) and Stanković (1973) among others. It is possible to obtain accuracy of 0·1°C over a 30°C range using these circuits.

Mean temperatures measured over a predetermined period are often required by the ecologist. A relatively simple method using a thermistor in a linearizing Wheatstone bridge circuit driving a miniature mercury coulometer of the Curtis type has been described by Brown (1973). Some earlier methods by other workers did not take into account the variation of internal resistance of the coulometer with temperature and current or even the non-linear characteristics of thermistors, resulting in inaccuracies especially over large temperature ranges.

In the method described by Brown the variations in internal resistance are buffered by including a high value resistance R_s in series with the coulometer.

Because of this series resistor a higher bridge supply voltage has to be used to obtain the required current flow through the coulometer. Care must be taken that self-heating effects do not occur in the thermistor. From the dissipation constant of the thermistor used the temperature rise per mW of power dissipated can be found. For maximum accuracy this rise should be less than half the required resolution of the thermometer.

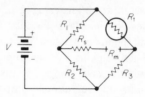

Figure 9.5   Integrating thermometer circuit

V is the battery
R_1, R_2, R_3, R_s, are fixed resistors
R_t is thermistor resistance
R_m coulometer resistance

In the circuit given by Brown (Fig. 9.5), the maximum voltage allowable across the thermistor is given by

$$E^2 = PR$$

where P = maximum power dissipation allowable
$\quad R = R_T$ = mid-range resistance of thermistor

The maximum supply voltage is thus given by $2E$ as $R_3 = R_T$
The other values of the bridge circuit are given by
1. Resistor R_1 equal to the thermistor resistance at the minimum of the temperature scale, hence the bridge is balanced at this point.
2. The linearization of the bridge output is achieved by making $R_2 = R_3$ = thermistor resistance at mid-point of temperature range.
3. The series resistor R_s should be greater than 10 kΩ and used to adjust the current flow and hence recording period of the coulometer.

3.3.2 Operational amplifier circuits

Operational amplifiers are used to perform various mathematical operations in the analogue computer field and to provide amplification impedance converting and level detecting functions in instrumentation systems. An operational amplifier is a high gain d.c. amplifier, the required response being obtained by the application of negative feedback between its output and input terminals. Operational amplifiers can be purchased constructed from discrete components in a modular form or more commonly they are made by bipolar monolithic circuit techniques, the integrated circuit. There are also hybrid devices employing both methods within the same package, in some power amplifiers for instance the driver stage uses an integrated circuit whilst the output stage uses discrete components.

When using an operational amplifier it is not normally necessary to know or understand the internal circuitry of the device, it may be considered as a 'black box'. What should be considered are the overall characteristics and the functions of the external connections. The most common form of operational amplifier is the differential input, single ended output type. This is because of the need in many applications to be able to swing the output voltage either side of zero and also to provide a floating input (isolated from earth) with high common mode rejection. When an amplifier has high common mode rejection the output voltage is dependent only on the input signal across the two inputs of the differential input amplifier, not on any common voltage between the input terminals and earth.

The most commonly used general purpose operational amplifier is of the 709 or the newer 741 type. The 709 lacks internal frequency compensation and external components as specified by the manufacturer have to be added to the circuit to obtain the desired frequency responses and to prevent instability at high gains. As with any high gain amplifier design care should be exercised with circuit layout and power supply decoupling to prevent instability, especially where two or more amplifiers are in close proximity. Decoupling capacitors of at least $0 \cdot 1$ μfd should be connected between the power supplies and earth as close as possible to the amplifier.

Certain precautions should be taken when using operational amplifiers in order not to damage them. The output of most amplifiers have built in protection against short circuit to earth or the power supply rails. The input to most amplifiers can also take a voltage up to that of the power supply. Damage will result though if a voltage greater than the power supply voltage is applied accidentally to the inputs or output. It should be remembered that when the power supply is turned off with a voltage still on the input or then applied to the input, damage may result. This is especially the case with low impedance inputs and large coupling capacitors used in a.c. amplifiers are such a case. Accidental reversal of the power supply polarity will also destroy the amplifier as heavy internal currents will flow. In situations where such accidents are liable to occur it is a wise precaution to place limiting resistances in the input and output circuits, where this is possible, and to connect high voltage diodes in series with the power supplies to the amplifier. Operational amplifiers usually require a dual or bipolar power supply, that is a stable d.c. supply of suitable voltage arranged symmetrically about earth potential. The operating voltage of most amplifiers is in the ± 5 V $\rightarrow \pm 15$V range. Some a.c. circuits can be run from a single power supply as can some single input d.c. amplifiers where common mode rejection is not required.

A recent introduction to the operational amplifier field is the micropower amplifier as manufactured by RCA, Silicon General, and others. These devices can deliver milliamps of output current yet only consume microwatts of standby power, compared to several hundred milliwatts for conventional types, Avery (1973). The economics in power supply provisions make these devices ideal for inclusion in field equipment especially as they can also be run from low supply voltages.

Another recent introduction is that of the F.E.T. input operational amplifier. The field effect transistor has a very high input impedance $(10^{-10}\,\Omega – 10^{-14}\,\Omega)$ and low input bias current and drift (in the order of 10^{-11} A). Two discrete F.E.T.s are used in the input stage of an operational amplifier to provide a hybrid device with the input characteristics of an F.E.T. These devices are useful in applications requiring high input impedance and low drift for instance in integrator circuits and amplifiers for ion selective probes.

Some amplifiers are made for special applications, the thermo-couple chopper amplifier is one such example. The input signal is 'chopped' into a square wave by the input stage and a.c. amplification is then used. This enables a high input impedance with very low offset voltage drift with reference to temperature and time to be obtained. This is an essential requirement where the low level signals from thermocouples have to be amplified and measured over long periods.

D.C. amplifiers

The differential input operational amplifier can be used in three basic circuit configurations, each with its own uses and limitations arising from the feedback networks employed. By assuming that the amplifier has ideal characteristics simplified expressions can be obtained for its closed loop gain in terms of the feedback network values alone. In most general applications where the closed loop gain required is much less than the open loop gain these expressions are sufficient to design a circuit with. It is, however, quite often necessary to consider the operational amplifiers input offset voltage and current characteristics and temperature coefficients when designing low drift d.c. amplifiers. A summary of the characteristics and applications of the three basic configurations is given in Table 9.2. For a full treatment of the theory of operational amplifiers, taking into account their practical characteristics and for descriptions of variations in these basic configurations for specific applications, some of the many books on the subject should be referred to, Clayton (1971), Hnatek (1973), Toby *et al.* (1971).

A.C. amplifiers

Operational amplifiers can be used to amplify a.c. signals with the added advantage that if a response down to d.c. is not required capacitor coupling

Table 9.2 Basic amplifier configurations.

Type	Circuit	Closed loop gain	Applications and characteristics
Inverting		$A_{cl} = -\dfrac{e_O}{e_i} = -\dfrac{R_2}{R_1}$	Gives gain with phase inversion. Adder applications.
Non-inverting		$A_{cl} = 1 + \dfrac{R_2}{R_1}$	Gives gain with no phase inversion. High input impedance. Buffer amplifier.
Differential		$A_{cl} = \dfrac{R_2}{R_1} = \dfrac{R_4}{R_3}$ $R_1 = R_3$ $R_2 = R_4$	For amplifying signals that are floated with respect to earth. Amplifying bridge circuit signals. Subtractor applications. Has high common mode rejection.

can be used so eliminating some of the problems associated with input offsets and drifts.

Voltage comparators

If an operational amplifier is used without negative feedback, i.e. open loop, the output will saturate and swing to the maximum positive or negative output voltage. A very small input voltage of the right polarity will cause the output to swing to the opposite maximum voltage so enabling the amplifier to be used as a switch. By using the summing junction and applying a reference voltage to it the amplifier can be made to switch only when the input voltage exceeds the reference voltage. Operational amplifiers especially designed for use as comparators are available commercially.

Non-linear feedback

The feedback network used with operational amplifiers need not be purely resistive. Non-linear feedback can be used to tailor the response of the amplifier to a particular purpose. Amplifiers with a logarithmic response can be designed exploiting the logarithmic response of a diode or transistor in the feedback path, several decades can be covered with careful choice of components. Precision rectification of low level a.c. signals can also be achieved by using a diode in the feedback path, Clayton (1971). The forward voltage drop across a diode is non-linear at low levels of current, making the usually rectification circuits inaccurate for low level-voltage. The inclusion of the diode in the feedback path reduces the effect of diode voltage drop at the output.

Integrators

Electronic integrators can be used in data logging where a measurement to be recorded is liable to large rapid changes, as in solarimetry. To continuously

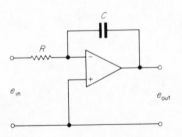

Figure 9.6   Basic integrator circuit.

record such measurement would produce a large amount of unnecessary data and subsequent processing problems. Infrequent short-duration sampling of the measurement can lead to errors and if a method of integrating the measurements over the sampling period is used, the average can then be recorded at the sampling time. An integrator circuit does have to be left running all the time but the power consumed is less than if a continuous recording technique was employed.

A basic integrator circuit for use with an operation amplifier is shown in Fig. 9.6. Negative feedback is provided through the capacitor C and it can be shown that for a perfect amplifier, Clayton (1971)

$$e_0 = -\frac{1}{CR} \int_0^t e_i dt$$

The output voltage is proportional to the voltage time integral of the input voltage. CR is the integrator time constant and $1/CR$ can be expressed as volts/second output per volt input.

In practice the amplifier will suffer from input bias current and offset voltage errors causing the output to drift with no input signal applied. In time the output will saturate either in a positive or negative direction depending on the polarity of the input offset voltages and this drift is the limiting factor in circuit design. The choice of amplifier is important especially when long time constants are used and amplifiers should be chosen which have low input bias current. The F.E.T. input amplifiers and chopper stabilized amplifiers are best for medium and long-term integration, up to one hour with care. Bipolar input amplifiers should only be used for very short-term integration. Drift in the output is also caused by leakage in the feedback capacitor C, the amount being proportional to the value of C and varies with different types of capacitor. Low leakage polycarbonate or polystyrene types should be used especially in integrators with long time constants. To equalize the input bias currents the non-inverting input should have a resistor in series with it equal to the inverting input resistor R. The initial input offset voltage should be balanced out by a recommended circuit to give minimum drift in the output. With careful design the main source of drift reduces to that caused by the temperature coefficients of the input offset voltage and bias currents.

A practical integrator circuit requires a method of setting the initial output to zero prior to the integration period. At the end of the period the output is recorded and then reset to zero. A circuit for providing reset and integrate facilities is shown in Fig. 9.7. A reed relay, actuated by the recording system, is used to discharge the feedback capacitor C before the integration period. The circuit shown in Fig. 9.7 has been used to integrate the output from a thermopile type dome solarimeter over quarter hour periods.

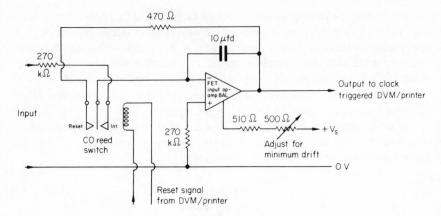

Figure 9.7   Practical integrator circuit.

3.3.3 Analogue−Digital converters

The standard recording voltmeter suffers from poor accuracy and the inability to read the instrument to better than about 0·5% of full scale. To overcome such difficulties the digital voltmeter was developed and is now incorporated in most data acquisition systems as a standard measuring unit. Its extremely good resolution with digital read-out makes this unit very adaptable for logging systems. Apart from its exceptional high accuracies of better than 0·01%, high input impedance of between 10 MΩ and 1,000 MΩ is now possible. The latter is extremely useful in logging systems since in many cases long leads between sensor and interface are required. The lead resistance, together with the source impedance presents only a very small fraction of the total input impedance and, therefore, does not limit the high accuracies which can be attained.

Automatic ranging is yet another useful feature with modern D.V.M.s. This allows mixed signals of low and high level to be fed direct from transducer to the unit without the aid of the interface but probably one of the best advantages of this development is that a lower or higher range can be automatically selected which will give the best sensitivity for the conditions which prevail.

Three well-known principles have been used in digital voltmeter developments. The first of these is known as the successive approximation method where the unknown d.c. input voltage is compared with a succession of voltages produced from a stable reference voltage source within the instrument itself. The operation can be explained with reference to Fig. 9.8. The input voltage is applied to the difference amplifier (comparator). The ring counter is operated by an internal clock circuit and advances to the first bi-stable switch, allowing

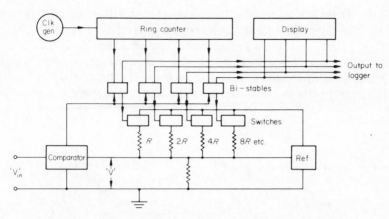

Figure 9.8 Block diagram of D.V.M. successive approximation method.

the lowest value of resistance R into the reference circuit, which produces the highest value of V. Should this be greater than the input applied, the output from the comparator switches off this bi-stable circuit and so advances the transistor ring counter which switches via the next bi-stable the next resistance $2R$ into circuit. This effectively reduced V to half its value due to the resistance weighting and so the comparator again compared the voltage input with the reference voltage. If now Vin is greater than V then the comparator output arranges for this bi-stable to remain on and so the ring counter advances once again. Resistance $4R$ is now switched into the comparator circuit providing a reference voltage of $\frac{3}{4}V$. Again the comparator decides whether Vin is greater or less than V and so takes appropriate action as before, allowing bi-stable 3 to remain in, providing V is less than Vin but switching it out if V is greater than Vin. When a null output from the comparator is established, the unknown value is then equal to that of the voltage reference. The number of switches now in circuit can be represented as a digital number which is equivalent to the input voltage. These resistors are, therefore, binary weighted (8421) and are arranged in groups of four—each representing one single decade number. All resistors in this instrument are precision types and a very stable reference voltage has to be used. However, very high operating speeds can be achieved with this type of instrument.

The second system uses the principle of slope versus time technique which converts an analogue quantity into a digital number by the following method. A ramp voltage wave-form is first produced the amplitude of which is directly proportional to time. The d.c. input signal is applied to a comparator. A start pulse first resets the counter and ramp voltage to zero and opens a gate circuit allowing pulses from a stable oscillator to be passed to a set of decade counters. Counting continues until the ramp wave-form equals that of the unknown signal. The comparator detects this condition and provides a stop pulse closing the gate; the counts registered within the counter, therefore, represent the input signal. The ramp wave-form is generated by charging a capacitor from a constant current source. Two comparators are sometimes used. The extra one detecting a zero reference point and allowing a counting gate to open. In higher resolution systems using this method the oscillator is of the crystal type having a frequency of 1 MHz. A block diagram of this type of instrument is shown in Fig. 9.9.

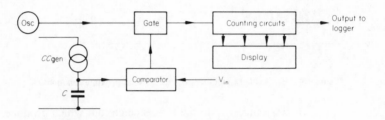

Figure 9.9   Block diagram ramp D.V.M.

The third type of D.V.M. is the integrator type or, as otherwise known, the voltage to frequency converter. The input signal is first amplified and passed via a series resistor to the input of an integrator which produces a ramp wave form, the slope of which varies with the input amplitude. As the ramp voltage output passes a predetermined trigger level, a precision pulse of known width and amplitude is fed back to the input of the integrator and resets it to its starting level. After completion, the ramp voltage is again allowed to rise. The frequency output of the wave-form produced by the integrator will now be a function of the input voltage. Two counters are used, one for timing and the second for display purposes. The frequency is fed through an accurately timed gate to the display counters. One of the main advantages of this type of instrument is its ability to reject mains hum. This is achieved because the resultant count is averaged over a fixed period, and if the count period is adjusted to that of the mains frequency (i.e. 20 milliseconds for 50 Hz) then the hum level will be averaged out, as will any harmonics at this frequency. Because of this integration technique low sensitivity to noise is achieved. A block diagram showing the principles of this technique is displayed in Fig. 9.10.

There are now on the market many low priced D.V.M.s suitable for use as analogue to digital converters. These are usually of the single input range digital panel meter type (D.P.M.s) and can be obtained with B.C.D. output and control logic for connection to external recording equipment. The $2\frac{1}{2}$ digit readout (0–199) that the cheaper D.P.M.s give is adequate for many purposes, increased accuracy being obtained in the more expensive $3\frac{1}{2}$ digit D.P.M.s (0–1999).

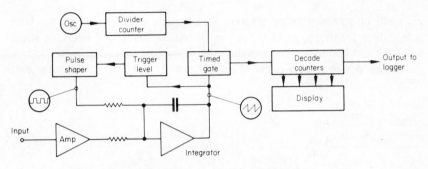

Figure 9.10   Block diagram integrating D.V.M.

3.4 TRANSDUCERS

These devices are used in electrical instrumentation systems and transform physical quantities or phenomenon to be measured into voltage or current outputs which are then suitable for data logger applications. Transducers fall roughly into two groups. Those which produce analogue signals as a function of a measured variable and secondly those which produce a frequency signal or series of events. The first types are normally fed to D.V.M. inputs, the latter to frequency counters or scalers.

The analogue types can be sub-divided into low and high level types. The latter requires no amplification and takes the form of the potentiometer type of sensor. They are used mainly where large analogue linear or angular displacement measurements are made. The low level type take the form of resistance thermometers, thermistors, strain gauges, thermocouples or voltaic devices and all need amplification before the input measurement can be referred to the analogue to digital converter.

Digital signals consist of pulse rates or a series of coded events. Here the signals are usually large enough to be presented direct to the measuring device used. This takes the form of a frequency meter or counter and since the signal is already in digital form, the need for conversion is not necessary. Sometimes a buffer store is required, especially if events are being continuously counted. Here the results are immmediately stored and the counter re-set for its next set of counts. The logger can then at its own speed accept the data already stored. Digital signals have a great advantage in that they offer only two conditions of either 'on' or 'off' and so can be recovered easily when used in noisy environments.

4 PRACTICAL MAGNETIC TAPE SYSTEMS

4.1 BASIC DATA LOGGING SYSTEM

Figure 9.11 shows a block diagram of a typical data logger system. Each sensor first transmits an electrical signal via its respective interface which amplifies and conditions the signal so that it may be measured by a common measuring device in the logger. The scanning unit is provided so that each

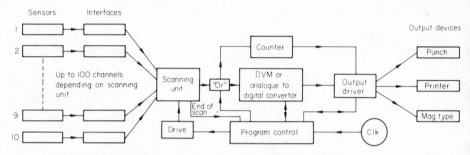

Figure 9.11   Block diagram typical data logger.

signal may be presented individually and in sequence to the measuring device of the logger. This measuring device is usually a D.V.M. or analogue to digital converter which converts the d.c. signal present into a digital output. In many logging systems a second measuring device may be incorporated, which takes the form of a frequency or periodic counter. This is a useful asset if signals from certain transducers being monitored are already in digital form such as pulse trains, events or frequencies. Both these measuring units present their output data in digital form to the output driver. This particular unit acts as an interface between the measuring unit and the output peripheral which, on command from the system control, transfers the output data from the measuring device to the input of the selected output machine. This machine prints, punches or records the input data and on completion, provides an instruction pulse which notifies the system control of its readiness to record more data. The system control unit acts as a programmer for the operation of the data logger and correlates the sequence of events for correct operation. It receives its instructions to start logging from a clock unit and therefore, enables the scan driver to advance the scanning unit one step so that the first transducer output may be connected to the D.V.M. or counter, whichever is appropriate. An instruction is then sent to the particular measuring unit to digitize its input signal. On completion it informs the control unit which then instructs the output device to transfer data to the output peripheral providing of course that this device is ready to receive data. The D.V.M., having completed its operation is now ready to receive input from another transducer and so the controller commands the scan drive unit to be advanced one more step connecting the next transducer to the D.V.M. The rate of switching is determined by the slowest device in use which is usually one of the output peripherals.

Since modern D.V.M. now have rapid digitizing times, it is not important to hold a transducer signal during a measurement period. The end-of-scan signal is generated from the last channel in the scan driver circuit and is routed to the system control so that the logger may be closed down until the next set of readings is required. The output data format which is usually set up within the logger, displays real time at the beginning of a logger scan followed by channel identification and sign which precedes each particular reading. These requirements are programmed into the system control and are offered to the output device at the appropriate time. The output device converts the output coded data from the logger and transforms it into print record, or coded punch paper tape which in the latter case, is suitable for further computer analysis. Speed is important so that data may be handled in comparatively short time periods and for this reason, high speed punch tape machines have been built and are now commonly used. However, printers are used for more directly read records.

Printers can take the form of electrical typewriters, strip printers, drum printers or teletypes and offer varying speed specification depending on the requirements needed. Firstly, teletypes and electrical typewriters offer greater flexibility with the additional facility of typing sub-headings if required but they have the disadvantage of being comparatively slow devices. This may be attributed to the fact that information is presented in serial form and is offered to the device one character at a time. Advantages of this system are that the data already in serial form can be conveniently used for data transmission purposes if required. The strip or drum printer is on the other hand a parallel device and prints all its information in one operation which considerably increases the print output capability. The drum printer is a development of the strip printer and offers high speeds. The device consists of a set of print characters distributed around the peripheral of a revolving drum which is driven at synchronous speeds. When, however, the required character is underneath the print page, a small print hammer operates to print the required character. This device is repeated across the line the required number of times and so a line of characters is printed for each drum revolution.

Tape punches are also used for output devices which are generally used for high speeds in the order of 110 characters per second. These units punch a series of coded holes in a 1 inch strip of paper with a maximum of eight holes across the width of the paper—each representing one character. These are driven by a synchronous or stepping motor while the punch hole actuation is normally of the electro-mechanical design. A smaller series of feeder holes is also punched between the data holes.

4.2 SYSTEMS IN USE

There are two types of logging system. Those which are used unattended in field experiments and the more sophisticated type used for laboratory or controlled experiments with an operator usually in attendance. Two typical systems of each type will be described in more detail.

4.2.1 Field systems

The majority of field data logging systems use magnetic tape for recording purposes. Two methods of recording are used (a) analogue and (b) digital. The latter is the most common form in use for data collection. Analogue recording is sometimes used in applications requiring control systems but now even this is being repeatedly replaced by more accurate digital techniques.

The design criteria for a field logging system calls for a compact, rugged, reliable, easily operated logger which has its own self-contained power supplies. These loggers are usually small and compact and are housed in water tight cases which are easily carried by individual personnel. Unattended field loggers may be installed in the field either in fibre glass boxes below the ground surface or above ground in wooden marine ply cases, mounted on poles 4 feet high. There is of course no real necessity to house such devices since their own case structure should be adequate for any field environment. However, housing such loggers in fibre glass boxes has the advantage of concealing these devices from vandals and the inquisitive public. Mounting in boxes on poles is recommended if the installation is on marshy or flooded sites. Pressurization of all loggers is recommended for satisfactory field operation and an important policy to adopt with these magnetic tape loggers is that no logger should ever be opened in the field.

Field logging systems operate on an individual clock basis; that is the logger is triggered at selective intervals by its own mechanical clock. Photograph (Fig. 9.12) shows the typical field data logging system and is the type at present in use by the Institute of Hydrology. It consists of two separate die cast aluminium boxes. One houses the interface cards necessary for amplifying or conditioning the signals received from the transducer and so arranging it to be acceptable to the data logger. The second houses the data logger itself, together with the rechargeable battery supplies which can be seen housed in the lid. Both lids are sealed effectively by the use of 'O' rings and when their lids are clamped to their respective cases, both units are then pressurized with dry air to a pressure of 5 lb per square inch so that ingress of moisture is not possible. The logger is connected to the interface by a suitable connecting cable which allows the complete logger to be exchanged in the field without having to expose the tape mechanism to the surrounding environment. Since the interface is not replaced then the calibration of the system is not

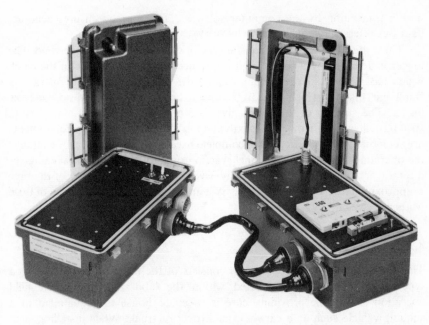

Figure 9.12 Field data logging system.

affected. Tape changing and battery replenishment can then be carried out in a dry environment without risking the mechanism to the hazards of the field conditions. With a suitable exchange of loggers a float system may be arranged to service a large number of loggers in use.

This equipment is a nominal 12-channel recorder with skip facilities on any channel. The first channel always records a precision reference voltage which also serves as an overscale number. The reason for this is two fold. Firstly the reference voltage is used to scale the rest of the recording within the particular scan and secondly it identifies the first channel so that sychronization of data can be maintained in the event of spasmodic errors which may occur from time to time.

The logger records in serial form an 8-bit binary word onto magnetic tape using the popular C.60 cassette. The tape transport mechanism is simple using an incremental stepping motor to advance the tape bit by bit with a packing density of 200 bits to the inch. No play back or re-wind facilities are incorporated in this system—the translation machine itself taking care of these requirements.

The analogue to digital converter in this logger is a cheap module working on the variable delay principle. The oscillator used is also a simple cheap unijunction type. Temperature drifts will naturally occur in such circuits but are catered for by recording a very stable reference voltage and correcting during translation by the computer softwear. This method ensures accurate data but offers a relatively inexpensive system.

These loggers use the mechanical type of clock with electrical re-wind. The logger is actuated by a reed switch positioned near the hour disc of the clock which has 12 pins spaced equally around the periphery. Each pin operates a small bar magnet which in turn operates the reed switch and so calls the logger. The spacing of these pins gives a 5-minute interval call but different spacing will give other period combinations if required. Power requirements are such that they well exceed the complete useage of the cassette; for example, the operating time for an 8-channel system using a 5-minute interval call gives a total working time of approximately 3 weeks. The total capacity of the tape is approximately 64,000 words. This system gives a mean power drain of 0·06 watts per 24 hours.

4.2.2 Laboratory system

This type of data aquisition system consists of the large more sophisticated type which is normally operated within the laboratory. However, field operation is possible providing care is taken to house the equipment in a suitable vehicle such as a caravan, land-rover or truck. When installing such devices in caravans, etc., consideration should be given to thermal installation and ventilation because such equipment especially peripheral types need a dry free environment to function properly. Layout of equipment is important with regards to weight distribution especially when towing. Heating and lighting must also be considered when installing such equipment and consideration towards the storage heater for heating purposes is well recommended. Power supply to the equipment is usually mains specified and this can be provided by a suitable diesel generating set. When considering power requirements heating and lighting loads should be accounted for and these loads help to stabilize the varying loads of the equipment so that the generator is always supplying at least 2/3rds of its maximum output. This is essential if a diesel engine is used otherwise fuel injection nozzles become blocked when working with light loads. No break power supplies offer safeguards against minor engine failures such as shortage of fuel and can save the embarrassment of loss of data at an important phase of the experiment. The system here is to supply all power from a storage battery and where mains voltages and frequencies are required by the equipment, then a transistor invert will adequately supply these necessities. The battery is constantly float charged from the diesel generator therefore if a failure should occur sufficient reserve power in the battery will maintain the equipment operational for a short time. Emergency lighting can also be supplied from the battery.

This equipment being the more sophisticated type is made up in modular form and is usually housed in a standard 19 inch rack system the complete system being housed in a metal cabinet. The modules are connected together to make individual units and usually consist of the following.

(a) Interfaces

These are plug-in modules that are selected for a particular sensor in use and amplify or condition the required signal so that it is presentable to the measuring unit via the multiplexer circuits. Each module represents an individual transducer and may be exchanged for any particular type of transducer in use for that particular channel. These are all plug-in units housed in a single rack unit. A further module supplies the required power supply voltages for amplification and is situated on the far right of all the units in use.

(b) Digital voltmeter

This is a standard commercial type of instrument which is adapted for rack mounting. It is placed in the logger system underneath the interface units. The digital display is provided on the front panel of this unit together with the necessary range controls. Filter and calibration controls are also mounted on the front panel. Most of these instruments also provide a polarity display of the signal being measured.

(c) Control unit

This is basically the logger itself and consists of all the previously described circuits which go together to make up the final logger. Displayed digitally on the front panel is real time which has been computed from the logger's own clock with the necessary switches provided to set correctly the actual time. The logger controls are very few and consist mainly of start preset, inch and output device selection. The lower unit houses all power supplies for the logger itself and because of the sheer bulk and weight is situated in the lower portion of the logger cabinet. Forced ventilation is sometimes provided but with modern loggers is now becoming obsolete.

The output peripherals such as teletypes or punched paper tape machines are free standing units housed in their own cabinets with connecting cables to the logger equipment. They house all their own control equipment and usually provide their own power supplies which are mains operated, fed from separate sockets.

4.2.3 Commercial equipment

Data logging equipment for most applications is offered on the commercial market and a particular system may be designed to accommodate such equipment which is at present available. Data logging systems should be designed to accommodate readily available commercial units. This saves considerable development time and offers economical advantages.

Many manufacturers specialize in particular projects, for instance digital voltmeters, counters, data loggers and output devices such as printers, punches and recorders. They offer considerable expertise within their specialized field and so provide the very latest techniques to their products. The system designer only needs to interface these particular devices to produce a suitable system for a particular application. Conditioning of lines, amplification and integration is readily obtained from operational amplifiers by the insertion of a few discrete components. Modular units are now used throughout in most logging systems and have the advantage of being easily changed for fault finding and maintenance purposes. This reduces 'down time' to a minimum if exchange units are readily available. Expansion for further channels is easily accommodated by the insertion of plug-in modular units which can increase logger capacity to 50 or a 100 channels. The back panel wiring is usually carried out initially so that the minimum of work is necessary for up-dating.

The latest developments within the data logging field include the application of computer control which offers considerable data compression facilities. This is where the computer's ability to handle large volumes of data is used to the full. Control by computer of the actual experiment may be handled by the required computer program. The latest range of 'mini-computer' now available will find their way into data logging systems. This will provide the scientist with a sophisticated tool which should improve and control our future experiments.

4.3 PRACTICAL APPLICATIONS

Data loggers are used in applications where a large number of measurements have to be made, whether it is required for experiment in the laboratory or field. Naturally the choice of logger is important, but the degree of sophistication necessary for a particular experiment should be considered. Having decided on the type needed, practical setting up of the complete system will have to be done whether it is required in the field or laboratory. The following paragraphs will give examples of logging systems set up for particular experimental projects and will show some of the practical points which should be considered when such a system is to be used.

4.3.1 Climatological stations

These stations are used to measure climatological variables at frequent intervals so that evaporation assessment may be obtained. These loggers will obviously be operated in field conditions and unattended, so that the choice here is the small magnetic type logger with low power consumption as shown in Fig. 9.12. This logger was designed specifically for such tasks. Interface cards are readily available for making such measurements as solar radiation, net radiation, temperature, depresssion, wind run, wind direction and rainfall. Further measurements such as soil heat flux, pressure, albedo and river level have all been successfully recorded with this logger. There is virtually no limit to the possible types of transducer that may be recorded with this type of logger and it is certainly the obvious choice for unattended field conditions.

All data loggers operating in the field at some time require tapes to be changed and new or recharged power supplies to be fitted. In adverse weather conditions it is dangerous to open these loggers and expose the tape transport mechanism to the mud, rain and moisture which will certainly damage this equipment. A safeguard policy adopted by the Institute of Hydrology is to use a 'float system' of loggers so that a complete logger may be changed in the field and the exchanged one passed back to the department for tape changing and battery replenishment. In the kinder environment of the laboratory, these operations are easily and safely carried out and the logger is then ready for its next operation. Disconnection of the logger from the interface only requires the removal of two plugs and since the interface is not disturbed, the calibration of the station is not upset.

4.3.2 Evaporation experiments

A further example of data logging application concerns an experimental project conducted by the Institute of Hydrology to study the factors involved in the evaporation from forests. A suitable site chosen for these investigations was Thetford Chase where large flat areas of forest are situated.

The requirement here was for a large number of meteorological measurements to be taken. The large volume of data resulting from such a project would involve considerable effort in analysis. The decision was therefore made to use a computer together with the logger making a complete data acquisition system with the advantages of considerable data compression and the facility of using computer control of the experiment when required. Outputs from the computer are presented on punched paper tape or, in many instances, profiles are plotted onto a graph plotter.

A photograph of the complete installation is shown in Fig. 9.13. On the extreme left in the foreground is the teletype used for input instructions to the computer as well as providing printed records of hourly averages. The left-hand top corner of the main frame houses the computer itself with all its controls and input reader. In the centre frame at the top, is housed the quartz thermometers and directly below the digital voltmeter of the logging system. The scanners for both these devices are situated respectively adjacent to its

Figure 9.13   Caravan installation at Thetford Chase showing computer and logging equipment.

measuring instrument in the top right-hand frame. The quartz scanner is capable of handling up to 60 temperature channels in multiples of ten, and the cross bar scanner below handles 200 channels of guarded two-pole analogue inputs or 600 single ones. Below this device is the anemometer interface unit handling 16 anemometers inputs. The rest of this panel is taken up with power supplies and monitor circuits. The two large punched paper tape output devices can be seen in the lower half of the centre rack. This equipment is operated in the forest and must therefore be housed in a suitable heated caravan. The power requirements are provided by a diesel generating set, surplus power being used for storage heaters which provide warmth and dry conditions necessary for the equipment to work satisfactorily in the caravan.

4.3.3 Bore hole logging

A similar data logging scheme was required to measure temperature profiles in bore holes together with conductivity and water levels. Up to ten bore holes were to be monitored in an area of 2 square kilometres. This scheme illustrates a different approach to a similar logging requirement as described for the Thetford Experiment. Because of their physical size and response the temperature sensors chosen were thermistors with seven being used in each bore hole. Three conductivity measurements were made using a conventional cell, and employing an operational amplifier combining a ring modulator circuit technique to prevent polarization. The tenth measurement is that of water level which is monitored by conventional techniques with float and potentiometer. All these ten analogue signals must be converted to a digital represented signal for transmission to a central logger. These circuits are all part of an outstation which must be controlled by the central logger. The capacity of the logger is to handle ten of these outstations which represent 100 channels of measurement. Since the outstation may be as far away as 1 km from the central logger and also to ease transmission problems the measured signals are all transmitted in digital form which greatly helps in the recovery of weak signals over long distances. The variation of this system is that the logger has a form of digital scanner rather than the conventional analogue type and so a control coded signal must be sent to each outstation when commanding measurements to be made. This system, nevertheless, is more complex but is more reliable over the distances with which measurements have to be recorded. The caravan installation is similar to the previous system except a no break power system to the logger is employed. The advantage here is that if there should be a power failure during a critical phase of the experiment, results will still be obtained for a limited period. Surplus power again is used for providing light and heating to the site caravan. The system described shows two major variations on a particular logger scheme to overcome field problems which are all important when the first designs are considered.

4.3.4 Meteorological site data collection

The last type of data logging to be described is used for monitoring meteorological data from the Institute's site. The data are mainly used for calibration of the field data loggers mentioned in Section 4.2.1. Here a standard commercial logger is used housed in a corner of the laboratory some 100 metres from the site. A multiway cable connects the sensors to the logger interfaces and since the signals are generally low-level then the choice of this cable is important for loss, screening and cross-talk. Suitable interfaces have to be incorporated between sensors and logger. These are rack mounted with their own power supplies. Each contains the operational amplifier necessary for suitable connection to the logger. Since each amplifier has an extremely high input impedance in the order of 10 MΩ then loss due to cable resistance will be comparatively small. Power requirements are not a problem here since mains power is readily available. The output can be either direct to a punch paper tape machine or a teletype and in this way failure of either device will allow the other machine to take over. Serial output to the teletype is used since connection to a distant computer via a G.P.O. line and modems can be made if required at a later date.

5 TRANSLATION AND DATA HANDLING

5.1 TRANSLATION OF DATA FROM CHARTS

The less sophisticated data recording systems suffer because the data cannot be translated automatically. Much time can be spent in extracting and tabulating data in a form suitable for analysis. Where chart recorders are used their graphical records can be especially difficult to translate accurately over any length of time. A semi-automatic reader that converts graphical readings into digital form (Fig. 9.14) can be constructed to ease this problem. A linear variable differential transformer (L.V.D.T.) of the type marketed by Electro Mechanisms Limited is used to measure the position of the chart reading. L.V.D.T.s have infinite resolution, their output being proportional to the

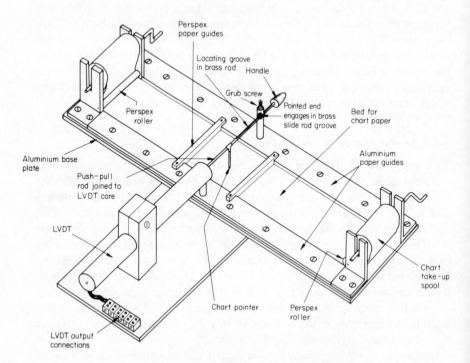

Figure 9.14   Semi-automatic chart reader.

displacement of the moveable core within the transformer. When used in a chart reader a pointer is attached to the rod which moves the core within the L.V.D.T. As the pointer is placed over a point on the graph the L.V.D.T. provides an electrical output proportional to the chart reading along the y axis for a particular point on the x axis.

L.V.D.T.s can be obtained with an exciter oscillator, phase sensitive demodulator and filter housed in the same unit so all that is required for operation is a stabilized 24 volt d.c. supply. The output voltage from the L.V.D.T. is then fed to a circuit which provides zero and scaling facilities, such a circuit for use with an L.V.D.T. (E.M.L. type 2,000 d.c.) and suitable for translating 6 cm wide charts is shown in Fig. 9.15. The circuit is designed to

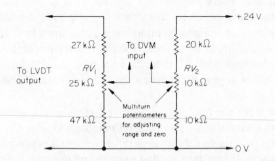

Figure 9.15   Chart reader scaling circuit.

provide the L.V.D.T. with the correct resistive load for linear operation and to provide an output from 0 to 10 volts over the chart width. Zero and scaling are set by $RV1$ and $RV2$ respectively. It should be noted that when this circuit is used, either the L.V.D.T. power supply or the D.V.M. input should be fully floated with respect to earth or else the output will be short circuited.

The output from the circuit is fed to a digital voltmeter and by selecting a suitable range a readout from 0–99 is obtained, which gives more than adequate resolution since chopper bar recorders are generally only accurate to within about 2% and small errors will arise in positioning the pointer over the chart reading. The readout from the D.V.M. may be recorded manually although a more efficient method is to use the D.V.M., providing it has a B.C.D. output, to drive a printer of punched tape unit.

The non-linearity of a chopper bar recorder, such as the Rustrak, requires a correction to be made to linear readings obtained with a L.V.D.T. chart reader. A suitable correction is

$y = x - a \sin (b.x)$
$y = $ corrected reading
$x = $ uncorrected reading
a and b are constants dependent upon recorder being used.

In the case of the Rustrak recorder $a = 1 \cdot 4$ and $b = 3 \cdot 6$. Further details of digitization and producing linear outputs from different types of chart recorders are given by Goncz (1974).

5.2 TRANSLATION OF DATA FROM MAGNETIC TAPE LOGGERS

Translation of the data obtained from a data logging system is necessary so that the coded results may be converted into their correct decimal equivalent and conversion to the correct physical units made. The correct scaling-factor must be introduced where appropriate and the final results tabulated in the required format. Data obtained from data logging systems are not always perfect, small errors or spasmodic faults may occur from time to time. These may be due to tape imperfection and so it is imperative that some form of parity checking be employed. All these data are best handled during translation by a computer where errors can be corrected and temperature drifts adjusted as required.

Translation from the field logger data first requires the use of equipment that will read the data already recorded on magnetic tape and re-arrange it in such a form that is suitable for direct parallel entry to a computer or punched paper tape machine. When data are first punched on to paper tape they may be later analysed by computer or printed and listed 'off-line' by a teletype machine.

The type of equipment shown in Fig. 9.16 shows a reading machine which translates magnetic tapes used by the field logger of Fig. 9.12. The miniature tape cassette is first placed on the machine and rewound to the start position. The playback head is brought into contact with the tape and pinch wheel which grips the tape between this and the capstan allowing the tape to advance past the reading head. The advancement of the magnetic tape past the recording head causes an induced voltage into the head windings when magnetic flux passes the small head gap, the polarizing of which will depend on the bit recorded. This signal is then amplified to a suitable level for logic operation. From each bit cell a clock pulse is also generated so that synchronization may be achieved.

The recorded data of the data logger is arranged as an 8-bit binary word in serial form. At the beginning of each word an identification word consisting of 0.0.0.1 is first produced. It is the duty of the translator to recognize such a code produced by the logger and arrange the next 8 bits of information in parallel form representing a binary coded 8-bit character. The computer or punch driver is now ready to receive these data and so the exchange is commenced. During this operation new data are already being coupled in parallel form providing the next character. A ready signal instructs the peripheral to take in the new data. This arrangement is controlled by the logic

Figure 9.16 Tape reading machine.

of the translator and so the speed is governed by the slowest device which in this case is the translator itself.

The data generated by the reading machine may be fed via a suitable control interface direct to the computer highway. This method has the distinct advantage of disregarding the coded punched paper tape stage which can involve large quantities of tape. However, by the use of a punch and drive unit, punched paper tape can be produced if a computer is not directly available.

Data outputs from many loggers are recorded directly via a teletypewriter machine which may be a serial input machine. The advantage of this method is that with suitable data modems data may be fed via a two-wire highway to a distant computer. The public telephone network or telex-lines may be used for such a system if required.

The data obtained from data loggers as previously mentioned has first to be checked for errors, and temperature drifts after which conversion for scaling factors, etc., must be made. These are best undertaken by computer and so the necessary programmes have to be written to obtain the final output required. The tapes used by data loggers, especially those recording in the field need the provision for updating records when necessary. This is best organized within the computer itself using its own storage devices for this purpose.

6 REFERENCES

ATKINS W.R.G. & POOLE H.H. (1930) Methods for the photo-electric and photo-chemical measurements of daylight. *Biol. Rev.* **5**, 91–113.

AVERY L.R. (1973) Recent advances in the design of micropower operational amplifiers. *Electronic Components* **14**, 326–330, 373–376.

BAYFIELD N.G. & PICKRELL B.G. (1971) The construction and use of a photoflux people counter. *Recreational news supplement, No.* **5**, 9–12.

BEAKLEY W.R. (1951) The design of thermistor thermometers with linear calibration. *J. Phys. E. Sci. Instrum.* **28**, 176–179.

BERTHET P. (1960) La mesure écologique de la température par détermination de la vitesse d'inversion du saccharose. *Vegetatio* **9**, 197–207.

BOWMAN M.J. (1970) On the linearity of a thermistor thermometer. *Radio and Electronic Engineer* **39**, 204–214.

BROWN J.M. (1973) A device for measuring the average temperature of water, soil or air. *Ecology* **54**, 1397–1399.

CLAYTON G.B. (1971) *Operational Amplifiers*. London, Butterworth.

DORE W.G. (1958) A simple chemical light meter. *Ecology* **39**, 151–152.

FEDERER C.A. & TANNER C.B. (1965) A simple pyranometer for measuring daily solar radiation. *J. Geophys. Res.* **70**, 2301–2306.

GONCZ J.H. (1974) Computer aided digitization of chart records. *J. Phys. E. Sci. Instrum.* **7**, 20–22.

HOLE V.H.R. (1971) Thermistor temperature measuring bridge circuits. *Electronic components.* **12**, 1167–1169, and **13**, 26–31.

HYDE F.J. (1971) *Thermistors*. London, Iliffe Books, Butterworth.

HNATEK E.R. (1973) *A Users Handbook of Integrated Circuits.* New York, John Wiley & Sons.

LEE R. (1969) Chemical temperature integration. *J. app. met.* **8**, 423–430.

LITUS J., NIEMIEC S. & PARADISE J. (1972) Transmission and multiplexing of analog or digital signals utilizing the CD4016A quad bilateral switch. *R.C.A. application note ICAN-6601.* R.C.A. Ltd., Sunbury-on-Thames, Middlesex, England.

MALMSTADT H.V. & ENKE C.G. (1969) *Digital Electronics for Scientists.* New York, W.A. Benjamin.

MAMMANO R.A. (1971) Using a dual polarity tracking voltage regulator. *Silicon General Application Bulletin No. 1.* Westminster, California, USA, Silicon General Inc.

PALLMAN H., EICHENBERGER E. & HASLER A. (1940) Eine neue methode der temperaturmessung bei ökologischen und bodenkundlichen untersuchungen. *Ber. Schweitz. bot. Ges.* **50**, 337–362.

PAMPLIN B.F. (1967) Unijunction transistors—how to use them. *Electronic Components* **8**, 59–63.

PEARSALL W.H. & HEWITT T. (1933) Light penetration in fresh water. II. Light penetration and changes in vegetation limits in Windermere. *J. Exp. Biol.* **10**, 306–312.

ROSS P.J. (1973) A low power voltage regulator for field instrumentation. *J. Phys. E. Sci. Instrum.* **6**, 969–970.

SMITH K. (1973) Latest trends in portable battery systems. *Electron* **19**, 17–24.

STANKOVIĆ D.K. (1973) Note on thermistor thermometer non-linearity. *J. Phys. E. Sci. Instrum.* **6**, 1237–1238.

S.T.C. INFORMATION NOTE TH/GEN (1960) Thermistors, general information. Standard Telephone and Cables Limited, Components Group, Footscray, Sidcup, Kent.

TANNER C.B., THURTELL G.T. & SWANN J.B. (1963) Integration systems using a commercial coulometer. *Soil Sci. Soc. Am. Proc.* **27**, 478–481.

THOMAS R. (1973) Multiplexing with TTL compatible F.E.T.s. *Electron* **39**, 38–40.

TOBY G.E., GRAEME J.G. & HUELSMAN L.T. (1971) *Operational Amplifiers, Design and Applications.* McGraw-Hill.

AUTHOR INDEX

Page where references appear in full shown in italics

Adams W.A.   346, *359*
Adamson R.S.   98, 125, *154*
Agnew A.D.Q.   111, 112, 120, *149*
Agricultural Research Council   346, *359*
Aitken M.J.   34, *69*
Ahn P.M.   323, *359*
Aimi R.   193, *219*
Alcock M.B.   176, 177, *219*
Alexandrov V.Y.   250, *285*
Al-Khafaf S.   354, *359*
Allan J.E.   445, *463*
Allen S.E.   209, 215, *219, 222,* 411, 414,
      415, 421, 428, 438, 460, *463*
Allen T.F.H.   146, *149*
Allison L.E.   442, 445, *463*
Alvim P. de T.   267, 268, *285*
Andelman J.E.   458, *463*
Andersen S.Th.   47, 52, 56, *69*
Anderson A.J.B.   135, 136, *154*
Anderson D.J.   89, 147, *149*
Anderson G.D.   326, *359*
Anderson J.L.   357, *359*
Anderson J.M.   181, 183, 190, 202, *219*
Anderson M.L.   24, *69*
Andersson F.   175, *219*
Andersson N.E.   265, *285*
Andrew R.   56, *69*
Arble W.C.   352, *366*
Archibald E.A.A.   100, 108, *149*
Arens K.   256, *285*
Armitage P.L.   300, *359*
Armstrong J.I.   268, *288*
Armstrong W.A.   89, *150*
Ashby E.   139, *149*
Ashby W.C.   189, *219*
Asher C.J.   278, *285*
Ashton P.S.   146, *149*
A. S. T. M.   355, *359*
Association of Official Analytical
      Chemists   421, *463*
Atkins W.R.G.   470, *506*
Atkinson D.   275, 277, 279, *285*
Aubert G.   319, *359, 360*

Audus L.J.   242, *285*
Austin M.P.   120, 132, 137, 146, 147, *149,
      152*
Austin R.P.   294
Avery B.W.   313, 323, 342, *360*
Avery L.R.   484, *506*

Bailey L.F.   265, *286*
Bailey N.T.J.   276, *286*
Baker D.   243, *289*
Balasubramaniam S.   264, *286*
Ball D.F.   297, 319, 328, 329, 331, 334, 335,
      337, 338, 346, 355, *360, 364,* 414,
      420, 442, *463*
Bandola E.   *224*
Bannister B.   35, 36, 38, *69*
Bannister P.   98, 146, *149,* 229, 231, 237,
      248, 249, 250, 251, 253, 254, 258,
      263, 265, 271, 273, 279, *286, 294*
Barber D.A.   285, *286*
Barber K.E.   5, 61, *69*
Barclay-Estrup P.   89, *149,* 168, *219*
Barkham J.P.   310, *360*
Barkley W.D.   44, *77*
Bartley D.D.   *70*
Barnes R.F.   *225*
Barratt B.C.   357, *360*
Barrs H.D.   255, 256, 259, *286*
Bascomb C.L.   445, *463*
Bates R.G.   *463*
Battarbee R.W.   40, *69, 78*
Baumgartner L.L.   204, *219*
Bavel C.H.M. van   268, *286*
Baver L.D.   340, *360*
Bawden M.G.   139, *149*
Bayfield N.G.   472, *506*
Bazilevich N.I.   207, *219, 226*
Beadle N.C.W.   215, *219*
Beakley W.R.   481, *506*
Beals E.   144, *149*
Beard J.S.   121, *150,* 389, *409*
Beattie D.J.   *288*

507

Beauchamp J.J.　176, *219*
Beaumont P.　420, *463*
Beckett P.H.T.　301, 329, 332, 335, *360, 361*
Becking R.W.　125, *150*
Beijerinck W.　43, 70
Beiswenger J.M.　*72*
Bell F.G.　*70*
Bell J.P.　351, *360*
Benedict J.B.　38, *70*
Benefield C.B.　186, *220*
Beresford M.　63, *70*
Bergh J.P. van den　284, 285, *286*
Berggren G.　44, *70*
Berglund B.　47, *73*
Berglund B.E.　*70*
Berry G. *366*
Bertelsen F.　56, *69*
Berthet P.　470, *506*
Bertsch K.　43, *70*
Beug H-J.　56, *70*
Beyer H.　285, *288*
Bidwell O.W.　319, *360*
Bie S.W.　329, *360*
Bierhuizen J.F.　268, *286*
Biggar J.W.　*364*
Biggs W.　376, *409*
Billings W.D.　89, *150,* 168, *220*
Bingham F.T.　*223*
Birch B.P.　63, 64, *70*
Birks H.H.　41, 60, 61, *70*
Birks H.J.B.　6, 7, 28, 46, 47, 48, 62, *70, 74*
Birse E.L.　86, *150*
Bisal F.　346, *360*
Bishop O.N.　113, *150*
Bissell V.C.　386, *409*
Black C.A.　334, 335, 339, 340, 341, 343,
　　　　　346, 347, 348, 349, 351, 353, 356,
　　　　　357, *360, 361,* 414, 459, *463, 464*
Bliss L.C.　168, 176, *220*
Blum G.　255, *294*
Blyth K.　381, *409*
Boaler S.B.　*150*
Boatman D.J.　89, *150*
Bocock K.L.　183, *220, 222*
Boggie R.　193, *220,* 279, *286*
Bohn H.L.　434, 435, *464*
Bommer D.　193, *226*
Bonny A.P.　54, *70, 78*
Booth R.S.　*226*
Bormann F.H.　423, *465*
Bornkamm R.　273, *287*
Botkin D.B.　168, *220*
Bouma J.　357, *359*
Bourdeau P.F.　168, *220*
Bourgeois W.W.　349, *361*
Bouyoucos G.J.　341, 354, *361*
Bowman G.E.　197, *220,* 241, *287*
Bowman M.J.　481, *506*

Boyce B.R.　*288*
Boyd I.W.　185, *226*
Bracher R.　1, *4*
Bradbury J.P.　29, *70*
Bradfield E.G.　450, *464*
Bradley R.I.　331, *365*
Bradshaw A.D.　236, 278, 280, 281, 283,
　　　　　287, 288
Bradstreet E.D.　*293*
Brady N.C.　298, *361*
Brandon P.F.　65, *70, 71*
Braun-Blanquet J.　98, 120, 123, 125, 128,
　　　　　138, *150*
Bray J.R.　143, *150,* 178, 188, 203, *220*
Bray R.H.　420, *464*
Brazier J.D.　42, *71*
Bremner J.M.　444, *464*
Brewer R.　357, *361*
Brewster L.A.　*72*
Bridge P.M.　*362*
Bridges E.M.　298, 319, *361*
British National Committee for Geology, *71*
British Standards Institution, 342, 346, 356,
　　　　　361
Brock K.　193, *228*
Brodin G.　*221*
Broeker W.S.　34, *71*
Brookes D.　58, *71*
Brooks R.R.　431, *464*
Brothwell D.　30, 34, 47, *71*
Brown A.　197, *220*
Brown A.H.F.　*220*
Brown C.A.　48, 52, *71*
Brown D.　97, *150*
Brown G.　358, *361*
Brown I.C.　420, *464*
Brown J.M.　469, 481, 482, *506*
Brown R.H.J.　261, *293*
Brown R.W.　353, *361*
Bruckmann G.　175, *227*
Brubaker L.A.　*72*
Brunsdon G.P.　*467*
Brunsven M.A.　204, *220*
Bryan R.B.　348, 355, *361*
Bryson R.A.　*82*
Buckman H.O.　298, *361*
Bunce R.G.H.　147, *150,* 176, 180, *225, 227*
Bunting B.T.　298, *361*
Burges A.　89, *150,* 195, *220,* 298, 355, 357,
　　　　　361, 363
Burris R.H.　211, *226, 229*
Buol S.W.　298, 318, 319, 327, *361*
Burleigh R.　31, 32, 36, *71*
Burnett J.H.　128, *150*
Burns R.C.　222, *289*
Burrough P.A.　329, 330, 332, *360, 361*
Burrows W.H.　181, *220*
Buscaloni L.　267, *287*

Caborn J.M. 395, 397, *409*
Cahoon G.A. 191, 192, *220*
Cain S.A. 100, 121, 138, *150*
Caldwell M.M. 194, *220*
Camp L.B. 194, *220*
Campbell D.J. 346, *361*
Campbell N.A. 319, *361*
Campbell R.M. 243, *287*
Capstick C.K. 186, *22*
Carlisle A. 209, *219, 220*
Carlson R.M. 458, 464
Carpenter J.H. 242, *287*
Carr A.P. 68, *71*
Carr D.J. 256, 258, 259, 261, *288*
Carritt D.E. 242, *287*
Carroll D.M. 330, *361*
Carson M.A. 355, *361*
Cary J.W. 347, *361*
Casparie W.A. 10, 11, 44, 46, *71*
Castro G.M. 121, 138, *150*
Catsky J. 255, *287*
Cescas M.P. 357, *361*
Chambers T.C. 56, *71*
Chapman E.G. *410*
Chapman H.D. 421, *464*
Chapman S.B. 1, 39, 54, *71,* 103, 107, 157,
        167, 168, 171, 179, 181, 183, 187,
        191, 195, 197, 198, 215, *220, 221,*
        470
Charles J.A. 192, *221*
Chater, E.H. 77
Chepil W.S. 347, 348, *361*
Childs E.C. 340, 347, 349, *361*
Chorley R.J. 148, *150*
Christian G.D. 431, 435, 450, *464*
Chrystal J. 242, *294*
Ciaccio L.L. 428, *464*
Clapham A.R. 94, 107, 230, *287*
Clark N.A. 442, *464*
Clarke G.R. 299, 301, 306, 311, 327, 329,
        362
Clarkson D.T. 279, *287*
Clausen J.J. 253, 254, 255, *287*
Claxton S.M. 210, *221*
Clayden B. 319, *365*
Clayton G.B. 481, 484, 486, 487, *506*
Clebsch E.E.C. *220*
Clements C.F. *71*
Clements F.E. 89, 96, 140, *150, 155*
Clifford M.H. 42
Clowes D.R. 74
Clymo R.S. 173, 183, 186, *221,* 278, *287*
Cole C.V. 421, *465*
Coleman D.C. 197, *221*
Coleman J.D. *362*
Coles J.M. 30, 61, *71*
Collins D. 34, *81*
Colquohoun D. 207, *221*

Colwell R.N. 139, *150*
Conolly A.P. *71*
Conway E.J. 197, *221*
Conway V.M. 44, 54, *71, 74*
Cooke G.W. 298, *362*
Cooke H.B.S. 27, 34, *71*
Coombe D.E. *287*
Coope G.R. 8, 33, 40, *70, 71*
Cooper C.F. *410*
Cope D.W. *362*
Corey J.C. *364*
Cormack E. *221*
Cornish P.M. 352, *362*
Cöster I. 46, *71*
Cottam G. 101, *150*
Coulter B.S. 434, *464*
Coupland R.T. 96, *150*
Coups E. 184, *225*
Courtney F.M. 327, *362*
Cowan I.R. 272, *287*
Cowie J.D. *363*
Cox A. 34, *71*
Cox E.A. *72*
Crabtree K. 40, 56, 60, *72*
Crafts A.S. 193, *228,* 261, *287*
Cragg J.B. 184, 186, *224*
Craig A.J. 35, *72*
Crapo N.L. 197, *221*
Crawford R.M.M. 230, *287*
Creer K.M. 34, *72*
Crisp D.T. 214, *221*
Crocker R.L. 195, *221*
Croker B.H. 204, *221*
Croney D. 353, *362*
Crossley D.A. Jnr. 183, 184, 204, *221, 225*
Crowder A.A. 47, *72*
Cruikshank J.G. 298, 319, 329, *362*
Cuanalo De La C H.E. 319, *362*
Cuddy D.G. 47, *72*
Cukor N. *366*
Cummings W.H. *286*
Currier H.B. *287*
Curtin W. 138, *150*
Curtis J.T. 101, 135, 141, 143, *150, 151*
Curtis K.E. 446, *464*
Curtis L.F. 299, *362*
Cushing E.J. 61, *72, 83*
Czekanowski J. 135, 144, *151*

Dagnall R.M. 435, *464*
Dahl E. *71,* 128, *151*
Dahlman R.C. 190, 193, 194, 195, *221*
Dale M.B. *72,* 135, *155*
Dalrymple G.B. *71*
Dalton F.N. *365*
Damblon F. 62, *72*
Daniels L.A. *225*
Dansereau P. 121, 137, *151*

Darby H.C.   63, *72*
Darwin F.   268, *287*
David D.J.   445, *464*
Davies E.G.   13, *72*
Davies T.A.W.   95, 105, *151*
Davis D.E.   63, *79*
Davis M.B.   9, 47, 54, 61, *72*
Davis R.B.   *72*
Davis W.E.   354, *362*
Davison A.W.   279, *285*
Deacon J.   62, *70, 72*
Dean L.A.   421, *465*
De Boodt M.   347, 348, *362*
Decker J.P.   241, 264, 265, 266, *287*
Deer W.A.   358, *362*
Deevey E.S.   21, 22, 61, *72*
De Geer G.   35, *72, 73*
Denton G.H.   38, *73*
Dept. of the Environment   427, 431, *464*
Devel L.   68, *73*
Dewar H.S.L.   30, *73*
Dewis J.   339, *362*
Dickson B.A.   195, *221*
Dickson C.A.   42, 43, 47, *73*
Dickson J.H.   42, 44, *73*
Dilworth M.J.   211, 221, 280, *287*
Dimbleby G.W.   12, 26, 41, 42, 43, 49, 56, *73*
Dittmer B.R.   *72*
Diver C.   68
Doell R.R.   *71*
Donkin R.A.   65, *73*
Doornkamp J.C.   *362*
Dore W.G.   470, *506*
Dos Santos J.M.   *410*
Drover D.P.   195, *220*
Drude O.   120, 121, *151*
Duchaufour Pl.   319, 323, *359, 362*
Duffield B.S.   107, *151*
Duncan U.K.   46, *73*
Dunne T.   300, *364*
Du Rietz G.E.   91, 92, 121, 149, *151*
Durno S.E.   12, 46, *81,* 186, *221*
Duvigneau H.   214, *221*
Dykeman W.R.   169, 170, *228*

Easter S.J.   353, *362*
Eckardt F.E.   242, 264, 265, *287*
Edgell M.C.R.   128, *151*
Edisbury J.R.   435, *464*
Edwards C.A.   183, *221*
Edwards R.S.   210, *221*
Eggink H.J.   214, *221*
Egner H.   208, 209, *221*
Ehrler W.L.   *286*
Eichenberger E.   *506*
Ellenberg H.   120, 121, 123, 124, 128, 138, *151,* 231, *287*

Ellern S.J.   193, *222*
Elton C.S.   86, 96, 109, 149, *151*
Elwell W.T.   435, *464*
Emerson W.W.   348, *362*
Enke C.G.   477, *506*
Epstein E.   340, *362*
Erdtman G.   47, 48, 56, *73*
Eriksson E.   *221*
Etherington J.R.   210, *222*
Evans C.C.   215, *219, 222*
Evans F.C.   182, *228*
Evans G.C.   234, 235, *287*
Evans G.H.   *73*
Evans J.G.   40, *73*
Evans R.   329, *362*

Faegri K.   9, 13, 14, 47, 48, 49, 52, 56, 60, *73*
Falck J.   338, *362*
Falk F.O.   265, *287*
Farrington A.   41, *76*
Farshad L.   *366*
Fearnsides M.   46, *73*
Federer C.A.   469, *506*
Feldman F.J.   431, 435, 450, *464*
Ferreira R.E.C.   277, *288*
Ferrill M.D.   193, *222, 228*
Firth J.G.   299, *362*
Fisher R.A.   103, 113, *151*
Fitzgerald G.P.   *229*
Fitzpatrick E.A.   298, 318, 319, *362*
Fitzsimons P.   *153*
Flack M.   *409*
Fleming G.A.   *292*
Flenley J.R.   47, *73*
Fletcher J.E.   176, *222*
Flint H.L.   249, 251, *288*
Flitters N.E.   392, *409*
Fogg G.E.   39, *73*
Follet E.A.C.   47, *81*
Ford A.S.   *365*
Ford E.D.   *222*
Forrest G.I.   173, 178, *222*
Forsythe J.F.   107, *151*
Fosberg F.R.   91, 120, 121, 122, 123, 128, 138, *151*
Fourt D.F.   349, *362*
Frank M.L.   197, 201, *228*
Frankland J.C.   184, 186, *222,* 336, *362*
Franklin G.L.   42, *71*
Franklin T.B.   *73*
Franks J.W.   8, *73*
Freeman G.F.   265, *287*
Freeman P.R.   234, *289*
Freitas F.   339, *362*
Frenkel R.E.   136, *151*
Frenzel B.   49, *73*
Frey D.G.   40, 56, *73*

Fritts H.C.   36, 37, *74*
Frohlich E.   *295*
Froment A.   202, *222*
Fry K.E.   267, *287*
Fujimaki K.   193, *219*
Funk J.R.   377, *409*
Fussell G.E.   *74*

Gaff D.F.   256, 258, 259, 261, *288*
Gallagher P.A.   336, *362*
Gardner W.H.   *360*
Gardner W.L.   *360*
Gardner W.R.   *365*
Garret S.D.   298, *362*
Gates D.M.   272, *288*, 370, *409*
Gawen D.   424, *464*
Gentry J.B.   *222*
Gessel S.P.   *223*
Ghilarov M.S.   188, *222*
Gibbs H.S.   335, *363*
Gidley J.A.F.   435, *464*
Gigon D.M.   278, *288*
Gigon A.   *288*
Gilbert O.J.B.   183, *220, 222*
Gilbert O.L.   370, *409*
Gillier J.E.   *225*
Gimingham C.H.   23, *74*, 89, *149, 153, 221*, 231, 285, *288*
Gittins R.   132, 146, 149, *151*
Globus A.M.   353, *363*
Glover J.   265, *288*
Godwin H.   8, 28, 30, 33, 41, 42, 43, 44, 47, 56, 61, 68, 69, *71, 73, 74*
Goedewaagen M.A.J.   189, *227*
Goldsmith F.B.   85, 121, 139, 140, 141, 145, 148, *151, 152*
Golley F.B.   185, *222*
Golterman H.   428, *464*
Goncz J.H.   504, *506*
Good R.   *74*
Goodall D.W.   98, 108, 130, 141, 149, *151, 155*
Goodier R.   140, *151*, 332, 355, *360, 363*
Goosen D.   329, *363*
Gordon A.D.   28, 62, *74*
Gordon R.B.   63, *74*
Gore A.J.P.   178, *222*, 428, *464*
Gorham E.   178, 181, 208, *220, 222*
Goring C.A.I.   459, *464*
Gortner R.A.   261, *289*
Graaf Fr. De.   40, *74*
Grace J.   234, *288*
Graeme J.G.   *506*
Grassland Research Institute Staff   177, *222*
Gray J.   49, *74*
Gray L.J.   *361*
Gray T.R.G.   *363, 365*
Green B.H.   44, 68, *74*

Green J.O.   203, *222*
Green H.E.   *363*
Greenhill W.L.   427, *464*
Greenland D.J.   *367*
Gregory F.G.   265, 266, 268, *288*
Gregory K.J.   *362*
Gregory K.S.   *409*
Gregory R.P.G.   281, 283, *288*
Greig-Smith P.   88, 97, 100, 101, 115, 116, 120, 132, 137, 141, 146, *149, 152*
Grieve B.J.   256, 258, 259, *288, 289*
Griffiths J.G.   374, 397, *410*
Griffiths O.W.   *409*
Grigg D.B.   *75*
Grim R.E.   358, *363*
Grime J.P.   244, 246, 281, 282, *288*, 340, *363*
Grimes B.H.   332, *363*
Grimshaw H.M.   *219*, 411, 413, 415, 446, *463, 464, 465*
Groenewoud H. van   86, *152*
Grover B.L.   214, *222*
Grubb P.J.   96, *152*, 335, *363*
Grummer G.   285, *288*
Guha M.M.   275, *288*
Gulliver R.L.   24
Guttenberg H. von   247, *288*
Gyllenberg G.   207, *222*

Haarlov M.   190, *222*
Hadley E.B.   168, *220*
Hafsten U.   49, *75*
Haggett P.   148, *150*
Haines F.M.   252, *288*
Hall J.B.   120, *152*
Hamilton E.L.   387, *409*
Hammel H.T.   *293*
Hammond R.F.   12, *75*
Handley W.R.C.   *288*
Hanks R.J.   354, *359*
Hansen B.   44, *75*
Hansen K.   215, *222*
Hardie H.G.M.   310, *364*
Hardman J.A.   *223*
Hardy R.W.F.   211, *222*, 280, *289*
Harley J.B.   68, *75*
Harley J.L.   279, *289*
Harnisch O.   *75*
Harper J.L.   88, *152, 222*, 230, 234, 235, *289*
Harris J.A.   261, *289*
Harris S.   348, *363*
Harris T.M.   203, *222*
Harrison C.M.   85, 90, 131, 132, 133, 134, 136, *151, 152*
Harrison D.   *365*
Harrold L.L.   400, *409*
Hasler A.   *506*

Hatch M.D. 246, *289*
Haveren B.P. 353, *361*
Havinga A.J. *75*
Havis J.R. 267, *285*
Hawksworth D.L. 370, *409*
Haworth E.W. 40, *75*
Haworth E.Y. *78*
Hayden C.W. 347, *361*
Hayes A.J. 184, *223*
Head G.C. 188, 189, *222, 226*
Heady H.F. 171, *222*
Heal O.W. 40, *75*, 186, *224*
Healey I.N. 180, 190, *219, 222*
Heath G.W. 183, *221*
Heath O.V.S. 237, 238, 241, 242, 246, 266, 267, 268, 272, *289, 294*, 376, *410*
Heinselman M.L. 68, *75*
Helbaek H. 47, *75*
Hellmuth E.O. 258, 259, *289*
Hemmingsen E.A. *293*
Hepple S. 357, *363*
Herlihy M. 336, *362*
Heron J. 209, *223*, 428, *464*
Hertz C.H. *285*
Hesketh J.D. 243, *289*
Hesse P.R. 339, *363*, 414, 434, 435, *464*
Hewitt E.J. 277, 278, *289*
Hewitt T. 470, *506*
Hewlett G. 26, *75*
Hewlett J.D. 255, *289*
Hibbert F.A. 28, 60, 61, 71, *75*
Hibble J. *221*
Higgs E. 29, 34, 47, *71*
Hill L. 397, *409*
Hill M.O. 146, *152*
Hills R.C. 349, *363*
Hinman W.C. 346, *360*
Hinson W.H. 354, *363*
Hirst J.M. 389, *409*
Hnatek E.R. 474, 484, *506*
Hoagland D.R. 277, *289*
Hodge C.A.H. *150*
Hodgson J.G. 281, 282, *288*
Hodgson J.M. 306, 342, *363*
Hofler K. 261, *289*
Hoglund M.P. 183, *221*
Hole F.D. 319, *360, 361*
Hole V.H.R. 481, *506*
Hollick F.S.J. 192, *226*
Holmes A. 301, *363*
Holmgren P. 272, *289*
Holstein R.D. *222, 289*
Hooper M.D. 38, 39, *75*
Hope-Simpson J.F. 88, *152*
Hopkins B. 108, *152*
Hornung M. *360*
Hoskins W.G. 27, 63, *75*
Hough W.A. 193, *223*

Howard J.A. 68, 139, *152*
Howard P.J.A. 184, 197, *223*, 442, *464*
Howell R.S. 300, *364*
Howie R.A. *362*
Huber B. 252, 265, *289*
Hubbard J.C.E. 22, 24, 36, *75*, *79*
Huckerby E. 44, *75*
Hudson N. 355, *363*
Huelsman L.T. *506*
Hughes A.P. 234, *289*
Hughes R.E. 161, 170, 171, 188, 202, *224*
Hunt R. 234, *289*
Hunter A.C.F. 283, *289*
Hunter R.F. 220, *286*
Huntley B. *74*
Hustedt F. *75*
Hutchinson T.C. 271, 281, 283, *289, 290*
Hyde F.J. 176, *223, 506*
Hygen G. 271, *290*

Idle D.B. 248, 272, *290*, 381, *409*
Ingram G. 442, *464*
Ingram H.A.P. 39, *75*
Institution of Water Engineers 407, *409*
International Society of Photogrammetry *75*
International Society of Soil Science 318, 323, *363*
Irving R.M. 250, *290*
Iversen J. 9, 13, 14, 48, 49, 56, 60, *73, 75*
Ivemey-Cook R.B. 135, *152*
Ivlev V.G. 216, *223*

Jackson E.K. *222, 289*
Jackson M.L. 342, 366, 414, *464*
Jackson W.A. 244, *290*
Jacques W.H. 193, *223*
James P.W. 25, *80*
James W.O. 242, *290*
Jane F.W. 42, *75*
Jankowska K. *224*
Janssen C.R. 6, 47, *75, 76*
Jarvis M.G. 329, 361, *363*
Jarvis M.S. 255, 256, 263, 271, 272, 273, *289, 290*
Jarvis P.G. 239, 241, 246, 255, 256, 263, 268, 271, 272, 273, 285, *289, 290, 291, 294*
Jarvis R.A. 329, *363*
Jefferies E.L. 264, *295*
Jeffers J.N.R. 160, *223*
Jeffrey D.W. 207, 208, *223*, 279, *290*, 346, *363*
Jehn K.H. 393, *409*
Jennings D.S. *223*
Jenny H. 162, 165, 188, *223*
Jenkins D.A. 357, *363*
Jenkinson D.S. 444, *464*

Jessen K.   41, *76*
Johansson O.   *221*
John M.K.   310, *363*
Johns G.G.   177, *223*
Johnson H.S.   *289*
Johnson R.E.   96, *150*
Johnston R.D.   387, *409*
Johnston W.B.   69, *76*
Jones B.M.   *224*
Jones E.L.   24, *81*
Jones E.W.   355, *367*
Jost L.   244, *290*
Jowsey P.C.   12, 19, *76*

Kaddah M.T.   342, *363*
Kanzow H.   285
Kaplan I.R.   *464*
Kappen L.   248, 250, *290*
Karlen W.   38, *72, 76*
Kassas M.   24, 38, *76*
Katz N.J.   43, *76*
Katz S.V.   *76*
Kaye S.V.   *226*
Keeney D.R.   458, *464*
Kelly O.J.   189, *223*
Kemnitzer R.   249, *292*
Kenworthy J.B.   215, *223*
Kershaw A.P.   61, *76*
Kershaw K.A.   88, 97, 101, 116, 118, 120,
     132, *152, 153*
Kerr P.F.   356, *363*
Ketcheson J.W.   346, *364*
King C.A.M.   301, 355, *363*
Kipiani M.G.   *76*
Kirkbright G.F.   435, *464*
Kirkby M.J.   355, *361*
Kirkham D.R.   197, 202, *223*
Kitching R.A.   354, *363*
Kittredge J.   175, *223*
Kleeman A.W.   416, *464*
Knapp B.J.   349, *363*
Knight A.H.   193, *220, 286*
Knight R.C.   268, *290*
Knipling E.B.   263, *290*
Knop W.   277, *290*
Koch W.   *291*
Kononova M.M.   188, 193, *223*
Kopf H.   199, *223*
Koppen W.   94, *153*
Korda V.A.   *222*
Kosonen M.   197, *223*
Koulter-Andersson E.   39, *77*
Kozlowski T.T.   36, *76,* 253, 254, 255, *287*
Kozub G.C.   *465*
Krajina V.J.   326, *367*
Kramer P.J.   255, *289*
Kreeb K.   257, 258, 261, *290*

Krogh A.   202, *223*
Krumbein W.C.   *363*
Kubiena W.L.   318, 323, 357, *363*
Kucera C.L.   190, 193, 194, 195, 197, 202,
     221, 223
Kuchler A.W.   68, 76, 88, 121, 125, 129,
     137, 138, *153*
Kuhnelt W.   298, *363*
Kulczynski S.   44, *76*
Kummel B.   48, *76*
Kunze G.W.   341, 357, 358, *365*
Kurtyka J.C.   382, *409*

Ladurie E. Le Roy   *76*
Lamb H.H.   *76*
Lambert A.M.   27, *76*
Lambert C.A.   40, *80*
Lambert J.M   120, 130, 133, 135, 136, *153,*
     155
Lambe E.   *153*
Lamborn R.E.   214, *222*
Lance G.N.   *155*
Lane B.E.   212, 213, *225*
Lane R.F.   138, *150*
Lang A.R.G.   353, *363*
Lange O.L.   168, *224,* 239, 242, 247, 248,
     249, 250, 264, 285, *290, 291*
Lange R.   247, 248, 264, *291*
Lanphear F.O.   250, *290, 294*
Larcher W.   243, 244, *291, 292*
Laryea K.B.   *362*
Latter P.M.   184, 186, *224*
Lavkulich L.M.   349, *361, 363*
Lawrence J.T.   176, *223*
Lebedov V.N.   *225*
Lee J.A.   201, *224*
Lee R.   470, 471, *506*
Leeuwen C.G. van   109, *153*
Legros J.P.   329, *364*
Leo M.W.M.   336, *364*
Leopold L.B.   300, *364*
Le Roux M.   264, *295*
Lewis M.C.   252, 272, *291*
Levitt J.   250, *291*
Leyton L.   387, 391, *409*
Libby W.F.   30, *76*
Lichti-Federovich S.   47, *79*
Lickens G.E.   423, *465*
Lieth H.   177, 189, 216, 218, *224,* 246, *291*
Lindley D.K.   413, *464*
Lindstrom G.R.   *367*
Lines R.   300, *364*
Lishman J.P.   39, *78*
Litus J.   477, *506*
Livne A.   261, *293*
Livingstone D.A.   *83*
Lloyd F.C.   266, *291*

Lloyd P.S. 340, *363*
Lock J.M 203, *224*
Loginov M.A. *225*
Lomnicki A. 182, *224*
Loneragan J.F. *285*
Long I.F. 392, 397, *409, 410*
Lopushinsky W. 267, 271, *291*
Loucks O.L. 141, *153*
Loveday J. 319, *364*
Lovett J.V. 177, *219*
Low A.J. 348, *364*
Ludlow M.M. 241, 246, *291*
Lunt O.R. *295*
L'Vov B.V. 435, *465*

Maarel E. van der 109, *153*
Macan T.T. 40, *76*
McAndrews J.H. 29, *76*
Macarthur R.H. 109, *153*
McArthur W.M. *361*
McCormack M.L. *223, 228*
McCormick J. *79*
McCracken E. *76*
McCracken R.J. *361*
Macfadyen A. 196, 197, 199, 200, 202, 203, *220, 224, 226*
McGuire J. 65, *81*
McIntosh R.P. 109, 140, 141, 143, *151, 153*
Mackay D.B. 250, *291*
McKell C.M. 191, 192, *224*
Mackenzie R.C. 358, *364*
Mackereth F.J.H. 21, 34, 35, 39, *72, 76*, 428, 431, *465*
McLaren A.D. 298, *364*
Macklon A.E.S. 259, 264, *291*
Macphee W.S.G. *360*
Macneil G.M. *360*
Macrae C. *222, 362*
McVean D.N. 128, *153*, 326, *364*
Madgwick H.A.I. 175, 209, *224, 225*
Maher L.J. 54, 60, *76*
Maignien R. 327, *364*
Major J. 160, 195, *221, 224*
Malloch A.J.C. 210, *224*
Malmstadt H.V. 477, *506*
Malone C.R. 216, *224*
Mammano R.A. 474, *506*
Manley G. *76, 77*
Mansfield T.A. 266, 268, 269, *289, 292*
Marchant R. *75*
Margalef R. 90, *153*
Mark A.F. 176, *224*
Marshall C.E. 358, *364*
Marshall T.J. 349, *364*
Martin A.C. 44, 77, 204, *219*
Martin D.J. 204, *224*
Martin M.H. 25, *77*, 199, 201, *224, 246*, *291*

Matelski R.P. 346, *366*
Mattson E. 39, 77
Maximov N.A. 252, 263, *291*
Medwecka-Kornas A. 180, 181, *224*
Mehlich A. 334, 335, *364*
Meidner H. 240, 246, 247, 262, 264, 266, 268, 269, *291, 292*
Mercer J.H. 77
Merrett P. 68, 77
Merrifield R.C.J. *363*
Merton L.F.H. 24, 36, 37, 77
Metcalfe G. 355, *364*
Metson A.J. 414, *465*
Mew G. 328, 329, 331, *360, 364*
Meyer B.S. 255, *292*
Michael G. 267, 273, *292*
Mick A.H. 354, *361*
Miller J. 242, *292*
Miller R. 265, *289*
Miller R.S. 86, 96, 109, 149, *151*
Milne R.A. *465*
Milner C. 161, 170, 171, 188, 202, *224*
Milner H.B. 358, *364*
Milthorpe F.L. 272, *287, 288, 292*
Minderman G. 187, 188, 190, 196, *224*
Mirreh H.F. 346, *364*
Mitchell A. 68, *77*
Mitchell C.W. 319, *360, 364*
Mitchell G.F. 28, 77
Mitchell J. 432, *465*
Mitchell R.L. 275, *288*
Mittre V. 56, 77
Molisch H. 267, *292*
Molneux L. *72*
Mongredien A. 68, 77
Montieth J.L. 170, 225, 236, 242, *292*, 370, 377, 392, 393, 404, *409, 410*
Montgomery E.G. 274, 285, *292*
Moody G.J. 458, *465*
Mooney H.A. 89, *150, 220*, 244, *292*
Moore A.W. 319, *364*
Moore J.J. 88, 123, 125, 136, 140, *153*
Moore N.W. 107, *153, 215, 225*
Moore P.D. 28, 77
Morgan A. *71*
Morrison I.R. 461, *465*
Morrison M.E.S. 44, 61, 77
Morton A.J. 119
Morton E.S. 191, 192, *220*
Moser W. *292*
Mosimann J.E. 61, 77
Moss R.P. 323, *364*
Mott G.O. 177, *225*
Mountford M.D. 144, *153*, 176, *225*
Muckenhausen E. 323, *364*
Muller-Dombois D. 120, 121, 123, *151*
Muir J.W. 310, 318, *364*
Mulcahy M.J. *361*

Mulkern G.B.   204, *220*
Mulqueen J.   277, *292*
Munnich K.O.   *83*
Murphy P.W.   184, *225*
Murray G.   189, *225*

Nakayama F.S.   *286*
Nassery H.   279, *292*
Nasyrov J.S.   170, *225*
National Soil Survey Committee of
        Canada   323, *364*
Neustupny E.   31, 77
Newbould P.J.   161, 162, 170, 174, 175,
        179, 188, 192, *222, 225, 226*
Newman E.I.   190, *225*
Newton J.D.   *225*
Nicholls P.H.   65, 77
Nichols H.   9, *77*
Nicol G.R.   *223*
Nielsen D.R.   335, *364*
Nielson J.A.   193, *225*
Niemiel S.   *506*
Nilsson S.   *73*
Nihlgard B.   209, *225*
Nimlos T.J.   193, *225*
Norcliffe G.B.   149, *153*
Norris J.M.   310, 319, 330, 332, *364*
Northcote K.H.   323, *365*
Norton W.M.   279, *286*
Nuffield Biology   246, 273, *292*

Oberdorfer E.   125, *153*
Odum E.P.   90, *153,* 178, *225*
Oertli J.J.   338, *365*
Ogata G.   259, *293*
Ogden J.   235, *289*
Ogg C.L.   442, *464*
Okada T.   *227*
Oldfield F.   5, 8, 9, 17, 28, 29, 48, 56, 69, *75,*
        77, 78
Olsen S.R.   421, 465
Olson J.S.   162, 165, 176, 178, 183, 184,
        188, 195, 196, *225, 222, 227, 228*
Olsson I.U.   31, *78*
Onal M.   258, *290*
O'Niel R.V.   *226*
Orloci L.   145, 146, 149, *153*
Orwin C.S.   63, *78*
Osbourne P.J.   40, *71, 78*
O'Sullivan P.E.   34, 47, 48, *78*
Osvald H.   44, *78, 89, 153*
Oudman J.   280, *292*
Overbeck F.   44, *78*
Ovington J.D.   175, 189, 209, *222, 224, 225,*
        362
Owen P.C.   257, *292*
Ozanne P.G.   *285*

Pack I.   *292*
Painter R.B.   369, 381, *409*
Pallman H.   470, *506*
Pamplin B.F.   478, *506*
Pankow H.   46, *71*
Paradise J.   *506*
Parizek R.P.   212, 213, *225*
Paris O.H.   204, *225*
Parkinson D.   184, *225, 298, 365*
Parkinson J.A.   411
Parson I.T.   234, *289*
Patten B.C.   160, *225*
Patten H.L.   60, *83*
Paulson B.   40, *78*
Payne-Gallwaye R.   *78*
Pearsall W.H.   125, *153,* 470, *506*
Pearse H.L.   268, *288*
Pearson L.C.   203, *226*
Pearson M.C.   68, *74,* 280, *294*
Pearson V.   279, *292*
Peck E.L.   386, *409*
Peck R.M.   47, 54, *78*
Pelton W.L.   400, *410*
Penman H.L.   272, *292,* 392, 403, 404, *410*
Pennington W.   28, 39, 41, 47, 54, *73, 78*
Penny L.F.   *77*
Perkins D.F.   139, *153*
Perrier A.   248, *292*
Perrin R.M.S.   82, 214, *226, 367*
Perring F.   78, 79, 141, *153,* 369, *410*
Perry P.J.   *79*
Pertz D.F.M.   268, *287*
Peterken G.F.   6, 23, 24, 25, 36, 63, *79,* 121,
        153, 170, *226,* 298, 319, *365*
Peterson G.H.   298, *364*
Pettijohn F.J.   *363*
Petruzewicz K.   202, *226*
Philip J.R.   401, *410*
Philips C.W.   *75*
Philips L.   *79*
Philips M.E.   119, *153*
Phillipson J.   216, 217, *226,* 298, *365*
Pielou E.C.   102, *134*
Pickrell B.G.   472, *506*
Pidgeon T.M.   139, *149*
Piggott C.D.   16, 24, 25, 36, 44, 52, 56, *79,*
        199, 201, *224,* 246, 279, *290, 291*
Pigott M.E.   16, 44, 52, 56, *79*
Pilcher J.R.   29, 33, *79, 80*
Piper C.S.   414, 421, *465*
Pisek A.   245, 246, 249, *292*
Pittwell L.R.   445, *465*
Pitty A.F.   300, *365*
Plater de C.V.   352, *365*
Platt R.B.   374, 397, *410*
Pohlen I.J.   306, *366*
Polwart A.   249, 250, 273, *292*
Pollacci G.   267, *287*

Pollard E.   25, 39, *79*
Pollard L.D.   358, *367*
Poole H.H.   470, *506*
Poore M.E.D.   108, 120, 125, 138, 140, *154*,
      326, *365*
Portsmouth G.B.   238, *292*
Post von L.   13, 47, 60
Potzger J.E.   63, *79*
Potzger M.E.   *79*
Powell M.C.   376, *410*
Praglowski J.   49, 52, *73, 79*
Pratt P.F.   421, *464*
Preis K.   95, *154*
Presley B.J.   *464*
Price W.J.   435, *465*
Prince H.C.   *79*
Pritchard D.J.   341, *365*
Pritchard N.M.   135, 136, *154*
Proctor J.   275, 276, 277, 281, 282, 283, *293*
Proctor M.C.F.   7, 46, *79*, 132, 135, *152*,
      154
Pullar W.A.   *363*
Pyatt D.G.   300, *365*

Quarmby C.   411, 446, *465*
Quirk J.P.   *367*

Rackham O.   65, *79*
Rafarel C.R.   *221*, 467
Rafes P.M.   203, *226*
Ragg J.M.   319, *365*
Rainford A.E.D.   260, *293*
Ramirez-Munoz J.   435, *465*
Ramsay J.A.   261, *293*
Randall R.E.   210, *226*
Rankin H.T.   63, *79*
Rankine W.F.   *79*
Rankine W.M.   *79*
Ratcliffe D.A.   44, *79*, 89, 109, 128, *153*,
      154, 326, *364*
Raunkiaer C.   91, 93, 94, 95, 123, 149, *154*
Raup O.   48, *76*
Raw F.   298, *361*
Rawes M.   203, 207, *226, 228*
Rawle P.R.   211, *227*
Rawlins S.L.   353, *365*
Rayner J.H.   319, *365*
Read H.H.   301, *365*
Reeve M.J.   346, *365*
Reichle D.E.   160, 207, *226*
Reiners W.A.   197, 201, *226*
Remezov H.P.   194, *226*
Renfrew C.   47, *71*
Renfrew J.M.   *79*
Rex R.W.   *465*
Reynolds E.R.C.   349, *365, 409, 410*
Reynolds S.G.   351, 356, *365*

Rhykerd C.L.   *225*
Rich C.I.   358, *365*
Richards L.A.   259, *293*, 351, 353, *365*
Richards P.W.   94, 95, 105, 121, *151, 154*
Richman S.   216, *226*
Richter H.   260, *293*
Ricklefs R.E.   166, *226*
Riley J.P.   446, *465*
Ritchie J.C.   47, *79*
Roberts B.K.   65, *79*
Roberts J.D.   411
Robertson J.S.   86, *150*
Robertson V.C.   138, *154*
Robinson M.E.   176, *222*
Robinson R.K.   *293*
Robson J.D.   349, *366*
Roden D.   65, *79*
Rodin L.E.   207, *219, 226*
Rogers J.S.   353, *365*
Rogers W.S.   189, *226*
Rorison I.H.   230, 231, 234, 237, 275, 278,
      279, *288, 293*
Rose C.W.   *286*, 340, 347, 349, *365*
Rose F.   25, *80*, 370, *409*
Ross D.J.   185, *226*
Ross P.J.   474, *506*
Rothacher J.S.   *286*
Rottenburg W.   260, *293*
Round F.E.   40, *80*
Rovner I.   41, *80*
Rowe P.B.   387, *409*
Rowley T.   *80*
Royal Society   27, *80*, 413, *465*
Rubel E.   121, *154*
Rudeforth C.C.   310, 331, *365*
Rufelt H.   *285*
Russell Sir E.J.   298, 358, *365*
Russell E.W.   298, *365*
Russell J.   268, *289*
Russell J.S.   *364*
Rutter A.J.   96, *154*, 189, 190, 199, *226*,
      227, 349, 350, 354, *365*, 387, 400,
      410
Rybníček K.   6, 9, 46, 48, *80*
Rybnickova E.   6, 9, 48, *80*
Rychnovská-Soudkova M.   273, *293*
Rychnovská M.   273, *293*

Sagar G.R.   *222*
Salisbury E.J.   68, *80*, 94, *154*, 195, 226
Salmi M.   48, *80*
Salt G.   192, *226*
Salter P.J.   343, 354, *366*
Sampson I.   266, *293*
Sandell E.D.   435, *465*
Sands C.H.S.   40, *71*
Sands K.   354, *365*

Sanglerat G.   356, *366*
Sarjeant W.A.S.   56, *80*
Satoo T.   172, 174, 175, *226*
Sator Ch.   193, *226*
Savraro M.   *80*
Scherdin G.   241, 242, *294*
Schimper A.F.W.   230, *293*
Schofield R.K.   272, *292*, 352, *366*
Scholander P.F.   259, *293*
Schöllhorn R.   211, *226*
Schulze E.D.   168, *224, 291*
Schumacker R.   62, *72*
Schuster J.L.   189, *227*
Schuurman J.J.   189, *227*
Schratz E.   263, *293*
Schwass R.H.   193, *223*
Scott D.   176, *227*
Searle S.R.   113, *154*
Selby M.J.   355, *366*
Sellick R.J.   63, *78*
Šesták Z.   237, 242, 243, *293*
Seybold A.   263, *293*
Shackleton N.J.   29, *80*
Shanks R.E.   183, *227*
Shardakov V.S.   259, *293*
Shaw K.   442, *465*
Shaw M.D.   352, *366*
Shaw M.W.   203, *227*
Sheail J.   *80*
Sheikh K.H.   189, 190, *227*, 278, *293*
Shimshe D.   261, *293*
Shimwell D.W. 88, 98, 101, 105, 108, 125,
            141, *154*
Shipp R.F.   346, *366*
Shontz J.P.   284, *366*
Shontz N.N.   284, *366*
Shotton F.W.   8, 33, 34, *77, 80*
Shropshire F.   244, *292*
Sikora A.   204, *225*
Simpson E.H.   109, *154*
Sims R.E.   54, *80*
Singh G.   *80*
Sivadjian J.   265, 267, *293*
Sjörs H.   430, *465*
Skene J.K.M.   336, *366*
Slack C.R.   *289*
Slater C.S.   354, *362*
Slatyer R.O.   255, 256, 257, 259, 260, 262,
            263, 268, *286, 293, 294, 295*
Small J.   *465*
Smiley T.L.   37, 38, *81*
Smith A.G.   21, 29, 33, *80*
Smith F.E.   160, *227*
Smith F.M.   386, *410*
Smith G.D.   318, 319, 323, *366*
Smith J.T.   *80*
Smith K.   474, *506*
Smith P.D.   *365*

Smith R.T.   27, 48, *81*
Smith S.E.   279, *294*
Sneath P.H.A.   120, 135, *154*
Sneddon J.I.   300, *366*
Snell C.T.   435, *465*
Snell F.D.   435, *465*
Soane B.D.   356, *366*
Society for Analytical Chemistry   424, *465*
Society for Promotion of Nature
            Reserves   109, *154*
Sokal R.R.   120, 135, 145, *154*
Sollins P.   *226*
Solomon M.E.   395, *410*
Sosebee R.E.   353, *362*
Southwood T.R.E.   144, *154*, 192, 202, *227*
Sorenson T.   112, 128, 144, 145, *154*
Spanner D.C.   259, 268, *294*, 353, *366*
Sparks B.W.   8, 27, 33, 40, *80, 82*
Sparling J.H.   281, *294*
Spence D.H.N.   242, *294*
Spencer H.L.   *288*
Spring D.   *80*
Stage A.R.   175, *227*
Stahl E.   267, *294*
Stalfelt M.G.   266, *294*
Stamp L.D.   139, *154*
Stankovic D.K.   481, *506*
Stebbings R.E.   22, *75*
Steele R.C.   25, *81*
Steere W.C.   *80*
Steponkus P.L.   250, *294*
Stevenson A.G.   285, *294*
Stewart D.R.M.   204, *229*
Stewart J.M.   12, 46, 47, *81*
Stewart W.D.P.   211, *229*, 280, *294*
Stewart W.S.   237, *294*
Steyn W.J.A.   422, *465*
Stocker O.   168, *227*, 253, 264, *294*
Stocking C.R.   *287*
Stokes M.A.   37, 38, *81*
Storr G.M.   204, *227*
Strangeways I.C.   393, *410*
Stuiver M.   35, *81*
Suess H.E.   *81*
Summerfeldt T.G.   458, *465*
Sutcliffe J.   273, *294*
Sutherland J.   *72*
Swain A.M.   39, 68, *81*
Swann J.B.   *506*
Swartout M.B.   216, *224*
Swift M.J.   180, *223*
Swinbank W.C.   403, *410*
Switsur V.R.   *71, 75*
Syers J.K.   *465*
Sykes J.M.   180, *227*
Szabo B.J.   34, *81*
Szafer W.   125, *154*
Szeicz W.   *154, 225,* 373, 374, 377, *409, 410*

Talbot L.M.   326, *359*
Tallis J.H.   65, *81*
Tandy J.D.   351, *367*
Tanner C.B.   342, 366, 397, *410,* 469, *506*
Tansley A.G.   1, *4,* 68, *81,* 89, 94, 96, 98,
        121, 125, *154,* 158, 227
Tate W.E.   *81*
Tauber H.   6, 47, *81, 83*
Taylor J.A.   138, *155*
Taylor N.H.   306, *366*
Tempel N.   *220*
Thake B.   211, *227*
Thom A.S.   272, *294*
Thomas A.M.   352, *366*
Thomas A.S.   100, 107, *155*
Thomas J.R.   458, *465*
Thomas K.W.   21, 58, *71, 81*
Thomas R.   477, *506*
Thomas T.M.   300, *366*
Thomasson A.J.   349, *365, 366*
Thompson F.B.   *409*
Thompson H.R.   117, *155*
Thompson R.   *72*
Thornthwaite C.W.   94, *155*
Thorpe J.   323, *366*
Thurston J.M.   276, *294*
Thurtell G.T.   *506*
Till O.   249, 250, *294*
Tinklin R.   256, 266, *294*
Tinsley H.M.   48, *81*
Tinsley J.   444, *465*
Tittensor R.M.   25, *81*
Tollan A.   386, *410*
Toby G.E.   484, *506*
Tolonen K.   19, 40, 41, 46, 54, *81*
Tracey J.G.   *155*
Treshow M.   275, *294*
Troels-Smith J.   10, 14, *81*
Truog E.   421, *465*
Tubbs C.R.   23, 24, 36, 65, *79, 81*
Turner C.   29, 61, *80, 82*
Turner F.   209, *228*
Turner J.   28, 35, 48, *79, 82*
Turner R.G.   283, *294*
Tutin T.G.   94, *150*
Tuxen R.   120, *155*
Tyner E.H.   *361*

Ueno M.   193, *227*
Ulmer W.   250, *294*
Ungerson J.   241, 242, *294*
Unterholzner R.   *292*
Urness P.J.   204, 205, *228*
Ursprung A.   255, *294*
United States Department of
        Agriculture   306, 311, 312, 313,
        318, 319, 323, 327, 329, 342, *366*

Van der Hammen T.   *82*
Van Dyne G.M.   171, *223, 227*
Van Haveren B.P.   *361*
Van Meter W.P.   *225*
Vickery P.J.   203, *227*
Vieweg F.H.   168, *227*
Vink A.P.A.   329, *366*
Vita-Finzi C.   33, 34, *82*
Visvalingam M.   350, 351, *367*
Volk R.J.   244, *290*
Voznesenskii V.L.   243, *294*

Waddington J.C.B.   29, *70*
Wadsworth R.M.   107, *155,* 370, *410*
Wagner G.H.   213, *227*
Wali M.K.   326, *367*
Walker D.   6, 10, 11, 44, *72, 74, 79, 82,* 89,
        154
Walker R.B.   267, *288*
Walker P.M.   10, 11, 44, *82*
Walker T.W.   *465*
Wallace A.   278, *295*
Wallace T.   275, *295*
Walling D.E.   *409*
Wallwork J.A.   298, *367*
Walshe M.J.   *292*
Walter H.   252, 260, 261, *295*
Walters C.S.   175, *227*
Walters S.M.   *79,* 369, *410*
Wanner H.   196, 197, *227*
Wanstall P.J.   *82*
Warburg E.F.   94, *150*
Ward P.F.   *79*
Ward W.T.   *364*
Warming E.   123, *155,* 230, *295*
Warren A.   90, *152*
Warren Wilson J.   98, *155,* 171, *227,* 381,
        410
Washburn A.L.   355, *367*
Watanabe F.S.   421, *465*
Watkin B.R.   *223*
Watson D.J.   234, *295*
Watson J.   301, *365*
Watson J.R.   341, *367*
Watt A.S.   27, 69, *82,* 88, 89, 90, 107, 121,
        154, 155, 355, *367*
Watts W.A.   29, 35, 41, 46, *82*
Waugham G.J.   211, *227, 228*
Weatherley P.E.   252, 255, 256, 259, 261,
        262, 264, 266, 268, *286, 291, 294,*
        295
Weaver C.E.   358, *367*
Weaver J.E.   96, *155*
Webb D.A.   86, 140, *155*
Webb L.J.   94, 96, *155*
Webb T.   48, *82*
Webster J.R.   199, *226*

Webster R.   301, 310, 318, 319, 329, 335, 353, *360, 362, 365, 367*
Weinemann H.   264, *295*
Weis-Fogh T.   190, *222*
Welbank P.J.   189, *228*
Welch D.   203, 207, *226, 228*
Wells C.B.   *367*
Wells T.   25
Went J.C.   186, *228*
Went F.W.   285, *295*
West R.G.   6, 7, 8, 9, 12, 18, 19, 27, 28, 29, 30, 31, 33, 34, 35, 40, 43, 47, 52, 54, 61, *70, 75, 77, 80, 82, 367*
Westenberg J   62, *82*
Westlake D.F.   158, *228*
Whipkey R.Z.   405, *410*
White E.J.   209, *219, 220, 228*
White J.   *153*
Whitmore T.C.   147, *152*
Whittaker E.   215, *228*
Whittaker R.H.   109, 141, 142, *155,* 174, 175, 176, *228*
Wiant H.V.   202, *228*
Wiegert R.G.   182, *228*
Wien J.D.   241, 264, 265, *287*
Wightman W.R.   65, *83*
Wijmstra T.A.   *82*
Wilcox W.W.   42, *83*
Willhite F.M.   420, *464*
Williams B.G.   348, *367*
Williams C.B.   109, *155*
Williams E.D.   189, *228*
Williams H.P.   446, *465*
Williams J.B.   343, 354, *366*
Williams J.D.H.   459, *465*
Williams M.   27, 63, 65, *83*
Williams O.B.   204, *228*
Williams S.T.   *363, 365*
Williams W.M.   334, 335, 337, 338, *360*
Williams W.T.   120, 130, 133, 135, 136, *153, 155,* 238, 267, *289*

Willis A.J.   86, *155,* 237, 264, 276, 279, *286, 295*
Willis E.H.   30, 31, *74, 83*
Wilmott A.J.   242, *295*
Wilson A.L.   461, *465*
Wilson A.M.   *224*
Wilson K.   195, *228*
Winkworth R.E.   98, *155*
Winter T.C.   41, 46, *82*
Wit C.T. de   284, *295*
Witkamp M.   197, 201, 202, *228*
Woods F.W.   193, *222, 223, 228*
Woodell S.R.J.   179, 180
Woodwell G.M.   168, 169, 170, 174, 175, 176, *220, 228*
Woolhouse H.W.   201, *224,* 230, 234, *288, 295*
World Meteorological Organization   387, *410*
Wormell P.   277, *288*
Wright H.E.   21, 28, 60, *72, 83*
Wright M.J.   334, *367*

Yabuki K.   *225*
Yamaguchi S.   193, *228*
Yarnell R.A.   47, *83*
Yarranton G.A.   102, 113, *155*
Yarwood S.M.   36, *83*
Yashihara K.   *227*
Yates E.M.   27, 65, *80, 83*
Yates F.   103, *151*
Yemm E.W.   237, *295*
Young A.   355, *367*
Young H.E.   207, *228*

Zalenskii O.V.   243, *294*
Zagwijn W.K.   *82*
Zoost M.A.   *363*
Zussman J.   *362*
Zyznar E.   204, 205, *228*

SUBJECT INDEX

(First page only cited when topic extends continuously over several pages)

Abundance 96, 100, 112, 120, 124
Accumulation 160, 162, 178
  curves 164, 165, 196
  definition 162
  in seral sites 195
Acetylene reduction 211, 280
Acer 24, 58, 59
  rubrum 203
Achillea ptarmica 111
Acid brown earth 324, 326, 335, 336
Acid digestion 424
Acid peat soil 325, 335, 336
Acidity of water 430
Actinograph 373, 374
Acrocladium cuspidatum 111
Aeolian deposits 302
Aerial photography 6, 8, 38, 107, 129, 138
  as historical records 68
  location of sampling sites 299
  soil mapping 328
  stereoscopic pairs 107, 139, 140
  snow cover 328
  types of image 140
  vegetation mapping 139
Aerodynamic resistance 404
Agathis australis 335
Agriculture 10, 38, 61, 67
  extension of 28
Agricultural and land use records 63
Agrostis canina 282
  setacea 279
  stolonifera 111, 282, 283
  tenuis 111, 120, 133, 134, 284
Airflow 395
Albedo 371
Alkalinity of water 430
Allerød interstadial 34, 41
Alliance, vegetation 123, 125
Allometric, law of growth 175
  regression, correction for bias 176
Alluvium 302, 321
Alnus 13, 28, 42, 52, 53, 57, 59
Alopecurus pratensis 284

Altimeter 300
Altitude 299, 300
Aluminium, in acid soil 280, 420
  extraction from soil 420
Ammonium-nitrogen, colorimetric
      method 456
  distillation method 454
Amphitrema flavum 45, 55
  wrightianum 45, 55
Amplifiers 485
  A.C. 484
  D.C. 484
  operational 482
Anemone nemorosa 25
Anemometer 396
Angelica sylvestris 111
Annual volume increment 175
Antagonism of nutrients 285
Anthoxanthemum odoratum 111
Analogue-digital converters 488
Analysis of vegetation 85, 109
  choice of method 87, 148
Ancient trackways 26, 30
Anthrone technique 237
Apparent growth increment 162
Arcella artocrea 55
  cantinus 55
  discoides 55
  rotundata 45, 55
Archaeology 27, 28, 30, 31, 34, 35, 38, 42,
      47, 61, 139, 192
Archives 23
Armeria maritima 145
Artemisia 53, 58, 59
Ash, crude 432
  silica free 434
  weight (see loss-on-ignition)
Ashing for chemical analysis 423
Aspect 299, 300
Association, analysis 91, 112, 120, 130, 136
  soil 327
  statistical 109, 110, 130, 132
  vegetation 91, 112, 120, 130, 136

Assulina muscorum 45, 55
   seminulum 45, 55
Aster tripolium 126
Atlantic period 28
Atmospheric fallout 208
Atomic absorption spectroscopy 435
   calcium 438
   iron 445
   magnesium 448
   manganese 449
Augers (see peat borers, soil augers and
   corers) 386
Aulocomnium palustre 111
Autecology 6, 86
Autoradiography 193
Available water capacity 352, 354
Avena sativa 275, 276

Basal area 101, 175
   increment 175
Basalt 304
Batteries, types and characteristics 475
Beechwood 180
Bellis perennis 58
Bench mark 300
Benzoic acid 216
Beta-ray attenuation 177
Betula 13, 41, 42, 53, 58, 59
   nana 41, 53
   pubescens 53
   verrucosa 53, 133, 134
Biostratigraphy 29
Bicarbonate indicators 201, 246, 273
Biomass (see Standing crop)
Bog (see also Mires) 9, 11, 40, 42, 170
   hydrology 46
   ombrogenous 9, 42
   raised 8, 10, 45, 48
   soil 350
   valley 9, 20
Bostrychia scorpioides 126
Botrychium 57
Boulder clay 16
Boundary layer resistance 271, 272
   theory 404
Bouyoucos, hydrometer 341
   resistance blocks 354
Bowen ratio 401
Brachythecium rutabulum 111
Braun-Blanquet cover scale 98
Braunerde 320, 326
Breast height 101, 175
Bridge circuits 480
Bristlecone pine calendar 31
British Ecological Society 1, 230
   Geomorphological Research Group 300
   Standards Institution 342, 346, 356

Bromoform separation 357
Bronze age 8
   trackways 30
Brown, calcareous soil 320, 324
   earth 320, 324, 326
   ferritic soil 320
   podzolic soil 321, 325, 326
   ranker 320, 324
Bryophytes 13, 16, 41, 42, 44, 45, 46, 111
Budgets 205, 206
   energy 215
   nutrient 157, 188, 205, 334
Bullinula indica 55
Buried soil profiles 26
Burning 44, 67
Burosem 321

Calcareous, gley 321, 325
   grassland 146
   marl 16
   soil 280, 324, 326
Calcicole 96
Calcifuge 96
Calcium 214
   atomic absorption method 438
   deficiency in magnesium soils 280
   EDTA titration method 440
   excess in soil 281
   radio-isotope (^{45}Ca) 193
Calluna vulgaris 10, 13, 17, 23, 27, 45, 57,
      89, 120, 131, 132, 134, 179, 181,
      183, 187, 195, 231, 251, 253, 263,
      265, 270, 335, 355
Calorie 216
Calorific, content 86, 205, 215
   value 215
Calorimeter, bomb 216
   corrections 216
Calorimetry 215
Cambridge University Collection of Air
      Photographs 69
Cannabis sativa 59
Carbohydrate, analysis 237
   production 237
Carbon 32, 39, 184, 185, 196
   determination 441
   losses from soil 185
   organic 441
      rapid titration method 444
      wet oxidation method 442
Carbon dioxide, assimilation 168, 170, 238,
      245
   buffers 242
   compensation point 245, 272
   concentration 169, 170, 199
   diffusion 272
   dissolved in water 431

Carbon dioxide—*cont.*
  evolution   184, 196, 244
  flux   170, 242
  loss from soil   185, 196
  measurement   241, 442
  radio-active   170, 193, 194, 243
  soil atmosphere   199, 246
Carbonate   33
  determination   445
Carex   17, 111
  echinata   111
  nigra   133, 134
  panicea   111
  pilulifera   131, 132, 133, 134
Carnivores   159
Carpinus betulus   57, 59
Carr   17
Carrying capacity   86
Caryophyllaceae   56, 57
Cation exchange   358
  capacity (C.E.C.)   417, 419
  column   210
Cell sap, extraction   260
Cellulase activity in soil   186
Centropyxis aculeata   55
  aerophila   55
  laevigata   55
Cereal   47
  pollen grains   56, 59
Chamaephytes   91
Chalk   303, 305
  grassland   27, 100, 106, 120, 141, 338
  soil   214, 335
Characeae   47
Character species   125
Charcoal   42
Chart, reader   502
  recorder   3, 405, 468, 471
Chemical analysis (see also under specific
      elements and nutrients)   171, 180,
      190, 213, 411
  accuracy and precision   413
  animal tissue   426
  calculation of results   436
  collection and storage of samples   333,
      414, 421, 426, 428
  drying and grinding samples   333, 415,
      422, 427
  general points   412
  instrumental techniques   435
  plant materials   275, 421
  preparation of sample solutions   416,
      423, 427, 431, 437
  relative to historical ecology   39
  soils   275, 414
    extraction   417, 431
    initial treatment   339, 416
    variability   335, 336, 337, 338

Chemical analysis—*cont.*
  standards and calibration curves   436
  waters   427
    alkalinity and acidity   430
    conductivity   429
    dissolved CO_2 431
    dissolved constituents   431
    dissolved oxygen   431
    dissolved solids   429
    pH   429
    suspended solids   429
    total organic matter   429
Chemical composition   205, 207
  litter   188
  plants   275, 421
  precipitation   208, 427
  soil   275, 414
Chemical inhibition of growth   285
Chenopodiaceae   57
Chernozem   320
Chestnut soil   320
Chi-squared (χ^2)   110, 116, 130
Chlorine, excess in soil   280
Chlorosis   278, 280
Chronozones   61
Cirsium palustre   111
Cistercians   65
Cladium mariscus   13, 17, 68
Classification, climate   94
  habitats   96
  peat stratigraphy   10
  rocks   303
  soils   318
  vegetation   107, 120
    agglomerative methods   123, 128,
      135
    Braun-Blanquet system   123, 125,
      128, 138
    divisive methods   121, 128, 130
    Ellenberg and Mueller-Dombois
      system   121, 123, 128
    Fosberg's system   121, 128
    physiognomic system   121, 123
Clay, content of soil   341
  definition   341
  minerals   358
Cliffs   129, 141, 145, 147, 210
Climate   7, 38, 46, 369
  changes of   44, 67
  classification   94
  and plant distribution   369
  and soil development   303
Climatological stations (see Weather stations)
Climatology   369
Climax vegetation   89
Clinometer   300
Clocks and timing units   476
Cluster analysis   120, 135

Cobalt chloride and thiocyanate paper 265, 267, 395
Cold resistance of plants 249
Colorimetric analysis 435
 ammonium 456
 iron 446
 manganese 450
 nitrate 457
 phosphate 458
 silica 461
Colour recording 11, 312
Community 125, 218, 278
 seral 89
 climax 89
Compaction 356
Compensation point 243, 244, 255, 272
Competition 88, 89, 231
 between ecotypes 284
 and ecological amplitude 231
 and mineral nutrition 274, 279, 283, 340
 and physiological response 243
 and soil phosphorus 279
Compositae 53, 57, 58, 59
Conductimetric, estimation of salt
  content 210
 soil respirometer 197, 198
Conductivity (see Electrical conductivity)
Conopodium majus 25
Conservation 298, 319
Constellation diagrams 111, 112
Constancy 125, 128
Consumption 203
Contagion 113
Continental drift 34
Contingency table 110, 130
Continuum analysis 143, 144
Convollaria majalis 25
Copper, deficiency 277
 toxicity 280
Coppicing 25, 65
Correlation 109, 112, 118
 coefficient 112, 118, 119, 135
Corylus avellana 42, 53, 58, 59, 60
Corythion dubium 45, 55
County archives 62
Covariation 118, 119
Cover 97
 —abundance 98, 124
 repetition 100
 scales 98, 124
Crop, marks 68
 meter 176
Cruciferae 58, 59
Crude ash (see Loss-on ignition)
Cryptodifflugia oviformis 45, 55
Cryptophytes 91
Culture solution 256, 277, 281
Curtis meter 469, 481

Cut shoot method 264
Cuticular, fragments 204
 resistance 271
Cyperaceae 44, 53, 57
Czekanowski's coefficient 144

Dachnowski peat borer 18, 19, 21
Dactylis glomerata 120, 285
Dark carboxylation 246
Data recording 467
 general design and construction 473
 integration methods 468
 interfacing 480
 magnetic tape loggers 472
 power supplies 474
 practical applications 498
 practical logging systems 492
 recorders (chart) and event counters 471
 timing and control 476
 translation and data handling 502
Data loggers 3, 371, 399, 467
 basic block diagram 492
 commercial systems 497
 field systems 494
 laboratory systems 496
Dating 6, 26
 absolute 27, 29, 34, 35
 chronometric 27, 29, 30, 31
 dendrochronological 35
 errors in stratigraphic 30
 hedges 26, 38, 39
 lichenometric 38
 palaeomagnetic 34
 pollen 27, 47
 radiocarbon 30
 radiometric 33
 relative 27
 stratigraphic 29
 varve 35
Daylight recorder 376
Decomposer 159
Decomposition 46, 120, 160, 178, 181
 definition 162
 exponential 163, 187, 188
 of roots 195
Decoy ponds 67
Deficiency of, nutrients 274, 275, 277, 281
 trace elements - 280
Dendrochronology 31, 35, 36
Dendroclimatology 31, 35, 36
Dendroecology 36
Density of, soil (see also Soil bulk
  density) 207, 334
 solutions 259
 vegetation 97, 101
Deschampsia flexuosa 111, 131, 279
 caespitosa 111, 133

Dessication 273
Design of experiments 276, 284
  factorial 276
Desorption curves 262
Description of vegetation 85, 90
  choice of method 86, 87
  floristic 96
  physiognomic 91
Devensian 8, 29, 41
Dew 381, 386
  balance 387
  measurement 395
  point 391
Diameter, basal 175
  breast height 101, 175
Diatoms 39, 40
Dickinson's cooling correction 217, 218
Differential, species 125, 135
  thermal analysis 358
Difflugia bacillifera 55
  leidyi 55
  oblonga 55
Diffusion pressure deficit 255
  of CO_2 272
Diffusive resistance 271
Digital volt meter 473, 477, 480, 492, 503
Digitalis purpurea 24
Direct harvest methods 171
Dissolution of the Monastries 28
Dissolved, CO_2 431
  oxygen 431
  solids 429
Distribution, contagious 113
  random 114
  regular 114
Disturbance 25
Diversity, indices 109
  species 109
Dolerite 304
Domain theory 385
Domesday Book 63
Domesticated plants 47
Domin scale 98, 128
Dominants 127
Drainage 25, 44, 335
  description 310
  ditches 350
  water analysis 212
Drosera 280
  rotundifolia 131
Drought resistance 263, 272
Dry, ashing 423
  weight determination 97, 177, 341, 415, 422, 427, 433
Dune heath, nutrient cycles 210
Du Rietz, life form system 92
Dy 13, 49

Earthworms 315
Ecology, definition 1
  limits of 3
Ecological amplitude 231, 251
Ecosystem 85, 205, 218
  concept 158, 160
  definition 159
  diagram 159
  modelling 160
  nitrogen fixation 211
Ecotypes 231, 274
  competition between 284
  on different soils 381
Elapsed time indicator 469
Electrical conductivity 143, 197, 198, 260, 429
  as measure of tissue damage 250, 273
Electrical power supplies 474
Electrochemical integrators 469, 482
Electrolytic CO_2 analyser 199, 200
Electromechanical counter 405
Electronic integrator 486, 488
Elm decline 54
Elutriation 191
Empetrum 57
  nigrum 231
Enclosure 24, 63, 65
Endymion non-scriptus 24, 25
Energy, balance 400
  budgets 157, 188, 205, 334
  solar 159, 371
  units of 216
Enteromorpha 126
  intestinalis 126
Environmental, gradients 140
  measurement 369
Epilobium 88
  montanum 279
  palustre 111
Epidermis 266
Equisetum 13
Erica 56, 57, 266, 270
  cinerea 131, 132, 133, 134, 231, 253, 263, 273, 335
  tetralix 45, 133, 134, 231, 253
Ericaceae 44, 53, 66, 271
Eriophorum 10
  angustifolium 119, 133, 134
  vaginatum 119, 133, 134
Euglypha alveolata 55
  laevis 55
  strigosa 55
Eurhynchium praelongum 111
Eutrophic, brown earth 324
  peat 325
Evaporation 271, 395, 399
  energy balance, and 401

Evaporation—*cont.*
    energy for 372
    latent heat of 401
    logging systems for 499, 500
    open water 404
    potential 404
    predictive equations 403
    water balance and 399
    vapour transfer method 403
Event counters 47
Exchangeable, bases , total 420
    cations (see Extractable cations)
    hydrogen 420
Exponential, decay 163, 187
    growth 234
Exposure 139, 299, 300
    classes 300
Extractable, anions 417
    cations 417
        choice of extractant 417
        variability in soil 337
Extraction techniques, nutrients 417
    pollen 40, 43
    roots 96, 189, 190
Factor analysis 143
Faecal analysis 204
Fagus sylvatica 7, 23, 28, 58, 59
Faithful species 135
Fen 13, 24, 66
    marl 321
    soil 320
Fertilizers 208
    predicting requirements 417
Festuca 144
    rubra 111, 133, 283
Fidelity 125
Field, capacity 212, 352, 354
    data logging systems 494
    experiments 275
    systems 26
Filipendula 58
    ulmaria 25, 111
Fine earth soil fraction 312, 333, 340
Fire 68, 89, 139
    loss of nutrients 211, 215
    temperatures 215
Fission track dating 30
Flame emission spectroscopy 435
    potassium 460
    sodium 462
    estimation of salt spray 210
Floras 67, 94
Floristics 120
Floristic analysis 96
Fluorine, dating 30
    excess in soil 280
Flushing 208, 211

Footpath, counting people 472
Forest (see also Woodland) 104, 121, 170
    dipterocarp 146
    fire 68
    litter accumulation 165
    litter decomposition 184
    rain 93, 95, 96, 120, 132
    soil 26
Fractional loss rate 163
Fraxinus excelsior 53, 58, 59
Freeze drying 216
Frequency 100
Frost 89
    erosion 355
    patterned ground 355
    resistance 250
Froth flotation 192
Fucus volubilis 126
Gabbro 304
Galinsoga ciliata 284
Galium 57
    hercynicum 111
    saxatile 133, 134
    uliginosum 111
    verum 25
Gas, analysis of soil atmosphere 199
    chromatography 211
Genotypic response 231
Geology 8, 27, 34
    of site 301, 311
Geomorphology 301
Germination 232
Girth at breast height (G.B.H.) 101
Glacial, deposits 302
    drift (see Boulder clay)
    interglacial cycle 27
    moraines 38, 195
    retreat stages 35
Gleyed soils 324, 326
Gleying 316
Glyceria fluitans 111
Gneiss 304
Gompertz growth curve 165, 166
Gradient analysis 140
Granite 303, 304
Grass heath 335, 355
Grassland, grazing 100, 202
    historical aspects 23, 24, 41, 66
    patterns of change 69, 88, 107
    production 170, 171, 176
    vegetation analysis 132, 146, 148
Gravel 9, 17, 303, 305
Grazing 24, 44, 67, 107, 202
    estimation of consumption 203
    exclosures 69, 202
    fire, source of new 215
    loss of nutrients 211

Grazing—*cont.*
    relation to plant competition 285
    relation to production 161, 181, 202
Ground water, flow 404
    gley 326
Growth, analysis 234
    cabinets 275
    curves 38, 117, 164, 196
    exponential 175, 234
    forms 300
    linear 175
    phases 94
    rate 165, 234
    response to radiant energy 234
    rings or increments 24, 27, 31, 35, 68,
        177, 181
Gulp dilution gauging 408
Gyttja 13

Habitation sites 26
Habitat, classification for animals 96
Haematite 29
Habrotrocha angusticollis 55
Halimione portulacoides 126
Heat, resistance of plants 248
    sensible and latent 400
    sensitive paints 215
    transfer 272
Heather check 285
Heathland 88, 96, 104, 171, 183, 231
    association analysis 131
    fires 215
    growth curve 167
    growth phases 89
    litter production 181
    lowland 26, 107, 177, 215
    nodal analysis 134
    nitrogen economy 280
    nutrient cycles 210
    root systems 96, 195
Heavy metals, loss in ashing 423
Hedera helix 58, 59
Hedges 25, 38
Heleopera petricola 55
    rosea 55
    sphagni 55
Helianthemum 53
Hemicryptophytes 91
Heracleum 53
Herbivores 159, 168, 202
Hiller peat borer 12, 18, 19, 21
Hippophae rhamnoides 280
Historical ecology 5
    dating techniques 27
    direct floral records 67
    discussion group 23
    documentary evidence 39, 62, 177

Historical ecology—*cont.*
    field techniques 7
    laboratory techniques 39
    maps and photographic records 68
    selection and location of sites 8
History, of landscape 27
    landuse 7
    vegetation 27
Hoagland's solution 277
Holcus lanatus 111
    mollis 111
Horneblende-gneiss 304
Hult-Sernander cover scale 98
Human interference and disturbance 11, 23,
        25, 29, 44, 61, 63, 181, 370
Humidity 378, 391, 401
Humification 10, 13, 45, 181
Hummock and hollow complex 12
Humic soils 324
Humus (see Soil organic matter)
Hyalosphenia elegans 55
    ovalis 55
    papilio 45, 55
    subflava 55
Hylocomium splendens 111
Hydrature 260
Hydraulic conductivity 357
Hydric raw soils 326
Hydrocotyle vulgaris 111
Hydrology 46, 212, 214
Hydro-photography 265, 267
Hydrosere 6, 67, 104
Hydrostatic potential 350
Hygrograph 394
Hygrometer 265, 394
    cobalt chloride 395
    dew point 395
    hair 394
    Shaw 394
Hygroscopic elements 394
Hyphae 45, 357
Hypnum 16

Identification of, animal remains in peat 55,
    40
    plant remains in peat 42
    pollen and spores 48, 56
    seeds and fruits 43
    Sphagna 46
Ilex aquifolium 24, 58, 59, 170
Increment cores 21, 174, 177
Indicator species 6, 10, 24
Infiltration, soil water 349
    method for stomatal aperture 267
Information analysis 135
Infra-red, absorption 265
    gas analyser 168, 197, 241, 246
    radiation 374

Inhibition of growth 285
Insects, phytophagous 203
    remains in peat 40
Insectivorous plants 280
Integration 3, 468, 482, 486
Integrators 469, 481, 486, 488
Interception, canopy 387, 389
Interglacial 8, 29, 35
International Biological Programme 4, 170, 218
    checksheets 121, 298
    handbooks 4, 170
Interstand distance 112, 145
Introduced plant species 24, 68
Iodine 39
    radio-isotope 193
    impregnation of sample bottles 209, 429
Ion-exchange resins 210, 431
Iron, age 30
    atomic absorption method 445
    colorimetric method 446
    deficiency in calcareous soil 280, 281
    ore mining 29
    soil extracts for analysis 420
    stone 305
Irrigation 213
Isohyets 385

Joule 216, 371
Juglans regia 59
Juncus 17, 86
    acutiflorus 111, 131, 134
    articulatus 111, 133
    conglomeratus 133
    effusus 111, 112, 120
    squarrosus 131, 184
Juniperus communis 53

Kastanozem 320
Kjeldahl analysis 211, 451, 455
Knop's solution 277
Kobresia simpliciuscula 279

Ladell can 192
Lagerberg-Raunkiaer cover scale 98
Lakes 8, 32, 39, 40, 46
    glacial 35
    meromictic 35
    sediments 6, 68
        declination changes in 34
Landform 299, 301
    unit 331, 355
Land surveys 62, 63
Late Devensian cold period 29, 41

Late-glacial 17, 29, 52, 54
Laterite 320
Leached brown earth 324
Leaching 159, 196, 208, 209, 211, 349, 357
    of litter samples 180
Lead 33
    toxicity in soil 280
Leaf area 203, 235
    ratio 236
    specific 236
Leaf chamber 193, 238, 246
    cooling 170
    effects on plant 239
    measurement of transpiration 264
Leaf, discs 238, 245, 254
    temperature 248
    wetness 389, 391
    weight ratio 236
Lesquerensia spiralis 280
Libby standard 31
Lichens 25, 26, 38
    atmosphere pollution 25, 370
    history of woodland 25
Life form 91, 121
    categories 91, 92, 93
    spectra 93, 94
Light, compensation point 245
    photochemical integrators 470
    relation to CO_2 exchange 245
    saturation 244
Lignin 49
Limestone 24, 25, 144, 148, 305
Limiting nutrients 276
Limiting plasmolysis 261, 262
Limonium humile 126
    vulgare 126
Linear variable differential transformer 502
Linossier's method 242
Lithology 303
Lithomorphic soils 326
Lithosol 321
Litter 40, 161
    accumulation 165, 181, 186
    bags 163, 183
    corers 186
    decomposition 181
    production 165, 178, 196
    traps 178, 179
Liverworts, distribution in Britain 132
Livingstone peat borer 18, 19, 21, 22
Loam 313
Local history 39, 63
Ligustrum vulgare 262
Loess 49
Logistic growth curve 165, 166
Lolium perenne 111, 120, 285
Lonicera 248
Lophocolea bidentata 111

Loss-on-ignition (L.O.I.), and bulk density of
  soils   207, 208
  determination   266, 433
  litter samples   187
  soil horizons   316
Lotus uliginosus   111
Luzula multiflora   131, 133, 134
Lychnis flos-cuculi   111
Lycopersicum esculentum   262
Lysimeter   212, 349, 400
  trench   212
  vacuum or suction   213
  weighing   400

Mackereth pneumatic sampler   12, 21, 34
Macrofossils   9, 11, 39, 41, 45, 46
Magnesium, atomic absorption method   448
  E.D.T.A. titration method   448
  soils   280
Magnetic, markers for permanent
        quadrats   107
  tape   405
      data loggers   472, 504, 505
Magnetism, remnant in lake deposits   34
Major, nutrients   276, 280
  soil group   323, 311
Manganese, atomic absorption method   450
  colorimetric method   450
Man made soil   326
Mapping and maps   8, 38, 62, 65
  geological   8
  individual plants   107
  land utilization   68
  permanent quadrats   107
  soil   327
  topographical   8, 68, 107, 138
  vegetation   68, 120
Marl   305
Mass spectrometer   211, 280
Mean unit leaf area   235
Mechanical analysis   340
Megaphyll   121
Menyanthes trifoliata   41, 58
Mentha aquatica   111
Mercron meters   469
Mercurialis perennis   25
Mercury coulometer   469, 481
Mesophyll   121
  resistance   268, 271, 272
Mesophytes   263
Meteorological recording site   501
Microfossils   19, 39, 41, 46, 56
Microphyll   121
Microtome   42
Mineral nutrition   274
  and competition   283, 340
  and ecological response   274

Mineral nutrition—*cont.*
  field experiments   275
  laboratory experiments   277
Mineralization   178, 181, 184
  after fire   215
Mineralogy   303
  of sand grains   357
  of soil   356
Minimal area   105, 107, 124, 129
Minor nutrients   277, 280
Mires   8, 11
  valley   9, 20
  growth   44
  nitrogen economy   280
Models, ecosystem   160
  exponential   188
  mathematical   160, 162, 272
  photosynthesis   246
  predictive   113, 160, 175
Moder   315
Moisture content (see also Dry weight
        determination)   432
Molinea caerulea   68, 111, 132, 133, 134
Molluscs   33, 40
Monastic records   7, 62
Monolith, peat   11, 21, 44, 212
  soil   189
Monomolecular curves   164, 196
Montgomery effect   274, 285
Moorland   214
Mor   315
  humus   48
Moraines   38, 195
Morgan's reagent   417
Morphology   120
  rhizome   119
  transpiration surface   266
Mountford's Index   144
Movement of soil water   349
Mull   315
Munsell colour system   11, 312
Mycorrhiza   194, 279, 285
Myriophyllum   57
Myxomatosis   100

Napoleonic War   24, 28, 65
Nardus stricta   144
National Archives Washington D.C.   62
Nature conservation   129, 140
Nebela carinata   55
  griseola   55
  marginata   55
  militaris   55
  parvula   55
  tincta   55
Neolithic   28, 30
Nessler's reagent   454, 456

Net assimilation rate 235
Neutron scatter 351
Nipher shield 385, 386
Nitrate-nitrogen, colorimetric method 457
  distillation method 456
  selective ion electrode 458
Nitrogen 207, 215, 216, 218
  ammonium 454
  excess in soil 280
  extraction for soil analysis 420
  fixation 208, 211, 280
  form in culture solution 278, 434
  in fertilizer experiments 276
  in waters 453
  Kjeldahl method 451
  limiting nutrient 280
  total organic 454
Nodal analysis 133, 134
Nodum 128
Non-calcareous gley 325
Non-gleyed raw mineral soil 324
Non-linear feedback 486
Norse 65
Nuphar 57
Nutrient, accumulation in plants 279
  antagonism 285
  availability 275
  budgets 157, 188, 205, 334
  inputs 206, 208
  content of soil 207, 334, 346, 439
  cycles 205, 218, 381
  deficiency 278
  estimation of content 207
  excess 280
  imbalance in soil 280
  levels in plants and soils 438, 439
  limitation 276
  losses 211
    by fire 215
    in soil solution 212, 214, 405
  recycling 206
Nutrients, minor 277

Oakwoods 25, 65
Obsidian dating 30
Ocean sediment cores 33
Olsen's extractant 421, 460
Orchis morio 25
Ordination 107, 140, 146, 147
Organic matter 45, 161
  total in water 429
  variability in soil 335, 336
Organic soil 320, 324, 326
Orthophyll 121
Osmotic potential 256, 260
Osmunda regalis 57

Oxygen 184, 215
  diffusion in soil 120
  electrodes 242, 431
  evolution 242
  isotope analysis 7
  Winkler's method 242, 431
  Linossier's method 242

Palaeoecology 47
Palaeomagnetism 6, 34
Pallman's method 470
Park grass experiment 276
Particle size analysis 333
  of rocks 303
  of soils 340, 354
Pattern 100, 101, 113
  analysis 116, 117, 118, 119
  causes of 114
  contagion 113
  detection 114
  regularity 114
  scale of 116
  soil 355
Peat, accumulation 9, 44, 186
  animal remains in 39, 40
  archaeological artifacts 30
  borers 12, 18, 19, 21, 42
  chemistry and vegetational history 39
  cutting 10, 42
  flows 8
  land 186
  monoliths 11, 21, 44
  ombrogenous 42, 44, 60, 205
  plant macrofossils in 43, 45
  pollen and spores in (see Pollen analysis)
  radiocarbon dating 30
  sections and stratigraphy 10, 16, 17, 30,
    40, 42, 44
  sediment symbols 10, 13, 14, 15
  soil 320, 324, 325, 326
Pelosols 326
Peltier effect 170, 250, 259
Pelvetia canaliculata 126
Penetrometer 356
Penman's equation 395, 403
People counter 472
Percentage base saturation 420
Performance 103
Peridotite 304
Periglacial, deposits 302
  features 26, 355
Permafrost 26, 355
Permeability 343
pF 352
Phanerophytes 91
Phenology 95, 177, 180, 232
Phenotypic response 231

pH, and soil classification 325
    of cultures 278
    measurement 429, 434
    of soil 275, 335
Phleum pratense 111
Phloem killing 238
Phosphorus, accumulation in plants 279
    addition in experiments 276
    availability in soil 278
    colorimetric method 458
    and competition 279
    extraction for analysis 420
    nutrition, sand dunes 275
    organic 459
    radioactive 193, 279
Photocells 376
Photochemistry 373, 470
Photographs 62, 68
Photography, aerial (see Aerial photography)
    balloon 107
    peat and soil sections 10, 306
    permanent quadrats 107
    sub-aerial 107
    stereo 107, 139, 140
    time lapse 189
    vegetation 107
Photomicrography 42, 56
Photogrammetry 386
Photorespiration 170, 237
Photosynthesis 233
    field measurement 242
    measures of 236
    net rate of 401
Phragmites communis 13, 17, 47
Phryganella 45
    hemisphaerica 55
Physiognomy 91, 120
Physiography 68, 299, 329, 332
Physiological, amplitude 231
    ecology 229, 232
Phytoliths 41
Phytosociology 123, 129, 135, 142
Picea 59, 285, 326
Pinus 13, 28, 42, 53, 57, 59, 169
    aristata 31, 32
    sylvestris 36, 53, 133
Plagiopyxis callida 55
Plant, communities 42
    damage and exposure 300
    macrofossils in peat 46
    mineral nutrition 274
    pigments 39
    resistance, to drought 272
        to extremes of temperature 249
    response to radiant energy 233
    water relations 251
Poa pratensis 133, 134
Podzol 26, 321, 324, 325

Poisson distribution 114
Polarimeter 470
Polarography 242
Pollarding 23, 24
Pollen and spores 41, 47, 206
    analysis 6, 22, 26, 28, 39, 47
    identification 48, 53, 56
    zones 28, 61
Pollution 25, 29, 275
Polygala vulgaris 133
    serpyllifolia 133
Polytrichum commune 111
Porometer 267, 269, 404
Post-glacial 28, 29, 52, 54
Potamogeton 41, 53, 57
Potassium 214
    —argon dating 34
    cycling 210
    fertilizer experiments 276
    flame emission method 460
Potometer 266
Potentilla 144
    erecta 111, 133, 134
    palustris 111
Power supplies 474, 494, 496
Prairie 64, 96, 195
Precipitation, gross and net 387, 388
    measurement 381
    source of nutrients 208
    spatial variation 381
    and water balance equation 399
Pressler increment corer 37, 174
Pressure, bomb apparatus 259, 260
    membrane apparatus 353
Principal component analysis 143, 146, 147
Probit analysis 251
Production, carbohydrate 237
    definitions 161
    ecology 86, 157
    primary 158, 161, 168, 182, 188, 218
    root 168
    secondary 202
    Sphagnum 173
    units 160, 167
Productivity 39
Pseudoscleropodium purum 111
Psychrometer 259, 378, 392, 393, 353
Psychrometric constant 392, 403
Pteridium aquilinum 57, 59, 66, 86, 89, 120,
        131, 133, 134, 184
Public Record Office 62
Puccinellia maritima 126
Pyranometer 469

Quadrat, method, for litter
        accumulation 186
    method, for litter loss 182

Quadrat—*cont.*
    for standing crop   171
    nested plot   105
    number   100, 105, 171
    permanent   69, 107, 229, 355
    point   97, 98, 99, 171
    random   103, 110, 129
    shape   107, 171
    size   100, 106, 124, 171
Quarter girth tapes   101
Quartz-gneiss   304
Quartz-schist   304
Quartzite   304
Quaternary   6, 27, 34, 43
    data bank of fossil records   62
Quercus   7, 23, 24, 53, 58, 59, 64, 65, 169, 203
    petrea   133

Rabbits   67, 100, 107
Radial wood increment   174
Radiant energy, flux   236, 242
    response of plants   233
Radiation, infra-red   372, 374
    long-wave   247, 371
    net   372, 376, 400, 403
        loss by plants   247
    short-wave   247, 372
    solar   371
    terrestrial   371
    ultra-violet   372, 374
    visible   372, 376
Radioactive, carbon dioxide   170, 193, 194, 243
    isotopes and tracers   27, 33, 184, 193
        and insect consumption   204
        half life   31, 163, 279
        nitrogen fixation   211, 280
        plant nutrition   279
        root studies   193
    phosphorus   279
    urea   193
Radioactivity of earth   386
Radiocarbon, dating   11, 21, 28, 30, 34, 38, 173
    dates   45
    years   31
Radiometers   248
Rainfall, mean areal   385
    nutrient input   206, 427
    samples for analysis   209, 428
    stemflow   209, 389
    throughfall   388
Raingauges   382, 383, 384, 388, 428
    fouling by birds   209
    recording   382, 383, 472
    siting   428

Random, distribution   114, 385
    numbers   103
    points   101
    quadrats   103, 110, 129
    samples   183, 331
Ranker   324, 326
Ranunculus   58
    flammula   111
    repens   111
Raunkiaer, life forms   91, 93, 95, 123
Raw mineral soil   324, 326
Reafforestation   28, 65
Reciprocal averaging   146
Recorders   405, 468, 471, 503
Redox potential   434
Reducing conditions in soil   435, 311
Reference collections   39, 42, 46, 56, 204
Refractive index   258
Regeneration   23
Regosol   321
Regression, allometric   175
    analysis   113, 166, 176, 389
    linear   175
    logarithmic   176
    multiple   113
Relative, growth rate   234
    humidity   391
    turgidity   252
    water content   252, 270
Relevee   124
Relief   299, 355
Rendzina   314, 320, 324
Reproductive strategy   234, 235
Resource survey   86
Respiration   159, 244
    dark   237
    photo   170, 237, 244
    rate   169
    root   196
Respirometer, Gilson   184
    soil   197
    Warburg   242
Results, expression of   60, 161, 187, 192
    calorimetry   216
    chemical analysis   413
    soil, analysis   334
        moisture   351
        nutrients   207
    tissue damage   250, 251
    transpiration   266
Rhamnus frangula   68
Rhinanthus minor agg.   25
Rhizocarpon   38
Rhizopods   40, 45, 55
Rhizosphere   285
Rhyolite   304
Rhytidiadelphus squarrosus   111
Richness   96

Rocks, lithological classification   303, 304,
        305
Root, activity   193
    biomass   187, 189, 194
    extraction   96, 189, 190
    exudates   194, 285
    frequency   100
    length   190
    morphology   119
    nodules   280
    observational pits   189
    pattern   193, 275
    penetration   346
    production   168, 188, 194, 196
    respiration   196, 197
    stratification   96, 315
    turnover time   194
    uptake   279
Rotatoria   55
Rotlehm   320
Rubus   133
    chamaemorus   248
Rumex   58, 59
    acetosa   58, 111, 275, 281
    acetosella   58
Running mean   106
Runoff   404
    chemical analysis   427
    water balance   399
Russian peat borer   12, 18, 19, 20, 21

Sach's half leaf method   238
Salicornia herbacea   126
    perennis   126
Saline soil   280, 320
Salix   53, 58, 59
Salt, marsh   104, 126, 355
    spray   141, 143, 209
        source of nutrients   209
Sample division   416
Sampling, bore holes   12
    chemical analysis   414, 421, 426, 428
    constant volume   333
    description and analysis of
            vegetation   103, 124, 129
    frequency   177, 180
    intensity, soil survey   328, 330
    nested plot   104
    numbers   100, 105, 171, 178, 180, 183,
            186, 337
    peat sections   10, 11
    plotless methods   101
    random   103, 331
    recording location   298
    regular or systematic   104
    representative   103

Sampling—*cont.*
    restricted random   104
    salt spray   209
    soil solution   212, 405
    stratified   105, 332
    transect   104
Sand   9, 12, 16, 17
    culture   277
    definition   312, 358
    dunes   23, 104, 106, 120, 173, 195, 338,
            355
        nutrition   210, 275, 279, 280
    grain mineralogy   357
    stone   305
Sandy soils   313
Savanna   132, 215
Saxifraga sarmentosa   262
Scabiosa columbaria   282
Scale of, classification   120, 129
    investigation   168
    spatial pattern   90
    temporal change   90
Scanning electron microscope   44, 56
Scheuchzeria palustris   47
Schist   304
Sclerophyll   121
Scutellaria minor   111
Sea level changes   7
Secale   59
Sections, pollen grains   48
    soil   189, 356
Sediments, disturbance of   10
    laminated   35
    limnic   13, 39, 40, 46, 48
    symbols   10, 13, 14, 15
Seeds and fruits   41, 43
    extraction from peat and soil   43, 192
    identification   43
Selaginella selaginoides   53
Selective ion electrode   458
Selenium cells   376
Senecio aquaticus   111
    vulgaris   235
Serosem   321
Serpentine   275, 276, 277, 280, 304
Sesleria caerulea   144
Sesquioxides (Fe and Al)   316
    soils   320
Sequoia gigantea   31
Shade plants   244
Shale   303, 305, 326
Sheep   202
Shelter belts   395
Shingle   24
Sieglingia decumbens   133
Sieves   43, 44, 49, 191, 415
    soils for analysis   341, 348, 415
Silica   41, 45, 49

Silicon, approximate gravimetric
    method 461
  colorimetric method 461
  solar cells 376
Silt 9, 16, 17, 305, 358
  definition 312
  determination 341
  mineralogy 358
  soils 313
Similarity 110
Site 297
  choice 8
  description 308
  drainage 310
  location 8
  physical characteristics 298
Slate 305
Slope 299, 300
Snow 385
Sociability classes 124
Sodium 210, 214
  excess in saline soil 280
  flame emission method 462
Soil 29, 39, 297
  accretion 355
  acidity 420
  aggregates and structural units 347
  analysis (see also Chemical analysis of
    soils) 275, 332
    preparation of samples 339, 341, 416
    sampling for 332
    variability 334, 336, 337
  animals 183, 190, 298, 315
  association 327
  atmosphere 199
  augers and corers 186, 189, 306, 330,
    333
  bulk density 207, 334
    measurement 343
    and compaction 356
    and loss-on-ignition 207, 346
  categories, key to 322
  chemical potentials 338
  classification 318
  colour 312
  compaction 356
  cores 189, 306
  creep 26
  depth 26
  distribution 298
  disturbance 24, 68
  dune 275, 277
  erosion 212, 355
  field description 306
  heat flux 401
  horizons 190, 307, 308, 309
    boundaries 312
    buried 355

Soil—*cont.*
  horizons—*cont.*
    chemical comparisons 337
    elluvial and illuvial 317
    sampling for analysis 333, 339
    symbols 316, 317
  humus classification 357
  major group 311, 323
  mapping and survey 306, 327
  mechanical strength 356
  metabolism 196
  mineralogy 356
  moisture 212, 231, 404
    availability 351, 352, 354
    content 351
    determination 214, 348
    field assessment 315
    movement 340, 347, 349
    and pore size distribution 347, 352
    resistance blocks 354
    and soil respiration 202
    tension (or suction) 213, 347, 352
    variability 335
    and water balance 399
  monoliths 189, 334
  nutrient 120, 207, 215
    availability 338
    content 207, 334, 346, 439
    status 419
  organic matter 163, 188, 196, 315
  oxygen diffusion rates 120
  parent material 306
  particle size 333, 340
  pattern 355
  phase 323
  phosphorus 278
  phytoliths 41
  pollen analysis 26
  pore space 190, 347
  porosity 314, 343, 352, 356
  profile 23, 26, 189, 214, 306
    classification 307, 318
    description 307
    drainage 311
    features 310
    modal 327
    permanent display 12, 334
    pits 306, 307, 333
  respiration 196
  saline 280
  salinity 145
  series 311, 323, 326, 332, 338, 359
  solution 212
  structure 313, 347
  sub-group 323, 329, 331, 338, 359
    temperature 500
  textural classes 313, 343
  texture 312, 340, 343

Soil—*cont.*
  thin sections   189, 356
  volume   207
  water, availability   351, 352, 354
    capacity   351
    weathering   208, 357
Solarimeter   372, 487
Solar, constant   371
  radiation   159, 371
Solifluction   29, 212
Sol Brun   320, 321, 326
Sol lessivé   321, 324
Solonchak   320
Solonetz   320
Sorenson's, coefficient   112, 144
Sparganium   57
Spartina anglica   126
  townsendii   68
Species-area curve   105, 108
Specific, conductance   429
  conductivity   429
  gravity   173, 175
Sphagnum acutifolium   45
  cuspidatum   13, 45, 46
  fuscum   13
  imbricatum   13, 16, 44, 45, 46
  magellanicum   45, 45
  palustre   111
  papillosum   45, 46
  recurvum   111
Stability of vegetation   109
Standing dead   171, 178, 188, 196
Standing crop   97, 120, 205
  definition   161
  measurement   170
  meter   176
Steady state   163, 194
Stellaria media   111
Stemflow   209, 387, 390, 427
Stomatal, aperture   233, 265, 266
  closure   263, 268, 270, 271
  opening   267
  resistance   265, 268, 271, 404
Stones   312, 333, 340
Storage of samples, animal tissue   426
  peat   11, 22
  plant material   422
  soil   414
  water   428
Stratification   95, 120
Stratigraphy   302
Streamflow   214
  analysis and soil development   214
  measurement   406
  nutrient losses   214
  samplers   214
Stress and shear strength   356
Suaeda maritima   126

Sub-boreal period   28
Sub-surface flow   159, 404, 405, 406
Succession   89, 140, 195
Succisa   58
  pratensis   111
Sucrose inversion method   470
Suction force   255
Sulphur   216, 218
Sun plants   244, 271
Sunshine recorders   375, 376
Surface flow   404
Surface water gley   326
Symbiosis   275, 285
Synecology   86
Systems analysis   160

Tantalum-182   184
Taraxacum   53, 57
Taxus baccata   37, 53, 59
Techniques choice of   2
  climate   370
  data logging   468
  vegetation analysis   86, 87, 148
Temperature   377, 401
  coefficient (Q_{10})   202
  compensation points   245
  and soil respiration   201
  gradients   401
  inversions   169, 170
  mean   169
  plant organs   247, 281
  and rainfall chemistry   208
  resistance   248
  vegetation fires   215
Temperature relations of plants   247
Tensiometer   353
Terra, fusca   320
  rossa   320
Terrain units   301
Tetrazolium salts   193, 250
Thermistors   247
  linearizing response   481
Thermocouple   215, 247, 372
  psychrometers   259
  amplification of signal   484
Thermograph   379
Thermoluminescence   30
Thermometer, deformation   378, 379
  integrating   481, 482
  Kata   397
  liquid in glass   247, 378
  resistance   378, 380
  sucrose inversion   470
  thermistor   247, 378, 380, 481
  thermocouple   247, 378, 381
  wet and dry bulb   392
Thermopile   372

Therophytes 91
Thiessen polygons 385
Tilia 25, 56, 58, 59
  cordata 53
Timber yield tables 101
Tipping bucket recorders 390, 405, 472
Tissue damage 250, 273
Tithe maps 67
Tolerance 280
Topographical exposure (Topex) value 300
Toxicity 274, 277, 280
Trace elements 277, 280, 423
Trampling 104, 356
Transducers 468, 473, 476, 491
Transects 104, 107
Translocation 180, 197, 237, 238, 252, 280
Transpiration 233, 247, 263, 387, 391
Tree, age and age classes 23
  girth/age relationships 36
  rings (see also Growth rings) 31, 35
Trend surface analysis 148, 331
Trichophorum cespitosum 133, 134
Trifolium repens 111, 118, 120
Triglochin maritima 126
Trigonopyxis arcula 45, 55
Trinema enchelys 55
Troells-Smith system 10, 14, 15
Truog's reagent 421
t-test 115
Tundra 26, 52, 170
Turgidity 252, 273
Turgor pressure 256, 261
Turnover time 194
Typha latifolia 57
  angustifolium 57

Ulex europaeus 24, 133, 134, 335
  minor 131, 133, 134
Ulmus 28, 53, 57, 59
Ultrasonic vibration 191, 340
Ultra-violet radiation 374
Units, of energy 216
  production ecology 160, 167
  water potential 255
Uranium series dating 33
Urtica dioica 279

Vaccinium myrtillus 120, 253, 270
  uliginosum 249
Vacuum, dessicator 353
  oven 216
Vapour pressure 257, 272, 391, 403
Variance 176, 207
  analysis 116
  :mean ratio 114
Varve dating 35

Vegetation, analysis (see Analysis of
    vegetation)
  as animal habitat 86, 90
  classification (see Classification of
    vegetation)
  cyclical phases, 89, 140
  description (see Description of vegetation)
  dynamics 88
  history of (see also Historical ecology) 5
  mapping 120, 137, 332
  pattern 26, 88, 89, 100
  profile diagrams 95, 104
  seasonality 88, 94, 121
  stratification 95, 120
  survey 67
  types 94
Vegetational succession 35
Velocity-area method 408
Verbascum 266
Vital staining 250, 273
Viola palustris 111
Volcanic deposits 29, 34
von Post's scale of humification 13

Warp soils 321
Water, balance 381, 399, 405
  chemical analyses 427
  deficit 273
  loss from plants 263
  potential 255, 353
  potential deficit 255, 268
  relations of plants 251
  status of plant tissue 233, 252, 272
  status of soils 272, 350, 352, 353
  table 350
Water content, of air 265
  of soil 350
  relative 252, 262
Water culture 277, 281
Water vapour 241, 271, 402
Weather stations 303, 370, 393, 498
Weighted similarity coefficient 146, 147
Weirs 214, 407
Wetness of foliage 387, 389
Wheatstone bridge, conductivity 429
  in logger systems 480
  porometer 268
  soil moisture determination 354
  temperature measurement 380, 481
Wilting point 352, 354
Wind 89, 208, 395
  direction 397
  heath fires 215
  erosion 355
  run 395, 403
  speed 396
Winkler's method 242, 431
Wood, sub-fossil remains 42

Woodland, ancient 25
  clearance 41, 63, 65
  history 23, 65
  standing crop 172, 174
  structure 23

Yate's correction 110
Yield 97
Younger Dryas period 29

Xerophytes 263
X-ray diffraction 358

Zinc toxicity 280
Zonation 141